Modern Acoustics and Signal Processing

Series Preface for Modern Acoustics and Signal Processing

In the popular mind, the term "acoustics" refers to the properties of a room or other environment—the acoustics of a room are good or the acoustics are bad. But as understood in the professional acoustical societies of the world, such as the highly influential Acoustical Society of America, the concept of acoustics is much broader. Of course, it is concerned with the acoustical properties of concert halls, classrooms, offices, and factories—a topic generally known as architectural acoustics, but it also is concerned with vibrations and waves too high or too low to be audible. Acousticians employ ultrasound in probing the properties of materials, or in medicine for imaging, diagnosis, therapy, and surgery. Acoustics includes infrasound—the wind driven motions of skyscrapers, the vibrations of the earth, and the macroscopic dynamics of the sun.

Acoustics studies the interaction of waves with structures, from the detection of submarines in the sea to the buffeting of spacecraft. The scope of acoustics ranges from the electronic recording of rock and roll and the control of noise in our environments to the inhomogeneous distribution of matter in the cosmos.

Acoustics extends to the production and reception of speech and to the songs of humans and animals. It is in music, from the generation of sounds by musical instruments to the emotional response of listeners. Along this path, acoustics encounters the complex processing in the auditory nervous system, its anatomy, genetics, and physiology—perception and behavior of living things.

Acoustics is a practical science, and modern acoustics is so tightly coupled to digital signal processing that the two fields have become inseparable. Signal processing is not only an indispensable tool for synthesis and analysis, it informs many of our most fundamental models for how acoustical communication systems work.

Given the importance of acoustics to modern science, industry, and human welfare Springer presents this series of scientific literature, entitled Modern Acoustics and Signal Processing. This series of monographs and reference books is intended to cover all areas of today's acoustics as an interdisciplinary field. We expect that scientists, engineers, and graduate students will find the books in this series useful in their research, teaching and studies.

William M. Hartmann
Series Editor-in-Chief

More information about this series at http://www.springer.com/series/3754

Frédéric Cohen Tenoudji

Analog and Digital Signal Analysis

From Basics to Applications

 Springer

Frédéric Cohen Tenoudji
Pierre and Marie Curie University, UPMC
Paris
France

Original Edition: Translation of Analyse des signaux analogiques et numériques
© Ellipses-Édition Marketing - Paris 2012

ISSN 2364-4915   ISSN 2364-4923   (electronic)
Modern Acoustics and Signal Processing
ISBN 978-3-319-82565-6   ISBN 978-3-319-42382-1   (eBook)
DOI 10.1007/978-3-319-42382-1

This Springer imprint is published by Springer Nature
The registered company is Springer International Publishing AG Switzerland

To Anne-Marie

Preface

The aim of this book is to explain and present naturally in a didactic manner the principles and methods of signal analysis. It is intended both for students who have no prior knowledge of the topic as well as for those who, having received introductory training, have only retained a disheartening ensemble of mathematical formulas with little or no appreciation for their underlying scientific basis. The goal of the author is to lay the foundations and to develop logically and progressively the mathematical tools in order to associate knowledge, intuition, and understanding. By focusing at every stage of the presentation on the essential aspects, it is then easy to make progress in the establishment of the theory and to build upon those fundamentals to simply expose and derive the most current techniques of signal processing.

A prerequisite is a first-year multivariable calculus course at the university level with the basic concepts used to solve differential equations, perform integration, and solve problems in linear algebra.

Students will come away from the book equipped with the handling of Dirac distribution, integration in the complex plane, applications of linear algebra, and the opportunity to link the abstraction of mathematical formulas with practical applications, to conceive and perform the fundamental operations in deterministic and random signal processing.

The notions and techniques exposed in this book are essential in different engineering fields: telecommunications, teledetection, acoustics, imaging, nondestructive evaluation, and defence.

Such techniques are used in:

Spectral analysis
Design of analog and digital filters
Amplitude and phase modulations in telecommunications
Voice recognition and speech synthesis
Sonar and radar ranging
Signal detection in the presence of noise
Echo cancelation on transmission lines

Noise cancelation, reduction
Data compression by parametric modeling
Data compression by multi-resolution
Seismic exploration
Noise source identification
Noise reduction
Control of antimissile systems (Only first concepts are given here)
Detection of first signs of mechanical failure
Sound analysis, musical instruments, music synthesis
Audio noise reduction

Selected problems, many of them with worked solutions, at the end of each chapter support the content with examples.

To enable the transition to applications, an overview of the MATLAB programming language is given in Appendix 3 along with an example program. As they work through the book, students are strongly recommended to write programs to derive the results presented in the text. They will discover with wonder how the treatments whose notions they had spent several hours, maybe even days, to master, can be performed in a few tens of a second using preprogrammed functions embedded in the language. Browsing through the information given by the help command within MATLAB is a fascinating journey in the signal processing territory.

The first part of the book primarily discusses continuous-time systems and signals because they provide intuitive access to basic concepts. The nature of a signal is inseparable from that of the systems that create or receive it. We first show in Chap. 1 that, for the linear and time-invariant (LTI) systems that are often encountered in physics, the exponential signals e^{st} have a remarkable property: The action of LTI systems on these signals leaves their shape unchanged. Only the amplitude and the temporal location of these signals are affected. The action of an LTI system comes down to the multiplication of the input signals e^{st} by a function, called the transfer function, which depends only on the complex parameter s. This situation is encountered for filtering harmonic signals $e^{j\omega t}$ (monochromatic), which are a special form of exponential signals. The rule is simple: The frequency of the output signal of the LTI system is the same as the frequency of the input signal.

Using the case of R, L, C electrical circuits, a thorough analysis of the first- and second-order systems is given in Chap. 2. It is shown that their properties are completely conditioned by the position of the poles and zeros of their transfer functions in the complex plane. These two filters are the building blocks of the vast majority of filters.

In Chap. 3 on Fourier series, we find the first superposition of elementary signals, i.e., the sum of an exponential at a base frequency and the exponentials whose frequencies are multiples of the base frequency. This type of periodic sum signals is encountered particularly in music. We derive and discuss the rule of decomposition and reconstruction of these signals on the basis of harmonic functions. The theoretical aspects are deepened by the introduction of the concept of

Hilbert space. We define Hermitian operators in this space. We show that the eigenvalues of these operators are real and that the eigenvectors related to two non-equal eigenvalues are orthogonal. We show that the eigenvectors of the operator $i\frac{d}{dt}$ have the form $e^{jn\omega_1 t}$ and may constitute a basis of the Hilbert space L^2 of periodic functions of period T_1. We also encounter the first example of optimal decomposition of a signal by a finite sum of functions. In that case, we show with an example the appearance of the Gibbs phenomenon on the reconstructed signal. In the last part of the chapter, we show the decisive advantage provided by Fourier analysis to characterize the physical properties of a signal.

The Dirac distribution plays an essential role in signal analysis. We define it in Chap. 4 as the infinite sum of monochromatic functions. This definition is best suited to signal theory because it leads naturally to the relationship between the impulse response of a system and its frequency response by Fourier transformation. This transformation is the cornerstone of signal analysis. It decomposes any signal into its monochromatic components as Newton's prism splits light. We simply deduce the response of a system to a signal of any shape.

The theoretical and practical aspects of Fourier transform of analog signals are developed in Chaps. 5–7. Chapter 5 introduces Fourier transform and its close relationship with the LTI systems. It is natural to decompose any signal in a series of harmonic components, to compute the action of the system on each component, and then reconstruct the results of those actions to recover the signal at the output of the system.

The discussion here emphasizes the essential nature of the Fourier integral, a key insight for students and practitioners: The projection of a function on sine functions. Simply put, it measures the proximity of this function with a sine wave according to the frequency of that sinusoid. This understanding then allows us to anticipate the effect of further treatments with a qualitative assessment of the situation.

Chapters 6 and 7 provide a range of detailed formulas and worked examples. It is strongly recommended that the reader work through these examples as an exercise. The ease of calculation that he will thereby acquire will be useful in a range of areas, from causal or analytical signals to modulations and time–frequency analysis, for example.

Chapter 8 is dedicated to the calculation of the impulse response of first- and second-order systems. The integration techniques in the complex plane used in these calculations are detailed in Appendix 1. We show that the causality of the system depends upon the position of the poles of its transfer function in the complex plane.

We explore in Chap. 9 the relationship between the two-sided Laplace transform and the Fourier transform. Attention is given to the domain of definition of the transfer function of a system and the consequence on stability and causality of that system. This property is an educational, striking example of the correspondence of a mathematical expression with a physical property.

Three main types of analog filters, Butterworth, Chebyshev and Bessel, are studied in Chap. 10. Their chief characteristics are given using the results of Chap. 2.

We explain qualitatively the differences in the properties of these three different classes of filters by the relative positions of the poles of their transfer functions in the Laplace plane. The study of these properties based on simple geometric arguments allows a general comprehension of the behavior filters. It is found in a slightly different form in the study of digital filters carried out in Chaps. 14 and 15 in the second part of the book.

In Chap. 11 we study the properties of causal and analytic signals. We demonstrate the formula giving the Fourier transform of the Heaviside function and prove the link by the Hilbert transform between the real part and the imaginary part of a signal in a domain when it is zero for negative values of the variable in the conjugate domain. Causal signals are the natural output of physical systems. In consequence, signal processing deals mainly with causal signals. Analytic signals have zero values at negative frequencies. They are a mathematical trick to allow an easy treatment of signal modulations.

While Fourier analysis is unrivaled to analyze the properties of linear systems and stationary signals, it is insufficient to account in an intelligible manner for the variation of signal properties over time. This is the case when dealing with the localization of echoes in radar or in seismic analysis. We are led to use a short-time Fourier transform and, more generally, to use the methods for time–frequency analysis developed in Chap. 12. A representation of the signals on alternative basis functions localized in time, as in continuous wavelet decomposition and in analysis with filter banks, is developed in this chapter.

Nowadays, signal recording and treatments are mainly digital. For this reason, the second part of the book is devoted to the presentation of digital processing methods. Claude Shannon has proven that we could sample a signal at each tick of a clock without loss of information. One can perfectly reconstruct the signal value at any time from the recorded samples if certain conditions are met. Of course, a condition on the frequency of the clock must be respected: The faster the signal variations are, the more frequent the samples will need to be in order to properly describe these variations, i.e., the greater the clock frequency must be. These notions are presented simply in Chap. 13 by qualitative arguments.

It is thus possible to sample a signal, process it digitally, and reconstruct the resulting processed analog signal, if desired. The prevalence of digital processing today is due to advances in electronics and computer technology, and to the algorithm of fast Fourier transform of Cooley and Tukey which has revolutionized signal processing. Because of this algorithm, it became possible to perform Fourier analysis in real time. It quickly became apparent to users that digital treatments were much more flexible and that they also allow treatment inapplicable in analog. In this second part, in parallel to the presentation of the analog processing, we define the numerical Fourier transform and the z-transform which is analogous to the Laplace transform for time-continuous signals. The eigenfunction z^n of digital LTI systems plays a role similar to the function e^{st} for analog systems developed in Chaps. 14 and 15. We define the digital moving average filters (MA).

Chapter 16 presents the Fourier transform of digital signals. The Shannon aliasing theorem and Shannon–Whittaker sampling theorem are demonstrated. Specific numerical transforms are discussed: the discrete Fourier transform and its use as the algorithm of fast Fourier transform (FFT). Fourier transform of time-limited signals is detailed, and the advantage of apodization windows is highlighted.

We find in Chap. 17 the properties of Autoregressive filters and ARMA filters. The pros and cons of these filters are compared to those of the MA filters encountered in Chap. 14.

Chapter 18 deals with minimum-phase filters and inverse filtering. The decisive advantage of numerical methods is also reflected in the calculation of inverse filters and in the treatment of nonstationary signals. The deconvolution techniques of a signal used in particular for the seismic signals are discussed.

We use the Haar transform as a first step for the description of nonstationary signals processing in Chap. 19. It allows a simple access to the concepts of filter banks and mirror filters. The Le Gall Tabatabai 5-3 filter used in the JPEG-2000 image compression standard is used to illustrate multiresolution methods. It becomes possible to decompose a signal using a simple filtering operation and return exactly to the signal using a second filter. The discrete wavelet transform is discussed using the example of the Daubechies wavelets. Their use is widespread today in signal processing and data compression of sound signals and images. The analogy between the filter bank processing and multiresolution analysis is emphasized.

Chapter 20 treats the parametric modeling of a signal as given by the impulse response of a digital system. The limits of Padé modeling are explained and the advantages of Prony's method are given. Prony's sytem of equations allows, for example, the modeling of a voice signal. It is called Linear Prediction Coding (LPC) in speech analysis. The chapter ends with the important concept of adaptive filters proposed by Widrow, which is a tracking algorithm in the least square sense that is able to subtract a spurious signal from the signal of interest. It provides an efficient noise canceler technique.

The third part of the book is devoted to the presentation of the properties of random signals and their treatments. After a refresher in the essential concepts of statistics on a single random variable and the normal law, Chap. 21 proceeds to an in-depth discussion of the statistics of two random variables.

The treatment of multiple r.v. is found in Chap. 22. The chi-square law used widely in statistics is presented and its use for the test of hypothesis of a probability law is illustrated by the example of testing the central limit theorem. The linear regression of a collection of data is studied by a simple method and by the use of results of linear algebra. We expose the Tikhonov regularization method which greatly improves the results when dealing with noisy data and ill-conditioned matrices. The maximum likelihood method of parameter estimation is discussed in several examples.

In Chap. 23 the correlation of two r.v., the correlation and covariance matrices, are defined. We show the optimality of Karhunen–Loève, principal components development, of a collection of random variables on a deterministic functions basis.

Chapter 24 is dedicated to the analysis of wide sense stationary signals (wss). We study the properties of their correlation functions, coherence, and power spectral densities. Filtering of random, digital, and analog signals is described. We study the role of filtering to improve the signal-to-noise ratio.

Spectral analysis of a random signal is often confronted with the fact that only one record of the signal is available which cannot claim to represent the statistical properties of the signal. However, when a signal is ergodic, it is possible to estimate the spectral properties from a single record using regularization methods. Different estimators of the autocorrelation function, the power spectral density, and methods to reduce the variance of these estimators are studied in Chap. 25.

Chapter 26 is dedicated to the parametric estimation of random signals. The Yule–Walker equations which enable the modeling of a regular process by an ARMA filter are established. Modeling a finite number of data is studied. The methods of extraction of significant components of Capon and Pisarenko are described.

Chapter 27, the final chapter in the book, develops the application of stochastic orthogonality on estimation and optimal filtering of random signals. The concepts have been established by Wiener. We present several Wiener filters for estimation and prediction using FIR, causal, and noncausal filters. In 1960 R. Kalman proposed a recursive algorithm for noise reduction and state system estimation. Its reach is beyond that of Wiener's filter as it is able to deal with nonstationary signals. It has the advantage of being highly computationally efficient which brings the possibility to make real-time estimations. We discuss its principle and provide a simple example of application.

Three appendices are included at the end of the book. The first two contain essential mathematical concepts. Appendix 1 is dedicated to integration in the complex plane and the residue theorem, which are used in the Fourier, Laplace, and z-transforms calculations.

Appendix 2 contains a review of matrices and linear algebra. The concepts discussed in this appendix are essential to the understanding of current digital processing methods.

Appendix 3 is devoted to the description of the MATLAB software and its use in signal analysis programming.

This book is translated, expanded, and updated from a book published in 2012 in French. I took the opportunity, while doing the translation in English, to bring improvements to the initial text and develop some aspects which seemed missing.

Acknowledgments

The writing of this book was made possible by my presence as an Emeritus Professor in the team Modeling, Propagation and Acoustic Imaging (MPIA) at the Institut Jean le Rond d'Alembert of the Pierre and Marie Curie (Paris VI) University. I want to thank my colleagues for their advice.

Thanks to Alice de Botton for her help in the translation of a large part of the manuscript.

I want to thank particularly Prof. William M. Hartmann for his precious suggestions and his numerous corrections to the manuscript. The rigor and clarity of the text owe him a lot.

I want to thank my editor, Sara Kate Heukerott, for her confidence and her warm support.

I thank Anne-Marie for her constant, patient support, for her help in the translation and for her care in reviewing the manuscript.

Paris, France   Frédéric Cohen Tenoudji

Contents

1	**Notions on Systems**	1
	1.1 Linear Systems	1
	1.2 Stationary Systems	2
	1.3 Continuous Systems	3
	1.4 Linear Time Invariant Systems (LTI)	3
	1.4.1 Eigenfunctions of LTI Systems	3
	1.4.2 Transfer Function and Frequency Response	6
	1.5 Linear Differential Equations with Constant Coefficients	7
	1.6 Linearity of Physical Systems	8
2	**First and Second Order Systems**	11
	2.1 First Order System. R, C Circuit	11
	2.1.1 Transfer Function	12
	2.1.2 Frequency Response	13
	2.1.3 Graphic Representation of the Frequency Response	15
	2.1.4 Geometric Interpretation of the Variation of the Frequency Response	17
	2.1.5 R, C Circuit with Output on the Resistor Terminals	19
	2.2 Second Order System. R, L, C Series Circuit	21
	2.2.1 Transfer Function	21
	2.2.2 Second Order System Frequency Response	23
	2.2.3 Geometric Interpretation of the Variation of the Frequency Response	23
	2.2.4 Bode Representation of the Gain	28
	2.3 Case of Sharp Resonance	29
	2.4 Quality Factor Q	30

3   Fourier Series   35
   3.1   Decomposition of a Periodic Function in Fourier Series   37
   3.2   Parseval's Theorem for Fourier Series   42
   3.3   Sum of a Finite Number of Exponentials   45
   3.4   Hilbert Spaces...   48
   3.5   Gibbs Phenomenon.....................................   51
   3.6   Nonlinearity of a System and Harmonic Generation   53

4   The Dirac Distribution   59
   4.1   Infinite Sum of Exponentials. Cauchy Principal Value   60
   4.2   Dirichlet Integral.....................................   61
   4.3   Dirac Distribution....................................   67
       4.3.1   Definition..   67
       4.3.2   Properties of the Dirac Distribution   69
       4.3.3   Definition of the Convolution Product............   70
       4.3.4   Primitive of the Dirac Distribution. Heaviside
                Function   72
       4.3.5   Derivatives of the Dirac Distribution............   73

5   Fourier Transform   77
   5.1   Impulse Response of an LTI System   77
   5.2   Fourier Transform of a Signal.........................   79
       5.2.1   Direct Fourier Transform.....................   79
       5.2.2   Inverse Fourier Transform   79
   5.3   Properties of Fourier Transform   82
       5.3.1   Symmetry Properties of the Fourier Transform
                of a Real Signal............................   82
       5.3.2   Time-Delay Property of the Fourier Transform   83
   5.4   Power and Energy of a Signal; Parseval–Plancherel
       Theorem   84
   5.5   Deriving a Signal and Fourier Transform   86
   5.6   Fourier Transform of Dirac Distribution
       and of Trigonometric Functions   87
   5.7   Two-Dimensional Fourier Transform   89

6   Fourier Transform and LTI Filter Systems   93
   6.1   Response of a LTI System to Any Form of Input Signal   93
   6.2   Temporal Relastionship Between the Input and Output
       Signals of an LTI Filter   95
   6.3   Fourier Transform and Convolution in Physics............   96
   6.4   Fourier Transform of the Product of Two Functions.........   97
   6.5   Fourier Transform of a Periodic Function.................   98
   6.6   Deterministic Correlation Functions   99
   6.7   Signal Spreads. Heisenberg–Gabor Uncertainty
       Relationship 102

7   Fourier Transforms and Convolution Calculations   111
  7.1   Fourier Transformation of Common Fonctions   111
    7.1.1   Fourier Transform of a Rectangular Window   111
    7.1.2   Fourier Transform of a Triangular Window   113
    7.1.3   Fourier Transform of Hanning Window   114
    7.1.4   Fourier Transform of a Gaussian Function   115
  7.2   Behavior at Infinity of the Fourier Amplitude of a Signal   120
  7.3   Limitation in Time or Frequency of a Signal   120
    7.3.1   Fourier Transform of a Time-Limited Cosine   120
    7.3.2   Practical Interest of Multiplying a Signal
             by a Time Window Before Calculating
             a Spectrum   122
    7.3.3   Frequency Limitation; Gibbs Phenomenon   122
  7.4   Convolution Calculations   124
    7.4.1   Response of a First Order System to Different
             Input Signals   124
    7.4.2   Examples of Calculations of Convolution   128

8   Impulse Response of LTI Systems   137
  8.1   Impulse Response of a First-Order Filter   138
  8.2   Impulse Response of a Second Order Filter   142

9   Laplace Transform   149
  9.1   Direct and Inverse Transforms   149
    9.1.1   Study of Convergence with an Example   151
    9.1.2   Another Example   153
  9.2   Stability of a System and Laplace Transform   153
    9.2.1   Marginal Stability   154
    9.2.2   Minimum-Phase Filter   155
  9.3   Applications of Laplace Transform   155
    9.3.1   Response of a System to Any Input Signal   157

10   Analog Filters   159
  10.1   Delay of a Signal Crossing a Low-Pass Filter   159
  10.2   Butterworth Filters   161
  10.3   Chebyshev Filters   166
  10.4   Bessel Filters   168
  10.5   Comparison of the Different Filters Responses   170

11   Causal Signals—Analytic Signals   177
  11.1   Fourier Transform of the Pseudo-Function $\frac{1}{t}$ 177
  11.2   Fourier Transform of a Causal Signal; Hilbert Transform   181
  11.3   Paley-Wiener Theorem   185
  11.4   Analytic Signal   186

        11.4.1   Instantaneous Frequency of a Chirp.   190
        11.4.2   Double-Sideband (DSB) Signal Modulation   190
        11.4.3   Single-Sideband Signal Modulation (SSB).   193
        11.4.4   Band-pass Filtering of Amplitude Modulated
                  Signal   195
  11.5 Phase and Group Time Delays   197
  11.6 Decomposition of a Voice Signal by a Filter Bank.   199

12 **Time–Frequency Analysis**.   207
  12.1 Short-Time Fourier Transform (STFT) and Spectrogram   208
  12.2 Wigner–Ville Distribution.   212
  12.3 Continuous Wavelet Transform.   217
        12.3.1   Examples of Wavelets   217
        12.3.2   Decomposition and Reconstruction of a Signal
                  with Wavelets   219
        12.3.3   Shannon Wavelet.   224

13 **Notions on Digital Signals**.   227
  13.1 Analog to Digital Conversion   228
  13.2 Criterion for a Good Sampling in Time Domain.   230
  13.3 Simple Digital Signals   232

14 **Discrete Systems—Moving Average Systems**.   235
  14.1 Linear, Time-Invariant Systems (LTI).   236
  14.2 Properties of LTI Systems   236
  14.3 Notion of Transfer Function   237
  14.4 Frequency Response of a LTI System   239
  14.5 Moving Average (MA) Filters   240
  14.6 Geometric Interpretation of Gain Variation with Frequency . . .   241
  14.7 Properties of Moving Average (MA) Filters, also Called
        Finite Impulse Response (FIR)   244
  14.8 Other Examples of All-Zero Filters (MA).   246

15 **Z-Transform**   253
  15.1 Definition   253
  15.2 Inversion of z-Transform.   256
  15.3 z-Transform of the Product of Two Functions   258
  15.4 Properties of the z-Transform   259
  15.5 Applications   259

16 **Fourier Transform of Digital Signals**.   263
  16.1 Poisson's Summation Formula   264
  16.2 Shannon Aliasing Theorem.   265
  16.3 Sampling Theorem of Shannon–Whittaker   267
  16.4 Application of Poisson's Summation Formula: Fourier
        Transform of a Sine   269
  16.5 Fourier Transform of a Product of Functions of Time   270

16.6 Parseval's Theorem. 271
16.7 Fourier Transform of a Rectangular Window 272
16.8 Fourier Transform of a Sine Function Limited in Time 273
16.9 Apodization Windows. 275
16.10 Discrete Fourier Transform (DFT) 279
    16.10.1 Important Special Case: The DFT
           of a Bounded Support Function. 281
16.11 Fast Fourier Transform Algorithm (FFT) 281
16.12 Matrix Form of DFT. 284
16.13 Signal Interpolation by Zero Padding 284
16.14 Artifacts of the Fourier Transform on a Computer 286

17 Autoregressive Systems (AR)—ARMA Systems 291
17.1 Autoregressive First-Order System 292
    17.1.1 Case of a Causal System 292
    17.1.2 Analysis of the Anticausal System. 295
17.2 Autoregressive System (Recursive) of Second Order 297
    17.2.1 Calculation of the System Transfer
           Function $H(z)$ 297
    17.2.2 Geometric Interpretation of Variation
           of Frequency Gain Magnitude. 299
    17.2.3 Impulse Response of Second-Order System. 302
    17.2.4 Functional Diagrams of the Digital System
           of Second Order 303
17.3 ARMA Filters. 305
17.4 Transition from an Analog Filter to a Digital Filter 310
    17.4.1 Correspondence by the Bilinear Transformation 310
    17.4.2 Correspondence by Impulse Response
           Sampling 312
    17.4.3 Correspondence by Frequency Response
           Sampling 313

18 Minimum-Phase Systems—Deconvolution 321
18.1 Minimum-Phase Systems 321
    18.1.1 Notion of Minimum-Phase System 321
    18.1.2 Properties of Minimum-Phase Systems 326
18.2 Deconvolution. 327
    18.2.1 Interest of Deconvolution 327
    18.2.2 Deconvolution Techniques. 328

19 Wavelets; Multiresolution Analysis 337
19.1 Dyadic Decomposition-Reconstruction of a Digital
     Signal; Two Channels Filter Bank 338
19.2 Multiresolution Wavelet Analysis. 346
19.3 Daubechies Wavelets 353

20   Parametric Estimate—Modeling of Deterministic
     Signals—Linear Prediction   375
     20.1   Least Square Method   376
     20.2   Padé Representation   378
            20.2.1   Padé Approximation   381
            20.2.2   All-pole Modeling   381
            20.2.3   Examples.   382
     20.3   Prony's Approximation Method. Shanks Method   385
            20.3.1   Prony's Method.   385
            20.3.2   Shanks Method   389
     20.4   All-pole Modeling in the Context of the Prony's
            Method   392
     20.5   All-pole Modeling in the Case of a Finite Number
            of Data   394
            20.5.1   Autocorrelation Method.   394
            20.5.2   Covariance Method   395
     20.6   Adaptive Filter   396
     References.   405

21   Random Signals: Statistics Basis   407
     21.1   First-Order Statistics   408
            21.1.1   Case of a Real Random Variable.   408
            21.1.2   Gaussian Distribution (Normal Law)   413
            21.1.3   Probability Density Function of a Function
                     of a Random Variable   419
     21.2   Second-Order Statistics   421
            21.2.1   Case of Two Real Random Variables   421
            21.2.2   Two Joint Gaussian r.r.   428
            21.2.3   Properties of the Sum of Two r.v   431
            21.2.4   Complex Random Variables   435

22   Multiple Random Variables—Linear Regression Maximum
     Likelihood Estimation   445
     22.1 χ_v^2 (Chi-Square) Variable with v Degrees of Freedom.   445
     22.2   Least Squares Linear Regression.   449
            22.2.1   Simple Method   449
            22.2.2   Elaborate Method   451
     22.3   Linear Regression with Noise on Data—Tikhonov
            Regularization.   452
     22.4   Parametric Estimation   455
            22.4.1   Issues of the Estimation.   455
            22.4.2   Maximum Likelihood Parametric Estimation   458
            22.4.3   Cramér-Rao Bound   460

23 Correlation and Covariance Matrices of a Complex
Random Vector 467
   23.1   Definition of Correlation and Covariance Matrices 467
        23.1.1   Properties of Correlation Matrix 468
   23.2   Linear Transformation of Random Vectors................ 469
   23.3   Multivariate Gaussian Probability Density Functions 471
   23.4   Estimation of the Correlation Matrix from Observations...... 474
   23.5   Karhunen-Loève Development 476
        23.5.1   Example of Using the Correlation
                 and Covariance Matrices 476
        23.5.2   Theoretical Aspects 478
        23.5.3   Optimality of Karhunen-Loève Development. 480

24 Correlation Functions, Spectral Power Densities
of Random Signals 483
   24.1   Correlation Function of a Random Signal................. 483
        24.1.1   Correlation Function of a Wide Sense
                 Stationary (WSS) Signal 484
        24.1.2   Properties of the Correlation Function 485
        24.1.3   Centered White Noise 486
   24.2   Filtering a Random Signal by a LTI Filter 486
        24.2.1   Expected Values 486
        24.2.2   Correlation Functions of Input and Output
                 Signals...................................... 487
   24.3   Power Spectral Density of a WSS Signal 488
   24.4   Filtering a Centered White Noise with a First Order Filter.... 491
   24.5   Coherence Function 493
   24.6   Autocorrelation Matrix of a Random Signal................ 496
   24.7   Beamforming 497
   24.8   Analog Random Signals................................ 498
   24.9   Matched Filter....................................... 501

25 Ergodicity; Temporal and Spectral Estimations 511
   25.1   Estimation of the Average of a Random Signal 512
        25.1.1   Expectation of the Average Estimator 512
        25.1.2   Variance of the Average Estimator 512
        25.1.3   Ergodicity Conditions 514
   25.2   Estimation of the Correlation Function 515
   25.3   Spectral Estimation.................................... 517
        25.3.1   Raw Estimator of the Power Spectral Density
                 or Periodogram 518
        25.3.2   Statistical Properties of the Periodogram 519
   25.4   Improvement of the Spectral Estimation................. 522

25.5   Search for Harmonic Components 524
        25.5.1   Capon Method ("Maximum Likelihood") 524
        25.5.2   Pisarenko Method 526

26 **Parametric Modeling of Random Signals** 529
    26.1   Paley–Wiener Condition 529
    26.2   Parametric Modeling of Random Signals 532
            26.2.1   Yule-Walker Equations 532
            26.2.2   Search of the ARMA Model Coefficients for a
                     Regular Process............................. 535
            26.2.3   AR Modeling of a Regular Random Signal........ 536
            26.2.4   MA Modeling of a Regular Random Signal 538

27 **Optimal Filtering; Wiener and Kalman Filters**................. 543
    27.1   Optimal Estimation.................................. 544
            27.1.1   Stochastic Orthogonality 544
            27.1.2   Optimal Least Squares Estimate................ 544
    27.2   Wiener Optimal Filtering 545
            27.2.1   FIR Wiener Filter 545
            27.2.2   Linear Prediction of a Random Signal........... 549
    27.3   IIR Wiener Filter 552
            27.3.1   Non-Causal Filter 552
            27.3.2   Causal Filter 553
    27.4   Kalman Filter 555
            27.4.1   Recursive Estimate of a Constant State 555
            27.4.2   General Form of the Kalman Recursive
                     Equation 556

Appendix A: Functions of a Complex Variable 563

Appendix B: Linear Algebra 573

Appendix C: Computer Calculations............................ 591

Bibliography 601

Index 605

About the Author

Frédéric Cohen Tenoudji is Emeritus Professor at Pierre and Marie Curie University in Paris. His research field is nondestructive evaluation by ultrasonics, defect characterization, and linear and nonlinear sound propagation in heterogeneous materials with applications to civil engineering: concrete cure monitoring, materials structural integrity. He develops ultrasound instrumentation for NDE with graphic user interface and embedded signal processing.

Cohen Tenoudji teaches signal processing, sensors, ultrasonics, and object-oriented programming. From 1985 to 1986 he was Member of the Technical Staff at the Science Center of Rockwell International, NDE Department, in Thousand Oaks, CA.

Chapter 1
Notions on Systems

In this chapter, we present the general properties of linear systems found in the physical world, particularly linear systems with time independent properties (LTI systems). After defining the concepts of eigenfunctions and eigenvalues of an operator, we show that the operator associated with a LTI system commutes with the time translation operator. We show that, as a consequence, the complex exponential functions of time are eigenfunctions of LTI systems. Thus, we attain the remarkable property that if the input signal of these systems is monochromatic, the signal at the output is also monochromatic and has the same frequency. We arrive in this way at the fundamental notions of transfer and frequency response functions of a system.

We define a system as a device producing a signal $y(t)$ (generally a physical quantity that can be transformed into an electrical signal) in response to an input signal $x(t)$. The system can be described mathematically by an operator O acting on the function $x(t)$ to provide the output function $y(t)$:

$$y(t) = O(x(t)). \qquad (1.1)$$

1.1 Linear Systems

Let $x_1(t)$ and $x_2(t)$ be any two signals. System output signals corresponding to the inputs $x_1(t)$ and $x_2(t)$ denote, respectively, $y_1(t)$ and $y_2(t)$. The system is linear if, given any two constants a_1 and a_2, to the input linear $a_1x_1(t) + a_2x_2(t)$, it makes the corresponding linear combination of the signals $y_1(t)$ and $y_2(t)$ with the same coefficients a_1 and a_2: $a_1y_1(t) + a_2y_2(t)$.

© Springer International Publishing Switzerland 2016
F. Cohen Tenoudji, *Analog and Digital Signal Analysis*,
Modern Acoustics and Signal Processing, DOI 10.1007/978-3-319-42382-1_1

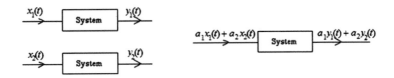

$$a_1 y_1(t) + a_2 y_2(t) = O(a_1 x_1(t) + a_2 x_2(t)). \tag{1.2}$$

In particular, it appears that if the system is linear, the doubling of the input signal results in a doubling of the output signal. The nonlinearity threshold of an operational amplifier is visible on an oscilloscope by the clipping of large values of the signal when the amplitude of the input signal is strongly increased.

1.2   Stationary Systems

A system is said to be stationary if its properties are invariant in time. Let $y(t)$ be the output corresponding to any given signal $x(t)$. A stationary system will respond the delayed output $y(t - \tau)$ to the delayed input $x(t - \tau)$.

In symbolic form we write: If $y(t) = O(x(t))$, then

$$y(t - \tau) = O(x(t - \tau)). \tag{1.3}$$

By definition, the translational operator in time T_τ performs the translation in time of a function $f(t)$ of an amount τ:

$$T_\tau f(t) = g(t) = f(t - \tau). \tag{1.4}$$

The left side of the relationship (1.3) can be read as follows: $y(t - \tau) = T_\tau y(t) = T_\tau O(x(t))$.

A system is said to be stationary if its properties are invariant in time. Let $y(t)$ be the output corresponding to any given signal $x(t)$. A stationary system will respond the delayed exit $y(t - \tau)$ to the delayed input $x(t - \tau)$.

While the right side of this relationship can be read as follows: $O(x(t - \tau)) = O(T_\tau x(t))$.

In other words, we can rewrite (1.3) in the form:

$$T_\tau O x(t) = O T_\tau x(t). \tag{1.5}$$

It is said that the two operators commute.

1.3   Continuous Systems

Let $x_n(t)$ be a sequence of input signals, and $x(t)$ the limit of this sequence when n tends to infinity. We note $y_n(t)$ and $y(t)$, respectively, the system responses to the signals $x_n(t)$ and to its limit $x(t)$. The system is continuous if

$$y(t) = \lim_{n \to \infty} y_n(t). \tag{1.6}$$

1.4   Linear Time Invariant Systems (LTI)

Simple physical systems generally have the property of being linear, time invariant, and continuous. Only these systems are studied later in this course.

1.4.1   Eigenfunctions of LTI Systems

A function $f(t)$ is said to be an eigenfunction of an operator O if the result of the action of the system on the function $f(t)$ is a function proportional to $f(t)$:

$$O(f(t)) = \lambda f(t), \tag{1.7}$$

where λ is a complex constant called the eigenvalue corresponding to the eigen-function $f(t)$.

Role of the Exponential Function e^{st}
The purpose of this section is to show that the functions of time with exponential form e^{st} are eigenfunctions of linear, time invariant operators.

The operator T_τ defined above performs translation in time of a function $f(t)$ by an amount τ:

$$T_\tau f(t) = g(t) = f(t - \tau). \tag{1.8}$$

When τ is positive, the shift is toward greater values, to the right. In this case, the value of the function $g(t)$ at the time t is the value that the function $f(t)$ had at the previous time $t - \tau$.

There is a relationship between the translation operator T_τ and the derivative operator $\frac{d}{dt}$:

It is assumed in what follows that the function $f(t)$ and its derivatives are sufficiently regular. Taylor development of $f(t - \tau)$ is then:

$$f(t - \tau) = f(t) - \tau \frac{df}{dt} + \frac{\tau^2}{2}\frac{d^2 f}{dt^2} - \frac{\tau^3}{6}\frac{d^3 f}{dt^3} + \cdots + \frac{(-1)^n \tau^n}{n!}\frac{d^n f}{dt^n} + \cdots \qquad (1.9)$$

One can formally write the translation operator as follows:

$$T_\tau = e^{-\tau\frac{d}{dt}} = 1 - \tau\frac{d}{dt} + \frac{\tau^2}{2}\frac{d^2}{dt^2} - \frac{\tau^3}{6}\frac{d^3}{dt^3} + \cdots \qquad (1.10)$$

Indeed:

$$T_\tau f(t) = e^{-\tau\frac{d}{dt}} f(t) = f(t) - \tau\frac{df}{dt} + \frac{\tau^2}{2}\frac{d^2 f}{dt^2} - \frac{\tau^3}{6}\frac{d^3 f}{dt^3} + \cdots = f(t - \tau). \qquad (1.11)$$

The Taylor formula giving the value of a function in the neighborhood of a point is recognized.

It is further noted that the operator T_τ commutes with the derivative $\frac{d}{dt}$.

Indeed,

$$\frac{d}{dt} e^{-\tau\frac{d}{dt}} f(t) = f'(t) - \tau\frac{d^2 f}{dt^2} + \frac{\tau^2}{2}\frac{d^3 f}{dt^3} - \frac{\tau^3}{6}\frac{d^4 f}{dt^4} + \cdots, \qquad (1.12)$$

and also:

$$e^{-\tau\frac{d}{dt}}\frac{d}{dt} f(t) = f'(t) - \tau\frac{d^2 f}{dt^2} + \frac{\tau^2}{2}\frac{d^3 f}{dt^3} - \frac{\tau^3}{6}\frac{d^4 f}{dt^4} + \cdots. \qquad (1.13)$$

This proves the commutativity of the two operators:

$$\frac{d}{dt} T_\tau f = T_\tau \frac{d}{dt} f. \qquad (1.14)$$

The operator $\frac{d}{dt}$ plays a fundamental role in the description of the evolution of physical systems with time (e.g., in differential equations with constant coefficients encountered in electricity).

It remains to show that T_τ and $\frac{d}{dt}$ have the same system of eigenfunctions. Noting $O_1 = \frac{d}{dt}$ and $O_2 = T_\tau$, we have formally:

$$O_1 O_2 = O_2 O_1. \qquad (1.15)$$

Let f_1 be an eigenfunction of O_1 with the eigenvalue λ_1. Assuming also that the operators are linear, we have

$$O_1 O_2 f_1 = O_2 O_1 f_1 = O_2 \lambda_1 f_1 = \lambda_1 O_2 f_1. \tag{1.16}$$

So it appears that $O_2 f_1$ is also an eigenfunction of O_1 with the same eigenvalue λ_1 as f_1. We necessarily have proportionality between $O_2 f_1$ and f_1 as they represent eigenvectors with the same eigenvalue. Then we see that, due to the commutativity of the operators, f_1 is also an eigenfunction of O_2. This result is of general application.

In the present case, we look first an eigenfunction f_1 of the operator $\frac{d}{dt}$. That function must be a solution of the differential equation:

$$\frac{d}{dt} f_1 = s f_1. \tag{1.17}$$

This equation is a first-order differential equation with one constant coefficient. Its general solution is

$$f_1(t) = A e^{st}. \tag{1.18}$$

We see that the eigenfunctions of the operator $\frac{d}{dt}$ have the form e^{st}, where s is any complex constant. We note that

$$\frac{d}{dt} e^{st} = s e^{st}. \tag{1.19}$$

We can check to complete that the exponential are eigenfunctions of the translation operator in time:

Let $f_1(t) = e^{st}$, then, $T_\tau e^{st} = e^{s(t-\tau)} = C e^{st}$. The eigenvalue is: $\lambda_1 = C = e^{-s\tau}$.

In summary, the eigenfunctions of the translation operator are exponential functions of time.

We note the following property, valid for any LTI system:

Let O be the system operator. To say that the operator O is translational invariant in time comes to write the commutation relation of the system operator with the translation operator $OT_\tau = T_\tau O$. The eigenfunctions of O will be to search through the eigenfunctions of T_τ, i.e., among the functions of the form e^{st}.

Thus

$$O(e^{st}) = \lambda e^{st}. \tag{1.20}$$

The major role to be played by the functions $f(t) = e^{st}$ for physical LTI systems appears here. As is made clear in the following, in electronics and signal processing, the eigenvalue λ of the system operator is denoted $H(s)$ as it is a function of s in the general case.

1.4.2   Transfer Function and Frequency Response

As shown above, when a signal of the form $x(t) = e^{st}$ is presented as input of a linear, time invariant system, the output signal will have the form

$$y(t) = H(s)e^{st}. \qquad (1.21)$$

The operator's complex eigenvalue $H(s)$ is called transfer (system) function of the system. It is also known as the transmittance of the filter.

$$x(t) = e^{st} \longrightarrow \boxed{\text{System}} \longrightarrow y(t) = H(s)\, e^{st}$$

In case where $x(t) = e^{j\omega t}$ that is to say that s is pure imaginary, the signal $x(t)$ is a monochromatic signal (also called harmonic signal) with pulsation ω written in complex notation following Euler's formula

$$x(t) = e^{j\omega t} = \cos \omega t + j \sin \omega t. \qquad (1.22)$$

Note that we use $j = \sqrt{-1}$ to represent the imaginary part instead than i. This notation is common in electricity and in signal analysis to avoid confusion with the current i flowing within a circuit.

At the filter output, we have:

$$y(t) = H(j\omega)e^{j\omega t}. \qquad (1.23)$$

It should be noted that the filter output signal $y(t)$ is also monochromatic with the same angular frequency as the input signal to the filter.

$H(j\omega)$ is named the frequency response. The common use in electronics and in signal analysis is to write $H(\omega)$ instead of $H(j\omega)$. One should be careful to avoid the difficulties caused by this change of notation.

In summary, when the input signal is monochromatic, the output signal of a linear, time invariant filter is monochromatic and has the same frequency as the input signal. Practically, if a filter is used in a nonlinear regime, as is the case of an operational amplifier whose output saturates for large values of the input signal, the output is no longer harmonic. Even when the input signal is monochromatic, one sees new frequencies in the output, generally multiples of the fundamental

frequency of the input signal. In the example of an operational amplifier in satu-
ration, when the input signal is sinusoidal, its development in the Fourier series has
only one coefficient. It corresponds to the frequency of the sine. The output signal
has a shape close to a periodic rectangular signal. As will be detailed in Chap. 3, its
development into a Fourier series have an infinite number of coefficients for the
frequencies corresponding to odd multiples of the fundamental frequency.

Note: A linear combination of two eigenfunctions is usually not an eigenfunction. Indeed, let $f_1(t)$ and $f_2(t)$ be two eigenfunctions of the operator with different
eigenvalues:

$$O(f_1(t)) = \lambda_1 f_1(t) \quad \text{and} \quad O(f_2(t)) = \lambda_2 f_2(t).$$

Then

$$O(a_1 f_1(t) + a_2 f_2(t)) = a_1 \lambda_1 f_1(t) + a_2 \lambda_2 f_2(t) \neq \lambda(a_1 f_1(t) + a_2 f_2(t)) \quad \text{if} \quad \lambda_1 \neq \lambda_2.$$

$$(1.24)$$

Thus, while $e^{j\omega t}$ and $e^{-j\omega t}$ are eigenfunctions of a system (e.g., an *RC* filter as an
electric circuit) $\cos \omega t = \frac{e^{j\omega t} + e^{-j\omega t}}{2}$ is not one in the general case.

We thus clearly see the benefits of using the complex exponential rather than
trigonometric sine and cosine functions in calculations.

It should be emphasized here that the physical signals are real as is $\cos \omega t$ and
not as the type of the complex exponential $e^{j\omega t}$ or $e^{-j\omega t}$. Besides, how can we
imagine a negative frequency $-\omega$ for a signal? Our answer is that negative frequencies are a mathematical fiction introduced to make calculations easier. We will
generally proceed as this in the calculations: we perform calculations with complex
exponential then, at the end, we return to real signals by extracting the real parts of
the results.

1.5   Linear Differential Equations with Constant Coefficients

Many physical systems, electrical (described by generalized Ohm's law) or
mechanical (fundamental relation of dynamics) satisfy the following general
equation:

$$\frac{d^m y(t)}{dt^m} + a_1 \frac{d^{m-1} y(t)}{dt^{m-1}} + a_2 \frac{d^{m-2} y(t)}{dt^{m-2}} + \cdots + a_m y(t)$$
$$= b_0 \frac{d^r x(t)}{dt^r} + b_1 \frac{d^{r-1} x(t)}{dt^{r-1}} + \cdots + b_r x(t) \qquad (1.25)$$

The coefficients of this equation $a_1, a_2, \ldots, a_m$ and $b_0, b_1, b_2, \ldots, b_r$ are constant. They contain the characteristics of the system which does not evolve with time. It only appears in this equation invariable linear combinations in the time of the input and output functions $x(t)$ and $y(t)$ and their derivatives. As before, the study of such a LTI system relies on its transfer function $H(s)$, that is to say, it relies on the response of the system $y(t) = H(s)e^{st}$ to the input $x(t) = e^{st}$.

The derivative of order r of $x(t)$ that appears in the (1.25) in this case is

$$\frac{d^r x(t)}{dt^r} = s^r e^{st}. \tag{1.26}$$

Similarly, the derivative of order m of $y(t)$ is

$$\frac{d^m y(t)}{dt^m} = s^m H(s) e^{st}. \tag{1.27}$$

After replacing in Eq. (1.25) and simplification by e^{st}, we obtain the following expression for the system transfer function $H(s)$:

$$H(s) = \frac{b_0 s^r + b_1 s^{r-1} + \cdots + b_r}{s^m + a_1 s^{m-1} + a_2 s^{m-2} + \cdots + a_m}. \tag{1.28}$$

As will be detailed in Chap. 2, the system properties are fully contained in the properties of the function $H(s)$ conditioned by the positions of the roots of its numerator (zeros of the transfer function) and the denominator (poles of the transfer function).

1.6   Linearity of Physical Systems

In the last paragraph, we discuss the case of linear and nonlinear systems. Generally a signal amplifier is expected to be linear, that is to say, satisfies the property given by the formula (1.2). A special case of this approach is that if the input signal is multiplied by a factor 2 (or 10, or any number) linearity causes the output signal to be also multiplied by 2 (or 10, or the same any number).

For example, is a circuit consisting of an operational amplifier of gain 50 linear? If the amplitude of the input signal is multiplied by 10, will the output signal also be multiplied by 10? Yes, as long as the amplitude of the output signal does not reach the power supply voltage of the op amp (± 12 V for example). Beyond that threshold the output signal is saturated to ± 12 V.

This operational amplifier circuit will therefore be considered as linear, as long as the output signal does not exceed ± 12 V, for example in the case of a 50 gain, as long as the input signal does not reach $\pm \frac{12}{50} = 240$ mV.

Generally, the nonlinearities of a system occur when the amplitude of the system input signal is important. Two physical examples of nonlinearities are the audio amplifier used with a strong input signal (a saturated guitar amplifier, prized by some rock groups is an extreme example) or a high intensity laser light passing through a transparent medium.

As will be discussed in Chap. 3 on the Fourier series, the nonlinearities are accompanied by generation of double, triple the fundamental frequency components, or more (these components are called harmonics of the fundamental frequency). The possibility of generating high frequencies by the harmonic is used in many applications' design in physics.

It will be specified on an example in Chap. 3, how the analysis of harmonics in the output signal of a system can be used to study the physical mechanism responsible for the nonlinearity of the system.

Summary

We have proved in this chapter that linear, time invariant systems operators, have eigenfunctions of the form of exponential functions. This has been shown to be the result of the commutativity of these operators with the time translation operator. We have explained the concepts of transfer and frequency response functions and demonstrated the fundamental property that the frequency of a monochromatic signal remains unchanged at the throughput of these systems. The next chapter will verify these concepts in the canonical examples of first and second-order systems, which are the cornerstone of electronic filter systems.

Exercises

I. Consider the system defined by the differential equation $y(t) = b_0 \frac{dx(t)}{dt} + b_1 x(t)$.

Show that this system operator noted O is linear, time invariant.

Solution:
Linearity: We note $y_1(t) = O(x_1(t))$ and $y_2(t) = O(x_2(t))$. Let $x(t) = c_1 x_1(t) + c_2 x_2(t)$ be a linear combination with any coefficients of the input functions.
Calculation of the output function $y(t) = O(x(t))$.

$$y(t) = b_0 \frac{d(c_1 x_1(t) + c_2 x_2(t))}{dt} + b_1(c_1 x_1(t) + c_2 x_2(t)),$$

$$y(t) = b_0 \frac{dc_1 x_1(t)}{dt} + b_0 \frac{dc_2 x_2(t)}{dt} + b_1 c_1 x_1(t) + b_1 c_2 x_2(t) = c_1 y_1(t) + c_2 y_2(t).$$

The latter relation corresponds to the definition of the system O linearity.
Translation invariance in time: Given $y(t) = O(x(t))$. What is $O(x(t - \tau))$?
$$O(x(t - \tau)) = b_0 \frac{dx(t-\tau)}{dt} + b_1 x(t - \tau) = b_0 \frac{dx(t')}{dt} + b_1 x(t') \text{ (with } t' = t - \tau).$$

We recognize in the right-hand side $b_0 \frac{dx(t')}{dt} + b_1 x(t') = y(t')$, which may be written $y(t - \tau) = b_0 \frac{dx(t-\tau)}{dt} + b_1 x(t - \tau)$ which corresponds to the definition of the stationarity of the system O.

II. Let the system defined by the equation: $y(t) = bx^2(t)$. Is this system linear? Time invariant?

   Solution:
   Linearity: We note $y_1(t) = O(x_1(t))$ and $y_2(t) = O(x_2(t))$. Let $x(t) = c_1 x_1(t) + c_2 x_2(t)$ be a linear combination with any coefficients of the input variables.
   Calculation of the output signal:

   $$y(t) = b(c_1 x_1(t) + c_2 x_2(t))^2 = b\left(c_1^2 x_1^2(t) + c_2^2 x_2^2(t) + 2c_1 x_1(t) c_2 x_2(t)\right)$$
   $$y(t) = c_1^2 y_1(t) + c_2^2 y_2(t) + 2bc_1 c_2 x_1(t) x_2(t) \neq c_1 y_1(t) + c_2 y_2(t).$$

   In the general case, the system is not linear.
   Translation invariance over time: The system is not translation invariant in time. This result stems from the fact that the multiplier of input signal is a function of time.
   Translational invariance in time: We have $y(t) = O(x(t))$. $O(x(t - \tau)) = bx^2(t - \tau)$.
   We recognize $y(t - \tau)$ in the right hand side. The system is time invariant.

III. Is the system defined by the equation: $y(t) = \cos t \, x(t)$ linear? Is it time invariant?

   Solution:
   Linearity: We study the system operation on a linear combination: let $x(t) = c_1 x_1(t) + c_2 x_2(t)$.
   $y(t) = \cos t(c_1 x_1(t) + c_2 x_2(t)) = c_1 \cos t x_1(t) + c_2 \cos t x_2(t) = c_1 y_1(t) + c_2 y_2(t)..$
   The system is linear.
   Translation invariance in time: $O(x(t - \tau)) = \cos t \, x(t - \tau)$.
   It is different from $y(t - \tau) = \cos(t - \tau) x(t - \tau)$. The system is not translation invariant in time.
   This result ensues from the fact that the multiplier of the input signal is a function of time.

Chapter 2
First and Second Order Systems

In this chapter, the properties of the transfer function and frequency response of first and second order systems are studied on some examples from electrical circuit laws. We show that their properties are governed by the poles (i.e., the zeros of the denominator) of the transfer function which is a rational fraction. A geometric argument based on the location of the poles of the transfer function in the complex plane allows a qualitative interpretation of the behavior of the frequency response with varying frequency. This geometric interpretation is easily generalized to situations with any number of zeros and poles. It proves useful for the understanding of the general behavior of filters. The study begins here with the simplest system, the first order system. Then the second order circuit system is presented thoroughly. The logarithmic Bode representation of the frequency gain is introduced and its advantages demonstrated. The quality factor Q of a resonant circuit is defined.

2.1 First Order System. R, C Circuit

Consider the electrical circuit consisting of a resistor and a capacitor in series (Fig. 2.1). The circuit is powered by an internal resistance-free generator of electromotive force $e(t)$. The charge on one plate of the capacitor is written q, and the voltage across the capacitor noted $v(t) = \frac{q(t)}{C}$.

The generalized Ohm law writes:

$$R\frac{dq}{dt} + \frac{q}{C} = e(t). \tag{2.1}$$

With a system point of view, we write $e(t)$ as the input variable and $v(t)$ as the output variable.

© Springer International Publishing Switzerland 2016
F. Cohen Tenoudji, *Analog and Digital Signal Analysis*,
Modern Acoustics and Signal Processing, DOI 10.1007/978-3-319-42382-1_2

Fig. 2.1 First order system;
R, C circuit

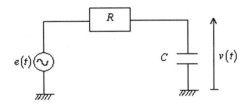

2.1.1 Transfer Function

The Eq. (2.1) can be written in the form of an operator acting on q:

$$\left(R\frac{\mathrm{d}}{\mathrm{d}t} + \frac{1}{C}\right)q = e(t). \tag{2.2}$$

This system is linear and time invariant. According to the fundamental result shown in Chap. 1, when $e(t)$ has the form e^{st}, the charge $q(t)$ on a plate of the capacitor and the voltage $v(t)$ across it will have the same exponential form. This can be checked:

Posing $e(t) = \mathrm{e}^{st}$ and looking for $q(t)$ in the form: $q(t) = B\mathrm{e}^{st}$.
Replacing its expression in Eq. (2.2) we have:

$$RB\frac{\mathrm{d}}{\mathrm{d}t}\mathrm{e}^{st} + \frac{B}{C}\mathrm{e}^{st} = \mathrm{e}^{st}. \tag{2.3}$$

By simplifying by e^{st}, we see that the proposed solution is valid if the following relationship is satisfied:

$$\left(Rs + \frac{1}{C}\right)B = 1. \tag{2.4}$$

or

$$B = \frac{1}{\left(Rs + \frac{1}{C}\right)} = \frac{C}{RCs + 1}. \tag{2.5}$$

The voltage across the capacitor (system output variable) is given by:

$$v(t) = \frac{q}{C} = \frac{1}{RCs + 1}\mathrm{e}^{st} = H(s)\mathrm{e}^{st}. \tag{2.6}$$

We notice that e^{st} is eigenfunction of the system and that $H(s)$ is its transfer function. The circuit transfer function $H(s)$ is thus written

2.1 First Order System. R, C Circuit

13

Fig. 2.2 Pole of the transfer function in the s plane

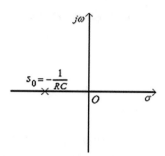

$$H(s) = \frac{1}{RCs + 1}.$$ (2.7)

$H(s)$ is a rational fraction with a simple pole (a simple zero of the denominator) in (See Fig. 2.2)

$$s_0 = -\frac{1}{RC}.$$ (2.8)

We can equivalently write $H(s)$ as

$$H(s) = \frac{-s_0}{s - s_0}.$$ (2.9)

The presence of a single simple pole is the reason for the first-order system name applying to this circuit.

2.1.2 Frequency Response

The frequency response is a particular case of the transfer function. In the function $H(s)$, the variable s is a complex number that will be written in the form: $s = \sigma + j\omega$. s belongs to the complex plane. With reference to the Laplace transformation detailed in Chap. 9, the s plane is also called Laplace plane. This plane is identified by the real axis σ and the imaginary axis $j\omega$.

$$v(t) = \frac{1}{RC(\sigma + j\omega) + 1} e^{\sigma t} e^{j\omega t}.$$ (2.10)

If $\sigma = 0$, that is, for a monochromatic input signal $e(t) = e^{j\omega t}$:

$$v(t) = \frac{1}{1 + jRC\omega} e^{j\omega t} = H(\omega) e^{j\omega t}$$ (2.11)

The frequency response is:

$$H(\omega) = \frac{1}{1 + jRC\omega} \tag{2.12}$$

The angular frequency ω is related to the frequency f by the relationship $\omega = 2\pi f$.

We see, of course, that $e^{j\omega t}$ is also eigenfunction of the system. The frequency response $H(\omega)$ (also called complex gain of the filter) is the transfer function H (s) evaluated on the imaginary axis $\sigma = 0$.

By showing the modulus $|H(\omega)|$ and the argument φ of the frequency response, we write:

$$v(t) = H(\omega)e^{j\omega t} = |H(\omega)|e^{j\varphi}e^{j\omega t} \tag{2.13}$$

Therefore, while the modulus of the input signal $e^{j\omega t}$ is 1, the output signal modulus is $|H(\omega)|$. It appears that the modulus of the frequency response is the gain in amplitude of the signal passing through the filter. The phase shift φ of the output signal relative to the input signal is the argument of the complex gain $H(\omega)$. The magnitude and phase are functions of ω in the general case.

Note: For convenience, the function $H(\omega)$ is called frequency response, although this function is expressed as a function of the angular frequency ω and not of the frequency f.

For the variation of the gain as a function of the frequency f, we replace ω by $2\pi f$ in the expression of $H(\omega)$.

As noted above, the function $e^{j\omega t}$ is the system eigenfunction but the function $\cos \omega t = \frac{e^{j\omega t} + e^{-j\omega t}}{2}$, linear combination of two eigenfunctions, is not. The system response for a cosine input is searched as follows:

If the electromotive force $e^{j\omega x}$ has the form $e(t) = \cos \omega t$, due to the linearity of the system, we can write the answer in the form:

$$v(t) = \frac{1}{2}\left(H(\omega)e^{j\omega t} + H(-\omega)e^{-j\omega t}\right) \tag{2.14}$$

$$v(t) = \frac{1}{2}\left(\frac{1}{1 + jRC\omega}e^{j\omega t} + \frac{1}{1 - jRC\omega}e^{-j\omega t}\right) = \frac{1}{2}\left(\frac{1}{1 + jRC\omega}e^{j\omega t}\right) + c.c. \tag{2.15}$$

c.c. is written to describe a complex conjugate of the previous term within the equation. The sum of a complex number and of its complex conjugate is equal to twice its real part, the following applies:

$$v(t) = \Re e\left(\frac{1}{1 + jRC\omega}e^{j\omega t}\right) \tag{2.16}$$

The notation $\Re()$ means that we must take the real part of the complex expression. It comes:

$$v(t) = \Re\left\{ \left(\frac{1 - jRC\omega}{1 + R^2C^2\omega^2}\right)(\cos \omega t + j \sin \omega t) \right\}$$
$$= \frac{1}{1 + R^2C^2\omega^2}(\cos \omega t + RC\omega \sin \omega t). \tag{2.17}$$

We can rewrite this result in the form:

$$v(t) = \frac{1}{\sqrt{1 + R^2C^2\omega^2}} \left(\frac{1}{\sqrt{1 + R^2C^2\omega^2}} \cos \omega t + \frac{RC\omega}{\sqrt{1 + R^2C^2\omega^2}} \sin \omega t\right), \tag{2.18}$$

or in another form:

$$v(t) = \frac{1}{\sqrt{1 + R^2C^2\omega^2}} \cos(\omega t + \varphi), \tag{2.19}$$

with

$$\cos \varphi = \frac{1}{\sqrt{1 + R^2C^2\omega^2}} \quad \text{and} \quad \sin \varphi = \frac{-RC\omega}{\sqrt{1 + R^2C^2\omega^2}}, \tag{2.20}$$

and then $\tan \varphi = -RC\omega$.

Behavior of the solution at low and high frequencies

At low frequencies, that is to say, when $RC\omega \ll 1$, we see on the solution (2.19) that $v(t) \cong \cos \omega t$. The output signal is in phase with the input signal and has equal amplitude.

At high frequency, when $RC\omega \gg 1$, the solution (2.19) becomes: $v(t) \cong \frac{1}{RC\omega} \sin \omega t$.

The output signal is in quadrature with the input signal with a phase shift $\varphi \cong -\frac{\pi}{2}$ and its amplitude decreases with frequency as $\frac{1}{\omega}$.

Note: Conciseness of the results when expressed in the form of complex exponentials will be compared to the heaviness from those expressed in sine and cosine.

2.1.3   Graphic Representation of the Frequency Response

Since $H(\omega) = \frac{1}{1 + jRC\omega}$, the modulus is:

$$|H(\omega)| = \frac{1}{\sqrt{1 + R^2C^2\omega^2}}, \tag{2.21}$$

and phase

$$\varphi = -\text{Arg}(RC\omega) \tag{2.22}$$

In Fig. 2.3, are represented the modulus and phase of $H(\omega)$.

For the value ω_c of ω such that $RC\omega_c = 1$, the value of the gain modulus is $\frac{1}{\sqrt{2}}$.

Using its value in decibels: $H_{dB} = 20\log_{10}(|H(\omega_c)|) = 20\log_{10}\left(\frac{1}{\sqrt{2}}\right) = -3\,\text{dB}$.

Frequency $f_c = \frac{\omega_c}{2\pi} = \frac{1}{2\pi RC}$ is called the -3 dB cutoff frequency.

It is seen in Fig. 2.3b that the phase variation range goes from $\frac{\pi}{2}$ to $-\frac{\pi}{2}$.

Bode representation

Scale in decibels

In the Bode representation the magnitudes logarithm are represented. As mentioned above, the decibel value of a quantity A is $A_{dB} = 20\log_{10} A$. This unit of measure was introduced by G. Bell to describe the acoustic sensitivity of the human ear (hence the name of this unit). The sensitivity of the ear is logarithmic: if the intensity of a sound is multiplied by 10, the ear feels a multiplication by 2. If the intensity is multiplied by 100, the ear feels a multiplication by 4. This physiological property allows the ear to hear correctly loud sounds, but remain sensitive to very low sounds. Moreover, as will be discussed in Chap. 3, the note of a musical instrument is accompanied by the presence of harmonics whose frequencies are multiples of the fundamental frequency. The amplitudes of these harmonics are specific to each instrument. They can be several tens of times lower than that of the fundamental component. As the ear analyzes the sounds from frequency, its logarithmic sensitivity somehow enhances the amplitude of low harmonics. This allows it to be physiologically sensitive to harmonics, so to the musicality of the instrument. It is important to remember that the representation in logarithm reinforces the low values of a variable relatively to strong values. This property is exploited in the Bode representation with which we may monitor small changes of

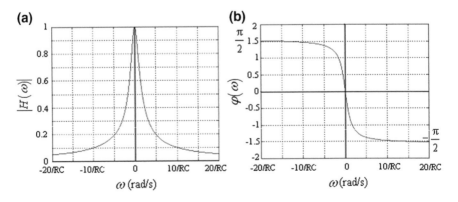

Fig. 2.3 Frequency response of R C circuit. **a** Modulus. **b** Phase

Fig. 2.4 Log–log plot of
gain magnitude (first order
system)

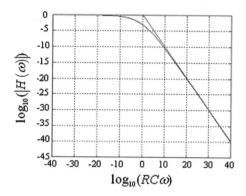

the variable values, whereas in linear representation, they would have been unde-
tectable. This representation has better dynamics. This explains why the gain of the
filters is most often plotted in dB.

Another quality of the logarithmic representation is that a variation with frequency in power law appears as a straight line whose slope gives the value of the
power law coefficient.

By definition, the decibel value of the frequency response is equal to 20 times
the base 10 logarithm of the frequency response modulus.

$$H_{dB} = 20 \log_{10}(|H(\omega)|). \tag{2.23}$$

Assuming that at high frequencies, the system has an asymptotic behavior of the
form $|H(\omega)| \simeq \omega^n$, then $H_{dB} = 20 \log_{10} \omega^n = n\, 20 \log_{10} \omega$. In a logarithmic rep-
resentation $H_{dB} = f(20 \log_{10} \omega)$, the variation is linear.

Figure 2.4 shows the gain in dB of the first order filter. Note the linear
asymptotic behavior of the high-frequency curve. The asymptote passes through the
point $(0, 0)$, that is to say, for the x-axis value $\omega = \frac{1}{RC}$. The slope of the line is -1,
reflecting the asymptotic gain as $\frac{1}{\omega}$ (Fig. 2.4). $|H(\omega)|$ decreases by 20 decibels per
decade (a decade corresponds to a multiplication of the frequency by a factor of 10).
This decrease is also -6 dB per octave (the octave is defined in music as the
interval between two notes when the frequency of a note is twice that of the other.
For example, the frequency of the note C is multiplied by 2 when going on a piano
keyboard from a C to a C immediately above).

2.1.4   Geometric Interpretation of the Variation
of the Frequency Response

It has been shown above that the transfer function is: $H(s) = \frac{-s_0}{s-s_0}$, with $s_0 = -\frac{1}{RC}$.
The frequency response is:

$$H(\omega) = \frac{-s_0}{j\omega - s_0}. \tag{2.24}$$

In the complex plane $s = \sigma + j\omega$. The point of the plane corresponding to the real pole $s_0 = -\frac{1}{RC}$ is noted on Fig. 2.5. The point M is the point $j\omega$ representative of the monochromatic signal to the frequency ω. The complex number in the denominator of $H(\omega)$ can be associated to the vector $\overrightarrow{PM}$. The modulus of $H(\omega)$ is inversely proportional to the length PM of that vector:

$$|H(\omega)| = \frac{|-s_0|}{PM} = \frac{1}{RC}\frac{1}{PM}. \tag{2.25}$$

Using the Pythagorean theorem we write $PM = \sqrt{\omega^2 + \frac{1}{R^2C^2}}$.

We find the variation in function of the frequency of the modulus of $H(\omega)$ according to the variation in length of the segment PM when the point M scans the vertical axis $\sigma = 0$ from $-j\infty$ (frequency $-\infty$) to $+j\infty$ (frequency $+\infty$).

For very high negative frequencies the segment PM is very large, and its inverse is very small. Thus $|H(\omega)|$ is very small. When the frequency decreases in absolute value to the zero frequency, the segment PM decreases, and $|H(\omega)|$ increases.

The segment PM is minimal for $\omega = 0$ and its inverse $|H(\omega)|$ is maximal. The gain will decrease continuously when ω increases from zero, the segment PM continuously growing. As shown on Fig. 2.6. Since the phase of the output signals is equal to the argument of $H(\omega)$,

$$\varphi(\omega) = \text{Arg}(H(\omega)) = \text{Arg}(-s_0) - \text{Arg}(j\omega - s_0).$$

s_0 being real and negative, we have $\varphi(\omega) = -\text{Arg}(j\omega - s_0)$.

The argument of $j\omega - s_0$ is equal to the angle formed by the vector $\overrightarrow{PM}$ with the horizontal axis. When the frequency is largely negative this angle is close to $-\frac{\pi}{2}$, the phase of $H(\omega)$ (the opposite to that angle) is then close to $\frac{\pi}{2}$. The change of phase with frequency is shown Fig. 2.6.

Fig. 2.5 Vector $\overrightarrow{PM}$ situation for a given frequency

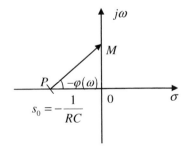

(a)

(b)

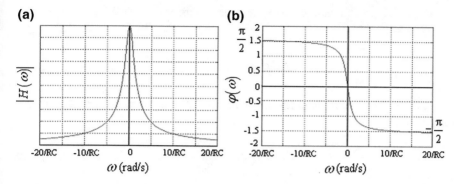

Fig. 2.6 Frequency response of R C circuit after geometric interpretation. **a** Modulus. **b** Phase

Fig. 2.7 R C Circuit with output taken at resistor terminals

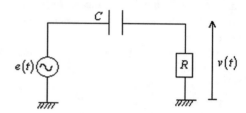

2.1.5 R, C Circuit with Output on the Resistor Terminals

This system is a second example of a first order system. The circuit is identical to that of Sect. 1.1 but the output voltage is taken at the terminals of the resistor (Fig. 2.7). We have the following diagram:

The calculation of the charge across the capacitor is the same as in Sect. 2.1.1. When $e(t) = e^{st}$ we have again:

$$q(t) = \frac{C}{RCs+1} e^{st}. \tag{2.26}$$

$$v(t) = R\frac{dq}{dt} = \frac{RCs}{RCs+1} e^{st} = H(s)e^{st}. \tag{2.27}$$

The transfer function is in this case:

$$H(s) = \frac{RCs}{RCs+1} = -s_0\frac{RCs}{s-s_0} = \frac{s}{s-s_0}. \tag{2.28}$$

The transfer function has a zero in $s = 0$ and a pole in $s_0 = -\frac{1}{RC}$.

Geometric interpretation of the variation of gain with frequency:
We have:

$$H(\omega) = \frac{j\omega}{j\omega - s_0}.$$ (2.29)

As can be seen in Fig. 2.5, the gain modulus is equal to the ratio of two segments:

$$|H(\omega)| = \frac{OM}{PM}.$$ (2.30)

As ω varies, the point M scans upward the axis $\sigma = 0$.

When $|\omega|$ is very large, the lengths of the segments OM and PM are very slightly different, the gain is close to 1. When ω is close to zero, the numerator becomes small while the denominator remains finite. The gain in amplitude $|H(\omega)|$ is close to zero.

The phase is the argument of the numerator of $H(\omega)$ minus the argument of its denominator:

$$\varphi(\omega) = \text{Arg}(j\omega) - \text{Arg}(j\omega - s_0).$$ (2.31)

$\text{Arg}(j\omega)$ equals $-\frac{\pi}{2}$ when $\omega < 0$ and equals $\frac{\pi}{2}$ if $\omega > 0$ (there is a π jump when ω passes through zero). As seen above, $-\text{Arg}(j\omega - s_0)$ varies from $\frac{\pi}{2}$ to $-\frac{\pi}{2}$ when ω varies from $-\infty$ to $+\infty$. The variations of the gain and phase with ω are shown in Fig. 2.8.

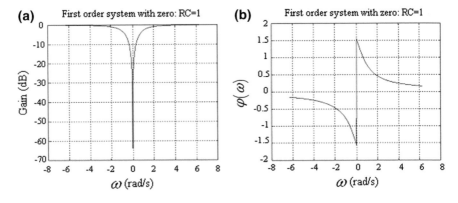

Fig. 2.8 Frequency gain for second R C circuit. **a** Modulus. **b** Phase

2.2 Second Order System. R, L, C Series Circuit

The emf $e(t)$ is applied to the terminals of a circuit composed of an inductor L, a resistor R and a capacitor C in series (Fig. 2.9). As above, the electric charge on a plate of the capacitor is denoted by q, and $v(t)$ is the voltage across the capacitor. Generalized Ohm's law takes the form:

$$L\frac{d^2q}{dt^2} + R\frac{dq}{dt} + \frac{q}{C} = e(t). \tag{2.32}$$

2.2.1 Transfer Function

This system is linear, invariant by translation in time. The circuit transfer function $H(s)$ is obtained by taking $e(t) = e^{st}$ for excitation and seeking $q(t)$ of the form $q(t) = Be^{st}$:

$$LB\frac{d^2}{dt^2}e^{st} + RB\frac{d}{dt}e^{st} + \frac{B}{C}e^{st} = e^{st}. \tag{2.33}$$

That is to solve the equation

$$\left(Ls^2 + Rs + \frac{1}{C}\right)B = 1. \tag{2.34}$$

It is necessary to have the equality $B = \frac{1}{Ls^2 + Rs + \frac{1}{C}} = \frac{C}{LCs^2 + RCs + 1}$.
The voltage across the capacitor is given by:

$$v(t) = \frac{q(t)}{C} = \frac{1}{LCs^2 + RCs + 1}e^{st} = H(s)e^{st}. \tag{2.35}$$

The system transfer function is therefore:

$$H(s) = \frac{1}{LCs^2 + RCs + 1}. \tag{2.36}$$

Fig. 2.9 Second order circuit; R L C in series

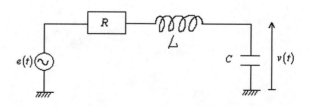

The transfer function is again a rational fraction which must initially determine the poles. The denominator is a polynomial in s. A general property of polynomials with complex coefficients is that they always have roots. These roots belong to the field of complex numbers. In addition, another property of polynomials is that when all the coefficients of the various powers of the variable s are real, the roots are either real or come in complex conjugate pairs.

Search of the Poles of the Transfer Function

The polynomial being of second degree, he always has two roots which will be distinct or multiple. For this reason, this circuit is called a second order filter.

The transfer function has the general form:

$$H(s) = \frac{1}{LCs^2 + RCs + 1} = \frac{1}{LC} \frac{1}{(s - s_1)(s - s_2)}. \tag{2.37}$$

Analysis of the roots of the quadratic polynomial $LCs^2 + RCs + 1$:
The discriminant of the polynomial is:

$$\Delta = R^2 C^2 - 4LC. \tag{2.38}$$

The roots of the polynomial are noted s_1 and s_2.

- If $\Delta > 0$, $s_{1,2} = \dfrac{-RC \pm \sqrt{R^2 C^2 - 4LC}}{2LC} = -\dfrac{R}{2L} \pm \sqrt{\dfrac{R^2}{4L^2} - \dfrac{1}{LC}}. \tag{2.39}$

The two roots are real.

- If $\Delta = 0$, $s_1 = s_2 = -\dfrac{R}{2L}$, the polynomial has a double real root. $\tag{2.40}$

- If $\Delta < 0$, $s_{1,2} = -\dfrac{R}{2L} \pm j\sqrt{\dfrac{1}{LC} - \dfrac{R^2}{4L^2}}$, the two roots are complex conjugate: $\tag{2.41}$

Writing

$$\omega_0 = \sqrt{\frac{1}{LC} - \frac{R^2}{4L^2}}, \tag{2.42}$$

we have:

$$s_{1,2} = -\frac{R}{2L} \pm j\omega_0 \tag{2.43}$$

2.2.2 Second Order System Frequency Response

The poles of the transfer function will condition the frequency response of the system $H(\omega)$, response to a monochromatic input signal of the form $e^{j\omega t}$.

Simply replacing s by $j\omega$ in the expression of $H(s)$, we have

$$H(\omega) = \frac{1}{-LC\omega^2 + jRC\omega + 1} = \frac{1}{LC}\frac{1}{(j\omega - s_1)(j\omega - s_2)}. \tag{2.44}$$

Note that $H(0) = 1$. Calculation programs like Matlab easily enable graphical representation of the modulus and phase of $H(\omega)$.

2.2.3 Geometric Interpretation of the Variation of the Frequency Response

It is interesting to further develop a geometric argument to interpret the variation of the frequency response. Its modulus is:

$$|H(\omega)| = \frac{1}{LC}\frac{1}{|j\omega - s_1||j\omega - s_2|}. \tag{2.45}$$

Having placed the poles s_1 (point P_1) and s_2 (point P_2) in the complex plane $(\sigma, j\omega)$, we see that the modulus of $H(\omega)$ is inversely proportional to the lengths of segments joining point M (representing $j\omega$) to the points P_1 and P_2.

$$|H(\omega)| = \frac{1}{LC}\frac{1}{MP_1 MP_2}. \tag{2.46}$$

The phase is given by the sum of the angles made by the vectors $\overrightarrow{P_1 M}$ and $\overrightarrow{P_2 M}$ with the x-axis:

$$\varphi(\omega) = \text{Arg}(H(\omega)) = -\text{Arg}(j\omega - s_1) - \text{Arg}(j\omega - s_2). \tag{2.47}$$

One can thus deduce qualitatively the following variations of gain and phase:

- If $\Delta > 0$, poles s_1 and s_2 lie on the real axis $\omega = 0$ (Fig. 2.10).

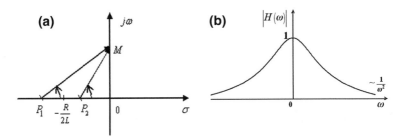

Segments P_1M and P_2M.   **Modulus of frequency response.**

Fig. 2.10 Geometric interpretation in case of two real poles. **a** Poles situation. **b** Gain modulus

Fig. 2.11 Phase variation
given by geometric
interpretation

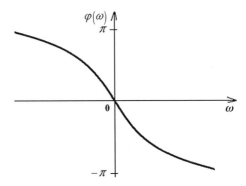

We see in Fig. 2.10 that the maximum gain value is obtained for zero frequency, value of ω for which the segments P_1M and P_2M are minimal. The gain decreases as $\frac{1}{\omega^2}$ when $|\omega| \rightarrow \infty$, each of the two segments P_1M and P_2M growing like $|\omega|$. The circuit behaves as a low pass filter.

When ω is largely negative, angles of the two vectors $\overrightarrow{P_1M}$ and $\overrightarrow{P_2M}$ with the horizontal are each approximately $-\frac{\pi}{2}$; phase will be π.

When ω increases, M scans vertically the axis $\sigma = 0$ and angles vary from $-\frac{\pi}{2}$ to $\frac{\pi}{2}$ (Fig. 2.11). They will be 0 for $\omega = 0$, the phase will be zero. Then as the angles increase toward $\frac{\pi}{2}$ the phase tends toward $-\pi$.

- If $\Delta = 0$, both poles are merged on the real axis (Fig. 2.12). The discussion is similar to the previous case and the system still has a low-pass filter behavior (Fig. 2.13).
- If $\Delta < 0$, the two poles are complex conjugates. We note H_1 and H_2 the projections of P_1 and P_2 on the axis $j\omega$ (Fig. 2.14). For large negative values of ω, we have the same behavior as before, the segments P_1M and P_2M are large and the modulus $|H(\omega)|$ very small, and the phase tends toward π (Figs. 2.14, 2.15 and 2.16).

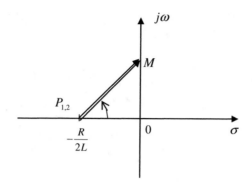

Fig. 2.12 Two poles merged on real axis

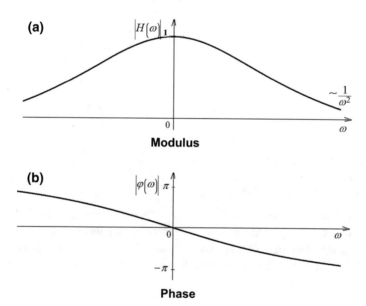

Fig. 2.13 Frequency gain in case of a double real pole. **a** Magnitude. **b** Phase

Fig. 2.14 Vectors P_1M and P_2M situation for a given frequency

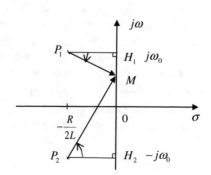

Fig. 2.15 Gain magnitudes for different damping situations

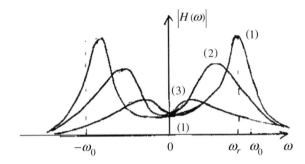

Fig. 2.16 Gain phases for different damping situations

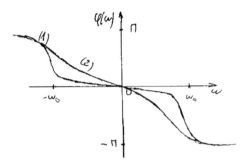

- When ω increases, a first maximum of $|H(\omega)|$ will occur when the product of the lengths of the segments P_1M and P_2M will reach a minimum. This will occur to a first intermediate position of M between H_2 and O.

The more points P_1 and P_2 will be close to the imaginary axis, that is to say, the more $\frac{R}{2L}$ will be small compared to ω_0, the more segments P_1M and P_2M can become smaller and $|H(\omega)|$ can become great. The resonance pulsation ω_r for which $|H(\omega)|$ is maximal will be closer to ω_0 when the points P_1 and P_2 are close to the imaginary axis.

$$|H(\omega)|_{\max} \cong |H(\omega_0)| = \frac{1}{LC}\frac{1}{H_1P_1}\frac{1}{H_1P_2}. \qquad (2.48)$$

The quantity $\frac{R}{2L}$ characterizes the damping of the circuit. Curves 1, 2 and 3 in Fig. 2.15 show the trend of the gain when the damping is increasing (with respect to ω_0). Thus, it is to remember that as the pole is closer to the vertical axis, the resonance is sharper and the resonance frequency nearer to ω_0.

Regarding the phase, it is found that when the pole is close to the vertical axis, the angle of the vector $\overrightarrow{P_1M}$ with horizontal changes abruptly from a value close to $-\frac{\pi}{2}$ to a value close to $\frac{\pi}{2}$ when ω passes through resonance (Fig. 2.16). In the case of strong resonance, phase starts from π and varies from π to 0 when ω passes the value $-\omega_r$ where the phase passes to $-\pi$. In the case of a damped system (pole farther from the vertical axis) angles vary more gradually.

It may be noted to put an end to this discussion that, as before, the gain decreases as $\frac{1}{\omega^2}$ when $|\omega| \to \infty$, both segments P_1M and P_2M in the denominator of the frequency response increasing as $|\omega|$.

On the following graphs showing the magnitude (Fig. 2.17) and phase (Fig. 2.18) of the gain, in the case where $L = 0.1$, $C = 0.1$ and where R was varied by taking the values 0.1, 0.3, 0.5, 0.7, 0.9.

The module is shown in linear scale:

$\frac{1}{RC}$ represents roughly the half width of the modulus of $H(\omega)$.

Phase varies from π to $-\pi$.

The resonance frequency ω_r which is the abscissa of the maximum of the frequency response modulus, is analytically determined by annulling the derivative

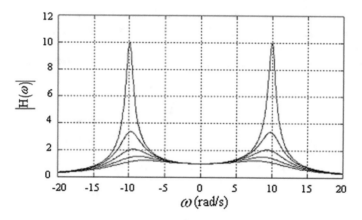

Fig. 2.17 Numerical simulations: gain magnitudes for different damping

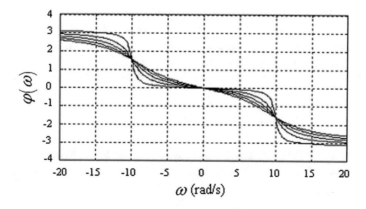

Fig. 2.18 Numerical simulations: gain phases for different damping

of $|H(\omega)|$. The denominator of the frequency responsemodule includes the sum of squares of the real and the imaginary parts:

$$|H(\omega)| = \sqrt{\frac{1}{(1 - LC\omega^2)^2 + R^2C^2\omega^2}}. \tag{2.49}$$

The resonant frequency ω_r is the frequency for which $\frac{d|H(\omega)|}{d\omega} = 0$.

This amounts to calculating the solutions of the equation canceling the derivative of the denominator. It comes:

$$\omega_r = \pm\sqrt{\frac{1}{LC} - \frac{R^2}{2L^2}}. \tag{2.50}$$

There are two solutions with opposite signs. Only the positive frequency is significant for real signals. It can be seen on the positive root that as the resistance R increases it causes the decrease of the resonant frequency, as it was anticipated qualitatively.

To calculate the filter gain at the resonance, this root is reported in the gain expression. It comes:

$$|H(\omega_r)| = \sqrt{\frac{1}{\left(1 - LC\omega_r^2\right)^2 + R^2C^2\omega_r^2}} = \frac{1}{RC\omega_0}. \tag{2.51}$$

Where ω_0 is given by (2.42). This result is remarkable for its simplicity.

2.2.4   Bode Representation of the Gain

Figure 2.19 shows the variation of $20\log_{10}(|H(\omega)|) = f\left(20\log_{10}\frac{\omega}{\omega_r}\right)$ for the following values of the system parameters: $R = 1\ \Omega$, $L = 10^{-4}$ H, $C = 10^{-6}$ F.

This gives the resonant frequency $\omega_r = 9.975 \times 10^4$ rad/s. The gain for the resonance frequency is approximately equal to 20 dB. The theoretical gain in decibels for the resonance frequency is calculated from the formula (2.51). It is

$$20\log_{10}|H(\omega_r)| = 20\log_{10}\frac{1}{RC\omega_0} = 20.01\ \text{dB}. \tag{2.52}$$

The system resonant frequency $\omega_r = 9.975 \times 10^4$ rad/s is slightly lower than the frequency $\omega_0 = 9.9875 \times 10^4$ rad/s, imaginary part of the positive pole frequency. Resonance is sharp.

2.2 Second Order System. R, L, C Series Circuit

29

Fig. 2.19 Log-log plot of gain magnitude (second order system)

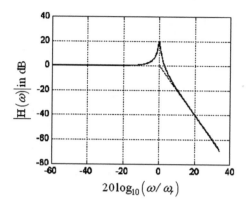

Note the linear behavior of the curve at high frequencies. As seen in Fig. 2.19, at high frequencies $|H(\omega)|$ decreases of 40 decibels per decade (when the frequency is multiplied by 10), as characteristic of the decay in $\frac{1}{\omega^2}$. This decrease corresponds to -12 dB per octave (when the frequency is multiplied by 2).

The asymptotic line to the high frequency curve (dotted line in Fig. 2.19) passes through the point $(0, 0)$, that is to say, for the abscissa value $\omega = \omega_r$. Please note that this is only true in the case of sharp resonance that is specified in the following paragraph.

2.3 Case of Sharp Resonance

We have seen that in the case of a sharp resonance, the resonance frequency which corresponds to the maximum of $|H(\omega)|$ is near ω_0. We can use in this case an approximate expression of $|H(\omega)|$ in the vicinity of the resonance. Geometrically, when ω is near ω_0, we allocate all of the variation of the modulus $|H(\omega)|$ to the variation of the segment MP_1. In the scheme of this approximation, the gain is maximum when M is in H_1:

$$|H(\omega)|_{\max} \cong |H(\omega_0)| = \frac{1}{LC} \frac{1}{H_1 P_1} \frac{1}{H_1 P_2}. \tag{2.53}$$

We have approximately $H_1 P_2 \cong H_1 H_2$, then:

$$|H(\omega)|_{\max} \cong |H(\omega_0)| = \frac{1}{LC} \frac{1}{H_1 P_1} \frac{1}{H_1 H_2}. \tag{2.54}$$

Under this approximation of sharp resonance, as $H_1 P_1 = \frac{R}{2L}$ and $H_1 H_2 = 2\omega_0$, we have then:

$$|H(\omega)|_{max} \cong \frac{1}{LC}\frac{2L}{R}\frac{1}{2\omega_0} = \omega_0^2 \frac{2L}{R}\frac{1}{2\omega_0} = \frac{L\omega_0}{R}. \tag{2.55}$$

Bandwidth at −3 dB of the resonator

Noting M_1 the point on the imaginary axis as $H_1M_1 = H_1P_1$ and ω_1 the corresponding angular frequency, we have

$$|H(\omega_1)| = \frac{1}{LC}\frac{1}{M_1P_1}\frac{1}{M_1P_2} \cong \frac{1}{LC}\frac{1}{\sqrt{(M_1H_1)^2 + (H_1P_1)^2}}\frac{1}{H_1P_2}$$
$$= \frac{1}{LC}\frac{1}{\sqrt{2(H_1P_1)^2}}\frac{1}{H_1H_2}. \tag{2.56}$$

Therefore

$$|H(\omega_1)| \cong \frac{1}{\sqrt{2}}|H(\omega)|_{max}. \tag{2.57}$$

Expressing this ratio in decibels:

$$|H(\omega_1)|_{dB} = 20\log_{10}|H(\omega_1)| \cong 20\log_{10}|H(\omega)|_{max} + 20\log_{10}\frac{1}{\sqrt{2}}, \tag{2.58}$$
$$|H(\omega_1)|_{dB} = |H(\omega)|_{max(dB)} - 3\,dB.$$

At point M_2 (pulsation ω_2) symmetrical of M_1 with respect to H_1, the attenuation is also 3 dB relatively to the maximum gain of the filter. Bandwidth at −3 dB is then as follows:

$$\Delta\omega = \omega_2 - \omega_1 = 2H_1P_1 = \frac{R}{L}.$$

2.4   Quality Factor Q

We name Quality factor Q the ratio

$$Q = \frac{\omega_0}{\Delta\omega} \tag{2.59}$$

The sharper the resonance, the smaller $\Delta\omega$ and the higher Q.

In this case:

$$\omega_0 = \sqrt{\frac{1}{LC} - \frac{R^2}{4L^2}} \quad \text{and} \quad Q = \frac{\sqrt{\frac{1}{LC} - \frac{R^2}{4L^2}}}{\frac{R}{L}}. \tag{2.60}$$

Since the damping is small, the second term in the root can be neglected when compared to the first term. Then

$$Q \cong \sqrt{\frac{1}{LC}\frac{L}{R}} = \sqrt{\frac{L}{C}\frac{1}{R}}. \tag{2.61}$$

Or, writing approximately

$$LC\omega_0^2 = 1, \quad Q = \frac{L\omega_0}{R}. \tag{2.62}$$

It is noted that, in the case of sharp resonance, the value of Q is equal to the maximum gain at resonance. Indeed, it has been seen that $|H(\omega_r)| = \frac{1}{RC\omega_0}$. As in the case of sharp resonance, the relationship $LC\omega_0^2 = 1$ is approximately satisfied, it finally comes $|H(\omega_r)| = \frac{L\omega_0}{R} = Q$, as it had been shown geometrically in the preceding paragraph (Eq. (2.55)).

Decrease over time in the amplitude of the eigenfunctions corresponding to the values of the poles

The eigenfunctions of the resonant system for values of s equal to those of the poles have the form

$$e^{s_{1,2}t} = e^{-\frac{R}{2L}t \pm j\omega_0 t} = e^{-\frac{R}{2L}t}e^{\pm j\omega_0 t} \tag{2.63}$$

The amplitude of these functions varies with time as $e^{-\frac{R}{2L}t}$. In a pseudoperiod $T_0 = \frac{2\pi}{\omega_0}$, this amplitude will vary by a factor

$$e^{-\frac{R}{2L}T_0} = e^{-\frac{R}{2L\omega_0}2\pi} = e^{-\frac{\pi}{Q}}. \tag{2.64}$$

When the Q-factor is great compared to 1, we can perform a limited expansion of the exponential and write: $e^{-\frac{\pi}{Q}} \cong 1 - \frac{\pi}{Q} + \ldots$

In a pseudoperiod, the amplitudes of functions $e^{s_1 t}$ and $e^{s_2 t}$ decrease by a factor $\frac{\pi}{Q}$. It will be shown in the following that the following linear combination of these functions $e^{s_{1,2}t}$ is the response of second order system in a very short pulse (Dirac pulse). This impulse response has the form:

$$h(t) = \frac{1}{LC}\frac{1}{(s_1 - s_2)}(e^{s_1 t} - e^{s_2 t})U(t). \tag{2.65}$$

In practice, in the case of complex conjugate poles, one measures the Q-factor from the decay of $h(t)$ during the pseudoperiod T_0.

Summary

The important first and second order electrical R, L, C circuit systems were studied in this chapter. The position of the poles of the transfer functions was used for qualitatively explaining the variation in frequency of the module and phase responses. This interpretation is fundamental in understanding the behavior of these filters and provides a generalized view of the frequency response of electronic systems. The Bode representation has been presented. The concept of quality factor used to characterize the properties of many physical systems was introduced.

Exercises

I. Consider the circuit composed of the series arrangement of a resistor $R = 100\ \Omega$, an inductor coil value $L = 0.01$ H, and a capacitance $C = 10^{-10}$ F. Note $e(t)$ the voltage across the assembly and $v(t)$ the voltage across the capacitor.

1. It is assumed that the emf $e(t)$ has the form $e(t) = e^{st}$ where s is a complex number capacitor $(s = \sigma + j\omega)$.

   (a) Give the expression of the voltage $v(t)$.
   (b) Give the expression of the filter transfer function. What are the poles of this transfer function? Represent the position of the poles.
   (c) Give the expression of the filter's frequency response. By a geometric argument based on the position of the poles, give the aspect of the variation of gain with frequency module.

2. Note that the transfer function can be written as a product of two terms of the first order $H(s) = H_1(s)H_2(s)$. From the variation of $|H_1(\omega)|$ with ω, give the -3 dB bandwidth of the first filter. By noticing that $|H_2(\omega)|$ remains approximately constant in the vicinity of the resonance, give the bandwidth at -3 dB of $|H(\omega)|$.

II. Consider again the circuit including the elements R, L and C placed in series with the output at the resistor terminals this time. Show that the transfer function is in this case:

$$H(s) = \frac{RCs}{LCs^2 + RCs + 1}.$$

Locate the zeros of $H(s)$ in the complex plane. Show that the circuit does not allow the continuous to pass (the frequency response is zero at zero frequency). Can this system keep a resonator character? Show that the resonance frequency is equal to $\omega_{00} = \sqrt{\frac{1}{LC}}$ whatever damping. Explain qualitatively that the presence of the zero of the transfer function pushes the positive resonance

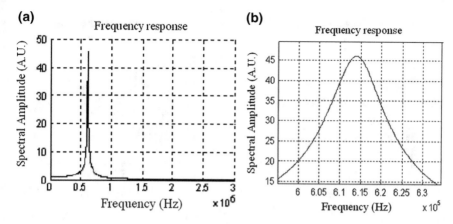

Fig. 2.20 Gain magnitude (**a**); with zoom in (**b**)

frequency toward higher frequencies, frequency as the negative pole frequency tends to decrease.

Which is the gain of the resonance filter? Show that the gain decreases at high frequencies.

Qualitatively, observe that the presence of the zero of the transfer function at $\omega = 0$ pushes the positive resonance frequency toward higher frequencies, while the negative pole tends to decrease that frequency. What is the filter gain at resonance? Show that the gain decreases as $\frac{1}{\omega}$ at high frequencies.

III. Create a circuit of the second order by arranging an inductor L, a resistor R and a capacitor C in series. The input signal is feeding the ensemble and the output signal is taken across the capacitance.

(A) The modulus of this filter frequency response is given by Fig. 2.20:

1. What is the value of the quality Q-factor of the circuit?
2. (a) Making the approximation of a sharp resonance, taking $R = 4.7\ \Omega$, evaluate L and C knowing that the resonant frequency is precisely $6.1389\ 10^5$ Hz.
   (b) Place the poles of the filter transfer function in the Laplace plane.

(B) The impulse response of that filter is given in Fig. 2.21.
Evaluate from these curves L and C the quality factor of the circuit, still taking $R = 4.7\ \Omega$.

Solution:

(A) In the graph of the frequency response we can estimate its maximum amplitude at about 46. In the course, it has been shown that the maximum amplitude is equal to the Q-factor. So we evaluate $Q = 46$.

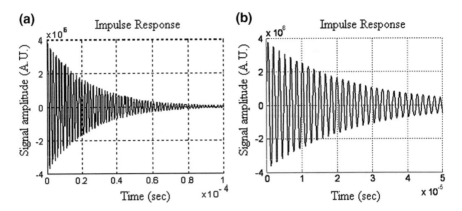

Fig. 2.21 Impulse response (**a**); with zoom in (**b**)

Second method for determining Q: $Q = \frac{\omega_0}{\Delta\omega} = \frac{f_0}{\Delta f}$ (Δf is the bandwidth at -3 dB). The amplitude at -3 dB is estimated to $\frac{46}{\sqrt{2}} = 32.5$.

On the graph of the frequency response the resonance frequency is seen to be $f_0 = 6.14 \times 10^5$ Hz. The frequencies for which the frequency response is attenuated by -3 dB are 6.21×10^5 Hz and 6.075×10^5 Hz, and It can be inferred that $Q = \frac{6.14}{(6.21-6.075)} = 45.5$, which value must be equal to the value given by the first method, the difference being due to uncertainties determinations on the graph. Since $Q = 45.5 = \frac{L\omega_0}{R}$, the resonance frequency in this case of sharp resonance is given by: $LC\omega_0^2 = 1$. It comes $L = 5.510^5$ H and $C = 1.2 \times 10^{-9}$ F.

(B) Determination from the impulse response: we measure graphically the pseudoperiod T_0 of the signal and we deduce $\omega_0 = \frac{2\pi}{T_0}$. In a pseudoperiod, the amplitude varies by the factor $e^{-\frac{\pi}{Q}}$. We deduce Q from it. We then calculate the constants L and C of the circuit knowing R = 4.7 Ω.

Chapter 3
Fourier Series

Fourier series have played an important role in the understanding and the development of signal analysis. The original interest was for music and the fact that the notes of many instruments are composed of frequencies which are multiples of a fundamental frequency. These tones are called harmonics as their combination is harmonious, pleasant to ear.

In the previous two chapters, we have highlighted the fundamental role played by exponential functions e^{st} and particularly by the exponential $e^{j\omega t}$ representing periodic, monochromatic signals. In this chapter, we study the decomposition-reconstruction of periodic signals in Fourier series. We study with a graphic example the idea behind the calculation of a coefficient which is the signal multiplication by a sine function followed by integration in time. We study the effect of limitation of the number of terms of the series on the signal reconstruction (Gibbs phenomenon) and the optimal coefficients of the reconstructing series. Hilbert spaces, which generalize these concepts, are introduced. We show that the functions $e^{j\omega t}$ are the eigenfunctions of the Hermitian operator $i\frac{d}{dt}$ and can be used as an orthogonal basis of development of periodic functions. At the end of the chapter, we illustrate the fact that frequency analysis and displays in logarithmic scale are favorable in the analysis of sounds and the nonlinearity of a system.

An example of a monochromatic signal is the sound generated by a tuning fork. Each note of the musical scale corresponds to a frequency well determined, for example,. A_4 corresponds to the frequency 440 Hz. The sound of a tuning fork which mainly consists of a sine wave is musically poor. A property common to all musical instruments is that the fundamental note is accompanied by harmonic frequencies, that is to say, by sound components whose frequencies are multiples of the fundamental frequency.

© Springer International Publishing Switzerland 2016
F. Cohen Tenoudji, *Analog and Digital Signal Analysis*,
Modern Acoustics and Signal Processing, DOI 10.1007/978-3-319-42382-1_3

For example, for a vibrating string under tension, fixed at its ends as is the case for a guitar or a piano, the natural frequencies of vibration of the string are given by the expression:

$$f_n = n\frac{c}{2l}, \tag{3.1}$$

where n is a positive integer, and c is the speed of the vibration along the string length. The speed is given by

$$c = \sqrt{\frac{T}{\rho}}, \tag{3.2}$$

with T the tension and ρ the mass per unit length of the string.

The excitation of the string by hitting or rubbing generates, in proportions that depend upon the instrument and the excitation mode, the fundamental frequency $f_1 = \frac{c}{2l}$ and its harmonics. It is its richness in harmonics which partially gives its specificity, timbre, to the note created by a musical instrument.

Similarly, in the spoken language, the vowels are compound sounds. They are the addition of signals whose frequencies are integer multiples of a fundamental frequency $f_1 = \frac{\omega_1}{2\pi}$ (this frequency is called the pitch). Physically, the sound generation of a vowel is explained as follows: The air comes out of the lungs in a continuous flow. The vocal cords (sorts of membranes located in the pharynx) periodically seal off the air. A pressure sensor placed downstream of the cords would measure a sequence of pulses in air pressure. The fundamental frequency of these pulses, the pitch, is about 100 Hz for men and approximately 200 Hz for women. These very short repetitive pulses—in an ideal mathematical modeling, we would speak of a Dirac comb which will be defined later in this course—are very rich in harmonics (with existence of harmonics up to the ranks 40 or 50). The sound of one vowel differs from that of another vowel by the relative importance of the different harmonics. The dimensions of sound resonators that are the larynx, nasal cavity and mouth determine the importance of these harmonics.

Mathematically, that composition of harmonic sounds can be noted as a sum of exponentials by the formula:

$$f(t) = \sum_{n=-\infty}^{+\infty} c_n e^{jn\omega_1 t}. \tag{3.3}$$

This composition is called a Fourier series. The development of a function in a Fourier series and the calculation of coefficients of the series are detailed in the next paragraph.

3.1 Decomposition of a Periodic Function in Fourier Series

Let $f(t)$ be a real or complex periodic function of time with period T_1. Its Fourier analysis is based on a sequence of functions $\varphi_n(t)$, periodicals, with period $\dfrac{T_1}{n}$, of exponential or sinusoidal shapes:

$$\varphi_n(t) = e^{jn\omega_1 t}, \text{ where } n \text{ is an integer and } \omega_1 = \frac{2\pi}{T_1} \tag{3.4}$$

The choice of these functions is dictated by the fact that they are eigenfunctions of most physical systems and that any two functions $\varphi_n(t)$ and $\varphi_m(t)$ are orthogonal when $m \neq n$; that is to say that their scalar product is zero. The scalar product of these two functions is defined by

$$\langle \varphi_n(t), \varphi_m(t) \rangle = \frac{1}{T_1} \int_0^{T_1} \varphi_n(t)\varphi_m^*(t)dt = \frac{1}{T_1} \int_0^{T_1} e^{j(n-m)\omega_1 t}dt. \tag{3.5}$$

In the integral appears the complex conjugate $\varphi_m^*(t)$ of the second term $\varphi_m(t)$ contained in the bracket in the left-hand side. By performing the integration we verify that:

$$\langle \varphi_n(t), \varphi_m(t) \rangle = \begin{vmatrix} 0 & \text{if} & n \neq m \\ 1 & \text{if} & n = m \end{vmatrix}. \tag{3.6}$$

Thus, the functions $\varphi_n(t)$ and $\varphi_m(t)$ are orthogonal if $m \neq n$.

We also write the last result in the condensed form $\langle \varphi_n(t), \varphi_m(t) \rangle = \delta(m - n)$, where $\delta(m - n)$ is the Kronecker symbol, equal to 0 if $n \neq m$ and to 1 if $n = m$. Thus, the functions $\varphi_n(t)$ are orthonormal.

The Fourier coefficients c_n of the function $f(t)$ are defined as the projections based of $f(t)$ on functions $\varphi_n(t)$. That is to say, they are given by the scalar product

$$\langle f(t), \varphi_n(t) \rangle = \frac{1}{T_1} \int_0^{T_1} f(t)\varphi_n^*(t)dt = \frac{1}{T_1} \int_0^{T_1} \sum_{n'=-\infty}^{+\infty} c_{n'}\varphi_{n'}(t)\varphi_n^*(t)dt$$

$$= \frac{1}{T_1} \sum_{n'=-\infty}^{+\infty} c_{n'} \int_0^{T_1} \varphi_{n'}(t)\varphi_n^*(t)dt = \frac{1}{T_1} \sum_{n'=-\infty}^{+\infty} c_{n'}\delta(n - n') = c_n.$$

$$c_n = \langle f(t), \varphi_n(t) \rangle = \frac{1}{T_1} \int_0^{T_1} f(t)e^{-jn\omega_1 t}dt. \tag{3.7}$$

Note that the integration interval extends over a signal period. It is easily shown that the coefficient c_n does not depend on the start time of the integration interval, but on the width of this interval only.

The fundamental theorem on Fourier series states that for a function $f(t)$ continuous and continuous derivative except for a finite number of points in a period, the infinite sum $\sum_{n=-\infty}^{+\infty} c_n \varphi_n(t)$ converges to half the sum of the limits to the left and right values of the function abscissa t:

$$\sum_{n=-\infty}^{+\infty} c_n \varphi_n(t) = \frac{1}{2}(f(t+0) + f(t-0)). \tag{3.8}$$

If the function $f(t)$ is continuous at time t, we will have:

$$\sum_{n=-\infty}^{+\infty} c_n \varphi_n(t) = f(t) \tag{3.9}$$

The expression in the development of exponential functions is then:

$$f(t) = \sum_{n=-\infty}^{+\infty} c_n e^{jn\omega_1 t}.$$

The ensemble of periodic functions $f(t)$ of period T_1 is a vector space. Indeed this set satisfies the following conditions of definition of a vector space: The scalar multiplication (by a complex or real) of a function of the space (periodic function of period T_1) also belongs to the space (it is also periodic with period T_1). Furthermore, linear combinations of any two functions of the space belong to it (any linear combination of periodic functions of period T_1 is also periodic with period T_1).

Since any function $f(t)$ of this space can be generated by a linear combination of functions $e^{jn\omega_1 t}$, we say that the infinite set of functions $e^{jn\omega_1 t}$ is a basis of the vector space of periodic functions of time with period $T_1 = \frac{2\pi}{\omega_1}$. The relations (3.6) show that this basis is orthonormal.

Development in the Particular Case of a Real Function
Since $f(t)$ is assumed real, then $f(t) = f^*(t)$. In the development of $f(t)$ we have then:

$$f(t) = \sum_{n=-\infty}^{+\infty} c_n e^{jn\omega_1 t} = f^*(t) = \sum_{n'=-\infty}^{+\infty} c_{n'}^* e^{-jn'\omega_1 t}. \tag{3.10}$$

By comparing the two sums, posing $n' = -n$, we get

$$\sum_{n=-\infty}^{+\infty} c_n e^{jn\omega_1 t} = \sum_{n=-\infty}^{+\infty} c_{-n}^* e^{jn\omega_1 t}. \tag{3.11}$$

We have then for all n:

$$c^*_{-n} = c_n, \quad \text{or} \quad c_{-n} = c^*_n, \forall n. \tag{3.12}$$

So, Fourier coefficients of terms of the series for negative frequencies are the complex conjugates of the coefficients of the terms for symmetric positive frequencies.

Discussion of the Complex Character of Coefficients c_n

It is seen that the coefficients c_n are complex in the general case, by expanding the exponential in the integral (3.7),

$$c_n = \frac{1}{T_1} \int_0^{T_1} f(t) e^{-jn\omega_1 t} dt = \frac{1}{T_1} \int_0^{T_1} f(t) \cos n\omega_1 t \, dt - \frac{j}{T_1} \int_0^{T_1} f(t) \sin n\omega_1 t \, dt. \tag{3.13}$$

As we can shift both boundaries of the integration interval without changing the value of the integral, we also have:

$$c_n = \frac{1}{T_1} \int_{-\frac{T_1}{2}}^{\frac{T_1}{2}} f(t) \cos n\omega_1 t \, dt - \frac{j}{T_1} \int_{-\frac{T_1}{2}}^{\frac{T_1}{2}} f(t) \sin n\omega_1 t \, dt. \tag{3.14}$$

Some interesting special cases are discussed in the following.

In the case where the function $f(t)$ is real, from (3.14) we can write: $c_n = a_n - jb_n$ with a_n and b_n real and given by

$$a_n = \frac{1}{T_1} \int_{-\frac{T_1}{2}}^{\frac{T_1}{2}} f(t) \cos n\omega_1 t \, dt \quad \text{and} \quad b_n = \frac{1}{T_1} \int_{-\frac{T_1}{2}}^{\frac{T_1}{2}} f(t) \sin n\omega_1 t \, dt. \tag{3.15}$$

Then:

$$f(t) = \sum_{n=-\infty}^{+\infty} (a_n - jb_n) e^{jn\omega_1 t} = \sum_{n=-\infty}^{+\infty} (a_n - jb_n)(\cos n\omega_1 t + j \sin n\omega_1 t). \tag{3.16}$$

As $f(t)$ is real, taking the real part of the right side of (3.16) we have:

$$f(t) = \sum_{n=-\infty}^{+\infty} a_n \cos n\omega_1 t + \sum_{n=-\infty}^{+\infty} b_n \sin n\omega_1 t, \tag{3.17}$$

sum of two series with a_n and b_n real coefficients given by (3.15).

The index n in the previous sums varies from minus infinity to plus infinity. Due to parity of cosine and sine functions contained in Eq. (3.15), we see that we have: $a_{-n} = a_n$ and $b_{-n} = -b_n$. One can reduce the summations interval to positive or zero values of n and write the relation (3.17) in the form:

$$f(t) = a_0 + 2\sum_{n=1}^{+\infty} a_n \cos n\omega_1 t + 2\sum_{n=1}^{+\infty} b_n \sin n\omega_1 t. \qquad (3.18)$$

Note the appearance of the factor of 2 resulting from the combination of positive and negative frequencies $e^{jn\omega_1 t}$ and $e^{jn\omega_1 t}$ exponential. The coefficient 2 could be removed in (3.18). By changing the notation, this equation becomes:

$$f(t) = \frac{A_0}{2} + \sum_{n=1}^{+\infty} A_n \cos n\omega_1 t + \sum_{n=1}^{+\infty} B_n \sin n\omega_1 t, \qquad (3.19)$$

with

$$A_n = \frac{2}{T_1} \int_{-\frac{T_1}{2}}^{\frac{T_1}{2}} f(t) \cos n\omega_1 t\, dt \quad \text{and} \quad B_n = \frac{2}{T_1} \int_{-\frac{T_1}{2}}^{\frac{T_1}{2}} f(t) \sin n\omega_1 t\, dt. \qquad (3.20)$$

If the function $f(t)$ is real and even, we have in addition $b_n = 0$, because it is the integral of an even function on a symmetrical interval around $t = 0$. Similarly, if the function $f(t)$ is real and odd, we will have $a_n = 0$, since the integral of an odd function over a symmetric interval around $t = 0$ is zero.

To conclude this discussion, the focus is on the development of Fourier of the function $f(t) = \cos(\omega_1 t + \varphi)$. By expanding the cosine:

$$f(t) = \cos \omega_1 t \cos \varphi - \sin \omega_1 t \sin \varphi = 2a_1 \cos \omega_1 t + 2b_1 \sin \omega_1 t,$$

with $2a_1 = \cos \varphi$ and $2b_1 = -\sin \varphi$.

The signal phase determines the distribution of the power of the signal (the magnitude of the signal; Power will be precisely defined in the next paragraph) between the real and the imaginary part of the Fourier coefficient c_1. If the phase is zero, the power is carried by the real part a_1 of the Fourier coefficient. If the phase is $\frac{\pi}{2}$, the power is carried by the imaginary part b_1 of the Fourier coefficient.

Illustration on an Example of the Calculation of the Fourier Coefficients

It is interesting here to illustrate how the dot product of $f(t)$ of period T_1 with the base functions $\cos n\omega_1 t$ and $\sin n\omega_1 t$ allows these functions to analyze the signal shape. In this example, the function $f(t)$ is a square wave, even, with zero mean and period T_1. On Fig. 3.1a Top) are shown $f(t)$ as well as the function $\cos \omega_1 t$. It is seen that the cosine follows the slow variations of the function. In Fig. 3.1b (Top) is the product $f(t) \cos \omega_1 t$ represented over one period. It is noted that this product is

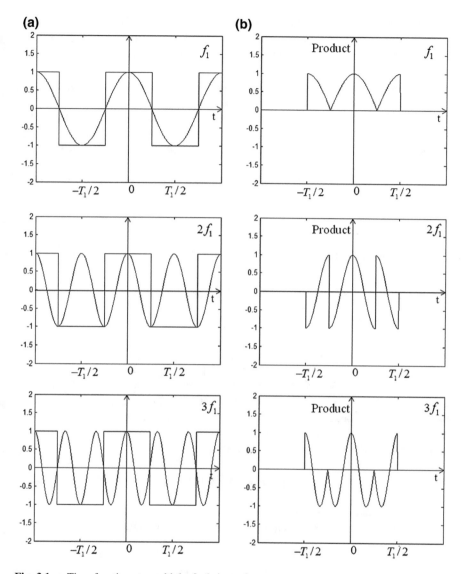

Fig. 3.1 a Time functions to multiply; **b** their products

always positive. Its integral over the period will have a relatively large positive value.

On Fig. 3.1a (Middle) the function $f(t)$ and $\cos 2\omega_1 t$. On Fig. 3.1b (Middle) is the product $f(t) \cos 2\omega_1 t$ over one period. We see that this product oscillates around zero. Its integral will be small. In fact, the integral will be zero because the negative oscillations compensate exactly in this case the positive oscillations. In consequence, the Fourier coefficient of harmonic 2 will be zero. On Fig. 3.1a (Bottom) are shown $f(t)$ as well as the function $\cos 3\omega_1 t$. On Fig. 3.1b (Bottom) is the product $f(t) \cos 3\omega_1 t$ over one period. It varies a lot around zero. The integral will be small. In fact, the integral will be negative because the negative part of the oscillations outweighs the positive ones.

Thus, the Fourier coefficient of harmonic 3 will be negative. We could do the same for higher orders n. For functions of the form $\sin n\omega_1 t$, the Fourier coefficients are all zero since they result from the integral of the product of an even by an odd function. The principle of Fourier analysis which is the integration of the product of the function to be analyzed by the trigonometric analysis functions is well understood in this example.

The exact calculation of the Fourier coefficients of a square signal will be made by means of an exercise later in this chapter.

3.2   Parseval's Theorem for Fourier Series

The power $P(t)$ of a complex signal $f(t)$ is defined as follows:

$$P(t) = f(t)f^*(t), \tag{3.21}$$

with $f^*(t)$ the complex conjugate of $f(t)$.

This definition is consistent with that of the power in an electrical circuit element which is equal to $P(t) = v(t)i^*(t)$, where $v(t)$ is the potential difference across the element and $i(t)$ is the current through that element. In signal analysis, $f(t)$ plays both the role of voltage and current. One can say that $f(t)$ appears as the voltage across an element with impedance $1\ \Omega$.

In the case of a real signal we obviously have

$$P(t) = f^2(t). \tag{3.22}$$

The power depends on time; we will speak of $P(t)$ as an instantaneous power. The signal energy is the integral of the power on the time axis:

$$E = \int_{-\infty}^{\infty} P(t)\mathrm{d}t = \int_{-\infty}^{\infty} f(t)f^*(t)\mathrm{d}t. \tag{3.23}$$

When the signal $f(t)$ is periodic of period T_1, the energy is infinite. For this type of signals, the focus is on the average signal power P_m over one period. It is defined by:

$$P_m = \frac{1}{T_1} \int_0^{T_1} f^*(t)f(t)\mathrm{d}t. \tag{3.24}$$

Replacing the functions by their Fourier developments, and setting $\varphi_n(t) = e^{jn\omega_1 t}$, with $\omega_1 = \frac{2\pi}{T_1}$, we write

$$P = \frac{1}{T_1} \int_0^{T_1} \sum_{n=-\infty}^{+\infty} c_n^* \varphi_n^*(t) \sum_{n'=-\infty}^{+\infty} c_{n'} \varphi_{n'}(t) dt. \tag{3.25}$$

Assuming that the mathematical conditions for reversing the order of the summations are verified, and using the fact that the functions $\varphi_n(t)$ are orthonormal over one period, we write:

$$P_m = \frac{1}{T_1} \sum_{n=-\infty}^{+\infty} \sum_{n'=-\infty}^{+\infty} c_n^* c_{n'} \int_0^{T_1} \varphi_n^*(t) \varphi_{n'}(t) dt$$

$$= \sum_{n=-\infty}^{+\infty} \sum_{n'=-\infty}^{+\infty} c_n^* c_{n'} \delta(n - n') = \sum_{n=-\infty}^{+\infty} c_n^* c_n = \sum_{n=-\infty}^{+\infty} |c_n|^2. \tag{3.26}$$

Thus we have the following relationship, which is the expression of the Parseval's theorem for Fourier series

$$P_m = \frac{1}{T_1} \int_0^{T_1} f^*(t) f(t) dt = \sum_{n=-\infty}^{+\infty} |c_n|^2. \tag{3.27}$$

This result tells that the average power of the signal over a period is the infinite sum of the squared moduli of the coefficients of the Fourier series. It is interesting to note in this formula that the quantities $|c_n|^2$ have the dimensionality of a power.

The squared modulus $|c_n|^2$ is the contribution to the average power of the harmonic signal $e^{jn\omega_1 t}$ of order n.

Example of Decomposition of a Periodic Signal
Let $f(t)$ be the square periodical signal (Fig. 3.2), of period T_1, formed by the repetition of the pattern $\Pi_T(t)$, which is a symmetrical rectangular window of width T, equal to 1 for $|t| < \frac{T}{2}$ and 0 zero elsewhere:

$$\Pi_T(t) = \begin{vmatrix} 1 & \text{for} & |t| < \frac{T}{2} \\ 0 & \text{elsewhere} \end{vmatrix}. \tag{3.28}$$

We may write:

$$f(t) = \sum_{n=-\infty}^{+\infty} \Pi_T(t - nT_1). \tag{3.29}$$

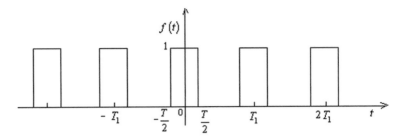

Fig. 3.2 Square periodic signal

It is necessary that $T < T_1$ in order that $f(t)$ is not always equal to 1.

Calculation of the coefficients of the development of $f(t)$: the function $f(t)$ is periodic, the integration being performed over a period, we may shift for convenience the boundaries in the integral appearing in (3.7), the interval being still equal to T_1.

Thus:

$$c_n = \frac{1}{T_1} \int_0^{T_1} f(t) e^{-jn\omega_1 t} dt = \frac{1}{T_1} \int_{-\frac{T_1}{2}}^{\frac{T_1}{2}} f(t) e^{-jn\omega_1 t} dt. \tag{3.30}$$

In this interval, $f(t) = \Pi_T(t)$. Since $\Pi_T(t)$ is zero outside the interval $\left\{-\frac{T}{2}, \frac{T}{2}\right\}$ and is 1 within the interval, we can write

$$c_n = \frac{1}{T_1} \int_{-\frac{T_1}{2}}^{\frac{T_1}{2}} f(t) e^{-jn\omega_1 t} dt = \frac{1}{T_1} \int_{-\frac{T_1}{2}}^{\frac{T_1}{2}} \Pi_T(t) e^{-jn\omega_1 t} dt = \frac{1}{T_1} \int_{-\frac{T}{2}}^{\frac{T}{2}} e^{-jn\omega_1 t} dt. \tag{3.31}$$

By performing the integration:

$$c_n = \frac{1}{T_1} \frac{e^{-jn\omega_1 \frac{T}{2}} - e^{jn\omega_1 \frac{T}{2}}}{-jn\omega_1} = \frac{1}{T_1} \frac{2 \sin\left(n\omega_1 \frac{T}{2}\right)}{n\omega_1}. \tag{3.32}$$

This result holds for $n \neq 0$, the denominator in Eq. (3.32) having to be different from 0 for the integration in this form to be possible.

For the case $n = 0$, integration (3.31) is carried out directly. In this case

$$c_0 = \frac{1}{T_1} \int_{-\frac{T}{2}}^{\frac{T}{2}} e^{-0} dt = \frac{1}{T_1} \int_{-\frac{T}{2}}^{\frac{T}{2}} dt = \frac{T}{T_1} \tag{3.33}$$

Since $\omega_1 T_1 = 2\pi$ the relationship (3.32) may be simplified, and we have

$$c_n = \frac{1}{n\pi} \sin\left(n\omega_1 \frac{T}{2}\right). \tag{3.34}$$

It is noted that the coefficients decrease as $\frac{1}{n}$. This type of decay is characteristic of the discontinuous nature of the function $f(t)$. As discussed later in this course, this decrease is considered slow.

Special case: If $T = \frac{T_1}{2}$, the signal is symmetrical. The coefficients for $n \neq 0$ become:

$$c_n = \frac{1}{n\pi} \sin\left(n\omega_1 \frac{T_1}{4}\right) = \frac{1}{n\pi} \sin\left(n\frac{2\pi}{T_1}\frac{T_1}{4}\right) = \frac{1}{n\pi} \sin\left(n\frac{\pi}{2}\right). \tag{3.35}$$

Expressing these results for different values of n, we get for the first terms:

$$c_0 = \frac{1}{2}, c_1 = \frac{1}{\pi}, c_2 = 0, c_3 = -\frac{1}{3\pi}, c_4 = 0, c_5 = \frac{1}{5\pi}, \dots$$

The function being even, the development of the Fourier series limited to positive values of n is $f(t) = a_0 + 2\sum_{n=1}^{+\infty} a_n \cos n\omega_1 t$.
For the first terms we have:

$$f(t) = \frac{1}{2} + \frac{2}{\pi} \cos \omega_1 t - \frac{2}{3\pi} \cos 3\omega_1 t + \frac{2}{5\pi} \cos 5\omega_1 t \dots \tag{3.36}$$

Except for $n = 0$ the coefficients of even orders are zero. The odd-order coefficients decrease as $\frac{1}{n}$. Only odd harmonics are present in the decomposition.

Note: The preceding square wave was positive or zero. One often uses in electronics a bipolar signal (symmetrically positive and negative). The Fourier coefficients of this signal are the same as above, except that the coefficient c_0 is zero in this case.

The coefficient c_0, given by the integral $c_0 = \frac{1}{T_1} \int_0^{T_1} f(t) dt$, represents the average value of the signal over one period.

3.3 Sum of a Finite Number of Exponentials

Optimal Development Coefficients of a Function

A problem may arise in practice when one is able only to use a finite number N of terms in the series (for example, in a numerical calculation). In this case, we seek to approach the better possible the function $f(t)$ by a linear combination of a finite number of basis functions (exponentials in the case of Fourier series). In the general case, we cannot hope to find the exact value of $f(t)$ for all time. We will have an estimate of $f(t)$. This estimator is noted $\hat{f}(t)$ by placing a cap on the function to be estimated. Its form is:

$$\hat{f}(t) = \sum_{n=1}^{N} c'_n \varphi(t) \tag{3.37}$$

The following question arises: How to choose the coefficients c'_n so that the error in estimating $f(t)$ is as small as possible? In other words how can we estimate optimally $f(t)$? Let us write the estimation error as:

$$e(t) = f(t) - \hat{f}(t). \tag{3.38}$$

We want $\hat{f}(t)$ to represent at best $f(t)$, so that the "distance" between $\hat{f}(t)$ and $f(t)$ should be the smallest possible. This distance is calculated from the scalar product $\langle e(t), e(t) \rangle$, the quadratic error ε which is defined by:

$$\varepsilon = \langle e(t), e(t) \rangle = \langle f(t) - \hat{f}(t), f(t) - \hat{f}(t) \rangle. \tag{3.39}$$

The error is searched to be minimal. That error is then called the standard error. $\sqrt{\varepsilon}$ is the distance between $f(t)$ and its estimator.

The coefficients c'_n that minimize the quadratic error are determined by the condition that the partial derivatives of ε with respect to the coefficients c'^*_n are zero

$$\frac{\partial \varepsilon}{\partial c^*_n} = 0. \tag{3.40}$$

The squared norm of the error is:

$$\|e\|^2 = \varepsilon = \frac{1}{T_1} \int_0^{T_1} \left| \sum_{n=1}^{N} c'_n \varphi_n(t) - f(t) \right|^2 dt. \tag{3.41}$$

The optimal coefficients will be obtained by canceling the following partial derivatives:

$$\frac{\partial \varepsilon}{\partial c'^*_n} = \left\langle e, \frac{\partial e}{\partial c'^*_n} \right\rangle = \frac{1}{T_1} \int_0^{T_1} \left(f(t) - \sum_{n'=1}^{N} c'_{n'} \varphi_{n'}(t) \right) \varphi^*_n(t) dt = 0. \tag{3.42}$$

Note that this equation can be written in the form of the cancelation of the scalar product

$$\langle e, \varphi_n(t) \rangle = 0. \tag{3.43}$$

In other words, the minimum error vector is orthogonal to the base vectors $\varphi_n(t)$. Since

$$\frac{1}{T_1} \int_0^{T_1} \left(\sum_{n'=1}^N c'_{n'} \varphi_{n'}(t) \right) \varphi_n^*(t) \mathrm{d}t = c'_{n'} \delta(n' - n) = c'_n, \tag{3.44}$$

and

$$\frac{1}{T_1} \int_0^{T_1} f(t) \varphi_n^*(t) \mathrm{d}t = c_n, \tag{3.45}$$

we have the result:

$$c'_n = c_n. \tag{3.46}$$

This is an important result. When the function $f(t)$ is approached by a linear combination of basis functions $\varphi_n(t)$ limited to a finite number of terms, the coefficients of the combination that minimize the quadratic error are the coefficients of the Fourier series of $f(t)$.

Important note: In the preceding derivation, it was assumed that the derivatives with respect to c'_n and over c'^*_n are independent. This may seem surprising, as c'_n and c'^*_n are complex conjugates. The underlying reason is that c'_n and c'^*_n are complex numbers composed of two independent real numbers. The derivation with respect to c'^*_n hides formally derivation versus these two numbers. It could be shown that in the final result of the cancelation of the derivative calculations with respect to the real numbers, we get the same results by considering formally c'^*_n as independent of c'_n.

Geometric interpretation: $f(t)$ can be considered as a vector belonging to an infinite dimensional space spanned by the infinitely many basis vectors $\varphi_n(t)$ (space of all linear combinations of the functions $\varphi_n(t)$). A finite number N of basis vectors generates a hyperplane Π_N in that space. The vector $f(t)$ is out of this plane in the general case. The estimator $\hat{f}(t)$ which consists of a linear combination of vectors which belong to this hyperplane Π_N will necessarily be in Π_N (Fig. 3.3).

Fig. 3.3 Vector $f(t)$ and its estimator $\hat{f}(t)$

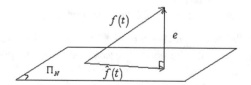

According (3.43) we have

$$e \perp \varphi_n \Rightarrow e \perp \hat{f}(t) \tag{3.47}$$

Thus, the error vector has to be perpendicular to the estimator. The optimal estimator $\hat{f}(t)$ is the projection of $f(t)$ onto the space Π_N generated by the N base vectors $\varphi_n(t)$.

3.4 Hilbert Spaces

The properties described above are within the scope of the general properties of Hilbert spaces. A Hilbert space is a vector space with a scalar product and provided with some additional properties.

Properties that define a Hilbert space:

- The sum of two elements of the set is an element of the set.
- The multiplication of an element of the set by a scalar belongs to the set.
- Definition of a scalar product:

  - The inner product of two elements of the set is a scalar.
  - A norm is defined from the scalar product. It is used for defining a distance between two elements of the set.
  - A Cauchy sequence of elements of the set converges.
  - The limit of a sequence of elements of the set belongs to the set. It is unique. We say that the space is complete.

Important properties of a Hilbert space:

- A subset composed of elements of the space generates a Hilbert space H_1, subset of H.
- One element of H is written as a linear combination of an element of H_1 and an element of its complement in H noted H_2.
- The subspaces H_1 and H_2 are orthogonal.

For example, the ensemble of even functions and that of odd functions, periodic with period T_1, and summable over a period form two orthogonal Hilbert subspaces of the ensemble of all periodic functions of time with period T_1.

We could separate the space into other orthogonal subspaces, such as separating the space generated by the N primary functions of the base and the one generated by the functions of the complementary set.

Theoretical Aspects of Exponential Fourier Series Expansion

Consider the set of periodic functions of period T_1 integrable over a period. Let $\psi_1(t)$ and $\psi_2(t)$ be any two functions belonging to that set. Following definition (3.5) their scalar product is given by:

$$\langle \psi_1(t), \psi_2(t) \rangle = \frac{1}{T_1} \int_0^{T_1} \psi_1(t) \psi_2^*(t) dt. \tag{3.48}$$

With this scalar product, the set is a Hilbert space.

Definition of a Self-Adjoint Operator:

Let O be an operator operating on the functions of this space. The adjoint operator $O^\dagger$ of O is defined by:

$$\left\langle O^\dagger \psi_1(t), \psi_2(t) \right\rangle = \langle \psi_1(t), O\psi_2(t) \rangle. \tag{3.49}$$

An operator O is said *self-adjoint* (or *Hermitian*) if it is such that:

$$O^\dagger = O \tag{3.50}$$

We have therefore in this case.

$$\langle O\psi_1(t), \psi_2(t) \rangle = \langle \psi_1(t), O\psi_2(t) \rangle. \tag{3.51}$$

The operator $O = i\frac{d}{dt}$ plays an important role in the context of Fourier series development of periodic functions with period T_1. Let us show that it is Hermitian. Following the definition (3.49) and the relationship (3.48), we have:

$$\left\langle O^\dagger \psi_1(t), \psi_2(t) \right\rangle = \frac{1}{T_1} \int_0^{T_1} \psi_1(t) \left(i\frac{d}{dt}\psi_2(t) \right)^* dt. \tag{3.52}$$

We integrate by parts:

$$\left\langle O^\dagger \psi_1(t), \psi_2(t) \right\rangle = -\frac{1}{T_1} i |\psi_1(t)\psi_2^*(t)|_0^{T_1} + \int_0^{T_1} i\frac{d}{dt}\psi_1(t)\psi_2^*(t) dt, \tag{3.53}$$

The first term of the second member is zero, since the functions are periodic. It comes:

$$\left\langle O^\dagger \psi_1(t), \psi_2(t) \right\rangle = \int_0^{T_1} i\frac{d}{dt}\psi_1(t)\psi_2^*(t) dt. \tag{3.54}$$

We see that $O^\dagger = O$. The operator $i\frac{d}{dt}$ is Hermitian.

Eigenvalues and eigenfunctions of a Hermitian operator

The eigenvalues are real

Let $\varphi(t)$ be an eigenfunction of a Hermitian operator O such that $O\varphi(t) = \lambda\varphi(t)$, a priori the eigenvalue λ is complex.

We use the fact that O is Hermitian to write [see (3.51)]:

$$\langle O\varphi(t), \varphi(t) \rangle = \langle \varphi(t), O\varphi(t) \rangle. \tag{3.55}$$

Since

$$\langle O\varphi(t), \varphi(t) \rangle = \langle \lambda\varphi(t), \varphi(t) \rangle = \lambda\langle \varphi(t), \varphi(t) \rangle,$$

and

$$\langle \varphi(t), O\varphi(t) \rangle = \langle \varphi(t), \lambda\varphi(t) \rangle = \lambda^*\langle \varphi(t), \varphi(t) \rangle,$$

then:

$$\lambda = \lambda^*. \tag{3.56}$$

We have demonstrated that the eigenvalues of a Hermitian operator are real.

The Eigenfunctions Relative to Nonequal Eigenvalues Are Orthogonal
To prove this property, we use $\varphi_1(t)$ and $\varphi_2(t)$ being any 2 eigenvectors of a Hermitian operator O.

We can write: $\langle O\varphi_1(t), \varphi_2(t) \rangle = \lambda_1\langle \varphi_1(t), \varphi_2(t) \rangle$.
Similarly: $\langle \varphi_1(t), O\varphi_2(t) \rangle = \lambda_2^*\langle \varphi_1(t), \varphi_2(t) \rangle$.
Since the eigenvalues are real we have $\lambda_2^* = \lambda_2$.
Taking in account the Hermiticity relationship (3.51), we must have

$$\lambda_1\langle \varphi_1(t), \varphi_2(t) \rangle = \lambda_2^*\langle \varphi_1(t), \varphi_2(t) \rangle = \lambda_2\langle \varphi_1(t), \varphi_2(t) \rangle.$$

If λ_1 and λ_2 are different, in order that the above equation could be satisfied, it is necessary that

$$\langle \varphi_1(t), \varphi_2(t) \rangle = 0. \tag{3.57}$$

We see that the eigenfunctions related to two different eigenvalues are orthogonal.

We apply now these results of a general scope to Fourier series development. First, we determine the eigenfunctions of the Hermitian operator $O = i\dfrac{d}{dt}$:
$O\varphi(t) = \lambda\varphi(t)$ [with λ real according to the property (3.56)],

$$i\frac{\mathrm{d}\varphi(t)}{\mathrm{d}t} = \lambda\varphi(t); \varphi(t) = Ce^{\frac{\lambda}{i}t},$$

or equivalently:

$$\varphi(t) = Ce^{i\omega t} \text{ with } \omega \text{ real and } C \text{ any constant.} \tag{3.58}$$

The argument in the exponential must be imaginary.

We have shown in Chap. 1 that the eigenfunctions of the operator $\frac{\mathrm{d}}{\mathrm{d}t}$, and by consequence, the eigenfunctions of LTI systems, were the functions Ce^{pt} with the eigenvalues p which may be any complex number. The operator $\frac{\mathrm{d}}{\mathrm{d}t}$ is non-Hermitian.

However, we may verify that the eigenfunctions of $i\frac{\mathrm{d}}{\mathrm{d}t}$, which are $Ce^{i\omega t}$, are also eigenfunctions of $\frac{\mathrm{d}}{\mathrm{d}t}$. In conclusion, among the eigenfunctions of $\frac{\mathrm{d}}{\mathrm{d}t}$, the eigenfunctions $Ce^{i\omega t}$ of $i\frac{\mathrm{d}}{\mathrm{d}t}$ may be chosen to form an orthogonal basis for a development of a periodic function. This is the principle of the Fourier series development of a function.

In Chap. 5 on Fourier transform, an other scalar product will be used for the development of square-integrable, nonperiodic functions, in the L_2 Hilbert space.

Remark: The above development on Hermitian operators derives from quantum mechanics where they have been used for many decades. The consideration of non-Hermitian operators in quantum mechanics is fairly recent. This subject is nowadays an active field of research.

3.5 Gibbs Phenomenon

This very important phenomenon in practice occurs when the order of the sum of exponentials is limited to a finite number N, that is to say, when limiting superiorly the harmonics frequency. This is the case when a signal passes through an amplifier with band limited to low frequencies. This phenomenon is observed, for example, on the screen of a 30 MHz bandwidth oscilloscope when viewing the square signal from a low frequency signal generator. In the case where the initial time function has a discontinuity (which is the case of the square wave), the limited bandwidth of the oscilloscope reduces the high-frequency content and gives a representation of the signal using a limited sum of exponentials. The visualized signal has an oscillation at the location of the discontinuity. One of the remarkable aspects of this phenomenon is that as great as is the maximum order N, the oscillations are always present and keep the same amplitude. Only the frequency of the oscillations increases with N. This phenomenon is quantitatively studied in the chapter on the applications of the Fourier transform.

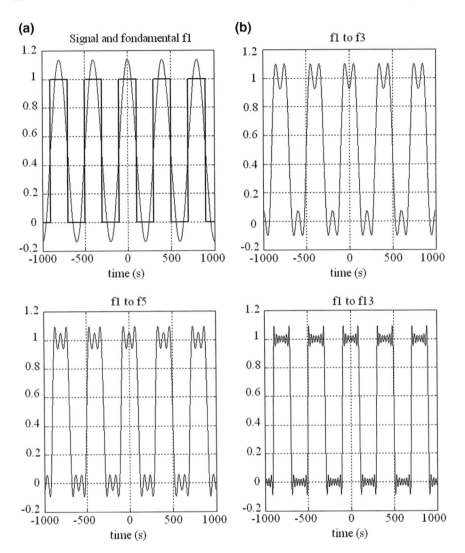

Fig. 3.4 Gibbs phenomenon for different sum number limitations

A numerical simulation is performed on the square signal studied in the previous example (3.29). In Fig. 3.4a (Top), the periodic square signal is shown as well as the partial sum $\frac{1}{2} + \frac{2}{\pi}\cos \omega_1 t$. In Fig. 3.4b (Top) is the sum limited to the first harmonic of odd row. On Fig. 3.4a (Bottom) is the sum of harmonics f_1–$5f_1$. The square wave aspect is beginning to appear. In Fig. 3.4b (Bottom) has harmonics from f_1–$13f_1$.

We can see an over-oscillation at the discontinuities of the function that expresses the Gibbs phenomenon. This phenomenon will continue to exist for arbitrarily large values of the highest frequency kept.

Gibbs phenomenon reflects both the prodigious nature of the development of a function as a sum of exponentials but also the default of this development: It is possible theoretically to obtain a constant value for $f(t)$ in an interval (the oscillation of functions $e^{jn\omega_1 t}$ cancel by interference in performing the summation) yet the sum should include an infinite number of terms. When seeking to represent the function $f(t)$ with a finite number of coefficients, this reconstitution presents spurious oscillations. This problem is encountered in compressing information in digital encoding (audio or video for example). Other compression techniques such as the decomposition of a signal on a wavelet basis do not have this drawback. This decomposition will be discussed in Chap. 12.

3.6   Nonlinearity of a System and Harmonic Generation

In the following example, we show using a numerical simulation that Fourier analysis provides information about a signal in the Fourier domain that a temporal analysis cannot detect. In particular, it can reveal the nonlinear character, even small, of a system.

Assume that a sine wave $x(t) = \sin(2\pi f_1 t)$ drives an amplifier whose gain is noted G. The amplifier is assumed to have a slight nonlinearity such that instead of having an output signal

$$y(t) = G_0 \sin(2\pi f_1 t), \tag{3.59}$$

where G_0 is a constant. The output signal of the system is:

$$y(t) = G_0 \sin(2\pi f_1 t) - \mathrm{sgn}(\sin(2\pi f_1 t))\alpha \sin^2(2\pi f_1 t). \tag{3.60}$$

The sign function $\mathrm{sgn}(x)$ equals 1 if its argument x is positive, and -1 if its argument is negative. α is the nonlinearity factor, smaller than 1.

The defect of this amplifier is that its gain decreases as the magnitude of the signal increases according to Eq. (3.60).

In the numerical application which follows, we take $G_0 = 1$. The signal frequency is chosen equal to $f_1 = 82$ Hz. A representation versus frequency in decibels is used to highlight the very low amplitudes (-150 dB corresponds to an amplitude of 3.2×10^{-8}, value extremely low). The value of the first Fourier coefficient corresponding to the fundamental frequency of 82 Hz is 1, which corresponds to a value equal to 0 dB. First, we represent in Fig. 3.5 the ideal case of an amplifier without nonlinearity for which $\alpha = 0$. Obviously there is just a single spectral line.

In the case of weak nonlinearity with $\alpha = 0.001$, as shown in Fig. 3.6, we are unable to detect the manifestation of this nonlinearity on the shape of the time signal (Fig. 3.6a). As far as we can judge, the signal maximum is always 1 and the sinusoidal shape seems unchanged. On the other hand, one sees in the Fourier domain additional lines whose amplitudes are given by the nonzero coefficients for

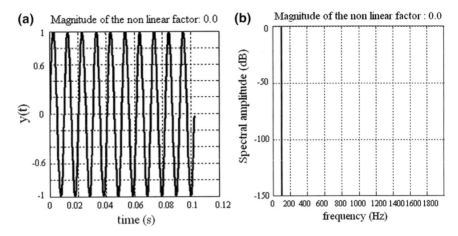

Fig. 3.5 Signal **a** and its Fourier magnitude **b** in case of zero nonlinearity ($\alpha = 0$)

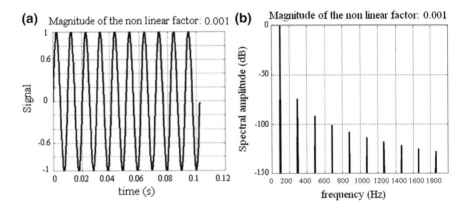

Fig. 3.6 Signal **a** and its Fourier magnitude **b** when $\alpha = 0.001$

odd multiple harmonic frequencies of 82 Hz (Fig. 3.6b). Here we see the interest of the Fourier analysis. And thanks to the representation in decibels (logarithmic representation), low values are less crushed by the dynamics that in linear scale and can be seen in the figure up to the 19th harmonic. The harmonic 3 is located approximately 75 dB below the fundamental. This corresponds to a ratio of 1.8×10^{-4} in linear scale, which is very little. One says that the harmonic distortion is -75 dB for the harmonic 3.

Very low harmonic distortion (~ -50 dB) is present in the audio amplifiers of high fidelity. Such a quality is sought after by musicians. It was noted above that the ear analyzes the signals in the Fourier domain and has a logarithmic sensitivity, which enables it to detect very weak signals mixed with strong signals, e.g., unwanted harmonics.

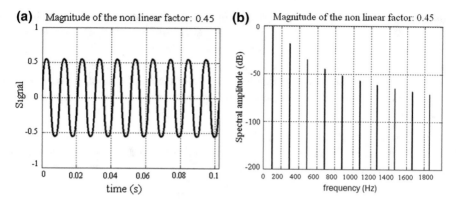

Fig. 3.7 Signal **a** and its Fourier magnitude **b** when $\alpha = 0.45$

The case of a very strong nonlinearity $\alpha = 0.45$ is discussed finally. This time the nonlinearity is also very apparent in the time domain (Fig. 3.7a). Signal peaks have weakened. In extreme cases, the signal begins to approximate a square wave. The harmonic Fourier coefficient c_3 of harmonic 3 is -20 dB about the fundamental c_1, a factor of 10 in amplitude (Fig. 3.7b).

We see how the Fourier domain allows quantifying the harmonic content. This analysis is particularly important in audio applications. We will detail later in this course how to record and process a signal to make a correct analysis of nonlinearities.

In a second example, the nonlinearity is assumed cubic, that is to say, that the output signal is given by:

$$y(t) = G_0 \sin(2\pi f_1 t) - \alpha \sin^3(2\pi f_1 t). \tag{3.61}$$

If $\alpha = 0.01$ the time and frequency representations of the output signal are plotted in Fig. 3.8. Again, the analysis in the time domain does not provide a tool to

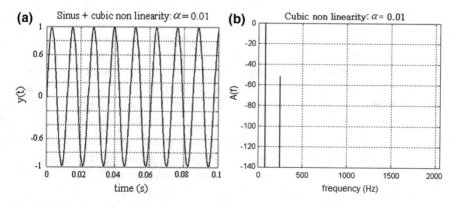

Fig. 3.8 Signal **a** and its Fourier magnitude **b** in case of a cubic nonlinearity

detect the nonlinearity. However, the Fourier domain is again very rich in information. We can be see on Fig. 3.8b that for such nonlinearity, only two Fourier coefficients, c_1 and c_3 are different from zero for the output signal, in contrast to what was observed for the quadratic nonlinearity of the previous example.

It appears from this example that the analysis of the amplitude of the harmonics of a signal is a valuable means of investigation of the physical system which creates this signal.

Total Harmonic Distortion of a Signal
This rate measures the ratio of the total power of higher order than 1 harmonics to the power of the fundamental component. We have therefore:

$$\text{THD} = \sqrt{\sum_{|n| > 1} \frac{|c_n|^2}{|c_1|^2}}. \tag{3.62}$$

It can also be written by (3.19):

$$\text{THD} = \sqrt{\sum_{|n| > 1} \frac{A_n^2 + B_n^2}{A_1^2 + B_1^2}}. \tag{3.63}$$

Assuming that the amplitude of the second harmonic relative to the fundamental is 1 %, and that of the rank 3 amplitude is 1.5 %, the THD will be:

THD $= \sqrt{0.01^2 + 0.015^2} = 0.018 = 1.8\%$, or in decibels -34.9 dB.

For high fidelity music, it is accepted that the THD of an amplifier should be smaller than 1 % (-40 dB).

Summary
This chapter has been dedicated to the study of complex periodic time signals. We have shown that theses signals can be expressed as a sum of harmonic signals whose frequencies are multiples of a fundamental frequency. The expansion coefficients are calculated by an integral which represents the projection of the signal on the basis of these harmonic functions. We qualitatively explain the magnitude of a projection on a given function. Parseval theorem on Fourier series expresses the power of a signal to be the sum of the squared moduli of the coefficients. We have shown that the coefficients of the optimal approximation of any signal as a linear combination of a limited number of basis functions are the Fourier series coefficients. A geometric interpretation of this behavior in Hilbert spaces has been given. We have shown that the functions $e^{j\omega t}$ are the eigenfunctions of the Hermitian operator $i\frac{d}{dt}$ and can be used as an orthogonal basis of development of periodic functions. The first manifestation of the Gibbs phenomenon is observed on the limited series. It has been shown on an example how the study of the amplitude of the Fourier coefficients can be used to study the nonlinearity of a system that may

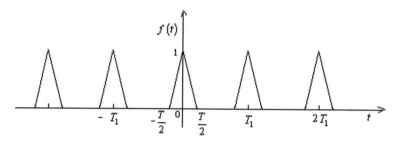

Fig. 3.9 Sawtooth signal

not be apparent in the time domain. The advantages of a logarithmic representation (dB) are shown to evaluate nonlinearity.

Exercise

Calculate the Fourier expansion of the sawtooth signal of the following form (Fig. 3.9). Show that the Fourier coefficients decrease with n like $\dfrac{1}{n^2}$. This decrease in $\dfrac{1}{n^2}$ of the coefficients will be interpreted later in the course as a feature of the Fourier expansion of continuous functions whose derivatives are discontinuous.

Chapter 4
The Dirac Distribution

We introduce in this chapter the concept of Dirac distribution. It is conceived as an infinitely brief pulse occurring at time zero. The Dirac distribution is a very powerful mathematical tool in signal analysis, especially in the Fourier transform. We demonstrate here the first golden formula of signal analysis:

$$\boxed{\int_{-\infty}^{\infty} e^{j\omega t} d\omega = 2\pi\delta(t)}$$

We give simple examples illustrating the rules of its use.

The response of a LTI system to a monochromatic signal input has been studied in Chaps. 1 and 2. In Chap. 3, it was shown that a periodic signal of period T_1 can be considered as the sum of monochromatic signals whose frequencies are multiples of a fundamental frequency $f_1 = \frac{1}{T_1}$. To go further, we must focus our attention on nonperiodic signals. They represent the general case and are richer in information than a simple monochromatic signal. The study of the response of LTI systems to signals of any form begins here by calculating the response of these systems to a special signal $x(t)$ resulting from the summation of infinite monochromatic signals with the same amplitude (here unit amplitude) for all frequencies:

$$\int_{-\infty}^{\infty} e^{j\omega t} d\omega = 2\pi\delta(t) \tag{4.1}$$

This particular signal, simple in its construction, is very interesting for the entirety of this course. However, it has no representation in the form of a function. The mathematical difficulty encountered to obtain the signal in the time domain as defined by the previous integral is that the integral does not converge.

© Springer International Publishing Switzerland 2016
F. Cohen Tenoudji, *Analog and Digital Signal Analysis*,
Modern Acoustics and Signal Processing, DOI 10.1007/978-3-319-42382-1_4

This chapter is devoted to the study of this important problem in signal analysis. It leads to the notion of the Dirac distribution, which is an extremely effective tool in calculations. We present in the following the notions of convolution product and of integral and derivatives of the Dirac distribution.

4.1   Infinite Sum of Exponentials. Cauchy Principal Value

Let us study the properties of the integral (4.1):

$$\int_{-\infty}^{+\infty} e^{j\omega t}\,d\omega. \tag{4.2}$$

Within the meaning of conventional integration of a function, this integral does not converge. Indeed, the convergence of this improper integral requires that the following limit exists

$$\lim_{\substack{A\to\infty\\B\to-\infty}} \int_{B}^{A} e^{j\omega t}\,d\omega = \lim_{\substack{A\to\infty\\B\to-\infty}} \frac{e^{jtA} - e^{jtB}}{jt}. \tag{4.3}$$

However, this limit does not exist since both exponentials appearing in the right hand side oscillate indefinitely when A and B tend to infinity, and therefore do not tend to a limit independently of each other. Thus, the integral in (4.1) does not converge.

One is led to focus on another summation of monochromatic signals corresponding to the Cauchy principal value of the integral. It is defined as

$$PV \int_{-\infty}^{\infty} e^{j\omega t}\,d\omega = \lim_{A\to\infty} \int_{-A}^{A} e^{j\omega t}\,d\omega \tag{4.4}$$

We now note the symmetry of the limits of integration. If the limit exists in the definition (4.4), the Cauchy principal value of the integral exists.

In the general case, when the integral of a function exists, the Cauchy principal value of that integral exists. But the converse is not true: the Cauchy principal value can exist without the convergence of the integral.

Fig. 4.1 Function $\frac{2\sin At}{t}$ as a function of time

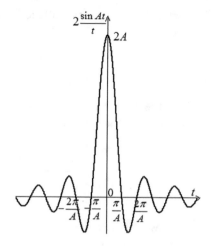

However, in the case studied here, even in the sense of the Cauchy principal value, there is no convergence. Indeed,

$$PV \int_{-\infty}^{\infty} e^{j\omega t}\,d\omega = \lim_{A\to\infty} \int_{-A}^{A} e^{j\omega t}\,d\omega = \lim_{A\to\infty} \frac{e^{jtA} - e^{-jtA}}{jt} = \lim_{A\to\infty} \frac{2\sin At}{t}. \qquad (4.5)$$

This last expression does not tend either to a finite limit $\forall t$ as $A \to \infty$ because the sine oscillates indefinitely between -1 and $+1$. However, we can see that the function oscillates more rapidly when A is great. The first zero of the function is obtained for $t_0 = \frac{\pi}{A}$. This value tends to zero as $A \to \infty$. The limit value at $t = 0$ of the function $\frac{2\sin At}{t}$ is $2A$. It tends to infinity with A. So, when A is large, the graph of the function $\frac{2\sin At}{t}$ (Fig. 4.1) shows a very pronounced maximum for $t = 0$ and oscillations for $t \neq 0$ whose amplitude decreases more quickly as A is greater.

4.2 Dirichlet Integral

However, when integrated over t the product of the function $\frac{2\sin At}{t}$, whose oscillations are fast when A is large, by any function $\varphi(t)$ with sufficiently slow variation, gives an approximately zero contribution to the integral for any value $t \neq 0$. Let us see this in more detail:

The integral of the product of a function $\varphi(t)$ by $\frac{2\sin At}{t}$ is called a Dirichlet integral.

Let $\varphi(t)$ be a continuous function at $t = 0$. The following relation holds:

$$\lim_{A\to\infty} 2 \int_{-\infty}^{+\infty} \varphi(t) \frac{\sin At}{t} dt = 2\pi\varphi(0). \tag{4.6}$$

Proof The function $\varphi(t)$ is assumed to be sufficiently regular, continuous and differentiable at many orders. The study focuses firstly on the integral

$$\int_a^b \frac{\varphi(t) - \varphi(0)}{t} \sin At\, dt \quad (\text{with } a < 0 < b). \tag{4.7}$$

By assumption, the function $\frac{\varphi(t) - \varphi(0)}{t}$ is continuous and has a continuous derivative around 0.

We note $f(t) = \frac{\varphi(t) - \varphi(0)}{t}$ and calculate the integral (4.7) by parts:

$$\int_a^b f(t) \sin At\, dt = -\frac{1}{A}[f(t) \cos At]_a^b + \frac{1}{A} \int_a^b f'(t) \cos At\, dt. \tag{4.8}$$

We note M a common upper bound to $f(t)$ and $f'(t)$ in the finite interval $[a, b]$. The modulus of the integral is less than $\frac{2M}{A} + \frac{M(b-a)}{A}$ which tends to zero when $A \to \infty$. Therefore it tends zero. Thus:

$$\lim_{A\to\infty} \int_a^b f(t) \sin At\, dt = \lim_{A\to\infty} \int_a^b \frac{\varphi(t) - \varphi(0)}{t} \sin At\, dt = 0. \tag{4.9}$$

This result can be rewritten as:

$$\lim_{A\to\infty} \int_a^b \frac{\varphi(t)}{t} \sin At\, dt = \varphi(0) \lim_{A\to\infty} \int_a^b \frac{\sin At}{t} dt. \tag{4.10}$$

We could have anticipated this result with a qualitative reasoning: As shown in Fig. 4.2, the function $\varphi(t)$ assumed continuous and regular in $t = 0$ can be considered constant and equal to $\varphi(0)$ in the vicinity of $t = 0$ (neighborhood smaller as A is greater). Only in this neighborhood, the function $\frac{2\sin At}{t}$ is significantly different from 0. This explains qualitatively the factorization of $\varphi(0)$ in the second member of (4.10).

Fig. 4.2 Functions $\frac{2\sin At}{t}$ and $\varphi(t)$

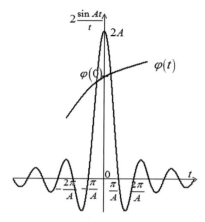

To demonstrate formula (4.6), it remains to be shown that

$$\lim_{A\to\infty}\int_{-\infty}^{+\infty}\frac{\sin At}{t}\,dt = \pi. \qquad (4.11)$$

We note $At = x; A\,dt = dx$ and $A > 0$,
We are led to evaluate the quasi-integral:

$$I = PV\int_{-\infty}^{+\infty}\frac{\sin x}{x}\,dx \qquad (4.12)$$

The function $\frac{\sin x}{x}$ which appears many times in signal analysis is called $\sin c$ (for cardinal sine). Then, one notes $\sin c(x) = \frac{\sin x}{x}$.

This integral has a singularity in $x = 0$. Summation will only be possible if the singularity is approached symmetrically around 0. This justifies the use of ½ circle of radius ε in the following calculation. The Cauchy Principal value has to be taken both for $x = 0$ and at infinite.

The integration is performed in the complex plane. The auxiliary function $\frac{e^{jz}}{z}$ is introduced, based on the complex variable $z = x + jy$. The principles of integration of a complex function are detailed in Appendix A1.

The function $\frac{e^{jz}}{z}$ is integrated on the closed contour (see Fig. 4.3) within which it is holomorphic (continuously differentiable).

Cauchy's theorem states that the path integral of a function on a closed contour within which the function is holomorphic is zero. In this case:

Fig. 4.3 Integration contour C

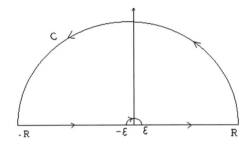

$$\int_{-R}^{-\varepsilon} \frac{e^{jx}}{x}dx + \int_{\alpha} \frac{e^{jz}}{z}dz + \int_{\varepsilon}^{R} \frac{e^{jx}}{x}dx + \int_{C} \frac{e^{jz}}{z}dz = 0.$$

$$\quad (1) \qquad\quad (2) \qquad\quad (3) \qquad\quad (4)$$

(4.13)

The integral (2) is carried clockwise along a semicircle α of radius ε, to avoid the singularity of the function $\frac{e^{jz}}{z}$ at $z = 0$. The integral (4) is carried along a semicircle C of radius R.

The sum I [formula (4.12)] is obtained from the summation of integrals (1) and (3) with the limits $\varepsilon \to 0$ and $R \to \infty$.

Firstly, one gets

$$\int_{-\infty}^{+\infty} \frac{e^{jx}}{x}dx = \int_{-\infty}^{+\infty} \frac{\cos x + j\sin x}{x}dx = j\int_{-\infty}^{+\infty} \frac{\sin x}{x}dx.$$

(4.14)

Indeed, since the function $\frac{\cos x}{x}$ is odd, its integration on a symmetric interval is zero:

$$\int_{-\infty}^{+\infty} \frac{\cos x}{x}dx = 0.$$

(4.15)

Therefore:

$$I = \int_{-\infty}^{+\infty} \frac{\sin x}{x}dx = \frac{1}{j}\operatorname{Im}\left(\int_{-\infty}^{+\infty} \frac{e^{jx}}{x}dx\right)$$

$$= \frac{1}{j}\lim_{\substack{R \to \infty \\ [\varepsilon \to 0]}}\left(\int_{-R}^{-\varepsilon} \frac{e^{jx}}{x}dx + \int_{\varepsilon}^{R} \frac{e^{jx}}{x}dx\right)$$

(4.16)

$$I = \int_{-\infty}^{+\infty} \frac{\sin x}{x} dx = \frac{1}{j} \lim_{\substack{R \to \infty \\ [\varepsilon \to 0]}} \{(1) + (3)\} \tag{4.17}$$

Taking into account Eq. (4.12), we can write

$$I = PV \int_{-\infty}^{+\infty} \frac{\sin x}{x} dx = -\frac{1}{j} \lim_{\substack{R \to \infty \\ [\varepsilon \to 0]}} \{(2) + (4)\} \tag{4.18}$$

The integrals $\int_{\alpha}$ and $\int_{C}$ are evaluated at the limits $\varepsilon \to 0$ and $R \to \infty$.
$(2)(4)$

The integral (4) on C is zero when $R \to \infty$, because the function $\frac{1}{z}$ appearing in the integral tends to zero, which allows the application of Jordan Lemma 3 (see Appendix 1).

Calculation of integral (2)

$$\int_{\alpha} \frac{e^{jz}}{z} dz : \tag{4.19}$$

The integral over the semi-circle can be set using the angle θ between the radius locating the point on the semi-circle and the horizontal.

We can note $z = \varepsilon e^{j\theta}$ and express the differential element dz on the circle of constant radius ε. The integral over α becomes:

$$\int_{\alpha} \frac{e^{jz}}{z} dz = \int_{\pi}^{0} \frac{e^{j\varepsilon e^{j\theta}}}{\varepsilon e^{j\theta}} \varepsilon e^{j\theta} j d\theta = j \int_{\pi}^{0} e^{j\varepsilon e^{j\theta}} d\theta, \tag{4.20}$$

where, as ε is very small, $e^{j\varepsilon e^{j\theta}} \cong e^{j0} = 1$.

Thus

$$j \int_{\pi}^{0} e^{j\varepsilon e^{j\theta}} d\theta \cong j \int_{\pi}^{0} d\theta = -j\pi. \tag{4.21}$$

Inserting this result in (4.18), we obtain:

$$I = \int\limits_{-\infty}^{+\infty} \frac{\sin x}{x} dx = \pi. \tag{4.22}$$

(Note that we have also $\int_0^{+\infty} \frac{\sin x}{x} dx = \frac{\pi}{2}$, as $\frac{\sin x}{x}$ is an even function).
Finally,

$$\lim_{A \to \infty} 2 \int\limits_{-\infty}^{+\infty} \varphi(t) \frac{\sin At}{t} dt = 2\pi\varphi(0). \tag{4.23}$$

Thus, returning to the notation of Eq. (4.5):

$$PV \int\limits_{-\infty}^{\infty} \int\limits_{-\infty}^{\infty} \varphi(t) e^{j\omega t} d\omega dt = 2\pi\varphi(0). \tag{4.24}$$

Symbolically we write:

$$\int_{-\infty}^{\infty} \delta(t)\varphi(t)dt = \varphi(0), \tag{4.25}$$

where it was noted:

$$PV \int\limits_{-\infty}^{\infty} e^{j\omega t} d\omega = 2\pi\delta(t). \tag{4.26}$$

This equation is often written (incorrectly as the main concept of principal value is omitted) in the form:

$$\boxed{\int\limits_{-\infty}^{\infty} e^{j\omega t} d\omega = 2\pi\delta(t).} \tag{4.27}$$

The formula (4.26) is very important. We could call it the golden formula for the calculations in signal processing because it can make easy calculations that would be very difficult without its use.

By a simple change of variable in (4.27), we see that we also have: $\int_{-\infty}^{\infty} e^{-j\omega t} d\omega = 2\pi\delta(t)$.

Similarly, exchanging roles of ω and t we can write: $\int_{-\infty}^{\infty} e^{\pm j\omega t} dt = 2\pi\delta(\omega)$.

4.3 Dirac Distribution

4.3.1 Definition

Sometimes, Eq. (4.25) is written as:

$$\int_{-\infty}^{\infty} \delta(t)\varphi(t)\mathrm{d}t = \varphi(0) \equiv \langle \delta, \varphi \rangle \qquad (4.28)$$

We have noted symbolically

$$\boxed{PV \int_{-\infty}^{+\infty} \mathrm{e}^{\mathrm{j}\omega t}\mathrm{d}\omega = 2\pi\delta(t)} \qquad (4.29)$$

These last two equations define the Dirac distribution $\delta(t)$. Sometimes one speaks of *Dirac function*. This is incorrect. Indeed, $\delta(t)$ has meaning only within an integral (4.28).

To avoid too much abstraction in the calculations, we write $\delta(t)$ out of an integral and visualize it (incorrectly) as a function of time with an arrow tending to infinity at $t = 0$ and with a zero value elsewhere. However, one should be aware of errors that could come from this oversimplification. In case of doubt about the behavior of $\delta(t)$ in a calculation, one should always return to the full writing (4.28).

We can find qualitatively this result by a numerical simulation:

Since $\mathrm{e}^{\mathrm{j}\omega t} = \cos \omega t + \mathrm{j} \sin \omega t$, the integration of the exponential consists in two separate integrals of cosine and sine functions of different frequencies.

If one represents graphically some cosine (Fig. 4.4), it is seen that, whatever the frequency, their common value at $t = 0$ is 1. While for $t \neq 0$ the cosine have

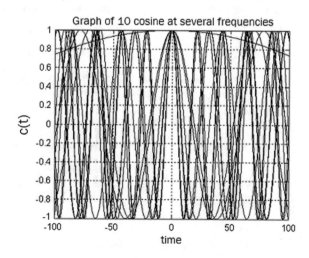

Fig. 4.4 Graph of ten cosine functions with different frequencies

various values depending on their frequency. It is clear that if we do the sum of all the cosine, for $t \neq 0$ the amount will be blurred by interference and we obtain 0. It is only at $t = 0$ that all cosine being equal to 1, the sum of the infinite number of terms will be infinite.

The sum of 5000 cosine with frequencies selected using a random number generator is shown in Fig. 4.5 we find the maximum amplitude value in $t = 0$ to be 5000 (each cosine is 1 at this point) and the general shape of a sinc function. It is understandable that if the number of cosines of different frequencies is infinitely large, the limit of this sum will be a Dirac distribution.

For the sum of imaginary terms in (4.27), we have the same phenomenon of cancellation for $t \neq 0$, but as all the sine functions are zero at $t = 0$, their sum will be zero also.

In summary, the infinite sum of exponentials will be real, infinite at $t = 0$ and will be zero elsewhere.

We must not forget that no time function can meet this definition. One talks of a distribution. The amount found is meaningful only within an integral as defined in Eq. (4.28).

Another view possible of the Dirac distribution is that of a rectangular pulse centered at $t = 0$ (Fig. 4.6). As it is not possible to define a function infinitely short, the Dirac distribution is defined by a passage to the limit of a rectangular pulse $\Pi_T(t)$ starting at time $t = -\frac{T}{2}$, with width T and height $1/T$. The area under the graph is 1, regardless of T.

Again, $\delta(t)$ cannot be considered as a function, since it is zero everywhere and infinite at $t = 0$, a behavior inconsistent with the definition of a function. In Fig. 4.7 we represent the Dirac distribution with the convention accepted by physicists to fix the imagination.

Fig. 4.5 Sum of 5000 cosine functions with random frequencies

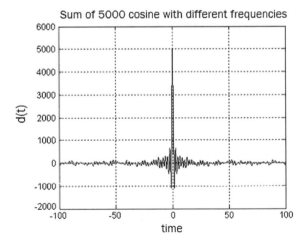

Fig. 4.6 Rectangular pulse
with area 1

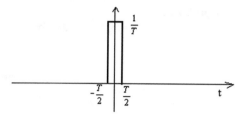

Fig. 4.7 Conventional
representation of Dirac
distribution

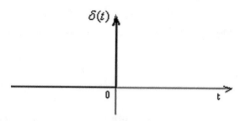

The Dirac distribution can be seen as the limit of different functions dependent upon a parameter when this parameter tends to zero (or infinite, depending on the function definition).

In particular, one shows that one may use a Gaussian function whose spread tends towards zero:

$$\delta(t) = \lim_{c \to 0} \frac{1}{c\sqrt{\pi}} e^{-\frac{t^2}{c^2}}. \tag{4.30}$$

We also have the following possible definition:

$$\delta(t) = \lim_{c \to 0} \frac{1}{c\sqrt{j\pi}} e^{j\frac{t^2}{c^2}}. \tag{4.31}$$

$f(t) = \frac{1}{c\sqrt{j\pi}} e^{j\frac{t^2}{c^2}}$ appears as a linearly frequency modulated signal.

4.3.2   Properties of the Dirac Distribution

Rigorous definition of the distribution $\delta(t)$ is given by its action on a function inside an integral. If the function $f(t)$ is continuous in $t = 0$, the following equation formally defines $\delta(t)$:

$$\int\limits_{-\infty}^{+\infty} f(t)\delta(t)dt = f(0). \tag{4.32}$$

Special case:
In the case where $f(t) = 1$, the formula (4.32) becomes:

$$\int_{-\infty}^{+\infty} \delta(t)\mathrm{d}t = 1. \tag{4.33}$$

Note:
In formula (4.29), the result remains unchanged if ω is replaced by $-\omega$ in the integral. We will have therefore:

$$PV \int_{-\infty}^{+\infty} e^{\pm j\omega t}\mathrm{d}\omega = 2\pi\delta(t). \tag{4.34}$$

Similarly, there will be a similar formula integrating over t instead of ω:

$$PV \int_{-\infty}^{+\infty} e^{\pm j\omega t}\mathrm{d}t = 2\pi\delta(\omega). \tag{4.35}$$

4.3.3   Definition of the Convolution Product

The convolution product of two functions $f(t)$ and $g(t)$, is defined by the following integral:

$$\boxed{y(t) = \int_{-\infty}^{+\infty} f(t')g(t - t')\mathrm{d}t'} \tag{4.36}$$

By the change of variables $t'' = t - t'$, we also have:

$$y(t) = -\int_{+\infty}^{-\infty} f(t - t'')g(t'')\mathrm{d}t'' = \int_{-\infty}^{+\infty} f(t - t')g(t')\mathrm{d}t', \tag{4.37}$$

t'' was replaced by t' in the last member of the equation.
The symmetry of the formulas (4.36) and (4.37) is noted.
The convolution integral is conventionally written by the following notation:

$$y(t) = f(t)\otimes g(t). \tag{4.38}$$

Properties: The convolution product is commutative, associative and distributive. These results come from the properties of integration.

Convolution of a function with the Dirac distribution:
This convolution is written:

$$f(t)\otimes\delta(t) = \int\limits_{-\infty}^{+\infty} f(t-t')\delta(t')dt' = f(t). \qquad (4.39)$$

We applied the formula (4.28) defining the Dirac distribution to get the result $f(t)$.

This result shows that the convolution of a function $f(t)$ with $\delta(t)$ gives back $f(t)$. It is said that $\delta(t)$ is the neutral element of the convolution product.

Translation property of the convolution of a function with $\delta(t - t_0)$**:**
The convolution of a function $f(t)$ with $\delta(t - t_0)$ leads to the translation of this function:

$$f(t) \otimes \delta(t - t_0) = \int\limits_{-\infty}^{+\infty} f(t - t')\delta(t' - t_0)dt' = \int\limits_{-\infty}^{+\infty} f(t - t'' - t_0)\delta(t'')dt''$$
$$= f(t - t_0). \qquad (4.40)$$

This last result is significant to remember:

$$f(t)\otimes\delta(t - t_0) = f(t-t_0). \qquad (4.41)$$

Scale change in the Dirac distribution:
Let the function $f(t)$ be continuous at $t = 0$ and a a real number different from zero. We have the following property:

$$\int\limits_{-\infty}^{+\infty} f(t)\delta(at)dt = \frac{1}{|a|}f(0). \qquad (4.42)$$

Indeed, by first treating the case $a > 0$:

$$\int\limits_{-\infty}^{+\infty} f(t)\delta(at)dt = \frac{1}{a}\int\limits_{-\infty}^{+\infty} f\left(\frac{x}{a}\right)\delta(x)dx = \frac{1}{a}f(0).$$

We have noted $at = x$.
In the case, where $a < 0$:

$$\int\limits_{-\infty}^{+\infty} f(t)\delta(at)dt = \frac{1}{a}\int\limits_{+\infty}^{-\infty} f\left(\frac{x}{a}\right)\delta(x)dx = -\frac{1}{a}\int\limits_{-\infty}^{+\infty} f\left(\frac{x}{a}\right)\delta(x)dx = -\frac{1}{a}f(0).$$

Again, we have noted $at = x$. Note the sign change caused by the change in the integral boundaries in x. Since $a < 0$, $x = \infty$ when $t = -\infty$ and $x = -\infty$ if $t = \infty$

The formula (4.42) assembles the two cases.

Similarly it can be shown that:

$$\int\limits_{-\infty}^{+\infty} f(t)\delta(at - t_1)\mathrm{d}t = \frac{1}{|a|}f\left(\frac{t_1}{a}\right). \tag{4.43}$$

4.3.4 Primitive of the Dirac Distribution. Heaviside Function

As seen in formula (4.33), $\int_{-\infty}^{+\infty}\delta(t)\mathrm{d}t = 1$. The argument exposed now lacks of mathematical rigor, but it leads to a result readily understandable and usable by a physicist.

The Dirac distribution being essentially localized in $t = 0$, the value of an integral of $\delta(t)$ will be different depending on whether the integration domain contains the point $t = 0$ or not. Thus we can write:

$$\int\limits_{-\infty}^{t}\delta(t)\mathrm{d}t = 0 \text{ if } t < 0 \text{ and } \int\limits_{-\infty}^{t}\delta(t)\mathrm{d}t = 1 \text{ if } t > 0. \tag{4.44}$$

Thus we see that the integral of the Dirac distribution is the Heaviside function $U(t)$ (Fig. 4.8).

This function is zero for negative times and is equal to 1 for positive times. Its value at $t = 0$ is not important in the calculations. Some authors take 0, others 1, others ½, for the value at the origin.

Conversely we have:

$$\boxed{\frac{\mathrm{d}U(t)}{\mathrm{d}t} = \delta(t)} \tag{4.45}$$

Fig. 4.8 Heaviside function

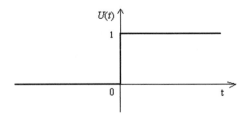

4.3.5   Derivatives of the Dirac Distribution

The first derivative is denoted:

$$\delta'(t) = \frac{d\delta(t)}{dt}.$$

(4.46)

Since $\delta(t)$ is a distribution, its derivation presents a problem. To give meaning to $\delta'(t)$, its action on a function $f(t)$ is evaluated through integration:

$$\int\limits_{-\infty}^{+\infty} \delta'(t)f(t)dt$$

(4.47)

By integrating by parts: $\int_{-\infty}^{\infty} \delta'(t)f(t)dt = [\delta(t)f(t)]_{-\infty}^{\infty} - \int_{-\infty}^{\infty} \delta(t)f'(t)dt$.
In the case where the function $f(t)$ <u>vanishes at infinity</u>, the integrated term vanishes and we get:

$$\int\limits_{-\infty}^{+\infty} \delta'(t)f(t)dt = - \int\limits_{-\infty}^{+\infty} \delta(t)f'(t)dt = -f'(0).$$

(4.48)

Thus $\delta'(t)$ is characterized in that its integration with a function $f(t)$ gives the value of that function at $t = 0$ with a reversed sign.

It was assumed that the derivative of the function $f(t)$ is continuous at $t = 0$.

Similarly, it is shown that if the function $f(t)$ (with a bounded support) is sufficiently regular in $t = 0$, we have:

$$\int\limits_{-\infty}^{\infty} \delta''(t)f(t)dt = f''(0).$$

(4.49)

More generally, noting $\delta^{(n)}(t)$ the nth derivative of the Dirac distribution, if the function $f(t)$ is sufficiently regular at $t = 0$, one has:

$$\int\limits_{-\infty}^{\infty} \delta^{(n)}(t)f(t)dt = (-1)^n f^{(n)}(0)$$

(4.50)

Summary
We have defined in this chapter the Dirac distribution $\delta(t)$ which plays a key role in the mathematics of signal analysis. The Dirichlet integral, which leads us here to the definition of the Cauchy's principal value was introduced for that purpose. Its evaluation was performed in the complex plane. We found the formula

$\int_{-\infty}^{\infty} e^{\pm j\omega t} d\omega = 2\pi\delta(t)$ which plays a very important role in calculations and which could be called the golden formula in signal analysis. The use of the Dirac distribution in calculations was exposed. Its primitive integral (the Heaviside function) and its derivatives were defined.

Exercises

I. Represent graphically the distributions $\delta(t-5)$ and $\delta(t+5)\otimes U(t)$.
   Answer: $\delta(t+5)\otimes U(t) = U(t+5)$.

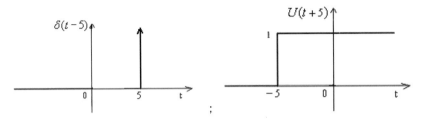

;

II. Show that we have:

$$\int_{-\infty}^{+\infty} \sin 5t\,\delta(t)dt = 0 \quad \int_{-\infty}^{+\infty} \sin 5t\,\delta(t+2)dt = \sin(-10) \quad \int_{-\infty}^{+\infty} \cos 5(t+2)\delta(t+2)dt = 1.$$

$$\int_{-\infty}^{+\infty} \cos 5(t+2)\delta(2t+4)dt = \frac{1}{2}.$$

III. Calculate the following derivatives: $\frac{d}{dt}U(t)\cos(\omega_0 t); \frac{d}{dt}U(t)\sin(\omega_0 t)$.

   Solution:

$$\frac{d}{dt}U(t)\cos(\omega_0 t) = \frac{dU(t)}{dt}\cos(\omega_0 t) - U(t)\omega_0\sin(\omega_0 t)$$
$$= \delta(t)\cos(\omega_0 t) - U(t)\omega_0\sin(\omega_0 t).$$

Otherwise, as $\delta(t)\cos(\omega_0 t) = \delta(t)$,

$$\int_{-\infty}^{+\infty} \delta(t)\cos(\omega_0 t)f(t)dt = \cos(0)f(0) = f(0) \text{ and } \int_{-\infty}^{+\infty} \delta(t)f(t)dt = f(0).$$

Finally:

$$\frac{\mathrm{d}}{\mathrm{d}t} U(t) \cos(\omega_0 t) = \delta(t) - U(t)\omega_0 \sin(\omega_0 t).$$

Similarly:

$$\frac{\mathrm{d}}{\mathrm{d}t} U(t) \sin(\omega_0 t) = \delta(t) \sin(\omega_0 t) + U(t)\omega_0 \cos(\omega_0 t) = U(t)\omega_0 \cos(\omega_0 t),$$

because

$$\int_{-\infty}^{+\infty} \delta(t) \sin(\omega_0 t) f(t) \mathrm{d}t = \sin(0) f(0) = 0.$$

Chapter 5
Fourier Transform

In the previous chapter the Dirac distribution has been introduced as a sum of exponentials $e^{j\omega t}$ with all possible frequencies and amplitude as one. Equation (4.35) can be rewritten as $\frac{1}{2\pi} PV \int_{-\infty}^{+\infty} e^{j\omega t} d\omega = \delta(t)$. Using this formula as a start, we introduce naturally the notion of impulse response of an LTI system. The impulse response appears to be the inverse Fourier transform of the frequency response of the system. This leads to the general definition of the inverse and direct Fourier transforms. We examine in this chapter the first main results given by Fourier transformation. The Parseval–Plancherel energy theorem is demonstrated. The important Poisson's summation formula which may be called the second golden formula in signal analysis is given as

$$\sum_{n=-\infty}^{\infty} e^{-j\omega n T_0} = \frac{2\pi}{T_0} \sum_{n=-\infty}^{\infty} \delta\left(\omega - n\frac{2\pi}{T_0}\right).$$

Finally, in this chapter, we present the elements of the two-dimensional Fourier transform.

5.1 Impulse Response of an LTI System

The impulse response of a system is defined as the system response to the input signal $\delta(t)$, when the system has not been prepared in advance. For instance, in the case of a first-order electrical system met in Chap. 2, non-preparation consists in the fact that the capacitor is not charged when the emf $e(t)$ is applied to the circuit.

In the following, the system is assumed to be linear time invariant (LTI). It is now shown that the impulse response which will be denoted as $h(t)$ is connected to the frequency response of the system $H(\omega)$ by integration. The reasoning is strengthened by the use of diagrams.

© Springer International Publishing Switzerland 2016
F. Cohen Tenoudji, *Analog and Digital Signal Analysis*,
Modern Acoustics and Signal Processing, DOI 10.1007/978-3-319-42382-1_5

In Chap. 1 it has been shown that the LTI system response to an input is $H(j\omega)e^{j\omega t}$. It was also said that the function $H(j\omega)$ was often $H(\omega)$ noted in signal analysis.

$$x(t)=e^{j\alpha t} \quad \boxed{\text{System}} \quad y(t)=H(\omega)e^{j\alpha t}$$

The system is linear, by hypothesis; its response to a sum of exponentials (represented here by integration in the sense of principal value) is the sum of each individual response according to the following scheme:

$$PV\int_{-\infty}^{+\infty}e^{j\alpha t}d\omega \quad \boxed{\text{System}} \quad PV\int_{-\infty}^{+\infty}H(\omega)e^{j\alpha t}d\omega$$

If, moreover, the input signal is divided by 2π, due to its linearity, the system response is also divided by 2π. The Dirac distribution may be recognized in the input signal. By convention, the system output is noted as $h(t)$.

$$\delta(t)=\frac{1}{2\pi}PV\int_{-\infty}^{+\infty}e^{j\alpha t}d\omega \quad \boxed{\text{System}} \quad h(t)=\frac{1}{2\pi}PV\int_{-\infty}^{+\infty}H(\omega)e^{j\alpha t}d\omega$$

To summarize, we note that if the input signal is

$$\delta(t) = \frac{1}{2\pi}PV\int_{-\infty}^{+\infty} e^{j\omega t}d\omega, \tag{5.1}$$

the output will be

$$h(t) = \frac{1}{2\pi}PV\int_{-\infty}^{+\infty} H(\omega)e^{j\omega t}d\omega. \tag{5.2}$$

The Cauchy principal value of the integral (5.2) defines the inverse Fourier transform. It is said that the impulse response of an LTI system is the inverse Fourier transform of the frequency response.

The principal value notation PV is often overlooked because it can embarrass students unfamiliar with this concept. This notation is omitted in the following while keeping in mind that the convergence of the PV is less restrictive than that of the improper integral. However, in practical calculations, we will have the right to let the boundaries go symmetrically to infinity in the evaluation of Eq. (5.2).

Note that, according to the above definition of the impulse response, it exists only if the sum (5.2) converges (in Cauchy's sense). In particular, it is necessary for the function $H(\omega)$ to be bounded for all ω. Accordingly, the transfer function $H(p)$ of the system should not have a pole on the imaginary axis (axis $j\omega$). It is also necessary that the modulus of the function $H(\omega)$ decreases quickly enough when $|\omega| \to \infty$ in order for integration to be possible.

5.2 Fourier Transform of a Signal

5.2.1 Direct Fourier Transform

In the previous section, the concept of inverse Fourier transform was encountered before the concept of direct Fourier transform. The reason is that the concept of superposition of exponential basis of formulas (5.1) and (5.2) appears to be fundamental in the understanding of the Fourier transform.

The Fourier transform $X(\omega)$ of a signal $x(t)$ is defined by the expression

$$X(\omega) = \int_{-\infty}^{+\infty} x(t)e^{-j\omega t}dt. \tag{5.3}$$

The existence of $X(\omega)$ is dependent on the integral convergence.

We see for example that it is necessary for the function to decrease rapidly enough at infinity so that the integral converges. A set of functions that play an important role is the set of square-summable (integrable)functions. This set generates a Hilbert space L_2.

In contrast, a periodic function, a sine example, does not decrease at infinity and will not have a Fourier transform in the sense of ordinary convergence of an integral. However, as noted above in the discussion of the Dirac distribution, it is possible to give a meaning to certain non-convergent summations. The sine function has a Fourier transform in the sense of distributions.

It is conventional to use lowercase to write functions in the time domain (for example $x(t)$) and to write the first letter in capital in the Fourier domain (example $X(\omega)$). It is recommended to follow this convention that allows better tracking during calculations on these functions. Scoring with a capital the first letter of the Heaviside function $U(t)$ is a rare case where the convention is not respected.

5.2.2 Inverse Fourier Transform

Having defined in the preceding paragraph the Fourier transform $X(\omega)$ of a function $x(t)$, it is now shown that it is possible to calculate $x(t)$, knowing $X(\omega)$.

Let $X(\omega)$ be the Fourier transform of $x(t)$, using the integration variable t' for demonstration purposes, we write

$$X(\omega) = \int_{-\infty}^{+\infty} x(t')e^{-j\omega t'}\, dt'. \tag{5.4}$$

Consider the integral

$$\int_{-A}^{A} X(\omega)e^{j\omega t}\, d\omega = \int_{-A}^{A} \int_{-\infty}^{+\infty} x(t')e^{j\omega t}e^{-j\omega t'}\, d\omega dt'. \tag{5.5}$$

We assume now without justification that the order of integration does not matter. Reversing the order of summations, however, requires that restrictive conditions on the convergence of the integrals are verified (the discussion of these conditions is beyond the scope of this book). These conditions are generally met in problems in signal analysis. In the following, the order of the summations will be systematically switched when necessary.

As demonstrated in Sect. 5.4,

$$\lim_{A\to\infty} \int_{-A}^{A} e^{j\omega(t-t')}\, d\omega = 2\pi\delta(t - t'). \tag{5.6}$$

By taking the limit of Eq. (5.5) when $A \to \infty$,

$$\lim_{A\to\infty} \int_{-A}^{A} X(\omega)e^{j\omega t}\, d\omega = \int_{-\infty}^{+\infty} x(t')\delta(t - t')2\pi dt' = 2\pi x(t). \tag{5.7}$$

We use equation

$$\int_{-\infty}^{+\infty} x(t')\delta(t - t')\, dt' = x(t). \tag{5.8}$$

The relationship (5.7) is written as

$$x(t) = \frac{1}{2\pi} \lim_{A\to\infty} \int_{-A}^{A} X(\omega)e^{j\omega t}\, d\omega. \tag{5.9}$$

Noting *PV* the Cauchy principal value of the integral, we have finally

$$x(t) = \frac{1}{2\pi} PV \int_{-\infty}^{+\infty} X(\omega)e^{j\omega t} d\omega. \tag{5.10}$$

This formula is that of the inverse Fourier transform, which calculates $x(t)$ from $X(\omega)$.

As stated previously, the convergence condition of the principal value is less restrictive than that of the integral (improper). Indeed, it is necessary to have convergence when the limits are reached independently when the variable goes to $-\infty$ and $+\infty$ in order that the integral can exist. In the summation leading to the Cauchy principal value, the two bounds $-A$ and $+A$ tend symmetrically to infinity. In that case, there may be a compensation phenomenon between contributions at infinity due to the symmetry of the bounds causing a finite limit of the sum.

As already mentioned, in practice one often omits *PV* in giving the formula and write symmetrically the pair of Fourier direct and inverse transforms:

Direct Fourier Transform:

$$X(\omega) = \int_{-\infty}^{+\infty} x(t)e^{-j\omega t} dt. \tag{5.11}$$

Inverse Fourier Transform:

$$x(t) = \frac{1}{2\pi} \int_{-\infty}^{+\infty} X(\omega)e^{j\omega t} d\omega. \tag{5.12}$$

As it appears in the inverse transformation formula, a signal $x(t)$ which has a Fourier transform can be considered as a sum of exponentials $e^{j\omega t}$ weighted by the factor $X(\omega)$. All frequencies are involved in the construction of $x(t)$.

Thus, even a non-periodic function of time appears as a sum of periodic functions (the exponential $e^{j\omega t}$). This result may seem surprising. On the other hand, how the sum of periodic functions which do not vanish at infinity could represent a non-periodic function $x(t)$ that could be null at infinite?

However, this is possible. Just think of the phenomenon of destructive interference encountered in optics, in which the sum of light vibrations can lead to dark areas on a screen.

The acceptance of the concept that Fourier developed in his famous paper on the propagation of heat proved difficult. The community of mathematicians was divided

at that time on that acceptance. The Fourier transform is now one of the pillars of
the modeling of physical phenomena.

Mathematically, we can consider the Fourier transform as the development of
the function $x(t)$ on the infinite, continuous basis of exponential $e^{j\omega t}$ with all possible frequencies. The function $X(\omega)$ is the coefficient of the development of $x(t)$
relative to the basis function $e^{j\omega t}$. It can be written as a scalar product:

$$X(\omega) = \int\limits_{-\infty}^{+\infty} x(t)e^{-j\omega t}dt = \langle x(t), e^{j\omega t}\rangle.$$

Since the scalar product $\langle e^{j\omega t}, e^{j\omega' t}\rangle = \int_{-\infty}^{\infty} e^{j(\omega-\omega')t}dt = 2\pi\delta(\omega - \omega')$ is zero for
$\omega \neq \omega'$, we say that these functions are orthogonal. The passage to the limit $\omega' \rightarrow$
ω acts as the normalization condition. The reader is encouraged to refer to the
discussion of Hilbert spaces in Chap. 3 to have a general perspective of the problem
addressed.

5.3   Properties of Fourier Transform

5.3.1   Symmetry Properties of the Fourier Transform of a Real Signal

Let $x(t)$ be a real signal and $X(\omega) = X_{Re}(\omega) + jX_{Im}(\omega)$ its Fourier transform:

$$X_{Re}(\omega) + jX_{Im}(\omega) = \int\limits_{-\infty}^{+\infty} x(t)e^{-j\omega t}dt = \int\limits_{-\infty}^{+\infty} x(t)\cos\omega t\,dt - j\int\limits_{-\infty}^{+\infty} x(t)\sin\omega t\,dt.$$

$$(5.13)$$

Identifying the real and imaginary parts of the two members of the equation, and
as $x(t)$ is real, it becomes

$$X_{Re}(\omega) = \int\limits_{-\infty}^{+\infty} x(t)\cos\omega t\,dt, \qquad (5.14)$$

and

$$X_{Im}(\omega) = -\int\limits_{-\infty}^{+\infty} x(t)\sin\omega t\,dt. \qquad (5.15)$$

Since $\cos \omega t$ and $\sin \omega t$ are, respectively, even and odd functions of ω, we have

$$X_{\text{Re}}(\omega) = \int_{-\infty}^{+\infty} x(t) \cos \omega t\, dt \quad \text{even function of } \omega$$

and

$$X_{\text{Im}}(\omega) = - \int_{-\infty}^{+\infty} x(t) \sin \omega t\, dt \quad \text{odd function of } \omega$$

Similarly, again using parities of $\cos \omega t$ and $\sin \omega t$ functions, it is seen that the modulus $|X(\omega)|$ is even and the phase $\varphi(\omega)$ odd. The modulus is given by

$$|X(\omega)| = \sqrt{X_{\text{Re}}^2(\omega) + X_{\text{Im}}^2(\omega)}$$
$$= \sqrt{\left(\int_{-\infty}^{+\infty} x(t) \cos \omega t\, dt \right)^2 + \left(\int_{-\infty}^{+\infty} x(t) \sin \omega t\, dt \right)^2 }. \qquad (5.16)$$

$|X(\omega)|$ is an even function because the squaring cancels the change of sign of $\sin \omega t$.

The phase

$$\varphi(\omega) = \text{Arg}\left(\frac{X_{\text{Im}}(\omega)}{X_{\text{Re}}(\omega)} \right) = \text{Arg}\, \frac{- \int_{-\infty}^{+\infty} x(t) \sin \omega t\, dt}{\int_{-\infty}^{+\infty} x(t) \cos \omega t\, dt}, \qquad (5.17)$$

is an odd function of ω since the sign of $\sin \omega t$ changes and that of $\cos \omega t$ stays the same in the change ω to $-\omega$.

Special cases:

If the signal $x(t)$ is real and even, the imaginary part of its Fourier transform is zero, as can be seen from Eq. (5.15) which becomes the integral of an odd function. Similarly, if $x(t)$ is real and odd, the real part of its Fourier transform is zero as shown by Eq.(5.14).

5.3.2 Time-Delay Property of the Fourier Transform

Let $f(t) = x(t - t_0)$ be the function $x(t)$ delayed by t_0. Its Fourier transform is

$$F(\omega) = \int_{-\infty}^{+\infty} x(t - t_0)e^{-j\omega t}\,dt = \int_{-\infty}^{+\infty} x(t')e^{-j\omega(t' + t_0)}\,dt'.$$

We have written $t' = (t - t_0)$.

$$F(\omega) = e^{-j\omega t_0} \int_{-\infty}^{+\infty} x(t')e^{-j\omega t'}\,dt' = e^{-j\omega t_0}X(\omega). \tag{5.18}$$

The Fourier transform of the delayed function is equal to the Fourier transform of the original function multiplied by the phase factor $e^{-j\omega t_0}$. This result is important. It plays a large role in signal analysis calculations.

5.4 Power and Energy of a Signal; Parseval–Plancherel Theorem

The power (instantaneous) $P(t)$ of a complex, certain signal $x(t)$ is defined as

$$P(t) = x(t)x^*(t) = |x(t)|^2. \tag{5.19}$$

If the signal $x(t)$ is real, the instantaneous power becomes

$$P(t) = x^2(t). \tag{5.20}$$

The energy E of a signal is the integral of the power for time varying from $-\infty$ to $+\infty$. In the case of a real signal, the energy is given by

$$E = \int_{-\infty}^{+\infty} x^2(t)\,dt. \tag{5.21}$$

The energy of a complex signal will be

$$E = \int_{-\infty}^{+\infty} |x(t)|^2\,dt. \tag{5.22}$$

Parseval–Plancherel theorem states how the energy can be calculated in the time domain or in the frequency domain. It is written here in the general case of a complex signal as

$$E = \int\limits_{-\infty}^{+\infty} |x(t)|^2 dt = \frac{1}{2\pi} \int\limits_{-\infty}^{+\infty} |X(\omega)|^2 d\omega. \tag{5.23}$$

Indeed,

$$\int\limits_{-\infty}^{+\infty} |x(t)|^2 dt = \int\limits_{-\infty}^{+\infty} x(t)x^*(t)dt. \tag{5.24}$$

Since we can write $x(t) = \frac{1}{2\pi} PV \int_{-\infty}^{+\infty} X(\omega)e^{j\omega t}d\omega$, and

$$x^*(t) = \frac{1}{2\pi}\left(PV \int\limits_{-\infty}^{+\infty} X(\omega)e^{j\omega t}d\omega \right)^* = \frac{1}{2\pi} PV \int\limits_{-\infty}^{+\infty} X^*(\omega)e^{-j\omega t}d\omega. \tag{5.25}$$

Using ω' as the variable of integration in one of the two integrals, we may write

$$\int\limits_{-\infty}^{+\infty} |x(t)|^2 dt = \frac{1}{(2\pi)^2} \int\limits_{-\infty}^{+\infty} PV \int\limits_{-\infty}^{+\infty} PV \int\limits_{-\infty}^{+\infty} X(\omega)X^*(\omega')e^{j\omega t}e^{-j\omega' t}d\omega d\omega' dt$$

We first evaluate the integral with respect to time in the second member. Since $PV \int_{-\infty}^{+\infty} e^{j(\omega-\omega')t}dt = 2\pi\delta(\omega - \omega')$, it appears necessary to evaluate the principal value of the integral in the left side.
We have

$$PV \int\limits_{-\infty}^{+\infty} |x(t)|^2 dt = \frac{1}{2\pi} PV \int\limits_{-\infty}^{+\infty} PV \int\limits_{-\infty}^{+\infty} X(\omega)X^*(\omega')\delta(\omega - \omega')d\omega d\omega'$$

After integration on ω', we finally have

$$PV \int\limits_{-\infty}^{+\infty} |x(t)|^2 dt = \frac{1}{2\pi} PV \int\limits_{-\infty}^{+\infty} X(\omega)X^*(\omega)d\omega = \frac{1}{2\pi} PV \int\limits_{-\infty}^{+\infty} |X(\omega)|^2 d\omega, \tag{5.26}$$

which is the expression of the Parseval theorem expressed by Eq. (5.23).
In practice, $x(t)$ is often a real function of time. Parseval's theorem in this case takes the following form:

$$E = PV \int\limits_{-\infty}^{+\infty} x(t)^2 dt = \frac{1}{2\pi} PV \int\limits_{-\infty}^{+\infty} |X(\omega)|^2 d\omega. \tag{5.27}$$

The quantity $|X(\omega)|^2$ appears as the spectral energy density. In practice, we note that we can calculate the energy of a signal by integration in the time domain or in the frequency domain. The notation PV is generally omitted in writing the theorem.

5.5   Deriving a Signal and Fourier Transform

In the following the FT $X_d(\omega)$ of the derivative of a signal $x(t)$ is inferred from the FT $X(\omega)$ of that signal:

$$X_d(\omega) = \int\limits_{-\infty}^{+\infty} \frac{\mathrm{d}x(t)}{\mathrm{d}t} \mathrm{e}^{-\mathrm{j}\omega t} \mathrm{d}t.$$

We integrate by parts: $X_d(\omega) = \left[x(t)\mathrm{e}^{-\mathrm{j}\omega t}\right]_{-\infty}^{\infty} + \mathrm{j}\omega \int_{-\infty}^{+\infty} x(t)\mathrm{e}^{-\mathrm{j}\omega t}\mathrm{d}t.$

When the function $x(t)$ vanishes at infinity, as is the case for a finite energy signal (this is not the case of a sine or cosine who oscillate indefinitely between -1 and $+1$), the first term cancels out, and we have

$$X_d(\omega) = \mathrm{j}\omega \int\limits_{-\infty}^{+\infty} x(t)\mathrm{e}^{-\mathrm{j}\omega t}\mathrm{d}t = \mathrm{j}\omega X(\omega). \tag{5.28}$$

Generalizing, and in the case where the derivatives of $x(t)$ are continuous up to order $n-1$ and tend to zero as t tends to infinity, we get

$$\frac{\mathrm{d}^n x(t)}{\mathrm{d}t^n} \xrightarrow{TF} (\mathrm{j}\omega)^n X(\omega). \tag{5.29}$$

Conversely, in the (very unlikely) case where primitives (antiderivatives) of $x(t)$ of order n are zero for $t \to \infty$, knowing the FT $X(\omega)$ of the signal $x(t)$, we can deduce the FT of its primitives as

$$(\text{Primitive of order } n \text{ of } x(t)) \xrightarrow{TF} \frac{1}{(\mathrm{d}\omega)^n} X(\omega). \tag{5.30}$$

We note that if the function $x(t)$ does not vanish for $t \to \infty$ the above formula is not applicable.

5.6 Fourier Transform of Dirac Distribution and of Trigonometric Functions

Fourier transform of $\delta(t)$:

$$\mathcal{F}(\delta(t)) = \int\limits_{-\infty}^{+\infty} \delta(t)\,e^{-j\omega t}dt = e^{-j\omega 0} = 1. \qquad (5.31)$$

Fourier transform of 1:

Since $PV \int_{-\infty}^{+\infty} e^{-j\omega t}dt = 2\pi\delta(\omega)$, the Fourier transform of 1 is $2\pi\delta(\omega)$.

Once again, only the PV of integrals appears, and strictly speaking, the constant function $f(t) = 1$ has no FT if the definition (5.11) is adopted.

Henceforth, we will often omit to include PV in the formulas on the Fourier transform to not overload the notation.

Fourier Transforms of $\cos\omega_0 t$ and $\sin\omega_0 t$:

By definition, the transform of $\cos\omega_0 t$ is

$$\int\limits_{-\infty}^{+\infty} \cos(\omega_0 t)e^{-j\omega t}dt = \int\limits_{-\infty}^{+\infty} \left(\frac{e^{j\omega_0 t} + e^{-j\omega_0 t}}{2}\right)e^{-j\omega t}dt. \qquad (5.32)$$

By the sum of two integrals, we have

$$\frac{1}{2}\int\limits_{-\infty}^{+\infty} e^{-j(\omega + \omega_0)t}dt + \frac{1}{2}\int\limits_{-\infty}^{+\infty} e^{-j(\omega - \omega_0)t}dt = \pi\delta(\omega - \omega_0) + \pi\delta(\omega + \omega_0). \qquad (5.33)$$

So

$$\mathcal{F}(\cos\omega_0 t) = \pi\delta(\omega - \omega_0) + \pi\delta(\omega + \omega_0). \qquad (5.34)$$

Similarly, we have

$$\mathcal{F}(\sin\omega_0 t) = -j\,(\pi\delta(\omega - \omega_0) - \pi\delta(\omega + \omega_0)). \qquad (5.35)$$

In Fig. 5.1 we see the real part of $\mathcal{F}(\cos\omega_0 t)$, and the imaginary part is zero. The imaginary part of $\mathcal{F}(\sin\omega_0 t)$ is shown in Fig. 5.2. The real part is zero.

Fig. 5.1 Fourier transform of $\cos \omega_0 t$

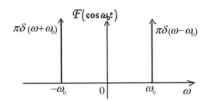

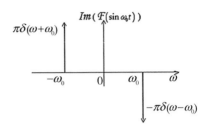

Fig. 5.2 Fourier transform of $\sin \omega_0 t$

Fourier Transform of a Dirac comb

A Dirac comb is a periodic sequence of Dirac impulses (Fig. 5.3). The period of these pulses is noted T_0:

$$\text{Ш}_{T_0}(t) = \sum_{n=-\infty}^{\infty} \delta(t - nT_0). \tag{5.36}$$

Calculation of the Fourier transform of this comb:

$$\mathcal{F}\left(\text{Ш}_{T_0}(t)\right) = \int_{-\infty}^{\infty} \sum_{n=-\infty}^{\infty} \delta(t - nT_0) e^{-j\omega t}\, dt = \sum_{n=-\infty}^{\infty} \int_{-\infty}^{\infty} \delta(t - nT_0) e^{-j\omega t}\, dt = \sum_{n=-\infty}^{\infty} e^{-j\omega nT_0}. \tag{5.37}$$

It was assumed that the conditions for interchanging the order of summations were met. This sum appears as the Fourier series of a periodic function of ω with period $\frac{2\pi}{T_0}$, because each exponential is periodic in ω. The result of this summation is noted $Y(\omega)$. As shown in Eq. (5.31), the coefficients of its Fourier series are all equal to 1.

Fig. 5.3 Dirac comb

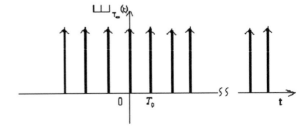

According to the general properties of Fourier series, they can be deduced from $Y(\omega)$ by the following formula:

$$c_n = 1 = \frac{T_0}{2\pi} \int_{-\frac{\pi}{T_0}}^{\frac{\pi}{T_0}} Y(\omega) e^{jn\omega T_0} d\omega. \tag{5.38}$$

We see qualitatively that the only way to get 1 regardless of n, i.e., regardless of the speed of variation with ω of the exponential functions, is that $Y(\omega)$ is zero for all frequencies except for $\omega = 0$. $Y(\omega)$ cannot be a function because it is too irregular. It is the Dirac distribution $\delta(\omega)$ (within a multiplicative coefficient). Between $-\frac{\pi}{T_0}$ and $\frac{\pi}{T_0}$, we must have

$$Y(\omega) = \frac{2\pi}{T_0} \delta(\omega). \tag{5.39}$$

Finally, as $Y(\omega)$ is periodic, we have the following:

Poisson summation formula

$$\sum_{n=-\infty}^{\infty} e^{-j\omega n T_0} = \frac{2\pi}{T_0} \sum_{n=-\infty}^{\infty} \delta\left(\omega - n\frac{2\pi}{T_0}\right). \tag{5.40}$$

In conclusion, it is seen that the Fourier transform of the Dirac comb with period T_0 is a Dirac comb with period $\frac{2\pi}{T_0}$:

$$\mathcal{F}\left(\text{Ш}_{T_0}(t)\right) = \frac{2\pi}{T_0}\left(\text{Ш}_{2\pi/T_0}(\omega)\right). \tag{5.41}$$

As will be seen in Chap. 6, the Poisson formula plays an important role in the calculations of the Fourier transform of periodic signals.

5.7 Two-Dimensional Fourier Transform

Let $x(t_1, t_2)$ be a function of two variables. Its two-dimensional FT is defined by

$$X(\omega_1, \omega_2) = \int_{-\infty}^{+\infty} \int_{-\infty}^{+\infty} x(t_1, t_2) e^{-j\omega_1 t_1} e^{-j\omega_2 t_2} dt_1 dt_2. \tag{5.42}$$

It may be shown as above (formula 5.12) that the inversion formula is

$$x(t_1, t_2) = \frac{1}{4\pi^2} \int\limits_{-\infty}^{+\infty} \int\limits_{-\infty}^{+\infty} X(\omega_1, \omega_2) e^{j\omega_1 t_1} e^{j\omega_2 t_2} d\omega_1 d\omega_2. \tag{5.43}$$

The principal value notations of these integrals have been omitted.

This transformation is widely used in the field of treatment of 2D images. In this case, the time variables t_1, t_2 are replaced by space variables x_1, x_2. The conjugate variables are noted as k_1, k_2 and are called spatial frequencies. One has, therefore,

$$F(k_1, k_2) = \int\limits_{-\infty}^{+\infty} \int\limits_{-\infty}^{+\infty} f(x_1, x_2) e^{-jk_1 x_1} e^{-jk_2 x_2} dx_1 dx_2 \tag{5.44}$$

and

$$f(x_1, x_2) = \frac{1}{4\pi^2} \int\limits_{-\infty}^{+\infty} \int\limits_{-\infty}^{+\infty} F(k_1, k_2) e^{jk_1 x_1} e^{jk_2 x_2} dk_1 dk_2. \tag{5.45}$$

Using of a vector notation and writing $\vec{x} = \begin{vmatrix} x_1 \\ x_2 \end{vmatrix}$ and $\vec{k} = \begin{vmatrix} k_1 \\ k_2 \end{vmatrix}$, we also have

$$F(k_1, k_2) = \int\limits_{-\infty}^{+\infty} \int\limits_{-\infty}^{+\infty} f(x_1, x_2) e^{-j\vec{k}.\vec{x}} dx_1 dx_2 \tag{5.46}$$

and

$$f(x_1, x_2) = \frac{1}{4\pi^2} \int\limits_{-\infty}^{+\infty} \int\limits_{-\infty}^{+\infty} F(k_1, k_2) e^{j\vec{k}.\vec{x}} dk_1 dk_2. \tag{5.47}$$

These formulas can be written in polar coordinates r, θ, and k, ϕ with

$$\begin{aligned} x_1 = r\cos\theta; x_2 = r\sin\theta \quad &\text{and} \quad dx_1 dx_2 = r dr d\theta. \\ k_1 = k\cos\phi; k_2 = k\sin\phi \quad &\text{and} \quad dk_1 dk_2 = k dk d\phi. \end{aligned} \tag{5.48}$$

$$F(k_1, k_2) = \int\limits_{0}^{+\infty} \int\limits_{0}^{2\pi} f(r, \theta) e^{-jk_1 r\cos\theta} e^{-jk_2 r\sin\theta} r dr d\theta \tag{5.49}$$

and

$$f(r,\theta) = \frac{1}{4\pi^2} \int\limits_{0}^{+\infty} \int\limits_{0}^{2\pi} F(k,\phi) e^{jkr\cos\phi\cos\theta} e^{jkr\sin\phi\sin\theta} k\,dk\,d\phi. \tag{5.50}$$

Special case:

If the function $f(r,\theta)$ depends only on r, the formula (5.49) becomes

$$F(k_1,k_2) = \int\limits_{0}^{+\infty} f(r)r\,dr \int\limits_{0}^{2\pi} e^{-jkr\cos\phi\cos\theta} e^{-jkr\sin\phi\sin\theta}\,d\theta, \tag{5.51}$$

or

$$F(k_1,k_2) = \int\limits_{0}^{+\infty} f(r)r\,dr \int\limits_{0}^{2\pi} e^{-jkr\cos(\phi-\theta)}\,d\theta.$$

The following important mathematical result is now used:

$$\int\limits_{0}^{2\pi} e^{jkr\cos\theta}\,d\theta = 2\pi J_0(kr), \tag{5.52}$$

where $J_0(x)$ is the first kind Bessel function of order 0.

The integration on θ being performed over the period of the cosine function, the integral no longer depends upon ϕ.

We then have $F(k_1,k_2) = 2\pi \int_0^\infty f(r)J_0(kr)r\,dr$.

It is noted that the result depends only upon $k = \sqrt{k_1^2 + k_2^2}$.

We have

$$F(k) = 2\pi \int\limits_{0}^{\infty} f(r)J_0(kr)r\,dr. \tag{5.53}$$

Conversely, the formula (5.50) becomes

$$f(r) = \frac{1}{2\pi} \int\limits_{0}^{+\infty} F(k)J_0(kr)k\,dk. \tag{5.54}$$

We say that the functions $g(r) = 2\pi f(r)$ and $F(k)$ are Hankel transforms (Fourier–Bessel transforms) of each other.

More generally, the Hankel transform of order $n \left(n \geq -\frac{1}{2}\right)$ of a function $f(r)$ is

$$F_n(k) = \int_0^\infty f(r)J_n(kr)r\mathrm{d}r. \quad \text{Inversely } f(r) = \int_0^\infty F_n(k)J_n(kr)r\mathrm{d}r. \quad (5.55)$$

$J_n(x)$ is the first kind Bessel function of order n.

Let us recall the orthogonality property of Bessel functions:

$$\int_0^\infty J_n(kr)J_n(k'r)r\mathrm{d}r = \frac{1}{k}\delta(k-k'), \quad (5.56)$$

with $k, k' > 0$.

Parseval–Plancherel theorem:

$$\int_0^\infty f(r)g(r)r\mathrm{d}r = \int_0^\infty F_n(k)G_n(k)k\mathrm{d}k. \quad (5.57)$$

Summary

Using the definition of the Dirac distribution, we have been able to show that the impulse response of an LTI system is the inverse Fourier transform of its frequency response. The direct and Fourier transform formulas result naturally. The important property of the phase factor induced by a time delay of a signal is given. The Parseval–Plancherel theorem is demonstrated. First important results were obtained: the Fourier transforms of a Dirac distribution, trigonometric functions, and a Dirac comb. The second golden rule of signal analysis, the Poisson's summation formula, was established. We have given elements of the two-dimensional Fourier transform.

Next chapter will study the use of Fourier transform in analyzing linear, stationary systems.

Exercises

1. Recall the value of the FT of function $h(t) = \frac{1}{RC}e^{-\frac{t}{RC}}U(t)$. Give the FT of $\frac{\mathrm{d}h}{\mathrm{d}t}$ by a direct calculation and by using Eq. (5.29).
2. Use the orthogonality property of Bessel functions given in Eq.(5.56) to verify the coherence of the two-dimensional transform pair given in Eqs. (5.53) and (5.54).
3. Using the integration property $\int J_0(x)x\mathrm{d}x = xJ_1(x)$, find the 2D Fourier transform of a circular disk (Result known as Airy pattern in optics).

Chapter 6
Fourier Transform and LTI Filter Systems

In the previous chapter, the Fourier transform and its inverse have been introduced in a natural way by studying the response of an LTI system using a sum of complex exponentials $e^{j\omega t}$ of equal amplitude. The response of an LTI system to any form of input signal is studied in this chapter. We demonstrate the relationships in the time and frequency domains between the input and output signals. A convolution in the time domain corresponds to a product in the frequency domain. We give the expression of the Fourier transform of the product of two functions. The formula of the FT of a periodic function establishes a bridge between Fourier series and Fourier transform. The deterministic correlation function is defined. The important application of measuring the delay between an impulsive signal and its replica is detailed. We give as an example the use in radar and sonar of a chirp signal. The spreads of a signal in time and frequency domains are defined. The Heisenberg–Gabor inequality is demonstrated. This inequality becomes equality in the case of a Gaussian signal. The chapter concludes with a discussion of the impossibility for a signal to have infinite supports simultaneously in time and frequency domains.

6.1 Response of a LTI System to Any Form of Input Signal

The reasoning is illustrated by a series of diagrams.

By definition of the frequency response of a system to an input of the signal with form $e^{j\omega t}$, the output signal will be written as $H(\omega)\,e^{j\omega t}$:

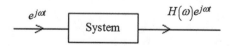

If the input signal $e^{j\omega t}$ is multiplied by an amplitude $X(\omega)$, the system being linear, the output will also be multiplied by $X(\omega)$:

$$X(\omega)e^{j\omega t} \longrightarrow \boxed{\text{System}} \longrightarrow X(\omega)H(\omega)e^{j\omega t}$$

The input signals are superimposed by an integration over ω. The system linearity leads to an integration of individual response outputs:

$$\int_{-\infty}^{\infty} X(\omega)e^{j\omega t}d\omega \longrightarrow \boxed{\text{System}} \longrightarrow \int_{-\infty}^{\infty} X(\omega)H(\omega)e^{j\omega t}d\omega$$

To complete, dividing by 2π, it shows the inverse Fourier transform of $X(\omega)$. Linearity results in the output division by the same factor 2π:

$$x(t)=\frac{1}{2\pi}\int_{-\infty}^{\infty} X(\omega)e^{j\omega t}d\omega \longrightarrow \boxed{\text{System}} \longrightarrow y(t)=\frac{1}{2\pi}\int_{-\infty}^{\infty} X(\omega)H(\omega)e^{j\omega t}d\omega$$

In summary, an input signal is given by

$$x(t) = \frac{1}{2\pi} \int_{-\infty}^{+\infty} X(\omega)e^{j\omega t}d\omega. \tag{6.1}$$

The system will provide the output signal $y(t)$ whose form is

$$y(t) = \frac{1}{2\pi} \int_{-\infty}^{+\infty} H(\omega)X(\omega)e^{j\omega t}d\omega. \tag{6.2}$$

Denoting $Y(\omega)$ the Fourier transform of $y(t)$, it is seen that we have the following property:

$$Y(\omega) = H(\omega)X(\omega). \tag{6.3}$$

Thus, the Fourier transform of the output signal of an LTI filter is the product of the Fourier transform of the input signal by the filter frequency response.

6.2   Temporal Relastionship Between the Input and Output Signals of an LTI Filter

One starts from Eq. (6.2): $y(t) = \frac{1}{2\pi} \int_{-\infty}^{+\infty} H(\omega)X(\omega)e^{j\omega t}d\omega$.

In this expression, we use $X(\omega) = \int_{-\infty}^{+\infty} x(t')e^{-j\omega t'}dt'$ in the integral giving $y(t)$, and then swaps the order of integration (assuming that the conditions for the validity of the permutation in integrations are met):

$$
\begin{aligned}
y(t) &= \frac{1}{2\pi} \int_{-\infty}^{+\infty} H(\omega) \int_{-\infty}^{+\infty} x(t')e^{-j\omega t'} dt' e^{j\omega t} d\omega \\
&= \int_{-\infty}^{+\infty} x(t')dt' \frac{1}{2\pi} \int_{-\infty}^{+\infty} H(\omega)e^{j\omega(t-t')}d\omega.
\end{aligned}
\tag{6.4}
$$

The integral on ω gives $h(t - t')$. We thus have

$$
y(t) = \int_{-\infty}^{+\infty} x(t') h(t - t')dt'.
\tag{6.5}
$$

One recognizes a convolution integral that we will note symbolically as

$$
y(t) = x(t) \otimes h(t).
\tag{6.6}
$$

> It is noteworthy that the output signal of an LTI system (not prepared) is the convolution of the input signal with the system impulse response.

Note The convolution product of $x(t)$ and $h(t)$ does not provide a complete solution to the problem in the case where the system is "prepared" by initial conditions. In the case of the first-order system found in Chap. 2, this preparation corresponds, for example, to an electric charge placed on the capacitor plates prior to application of the input signal. In this case, the general solution comprises additional terms.

Physical systems are generally damped; these additional terms disappear in the long run where only the term $x(t) \otimes h(t)$ is likely to remain. We will call the first term as transient, the term resulting from the convolution representing the stationary solution.

Direct proof of formula (6.6).

Using the property of the Dirac distribution, we write

$$
x(t) = \int_{-\infty}^{\infty} x(\tau)\delta(t - \tau)d\tau.
\tag{6.7}
$$

The output result of the action of the system operator O on the input signal is

$$y(t) = O(x(t)) = O\left(\int_{-\infty}^{+\infty} x(\tau)\,\delta(t - \tau)\,d\tau\right). \tag{6.8}$$

Due to the linearity, we can swap the integral operator and the system operator O:

$$y(t) = \int_{-\infty}^{+\infty} x(\tau)\,O(\delta(t - \tau))\,d\tau. \tag{6.9}$$

Note that the operator O did not act on $x(\tau)$ which is considered as the weight assigned to the Dirac pulse $\delta(t - \tau)$ which is a function of time t.

By definition of $h(t)$, we have

$$h(t) = O(\delta(t)). \tag{6.10}$$

Since the system is time invariant we have

$$O(\delta(t - \tau)) = h(t - \tau). \tag{6.11}$$

It now comes using the system linearity property:

$$y(t) = \int_{-\infty}^{+\infty} x(\tau)\,h(t - \tau)\,d\tau = x(t) \otimes h(t). \tag{6.12}$$

In summary, it has been shown that the Fourier transform of the convolution of two functions is the simple product of the Fourier transforms of these two functions:

$$\boxed{y(t) = x(t) \otimes h(t) \text{ is transformed into } Y(\omega) = X(\omega)H(\omega).}$$

6.3 Fourier Transform and Convolution in Physics

Many physical systems have the LTI property. The propagation of an electrical signal, light or sound, in a medium can be interpreted as the passage of a signal in a filter, most often LTI. As has been shown, after passing through the filter, the signal is the convolution of the input signal by the impulse response of the filter. Some convolution products are calculated in Chap. 7, and we will find that even in the simplest case, the calculation is thorny. In contrast, in the frequency domain, the Fourier transform of the output signal is simply the product of the input signal

Fourier transform by the system frequency response. Thus, most often in Physics, system analysis will be performed for a monochromatic input signal in the system. The response to any signal is obtained by inverse Fourier transform of the product of the monochromatic output function (frequency response) with the Fourier transform of the input signal.

6.4 Fourier Transform of the Product of Two Functions

The formula now demonstrated conveys great interest in practice.
Consider the product of two functions

$$g(t) = x(t)f(t). \tag{6.13}$$

Calculation of its Fourier transform:

$$G(\omega) = \int_{-\infty}^{+\infty} x(t)f(t)e^{-j\omega t}dt. \tag{6.14}$$

Expressing $f(t)$ from its inverse FT,

$$G(\omega) = \frac{1}{2\pi} \int_{-\infty}^{+\infty} x(t) \int_{-\infty}^{+\infty} F(\omega')e^{j\omega' t}e^{-j\omega t}\,d\omega'dt. \tag{6.15}$$

By interchanging the order of integration,

$$G(\omega) = \frac{1}{2\pi} \int_{-\infty}^{+\infty} F(\omega')d\omega' \int_{-\infty}^{+\infty} x(t)e^{-j(\omega-\omega')t}\,dt = \frac{1}{2\pi} \int_{-\infty}^{+\infty} F(\omega')X(\omega-\omega')\,d\omega'. \tag{6.16}$$

A convolution integral is recognized as calculated in the Fourier domain.
Thus, the Fourier transform of a simple product of two functions of time is the convolution of the Fourier transforms of the two functions divided by 2π:

$$g(t) = x(t)f(t) \text{ is transformed in } G(\omega) = \frac{1}{2\pi}X(\omega) \otimes F(\omega). \tag{6.17}$$

6.5   Fourier Transform of a Periodic Function

The development of a periodic function in Fourier series was studied in Chap. 3. The periodic function $f(t)$ of period T_1 is written in the form $f(t) = \sum\limits_{n=-\infty}^{+\infty} c_n e^{jn\omega_1 t}$. The Fourier coefficients are given by integrals over one period: $c_n = \frac{1}{T_1} \int\limits_0^{T_1} f(t) e^{-jn\omega_1 t}\, dt$.

We seek here the relationship between the Fourier transform of $f(t)$ and the coefficients of its Fourier series expansion. We denote $f_0(t)$ the function equal to $f(t)$ over a period and zero elsewhere. The function $f(t)$ appears as the infinite sum of the function $f_0(t)$ and its translated of all the multiples of the period T_1:

$$f(t) = \sum_{n=-\infty}^{+\infty} f_0(t - nT_1). \tag{6.18}$$

Using the property that the translated of a function can be written as a convolution $f_0(t - nT_1) = f_0(t) \otimes \delta(t - nT_1)$, we can rewrite Eq. (6.18) in the following form:

$$f(t) = \sum_{n=-\infty}^{+\infty} f_0(t) \otimes \delta(t - nT_1), \tag{6.19}$$

or using the distributive property of convolution:

$$f(t) = f_0(t) \otimes \sum_{n=-\infty}^{+\infty} \delta(t - nT_1). \tag{6.20}$$

Calculation of the FT of $f(t)$: The FT of this convolution is equal to the products of FTs.

We can write

$$F(\omega) = F_0(\omega) \int_{-\infty}^{+\infty} \sum_{n=-\infty}^{+\infty} \delta(t - nT_1) e^{-j\omega t} dt = F_0(\omega) \sum_{n=-\infty}^{+\infty} e^{-j\omega n T_1}, \tag{6.21}$$

with

$$F_0(\omega) = \int_{-\infty}^{+\infty} f_0(t) e^{-j\omega t} dt = \int_0^{T_1} f_0(t) e^{-j\omega t} dt. \tag{6.22}$$

Using the Poisson summation formula (see Chap. 5), we get

$$F(\omega) = F_0(\omega) \sum_{n=-\infty}^{+\infty} \delta(\omega - n\omega_1), \qquad (6.23)$$

and we can write

$$F(\omega) = \sum_{n=-\infty}^{+\infty} F_0(n\omega_1)\delta(\omega - n\omega_1). \qquad (6.24)$$

The spectrum of $F(\omega)$ is a line spectrum, a Dirac comb. The weights associated with each Dirac distribution are the values of the function at theirs abscissa:

$$F_0(n\omega_1) = \int_0^{T_1} f_0(t)e^{-jn\omega_1 t}dt = c_n. \qquad (6.25)$$

It was recognized in $F_0(n\omega_1)$ the coefficients c_n of the Fourier series of $f(t)$.

The results of this section are important since they provide a bridge between Fourier series and Fourier transform developments of a periodic function.

6.6   Deterministic Correlation Functions

The deterministic cross-correlation function of two signals $x(t)$ and $y(t)$ is defined by the integral

$$r_{xy}(\tau) = \int_{-\infty}^{+\infty} x(t+\tau)y^*(t)dt. \qquad (6.26)$$

Its Fourier transform is

$$R_{xy}(\omega) = \int_{-\infty}^{+\infty} r_{xy}(\tau)e^{-j\omega\tau}\,d\tau = \int_{-\infty}^{+\infty} y^*(t) \int_{-\infty}^{+\infty} x(t+\tau)e^{-j\omega\tau}\,d\tau dt,$$

$$\qquad (6.27)$$

$$R_{xy}(\omega) = \int_{-\infty}^{+\infty} y^*(t) \int_{-\infty}^{+\infty} X(\omega)e^{j\omega t}\,dt = X(\omega)Y^*(\omega).$$

The deterministic autocorrelation function of a signal $x(t)$ is defined by the integral

$$c(\tau) = \int_{-\infty}^{+\infty} x(t+\tau)x^*(t)\mathrm{d}t. \tag{6.28}$$

For a real signal this operation is equivalent to multiplying the signal by its translated version of the parameter value $-\tau$.

We can reveal a convolution product by the change of variable $t' = -t$. We have

$$c(\tau) = \int_{-\infty}^{+\infty} x^*(-t')x(\tau - t')\,\mathrm{d}t' = x^*(-\tau) \otimes x(\tau). \tag{6.29}$$

We note that in the case where the function $x(t)$ is real, the correlation $c(\tau)$ is equal to the convolution of $x(t)$ with its reversed in time $x(-t)$.

The Fourier transform of the autocorrelation function is denoted $C(\omega)$ as

$$C(\omega) = \int_{-\infty}^{+\infty} c(\tau)\mathrm{e}^{-\mathrm{j}\omega\tau}\,\mathrm{d}\tau. \tag{6.30}$$

Taking the FT of the convolution yields

$$C(\omega) = X(\omega)F(x^*(-\tau)) = X(\omega)X^*(\omega) = |X(\omega)|^2. \tag{6.31}$$

One would deduce directly this result from the relationship (6.27).

Referring to the Parseval theorem on energy, $|X(\omega)|^2$ is called the spectral energy density of the signal.

By extension $R_{xy}(\omega) = X(\omega)Y^*(\omega)$ is called the spectral energy density of interaction of the two signals $x(t)$ and $y(t)$. In the case where the function $x(t)$ is real, we have

$$c(\tau) = \int_{-\infty}^{+\infty} x(t)x(t+\tau)\mathrm{d}t. \tag{6.32}$$

When $x(t)$ is real, its FT verifies the relationship $X^*(\omega) = X(-\omega)$. We will also have in this case:

$$C(\omega) = X(\omega)X(-\omega). \tag{6.33}$$

Localization of the maximum of the autocorrelation function of a real signal
The following inequality is verified:

$$|c(\tau)| \leq c(0). \tag{6.34}$$

This property is demonstrated in the following in the case of a real signal $x(t)$. For any real parameter λ, the next integral of the squared quantity is always positive or null:

$$\int_{-\infty}^{+\infty} (x(t+\tau) + \lambda x(t))^2 \, dt \geq 0. \tag{6.35}$$

Developing the square within the integral:

$$\int_{-\infty}^{+\infty} \left(x^2(t+\tau) + 2\lambda x(t)\, x(t+\tau) + \lambda^2 x(t) \right) dt \geq 0,$$

or

$$c(0) + 2\lambda c(\tau) + \lambda^2 c(0) \geq 0. \tag{6.36}$$

The left inequality member is a quadratic polynomial in λ. To be always positive or zero it is necessary that, $c(0)$ being positive, it has no root. That is to say, its discriminant must be negative or zero. Thus, the polynomial coefficients must verify the following condition: $c^2(\tau) - c^2(0) \leq 0$, and because $c(0)$ is always positive or zero: $|c(\tau)| \leq c(0)$. Q.E.D.

We notice that the maximum of the autocorrelation function modulus is located in $\tau = 0$. This property is widely used in signal analysis, such as in radar or sonar, to calculate the delay of a replica of a signal relative to the signal transmitted to probe the environment. Indeed noting $y(t)$ the echo signal delayed by b, $y(t) = Ax(t-b)$ (b is the replica delay).

$$i(\tau) = \int_{-\infty}^{+\infty} x(t)\, y(t+\tau)\, dt = A \int_{-\infty}^{+\infty} x(t)\, x(t-b+\tau)\, dt. \tag{6.37}$$

The cross-correlation function $i(\tau)$ will have its maximum for $-b + \tau = 0$, so for $\tau = b$.

In the following simulation a chirp function was chosen, which has the remarkable property of having an autocorrelation function very localized in time. It is shown how one can determine the delay of the second signal, delayed by 400 μs by pointing to the maximum of the cross-correlation function (Fig. 6.1).

Due to the continuous variation of the frequency, for a slight time lag, the function oscillations of different frequencies are multiplied, which leads to cancelation of the integral of their product. The cross-correlation is very localized in time and the delay easily measured (see Exercise VI). This property is the reason of the wide use of chirp signals in radar.

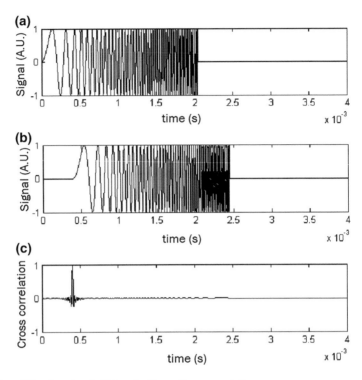

Fig. 6.1 **a** Chirp function; **b** Delayed chirp; **c** Cross-correlation

6.7 Signal Spreads. Heisenberg–Gabor Uncertainty Relationship

To measure the spreads in time σ_t and frequency σ_ω of a signal, the following definitions of these quantities are used:

$$\sigma_t^2 = \frac{1}{E} \int\limits_{-\infty}^{\infty} t^2 |f(t)|^2 \, dt; \quad \sigma_\omega^2 = \frac{1}{2\pi E} \int\limits_{-\infty}^{\infty} \omega^2 |F(\omega)|^2 \, d\omega. \qquad (6.38)$$

where E is the signal energy: $E = \int_{-\infty}^{\infty} |f(t)|^2 \, dt$.

The decay at infinity of functions $|f(t)|^2$ and $|F(\omega)|^2$ are assumed fast enough to ensure the convergence of integrals. According to Parseval's theorem (see Chap. 5),

$$E = \int\limits_{-\infty}^{\infty} |f(t)|^2 \, dt = \frac{1}{2\pi} \int\limits_{-\infty}^{\infty} |F(\omega)|^2 \, d\omega. \qquad (6.39)$$

Care is taken to choose the origins of time and frequency, such that

$$\int_{-\infty}^{\infty} t|f(t)|^2 \, dt = 0 \quad \text{and} \quad \int_{-\infty}^{\infty} \omega|F(\omega)|^2 \, d\omega = 0. \tag{6.40}$$

We show now that if the condition

$$\sqrt{t}f(t) \to 0, \tag{6.41}$$

is verified when $|t| \to \infty$, then

$$\sigma_t \sigma_\omega \geq \frac{1}{2}, \tag{6.42}$$

with equality occurring only when

$$f(t) = Ae^{-\alpha t^2}. \tag{6.43}$$

For ease of demonstration, it is assumed that function $f(t)$ is real.
According to Schwarz inequality, we can write

$$\left| \int_{-\infty}^{\infty} tf \frac{df}{dt} \, dt \right|^2 \leq \int_{-\infty}^{\infty} t^2 f^2 \, dt \int_{-\infty}^{\infty} \left| \frac{df}{dt} \right|^2 \, dt. \tag{6.44}$$

Integrating by parts the first member and using the assumption (6.41) and taking $E = 1$ to facilitate writing,

$$\int_{-\infty}^{\infty} tf \frac{df}{dt} \, dt = \int_{-\infty}^{\infty} t \frac{df^2}{2} = t \frac{f^2}{2} \Big|_{-\infty}^{\infty} - \frac{1}{2} \int_{-\infty}^{\infty} f^2 \, dt = -\frac{1}{2}. \tag{6.45}$$

According to the properties of the Fourier transform of the function $f(t)$ which is assumed regular,

$$\int_{-\infty}^{\infty} \left| \frac{df}{dt} \right|^2 \, dt = \frac{1}{2\pi} \int_{-\infty}^{\infty} \omega^2 |F(\omega)|^2 \, d\omega, \tag{6.46}$$

since FT of $\frac{df}{dt}$ is in this case $j\omega F(\omega)$.

By transferring these results in the inequality (6.44) we have the inequality

$$\frac{1}{4} \leq \int_{-\infty}^{\infty} t^2 f^2 \, dt \frac{1}{2\pi} \int_{-\infty}^{\infty} \omega^2 |F(\omega)|^2 \, d\omega, \tag{6.47}$$

so $\frac{1}{4} \leq \sigma_t^2 \sigma_\omega^2$, which demonstrates Eq. (6.42).

The Schwarz inequality becomes an equality if the two terms in the integral (6.44) are proportional, that is to say if $\frac{df}{dt} = ktf$, as can be seen by solving this differential equation, if

$$f(t) = Ae^{k\frac{t^2}{2}} = Ae^{-\alpha t^2}, \tag{6.48}$$

(with $\alpha > 0$ to make integration of $f(t)$ possible).

Thus, the Gaussian function has the property to have a product of spreads in time and frequency minimum. This property has been recognized by Gabor. It proposes the use of this function limited to its central part as a sliding multiplicative window for the calculation of the spectrogram (see definition in Chap. 12).

The inequality $\sigma_t \sigma_\omega \geq \frac{1}{2}$ implies that we cannot arbitrarily make the spread of a signal small in the time domain or in the frequency domain without an expansion in the conjugate domain. This inequality is identical to the Heisenberg uncertainty relation encountered in quantum mechanics. It is known in the field of signal analysis as the Heisenberg–Gabor inequality. Using the time and frequency variables, this inequality takes the form

$$\sigma_t \sigma_f \geq \frac{1}{4\pi}, \tag{6.49}$$

or, noting, respectively, T and B the spreads of the signal in the time and frequency domains,

$$BT \geq \frac{1}{4\pi}. \tag{6.50}$$

Impossibility for a signal to have simultaneously finite time and frequency supports

The above inequality does not inform about the possibility that a signal has limited supports both in time and frequency. We now show that it is impossible to have a limited time support width T when the frequency support is limited to a band B.

Inversion Fourier formula would be written otherwise as

$$x(t) = \frac{1}{2\pi} \int_{-\frac{B}{2}}^{+\frac{B}{2}} X(\omega)e^{j\omega t} \, d\omega = 0 \quad \text{when} \quad |t| > \frac{T}{2}. \tag{6.51}$$

If $x(t)$ was zero outside support $\left[-\frac{T}{2}, \frac{T}{2}\right]$, it would be the same for its derivatives, and its nth derivative could be written as

$$\frac{d^n x(t)}{dt^n} = \frac{1}{2\pi} \int_{-\frac{B}{2}}^{+\frac{B}{2}} (j\omega)^n X(\omega) e^{j\omega t}\, d\omega = 0 \quad \text{when } |t| > \frac{T}{2}. \qquad (6.52)$$

Taking any time s within the range $\left[-\frac{T}{2}, \frac{T}{2}\right]$, using the inversion of the Fourier formula, we can write

$$x(s) = \frac{1}{2\pi} \int_{-\frac{B}{2}}^{+\frac{B}{2}} X(\omega)\, e^{j\omega s}\, d\omega$$

We can also write

$$x(s) = \frac{1}{2\pi} \int_{-\frac{B}{2}}^{+\frac{B}{2}} X(\omega) e^{j\omega(s-t)} e^{j\omega t}\, d\omega. \qquad (6.53)$$

We can replace in the integral the exponential by its development in Taylor series:

$$e^{j\omega(s-t)} = \sum_{n=-\infty}^{\infty} \frac{[j\omega(s-t)]^n}{n!}. \qquad (6.54)$$

Substituting this expression for the exponential in Eq. (6.53),

$$x(s) = \frac{1}{2\pi} \int_{-\frac{B}{2}}^{+\frac{B}{2}} X(\omega) \sum_{n=-\infty}^{\infty} \frac{[j\omega(s-t)]^n}{n!} e^{j\omega t}\, d\omega$$

$$= \frac{1}{2\pi} \sum_{n=-\infty}^{\infty} \frac{[(s-t)]^n}{n!} \int_{-\frac{B}{2}}^{+\frac{B}{2}} X(\omega)(j\omega)^n e^{j\omega t}\, d\omega \qquad (6.55)$$

The last integral is null by hypothesis, according to Eq. (6.52). This logically leads to $x(s) = 0$, for any s in the range $\left[-\frac{T}{2}, \frac{T}{2}\right]$, which is contrary to the hypothesis.

We thus arrive at the result that the condition (6.51) is impossible to achieve for a band-limited signal. A signal may not have both a limited frequency bandwidth and be limited in time to a finite length interval.

Summary
We have demonstrated the relationships in the time and frequency domains between the input and output signals of an LTI system (convolution in the time domain, product in frequency domain). We showed that the Fourier transform of the product of two functions is the convolution of their FT divided by 2π. The use of distributions allows considering Fourier series as a special case of Fourier transform. We have defined and discussed the properties of the deterministic correlation function which is of uttermost importance for the comparison of two signals. We have shown the interest of using a sweeping frequency signal (chirp) for target localization by correlation radar. The spreads of a signal in time and frequency domains have been defined and the Heisenberg–Gabor inequality demonstrated. The advantage brought by a Gaussian signal has been shown. The chapter concludes by exposing the impossibility for a signal to have infinite supports, simultaneously in time and frequency domains.

In the next chapter, several Fourier transforms of useful functions and convolution calculations are given.

Exercises

I. 1a. Calculate the Fourier transform of the function $f(t) = \sin \omega_0 t$.

1b. Note $\Pi_T(t)$ the rectangular window centered at the origin with width $2T$: $\Pi_T(t) = 1$ if $|t| < T$ and 0 elsewhere. Calculate its Fourier transform. Specify the zeros of this function.

2a. Let the function $g(t) = f(t)\Pi_T(t)$ Deduct from Question 1 the Fourier transform of $g(t)$ (a convolution calculation).

2b. It is assumed that the half width T of the rectangular window is exactly three times the period T_0 of $f(t)$. Show that the function $g(t)$ is continuous. What can be expected on the decay of the high frequency spectrum of $g(t)$?

3. By a drawing show how the superposition of the two sincs in the frequency domain results in a partial compensation of the lobes at high frequencies.

II. An electronic multiplier circuit provides the product $y(t)$ of a function $x(t)$ with a Dirac comb $p(t) = \sum\limits_{n=-\infty}^{\infty} \delta(t - nT)$.

1. Calculate the Fourier transform $Y(\omega)$ of $y(t) = x(t)p(t)$.

2. It is assumed that the spectrum of $x(t)$ is limited to an interval $\{-f_m, f_m\}$. Give the frequency response of a filter that retrieves $x(t)$ from $y(t)$.

III. Spectrum analyzer.

1. Let the function $f(t) = \cos \omega_0 t$. Calculate its Fourier transform $F(\omega)$.

2. Let the rectangular window $\Pi(t) = 1$ for $0 \le t < T$, and equal to 0 elsewhere. Calculate the Fourier transform $W(\omega)$ of the function $\Pi(t)$.

Fig. 6.2 Bartlett window

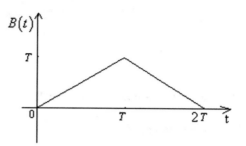

3. It is assumed that $T = 10T_0$ where $T_0 = \frac{2\pi}{\omega_0}$. Represent $F(\omega)$ and $W(\omega)$ functions moduli.

4. A spectrum analyzer calculates the FT of a signal $f(t)$ multiplied by the rectangular window $\Pi(t)$. We note $g(t)$ the product. $g(t) = f(t)\Pi(t)$.

   a. Calculate its FT $G(\omega)$.

   b. Show that this function is the sum of two functions, and represent their respective moduli.

   c. Explain why $G(\omega)$ will be an estimate of $F(\omega)$ which improves as the width T of the window is larger compared with the period T_0 of the cosine.

5. Bartlett window $B(t)$ is defined as the triangular window shown in Fig. 6.2.

  Assuming that this function $B(t)$ is given by $B(t) = \Pi(t) \otimes \Pi(t)$, calculate its FT $W_B(\omega)$.

6. Justify the interest, when performing the spectral analysis, to multiply the function $f(t) = \cos \omega_0 t$ by the Bartlett window rather than by the rectangular window to estimate the spectral amplitude.

IV. Lock-in amplifier.

1. Let the function $x_0(t) = \cos \omega_0 t$. Calculate its Fourier transform $X_0(\omega)$. Graph this FT.

2. Let the signal $x_2(t) = x_0^2(t)$. Calculate the FT $X_2(\omega)$ of $x_2(t)$ using the fact that $x_2(t)$ is the product of $x_0(t)$ by itself and the theorem giving the FT of a product of two functions. Graph $X_2(\omega)$.

3. Why can we say that $\int_{-\infty}^{\infty} x_2(t)\,dt = X_2(0)$?

4. Evaluate the energy E of the signal $x_0(t)$ in the time domain: $E = \int_{-\infty}^{\infty} x_0^2(t)dt$ and in the frequency domain (in the latter case, rely on Question 3. One can discuss starting from the graph of $X_2(\omega)$).

5. Let $f(t) = A_0 \cos(\omega_0 t + \varphi_0) + A_1 \cos(\omega_1 t + \varphi_1)$ be a signal where A_0 and A_1 are two positive real constants. Calculate the FT $F(\omega)$.

6. Represent individually the FT of each component in $F(\omega)$. Qualitatively infer the shape of the spectrum $F(\omega)$.

7. Consider the rectangular window $\Pi(t) = 1$ for $-\frac{T}{2} \le t < \frac{T}{2}$, and 0 elsewhere.

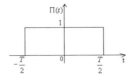

Calculate the Fourier transform $W(\omega)$ of the function $\Pi(t)$.
Calculate the Fourier transform $G(\omega)$ of the function $g(t) = \Pi(t)\cos(\omega_2 t + \varphi_2)$.
Represent this FT.

8. A lock-in amplifier is an apparatus which, to assess the magnitude of a monochromatic component in a signal $f(t)$, performs electronically the following operation (product and integration): $M = \int_{-\infty}^{\infty} f(t)g(t)\,dt$, where the function $g(t)$ is given in the previous question.

The following questions help to understand the operation if the function $f(t)$ has the form given in Question 5.

a. Calculate the FT of the product $f(t)g(t)$. Graphically explain the composition of the spectrum.
b. Building on Question 3, give the conditions upon ω_2 and φ_2 for M to be great. Show that if $\omega_1 \gg \omega_0$ we can evaluate A_0 or A_1.

Solution:

1. $X(\omega) = \pi(\delta(\omega - \omega_0) + \delta(\omega + \omega_0))$ refer to the drawing in the course.

2.
$$X_2(\omega) = \frac{1}{2\pi}X(\omega) \otimes X(\omega) = \frac{1}{2\pi}\pi(\delta(\omega - \omega_0) + \delta(\omega + \omega_0)) \otimes \pi(\delta(\omega - \omega_0) + \delta(\omega + \omega_0))$$
$$X_2(\omega) = \frac{\pi}{2}(\delta(\omega - 2\omega_0) + 2\delta(\omega) + \delta(\omega + 2\omega_0))$$

3. Since $X_2(\omega) = \int_{-\infty}^{\infty} x_2(t)e^{-j\omega t}\,dt$, for $\omega = 0$ we have $X_2(0) = \int_{-\infty}^{\infty} x_2(t)\,dt$
4. Since the signal is periodic, its energy is infinite and the integral of the square will not converge. In the frequency domain, the energy is $X_2(0)$ and is infinite due to the term $\pi\delta(\omega)$ which is infinite in $\omega = 0$.

5.
$$F(\omega) = A_0\pi(\delta(\omega - \omega_0) - \delta(\omega + \omega_0))\cos\varphi_0 - A_0\pi j(\delta(\omega - \omega_0) + \delta(\omega + \omega_0))\sin\varphi_0$$
$$+ A_1\pi(\delta(\omega - \omega_1) + \delta(\omega + \omega_1))\cos\varphi_1 - A_1\pi j(\delta(\omega - \omega_1) - \delta(\omega + \omega_1))\sin\varphi_1.$$

6. The real and imaginary parts of $F(\omega)$ are each constituted of two Dirac distributions.
7. $W(\omega) = T\,\mathrm{sin}\,c\left(\frac{\omega T}{2}\right); \quad G(\omega) = \frac{1}{2\pi}T\,\mathrm{sin}\,c\left(\frac{\omega T}{2}\right) \otimes \quad (\pi(\delta(\omega - \omega_2) - \delta(\omega + \omega_2))$
$\cos\varphi_2 - \pi j(\delta(\omega - \omega_2) + \delta(\omega + \omega_2))\sin\varphi_2).$

Convolutions with Dirac distributions result in translations of the sinc function.

8. a. The FT is $\frac{1}{2\pi}F(\omega) \otimes G(\omega)$. We have translations of the sinc function.
   b. M is also the value of FT in the previous question in $\omega = 0$. There must be a translated sinc with a maximum in $\omega = 0$. We must have $\omega_2 = \omega_1$ or

$\omega_2 = \omega_0$. If the angular frequencies ω_1 and ω_0 are very different, the sinc functions are widely spaced and overlapping is negligible. We can then determine A_0 (or A_1). Phase φ_2 is set to be equal to the phase φ_0 (or φ_1), to get M to be maximum. Therefore, we can measure the magnitude and phase of each component 1 or 2 by choosing the frequency ω_2 and phase φ_2.

V. Consider the Bartlett window $\Lambda_{2T}(t)$, shown in the following figure:

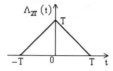

Using the results of the course, answer the following questions:

1. Calculate the energy of the signal $\Lambda_{2T}(t)$ by integration in the time domain, and by integration in the frequency domain. Recall the general theorem giving the equality of these calculation results.
2. Calculate the spreads of this signal in the time and frequency domains. Verify that their product satisfies the inequality Heisenberg–Gabor.

Reminder: $\int_{-\infty}^{\infty} \frac{\sin^2 x}{x^2}\,dx = \pi$; $\int_{-\infty}^{\infty} \frac{\sin^4 x}{x^2}\,dx = \frac{\pi}{2}$; $\int_{-\infty}^{\infty} \frac{\sin^4 x}{x^4}\,dx = \frac{2\pi}{3}$.

Solution:

1. The energy of this real signal is defined by $E = \int_{-\infty}^{\infty} \Lambda^2(t)\,dt$. After integration, $E = \frac{2}{3}T^3$. By Parseval–Plancherel theorem, the energy is also $E = \frac{1}{2\pi}\int_{-\infty}^{\infty} \Lambda^2(\omega)\,d\omega$.
   $\Lambda(\omega)$ is given by $\Lambda(\omega) = T^2 \sin c^2\left(\frac{\omega T}{2}\right)$.
   Then $E = \frac{1}{2\pi}\int_{-\infty}^{\infty}\left(T^2 \sin c^2\left(\frac{\omega T}{2}\right)\right)^2 d\omega = \frac{T^4}{\pi}\int_0^{\infty} \sin c^4\left(\frac{\omega T}{2}\right) d\omega = \frac{2T^3}{3}$

2. $\sigma_t^2 = \frac{1}{E}\int_{-\infty}^{\infty} t^2\Lambda^2(t)\,dt = \frac{T^2}{10}$; $\sigma_\omega^2 = \frac{1}{2\pi E}\int_{-\infty}^{\infty}\omega^2|\Lambda(\omega)|^2\,d\omega = \frac{3}{T^2}$.
   $\sigma_1\sigma_\omega = \sqrt{\frac{3}{10}} > \frac{1}{2}$. The Heisenberg–Gabor inequality is verified.

VI. A complex chirp is expressed as $x(t) = e^{j\left(\omega_0\left(t + at^2\right)\right)}$. ω_0 and a are two real constants.

   1. Calculate its autocorrelation function.
   2. In practice the integration interval is necessarily limited to $[-T, T]$. Calculate the autocorrelation of the time limited signal.

Solution:

1. $r_{xx}(\tau) = \int_{-\infty}^{\infty} x(t + \tau)x^*(t)\,dt$.

$$r_{xx}(\tau) = \int_{-\infty}^{\infty} e^{j\left(\omega_0\left(t+\tau+a(t+\tau)^2\right)\right)} e^{-j\left(\omega_0\left(t+at^2\right)\right)} dt = \int_{-\infty}^{\infty} e^{j\left(\omega_0\left(\tau+2at\tau+a\tau^2\right)\right)} dt$$

$$r_{xx}(\tau) = e^{j\omega_0\tau} e^{j\omega_0 a\tau^2} \int_{-\infty}^{\infty} e^{j2\omega_0 at\tau} dt = e^{j\omega_0\tau} e^{j\omega_0 a\tau^2} 2\pi\delta(2\omega_0 a\tau), \quad \text{or} \quad r_{xx}(t) = 2\pi\delta$$
$(2\omega_0 a\tau)$.

This result shows the efficiency of the use of a chirp in radar target localization.

2. By assumption, in practice, the signal is limited to $[0, T]$. Its autocorrelation is

$$r_{xx}(\tau) = \int_{-T}^{T} e^{j\left(\omega_0\left(t+\tau+a(t+\tau)^2\right)\right)} e^{-j\left(\omega_0\left(t+at^2\right)\right)} dt = \int_{-T}^{T} e^{j\left(\omega_0\left(\tau+2at\tau+a\tau^2\right)\right)} dt.$$

$$r_{xx}(\tau) = e^{j\omega_0\tau} e^{j\omega_0 a\tau^2} \int_{-T}^{T} e^{j2\omega_0 at\tau} dt = e^{j\omega_0\tau} e^{j\omega_0 a\tau^2} \frac{e^{j2\omega_0 aT\tau} - e^{-j2\omega_0 aT\tau}}{2j\omega_0 a\tau}.$$

$$r_{xx}(\tau) = e^{j\omega_0\tau} e^{j\omega_0 a\tau^2} \frac{\sin(2\omega_0 aT\tau)}{\omega_0 a\tau};$$

$$|r_{xx}(\tau)| = \left|\frac{\sin(2\omega_0 aT\tau)}{\omega_0 a\tau}\right| = 2T|\sin c(2\omega_0 aT\tau)|.$$

1/2 the width of the correlation peak is obtained by $\frac{\sin x}{x} = 0$, therefore for $x_0 = \pi$.
It comes $\tau_0 = \frac{\pi}{2\omega_0 aT}$. The width tends to zero for large T, ω_0, a.

Chapter 7
Fourier Transforms and Convolution Calculations

This third chapter on Fourier transforms deals with the application on practical cases of the theorems on Fourier transform and convolution established in previous chapters. The Fourier transform of common windows used in signal analysis are evaluated. Successively rectangular (also called boxcar), triangular (Bartlett), Hanning, and Gaussian are considered. We use the notions on complex integration given in Appendix 1 to calculate the Fourier transform of Gaussian functions We show on practical examples that for smooth, time-limited signals, the decrease of the Fourier transform at infinity is related to the continuity properties of the signal at its boundaries in the time domain. The smoother the junction is in the time domain, the faster the amplitude of the Fourier transform decreases at infinite frequencies. Examples of calculations of convolution are detailed at the end of this chapter to help diminish the risks of errors often made. The chapter ends with a table of Fourier transforms of common functions.

7.1 Fourier Transformation of Common Fonctions

7.1.1 Fourier Transform of a Rectangular Window

Note $\Pi_T(t)$ the even function, equal to 1 for $|t| < \frac{T}{2}$ and 0 elsewhere (Fig. 7.1):

$$\Pi_T(t) = \begin{vmatrix} 1 & |t| < \frac{T}{2} \\ 0 & \text{elsewhere} \end{vmatrix}, \tag{7.1}$$

This function is symmetrical around $t = 0$, with width T. Let us calculate its Fourier transform

© Springer International Publishing Switzerland 2016
F. Cohen Tenoudji, *Analog and Digital Signal Analysis*,
Modern Acoustics and Signal Processing, DOI 10.1007/978-3-319-42382-1_7

Fig. 7.1 Rectangular
window

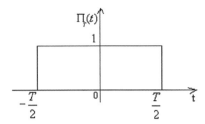

$$W_r(\omega) = \mathcal{F}(\Pi_T(t)) = \int_{-\infty}^{+\infty} \Pi_T(t)e^{-j\omega t}dt = \int_{-\frac{T}{2}}^{\frac{T}{2}} e^{-j\omega t}dt = \frac{e^{-j\omega\frac{T}{2}} - e^{j\omega\frac{T}{2}}}{-j\omega} \quad (7.2)$$

so

$$W_r(\omega) = \frac{2\sin\frac{\omega T}{2}}{\omega} = \frac{2T}{T\omega}\frac{\sin\frac{\omega T}{2}}{1} = T\frac{\sin\frac{\omega T}{2}}{\frac{\omega T}{2}} = T\,\mathrm{sinc}\left(\frac{\omega T}{2}\right). \quad (7.3)$$

We used the classic notation $\frac{\sin x}{x} = \mathrm{sinc}(x)$ that defines the cardinal sine function.

The shape of the Fourier transform is given in Fig. 7.2. The amplitude of the main maximum is evaluated by taking the limit for $\omega \to 0$ of $W_r(\omega)$. As $\sin\frac{\omega T}{2}$ is equivalent to $\frac{\omega T}{2}$ for small values of ω, we have $W_r(0) = T$.

We now wish to estimate the value of the second extremum (negative here). We seek an extremum of a function of the form $F(x) = T\frac{\sin x}{x}$.

The abscissa of the function extrema is obtained by cancelation of its derivative with respect to x

Fig. 7.2 Fourier transform of
a rectangular window

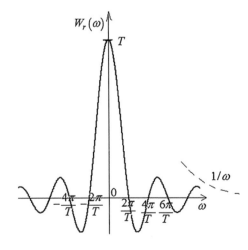

$$\frac{d}{dx}\left(\frac{\sin x}{x}\right) = \frac{x\cos x - \sin x}{x^2} = 0. \tag{7.4}$$

This derivative is equal to zero for values of x such that $\mathrm{tg}\,x = x$. The transcendental equation is solved numerically or graphically. The first solution after $x = 0$ is $x_{\mathrm{max}2} = 4.4934$.

The relative amplitude of the second extremum with respect to the main lobe is given by

$$\frac{|W(x_{\mathrm{max}2})|}{|W(0)|} = 20\log_{10}\left(-\frac{\sin 4.4934}{4.4934}\right) = -13.26\mathrm{dB}. \tag{7.5}$$

The abscissa of the following secondary extremum is $x = 7.7252$. Its relative magnitude is 0.1284, or in decibels, $-17.83\,\mathrm{dB}$. We also note that except for the first solutions, the solutions x_k are close to the values $\frac{k\pi}{2}$ with k odd. When approaching the value of the first secondary extremum by the function's value $-17.9\,\mathrm{dB}$ for the abscissa $x = \frac{3\pi}{2} = 4.7124$ making the numerator maximum, the error on the evaluation of this extremum is below 0.1 dB.

It will be noted to complete that the high frequency envelope decay of function $W_r(\omega)$ is as $\frac{1}{\omega}$. This decrease with ω is slow (6 dB per octave). This slow decrease reflects the discontinuous nature of the function $\Pi_T(t)$.

7.1.2   Fourier Transform of a Triangular Window

The convolution of the rectangular window $\Pi_T(t)$ with itself is a triangular window $\Lambda_{2T}(t)$ with width $2T$. It is called the Bartlett window in signal analysis (Fig. 7.3).

$$\Lambda_{2T}(t) = \Pi_T(t) \otimes \Pi_T(t). \tag{7.6}$$

The theorem of the Fourier transform of the convolution of two functions states that the Fourier transform of the convolution product is equal to the product of the functions Fourier transforms. The FT of the function $\Lambda_{2T}(t)$ will then be the square of the FT of $\Pi_T(t)$ seen in Fig. 7.4

$$W_{\mathrm{tri}}(\omega) = W_r^2(\omega) = T^2\,\mathrm{sin}\,c^2\left(\frac{\omega T}{2}\right). \tag{7.7}$$

Fig. 7.3 Bartlett window

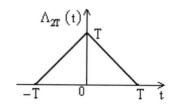

Fig. 7.4 Fourier transform of
Bartlett window

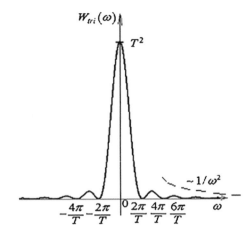

Note in the figure the decay as $\frac{1}{\omega^2}$ of the function's envelope. This rate of decay is faster than that of the rectangular window. It is due to the fact that the function $\Lambda_{2T}(t)$ is continuous but its derivative is discontinuous (in 0 and $2T$).

7.1.3 Fourier Transform of Hanning Window

This window is defined by (see Fig. 7.5).

$$
\left|
\begin{array}{ll}
w_H(t) = 0.5 + 0.5\cos\frac{2\pi}{T}t & \text{for } |t| < \frac{T}{2} \\
w_H(t) = 0 & \text{elsewhere}
\end{array}
\right. \tag{7.8}
$$

Note that the function and its derivative are continuous (horizontal tangent) in $|t| = \frac{T}{2}$. The decay at infinity of the FT will necessarily be faster than the previous two windows.

Calculation of FT of $w_H(t)$

$$
W_H(\omega) = 0.5\int_{-\frac{T}{2}}^{\frac{T}{2}} e^{-j\omega t}\,dt + 0.25\int_{-\frac{T}{2}}^{\frac{T}{2}} e^{j(\omega_e - \omega)t}\,dt + 0.25\int_{-\frac{T}{2}}^{\frac{T}{2}} e^{-j(\omega_e + \omega)t}\,dt, \tag{7.9}
$$

Fig. 7.5 Hanning window

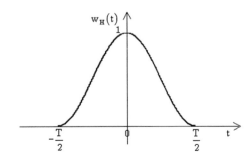

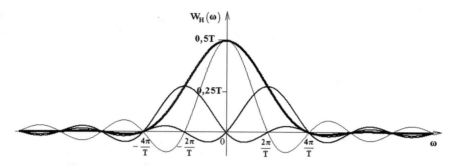

Fig. 7.6 Fourier transform resulting from a summation

$$W_H(\omega) = 0.5T\frac{\sin\frac{\omega T}{2}}{\frac{\omega T}{2}} + 0.25T\frac{\sin(\omega_e - \omega)\frac{T}{2}}{(\omega_e - \omega)\frac{T}{2}} + 0.25T\frac{\sin(\omega + \omega_e)\frac{T}{2}}{(\omega + \omega_e)\frac{T}{2}}. \qquad (7.10)$$

(We wrote $\omega_e = \frac{2\pi}{T}$). This Fourier transform is the sum of three cardinal sine (Fig. 7.6). The first is centered at $\omega = 0$. It is represented by the fine line in the figure, the other two are respectively centered in $+\omega_e$ and $-\omega_e$ with half-relative amplitude. The function $W_H(\omega)$ is shown in bold lines.

As seen in the figure, the central lobe of this sum is twice as wide as that of the function $\frac{\sin x}{x}$. We also see that the compensation of side lobes leads to lower values of the oscillations amplitudes of function $W_H(\omega)$ and therefore a more rapid decrease with ω than in the case of the FT of the rectangular window. The first secondary extremum of $W_H(\omega)$ is $-32.3\,\mathrm{dB}$ below that of main lobe.

This result is to be compared with that of the FT of the rectangular window. For the frequency of this extremum, the maximum amplitude of the second secondary lobe of the rectangular window of the FT is only $-17.833\,\mathrm{dB}$ below the main lobe. This shows that the oscillations of the Hann window FT decrease much faster than those of the FT of the rectangular window. This is because the Hann window function is continuous and its derivative is continuous.

7.1.4 Fourier Transform of a Gaussian Function

Gaussian functions are widely used in the field of signal analysis and in statistics. Before calculating the Fourier transform of a Gaussian, the properties of the integrals of Gaussian functions are described.

Integration of Gaussian functions
A Gaussian function is a function of the form

$$f(x) = e^{-\alpha x^2} \text{ with } \alpha > 0 \text{ real.} \qquad (7.11)$$

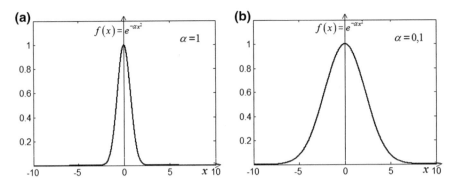

Fig. 7.7 Gaussian functions for parameter values: **a** $\alpha = 1$ and **b** $\alpha = 0.1$

The parameter α controls the Gaussian decay with x, and by consequence, the width of the bell curve. In Fig. 7.7a, we have $\alpha = 1$ and in 7.7b $\alpha = 0.1$.

The general methodology for calculating the Gauss integrals is now described. One first calculates the integral

$$I(\alpha) = \int\limits_{-\infty}^{+\infty} e^{-\alpha x^2}\, dx \text{ with } \alpha > 0 \text{ real.} \tag{7.12}$$

The calculation of this integral is not straightforward. Various techniques may be used. The method given here is through the calculation of the squared integral in order to perform the integration in polar coordinates

$$I^2(\alpha) = \int\limits_{-\infty}^{+\infty} \int\limits_{-\infty}^{+\infty} e^{-\alpha(x^2 + y^2)}\, dxdy = \int\limits_{0}^{\infty} \int\limits_{0}^{2\pi} e^{-\alpha r^2}\, rdrd\theta.$$

$$I^2(\alpha) = 2\pi \int\limits_{0}^{\infty} e^{-\alpha r^2}\, rdr. \tag{7.13}$$

We note $r^2 = u;\ du = 2rdr$

$$I^2(\alpha) = \pi \int\limits_{0}^{\infty} e^{-\alpha u}\, du = -\frac{\pi}{\alpha}[e^{-\alpha u}]_{0}^{\infty} = \frac{\pi}{\alpha}. \tag{7.14}$$

The desired integral is then

$$I(\alpha) = \sqrt{\frac{\pi}{\alpha}}. \tag{7.15}$$

In the calculation of the moments of the Gaussian distribution, it is interesting to use the derivatives of $I(\alpha)$

$$\frac{dI(\alpha)}{d\alpha} = - \int_{-\infty}^{+\infty} x^2 e^{-\alpha x^2} dx = \sqrt{\pi}\left(-\frac{1}{2}\right)\alpha^{-\frac{3}{2}}. \tag{7.16}$$

Thus we see that

$$\int_{-\infty}^{+\infty} x^2 e^{-\alpha x^2} dx = \frac{1}{2}\sqrt{\frac{\pi}{\alpha^3}}. \tag{7.17}$$

Similarly taking twice the derivative relatively to α

$$\frac{d^2 I}{d\alpha^2} = \int_{-\infty}^{+\infty} x^4 e^{-\alpha x^2} dx = \sqrt{\pi}\frac{3}{4}\alpha^{-\frac{5}{2}}. \tag{7.18}$$

It is thus seen that by successive differentiations we can calculate all integrals containing even powers, that is, calculate the various moments of the Gaussian distribution. The odd moments are zero, as the integrand is odd on the interval $-\infty$, $+\infty$.

Calculation of the Fourier transform of $e^{-\alpha x^2}$

$$I(k) = \int_{-\infty}^{+\infty} e^{-\alpha x^2} e^{-ikx} dx \text{ with } \alpha > 0 \text{ real.} \tag{7.19}$$

To calculate $I(k)$ the function $f(z) = e^{-\alpha z^2}$ is integrated in the complex plane on a positively oriented contour C, consisting of line L_1, segment of x axis limited by $-R$ and R, of the vertical segment L_2 with abscissa R limited by $y = 0$ and $y = a$, of the horizontal segment L_3 with ordinate $y = a$ limited by R and $-R$, and the vertical segment L_4 with abscissa $-R$ (Fig. 7.8)

Fig. 7.8 Integration contour

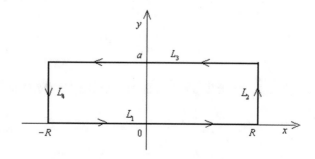

The function $f(z)$ is holomorphic inside contour C. The integral over this contour is zero according to Cauchy theorem.

$$\oint_C e^{-\alpha z^2} dz = \int_{L_1} e^{-\alpha z^2} dz + \int_{L_2} e^{-\alpha z^2} dz + \int_{L_3} e^{-\alpha z^2} dz + \int_{L_4} e^{-\alpha z^2} dz = 0. \quad (7.20)$$

In the first integral on L_1, we have $y = 0$. So

$$\int_{L_1} e^{-\alpha z^2} dz = \int_{-R}^{R} e^{-\alpha x^2} dx. \quad (7.21)$$

As shown above, when $R \to \infty$ this integral tends to $\sqrt{\frac{\pi}{\alpha}}$.

In the integral over L_2, $\int_{L_2} e^{-\alpha z^2} dz = \int_R^{R+ia} e^{-\alpha z^2} dz$. With $dz = idy$.

We will show that this integral tends to zero as $R \to \infty$. The modulus of the integral is less than or equal to the integral of the modulus. We have the inequalities

$$\left| \int_{L_2} e^{-\alpha z^2} dz \right| \leq \int_{L_2} \left| e^{-\alpha z^2} \right| |dz| = \int_0^a \left| e^{-\alpha(R+iy)^2} \right| dy = \int_0^a e^{-\alpha(R^2 - y^2)} dy. \quad (7.22)$$

So,

$$\left| \int_{L_2} e^{-\alpha z^2} dz \right| \leq \int_0^a e^{-\alpha(R^2 - y^2)} dy \leq e^{\alpha a^2} \int_0^a e^{-\alpha R^2} dy = e^{\alpha a^2} e^{-\alpha R^2} \int_0^a dy = a\, e^{\alpha a^2} e^{-\alpha R^2}.$$

$$(7.23)$$

The last quantity tends to zero when $R \to \infty$ (a is finite), it follows that

$$\int_{L_2} e^{-\alpha z^2} dz \to 0 \text{ when } R \to \infty.$$

It will be the same for the integral over L_4. So we have

$$\int_{L_3} e^{-\alpha z^2} dz = - \int_{L_1} e^{-\alpha z^2} dz. \quad (7.24)$$

We can therefore write this last equality in the form

$$\int_{-\infty}^{+\infty} e^{-\alpha(x+ia)^2}\,dx = \int_{-\infty}^{+\infty} e^{-\alpha x^2}\,dx = \sqrt{\frac{\pi}{\alpha}}. \tag{7.25}$$

Expanding the square contained in the left-hand side

$$\int_{-\infty}^{+\infty} e^{-\alpha x^2 - 2i\alpha ax + \alpha a^2}\,dx = e^{\alpha a^2}\int_{-\infty}^{+\infty} e^{-\alpha x^2} e^{-2i\alpha ax}\,dx = \sqrt{\frac{\pi}{\alpha}}. \tag{7.26}$$

It is recognized in the last integral the sought Fourier transform. Noting $k = 2a\alpha$ we can write

$$\int_{-\infty}^{+\infty} e^{-\alpha x^2} e^{-2i a\alpha x}\,dx = \int_{-\infty}^{+\infty} e^{-\alpha x^2} e^{-ikx}\,dx = e^{-\alpha a^2}\sqrt{\frac{\pi}{\alpha}} = e^{-\frac{k^2}{4\alpha}}\sqrt{\frac{\pi}{\alpha}}. \tag{7.27}$$

Finally

$$\int_{-\infty}^{+\infty} e^{-\alpha x^2} e^{-ikx}\,dx = e^{-\frac{k^2}{4\alpha}}\sqrt{\frac{\pi}{\alpha}}. \tag{7.28}$$

It is thus seen that the Fourier transform of a Gaussian is a Gaussian.

Returning to the conjugate variables t and ω used in signal analysis, we will have the formula for the Fourier transform of the function $e^{-\alpha t^2}$:

$$\int_{-\infty}^{+\infty} e^{-\alpha t^2} e^{-j\omega t}\,dt = e^{-\frac{\omega^2}{4\alpha}}\sqrt{\frac{\pi}{\alpha}}. \tag{7.29}$$

It will be accepted here that this result remains valid if α is complex, provided that $\mathrm{Re}(\alpha) > 0$.

Finally, we seek to determine the transform of a Gaussian whose maximum amplitude occurs at abscissa τ

$$f(t) = e^{-\alpha(t-\tau)^2}.$$

This function is translated by a quantity τ from the function $e^{-\alpha t^2}$. Using the time-delay property connecting the FT of a translated function to that of the non-translated one by the multiplication by a phase factor depending on the function, we will have

The FT of $f(t) = e^{-\alpha(t-\tau)^2}$ is expressed as

$$F(\omega) = e^{-i\omega\tau} e^{-\frac{\omega^2}{4\alpha}} \sqrt{\frac{\pi}{\alpha}}. \tag{7.30}$$

7.2   Behavior at Infinity of the Fourier Amplitude of a Signal

It has been previously found that the amplitude of the FT of a rectangular function was decreasing as $\frac{1}{\omega}$ when $\omega \to \infty$ and that of a triangular function decreases as $\frac{1}{\omega^2}$. The more regular the function is in the time domain (continuous and differentiable at order n), the more its decrease at infinity is fast.

Time property	Decay in frequency at infinity
Function with a discontinuity	$\frac{1}{\omega}$
Continuous function, discontinuous derivative	$\frac{1}{\omega^2}$
...	...
Discontinuity in the nth derivative	$\frac{1}{\omega^{n+1}}$

Smoothing performed by a convolution product
The convolution products are functions obtained by integration. In general, the integration provides a function more regular than the one integrated. For example, integration of a function with a finite discontinuity gives a function continuous at the discontinuity. The integral of a function with an angular point will be rounded at this point. Convolution generally operates a smoothing akin to a low-pass filtering. This property will be found in the following section, the analysis being made in the frequency domain.

Another interesting property that will be seen in Example 7.2, Sect. 7.3 is that the support of the convolution product is equal to the sum of the supports of the two functions in the product.

7.3   Limitation in Time or Frequency of a Signal

7.3.1   Fourier Transform of a Time-Limited Cosine

Calculating the Fourier transform of a cosine with angular frequency ω_0 limited in time by a rectangular window to the interval $\left\{-\frac{T}{2}, \frac{T}{2}\right\}$ (Fig. 7.9)

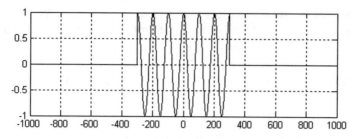

Fig. 7.9 Time limited cosine

$$y(t) = \Pi_T(t) \cos \omega_0 t. \tag{7.31}$$

The FT of a two functions product is equal to the convolution of FTs divided by 2π. (See formula (6.17)). It follows from the previous results that

$$Y(\omega) = \frac{1}{2\pi} T \frac{\sin \frac{\omega T}{2}}{\frac{\omega T}{2}} \otimes (\pi\,\delta(\omega - \omega_0) + \pi\,\delta(\omega + \omega_0)). \tag{7.32}$$

The convolution of a function with $\delta(\omega \pm \omega_0)$ resulting in its translation, we have

$$Y(\omega) = \frac{T}{2} \left\{ \frac{\sin(\omega - \omega_0)\frac{T}{2}}{(\omega - \omega_0)\frac{T}{2}} + \frac{\sin(\omega + \omega_0)\frac{T}{2}}{(\omega + \omega_0)\frac{T}{2}} \right\}. \tag{7.33}$$

Note that the FT of a time-limited cosine is given directly by the FT of the rectangular window limiting the cosine function (sinc function). The frequency ω_0 of the cosine, in turn, acts on the position of the sinc function on the ω axis.

The time limitation results in a spreading of the FT on the entire frequency axis (Fig. 7.10).

The temporal discontinuity caused by the rectangular window boundaries produces a spreading of the spectrum decreasing as $\frac{1}{\omega}$. We note from the figure that the spreading of the sinc centered in $-\omega_0$ (shown in thin line) is superimposed on the

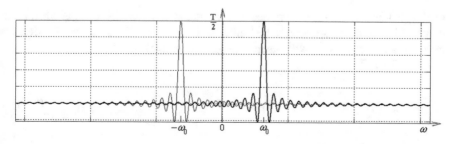

Fig. 7.10 Two constituents of the sum in FT

one centered in ω_0. It follows that the maximum of the positive frequencies is not located exactly in ω_0, causing an error when trying to measure an unknown frequency ω_0. To reduce this effect we will take care in practice to satisfy the condition $T \gg T_0$, which implies that $\omega_0 \gg \frac{2\pi}{T}$. The peaks are relatively far from each other, relatively to the period of oscillation in the frequency domain.

7.3.2 Practical Interest of Multiplying a Signal by a Time Window Before Calculating a Spectrum

In practice, when we want to compute the Fourier transform of a physical signal a spectrum analyzer is used or the calculation is done numerically. The operation can be done only on a time signal limited by the temporal analysis window, because one cannot wait forever to get the result.

Thus, looking to the analysis of a sinusoidal signal at the output of an amplifier for example, the actual analysis is performed on a time signal limited by a rectangular window (or another window if a numerical calculation is carried out). A sinc function which spreads the frequency information on the whole frequency axis is obtained.

In case a different frequency small signal is superimposed on a large signal, the spectral component of the small signal is embedded in the oscillations of the sinc of the large signal. It may happen that this small spectral component is not observable. However, if the data is multiplied by a Hann window (for example), the oscillations of the spectrum of the main signal decreases much more rapidly with frequency and thus a small spectral component is less likely to be embedded in the oscillations of large signal and can be more easily observable. Hann window is an example of an apodization window. This operation of multiplication of a signal by a window is particularly easy to make in numerical signal processing where they are commonly used.

7.3.3 Frequency Limitation; Gibbs Phenomenon

Gibbs phenomenon is the occurrence of oscillations on a function of time when the width of its spectrum is forced to be limited (this is the case for a rectangular signal crossing a low-pass filter with steep edges, for example).

The FT of $\Pi_T(t)$ is

$$W_r(\omega) = T \frac{\sin \frac{\omega T}{2}}{\frac{\omega T}{2}}. \tag{7.34}$$

Assume now that we limit the spectrum to low frequencies, for example by passing through an ideal low-pass filter with frequency response $H(\omega) = \Pi_{\omega_M}(\omega)$, where ω_M is the maximum frequency $\left| \begin{array}{ll} \Pi_{\omega_M}(\omega) = 1 & \text{for } |\omega| < \omega_M \\ \Pi_{\omega_M}(\omega) = 0 & \text{elsewhere} \end{array} \right.$.

The limited spectrum at low frequencies is

$$W_{LF}(\omega) = W_r(\omega)H(\omega). \tag{7.35}$$

Its inverse FT is

$$w_{LF}(t) = \Pi_r(t) \otimes h(t). \tag{7.36}$$

$h(t)$ is given by

$$h(t) = \frac{1}{2\pi} \int_{-\infty}^{\infty} H(\omega) \, e^{j\omega t} d\omega = \frac{1}{2\pi} \int_{-\omega_M}^{\omega_M} e^{j\omega t} d\omega = \frac{1}{2\pi} \frac{e^{j\omega_M t} - e^{j\omega_M t}}{jt} = \frac{1}{\pi} \frac{\sin \omega_M t}{t},$$

$$\tag{7.37}$$

so

$$w_{BF}(t) = \Pi_r(t) \otimes \frac{1}{\pi} \frac{\sin \omega_M t}{t}. \tag{7.38}$$

Assume now that $\omega_M T > > 1$, i.e., that the oscillations of $\frac{\sin \omega_M t}{t}$ are rapid compared to the duration of the rectangular signal. For t close to 0, the convolution integral includes the oscillations of the function $\frac{\sin \omega_M t}{t}$ on both sides of the maximum and therefore we have approximately $w_{BF}(t) = \int_{-\infty}^{\infty} \frac{1}{\pi} \frac{\sin \omega_M t'}{t'} dt' = \frac{1}{\pi} \int_{-\infty}^{\infty} \frac{\sin x}{x} dx = 1$.

In contrast, when t is close to $\frac{T}{2}$, the integration is done only on a part of the oscillations of the $\frac{\sin \omega_M t}{t}$ function. Compensation between the oscillations is no longer total and in the result of the convolution integral it appears values that hover around 1. The maximum amplitude of the oscillation comes for $t = \frac{T}{2} - \frac{\pi}{\omega_M}$, when the central peak is completely included in the integral. As seen on the graph, the maximum value is $\frac{1}{\pi} \int_{-\infty}^{\pi} \frac{\sin x}{x} dx = 1.0895$ (value calculated numerically).

In conclusion, when $\omega_M T \gg 1$ and when the function spectrum is limited to low frequencies, after return to the time domain, oscillations occur in the vicinity of the discontinuity. The amplitude of these oscillations is independent of the width of the frequency band kept. They persist even if the width of the frequency band kept is high.

The Gibbs phenomenon appears in (Fig. 7.11). The oscillation of the reconstituted signal after limitation to the low frequencies of the spectrum of the signal $\Pi_T(t)$ appears in the vicinity of the discontinuity of the function at $\frac{T}{2}$.

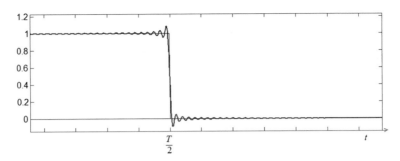

Fig. 7.11 Gibbs phenomenon at function discontinuity

Note: We have studied the manifestation of the Gibbs phenomenon in the time domain. Given the symmetry of the direct and inverse FT formulas, the phenomenon is of course also present in the frequency domain. One example is the FT of the rectangular window, discontinuous function in time. It is a sinc oscillating function.

7.4 Convolution Calculations

7.4.1 Response of a First Order System to Different Input Signals

The impulse response of the first order RC system in Chap. 2 is $h(t) = \frac{1}{RC} e^{-\frac{t}{RC}} U(t)$.

1. We now seek to evaluate the system response to an input signal of the form $e(t) = x(t) = U(t) \cos \omega_0 t$. We look for $y(t) = \int_{-\infty}^{+\infty} x(\tau) h(t - \tau) d\tau$. Using the decomposition $\cos \omega_0 t = \frac{1}{2} \left(e^{j\omega_0 t} + e^{-j\omega_0 t} \right)$, we can write

$$y(t) = \frac{1}{2} h(t) \otimes U(t) e^{j\omega_0 t} + c.c. \tag{7.39}$$

Calculating the first integral

$$I_1 = \frac{1}{2} h(t) \otimes U(t) e^{j\omega_0 t} = \frac{1}{2RC} \int_{-\infty}^{+\infty} U(\tau) e^{-\frac{\tau}{RC}} U(t - \tau) e^{j\omega_0 (t - \tau)} d\tau.$$

Since the function $U(\tau)$ is zero for $\tau < 0$ and 1 for $\tau > 0$ we have

$$I_1 = \frac{1}{2RC} \int_0^{+\infty} e^{-\frac{\tau}{RC}} U(t - \tau) e^{j\omega_0 (t - \tau)} d\tau. \tag{7.40}$$

The above integral is zero for $t \leq 0$ because in this case, $U(t - \tau)$ is zero in (7.40).

What is its value for $t > 0$? The upper limit of the integral is t because beyond this value of τ, $U(t - \tau)$ vanishes.

$$I_1 = \frac{1}{2RC} \int_0^t e^{-\frac{\tau}{RC}} e^{j\omega_0(t-\tau)} d\tau = \frac{1}{2RC} e^{j\omega_0 t} \int_0^t e^{-\left(j\omega_0 + \frac{1}{RC}\right)\tau} d\tau,$$

$$I_1 = \frac{1}{2} \frac{1}{(jRC\omega_0 + 1)} \left(e^{j\omega_0 t} - e^{-\frac{t}{RC}}\right).$$

(7.41)

By adding the complex conjugate integral, we get

$$y(t) = -\frac{1}{R^2 C^2 \omega_0^2 + 1} e^{-\frac{t}{RC}} U(t) + \frac{1}{R^2 C^2 \omega_0^2 + 1} (\cos \omega_0 t + RC\omega_0 \sin \omega_0 t) U(t).$$

(7.42)

The output signal of the filter is composed of two terms

- The first term is a transient term which becomes very small when time exceeds a few RC.
- The second term is the stationary term already found in Chap. 2, where the response of the first-order system to the input signal $\cos \omega t$ was calculated, this signal existing since time minus infinity.

Thus after times exceeding few RC, the output signal reaches its steady state, where everything happens as if the input signal filter had been present from very remote times.

2. In this other example (Fig. 7.12), the determination of the system response is performed as a convolution in the presence of the initial condition in the first-order system RC. It is assumed that at time $t = 0$, the capacitor is charged with the charge q_0.

When the switch is opened, the potential of the point A relative to ground is $\frac{q_0}{C}$. Ohm's law is written $R\frac{dq}{dt} + \frac{q}{C} = \frac{q_0}{C}$, with no current flowing in the resistor.

When the switch is closed, the potential of point A drops instantaneously to 0; we have

$$R\frac{dq}{dt} + \frac{q}{C} = 0.$$

(7.43)

We can gather these two equations in the form

$$R\frac{dq}{dt} + \frac{q}{C} = \frac{q_0}{C} U(-t).$$

(7.44)

Fig. 7.12 R C circuit

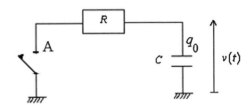

Fig. 7.13 a Time reversed of
$U(t)$. **b** Impulse response $h(t)$

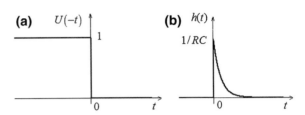

Everything happens as if the system was attacked by the electromotive force
$e(t) = \frac{q_0}{C} U(-t)$. The output of the system is determined by the relationship

$$y(t) = \int_{-\infty}^{+\infty} h(\tau)e(t-\tau)d\tau = \frac{1}{RC}e^{-\frac{t}{RC}}U(t) \otimes \frac{q_0}{C}U(-t). \tag{7.45}$$

It is seen that the function $y(t)$ is obtained by integrating the product of two
functions of τ: the function $h(\tau)$ and the function $e(t-\tau)$ obtained by reversing in
time the function $e(\tau)$ (to obtain the function $e(-\tau)$) then translation by t. The
graphic resolution that is used in the following calculation avoids calculation errors
often committed in the evaluation of convolution products. First we represent the
functions to be convoluted (Fig. 7.13)
    In (Fig. 7.14) are shown the case $t < 0$, the functions contained in the convo-
lution integral of the variable of integration τ and their product. Care is taken in
representing the functions one above the other to make things simple and minimize
errors of reasoning.
    When $t < 0$

$$y(t) = \frac{1}{RC}\frac{q_0}{C}\int_{0}^{+\infty} e^{-\frac{\tau}{RC}}d\tau = -\frac{q_0}{C}\left[e^{-\frac{\tau}{RC}}\right]_{0}^{+\infty} = \frac{q_0}{C} \tag{7.46}$$

For the case $t > 0$, the functions and their product are (Fig. 7.15)
For $t > 0$, we have

Fig. 7.14 Two functions to multiply (*top*). Their product (*bottom*)

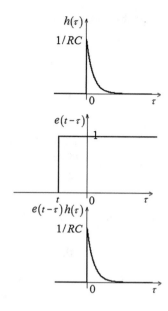

Fig. 7.15 Two functions to multiply (*top*). Their product (*bottom*)

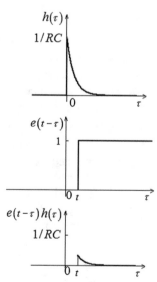

$$y(t) = \frac{1}{RC} \frac{q_0}{C} \int_t^{+\infty} e^{-\frac{\tau}{RC}} d\tau = -\frac{q_0}{C} \left[e^{-\frac{\tau}{RC}} \right]_t^{+\infty} = \frac{q_0}{C} e^{-\frac{t}{RC}} \tag{7.47}$$

In summary, the output signal will have the form (Fig. 7.16)

Fig. 7.16 Output signal

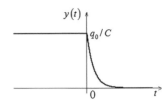

We find back with the convolution calculation the capacitor discharge law that occurs at the grounding of the circuit input.

7.4.2 Examples of Calculations of Convolution

Example 1 Let the functions $x(t)$ and $h(t)$ given by

$$x(t) = U(t) - U(t - 2) \text{ and } h(t) = e^{-at}U(t).$$

($U(t)$ Is the Heaviside function and a a real positive number). $x(t)$ is a rectangular window equal to 1 between 0 and 2.

The convolution has the form $y(t) = \int_{-\infty}^{+\infty} (U(\tau) - U(\tau - 2))e^{-a(t-\tau)}U(t - \tau)d\tau$.

The formal calculation of this integral involves a comprehensive analysis of functions supports. Again it is recommended that the complete graphical processing described now should be used. First of all, the functions to integrate are (Fig. 7.17)

We look for $y(t) = \int_{-\infty}^{+\infty} x(\tau)h(t - \tau)d\tau$. We represent graphically the functions $x(\tau) = U(\tau) - U(\tau - 2)$, $h(t - \tau) = e^{-a(t-\tau)}U(t - \tau)$ and their product.

The function $h(t - \tau) = e^{-a(t-\tau)}U(t - \tau)$ is the reversed in time of $h(\tau) = e^{-a\tau}U(\tau)$ translated by t.

Figure 7.18 gives the graph of the time reversed $h(-\tau) = e^{a\tau}U(-\tau)$

For the translation of the function by t, three cases are met

For $t < 0$ (Fig. 7.19)

It is noted that for $\tau < 0$, $x(\tau) = 0$. The product of the two functions is zero.

Fig. 7.17 Two functions to convolve

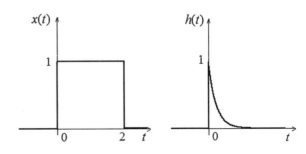

Fig. 7.18 Time reversed of $h(\tau)$

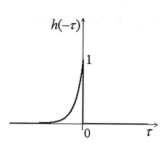

Fig. 7.19 Two functions to multiply (*top*); Their product (*bottom*)

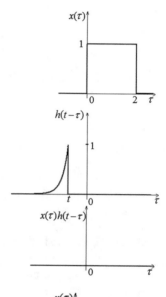

Fig. 7.20 Two functions to multiply (*top*); Their product (*bottom*)

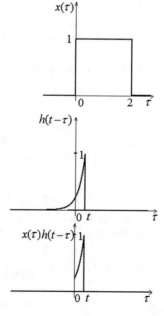

Fig. 7.21 Two functions to multiply (*top*); Their product (*bottom*)

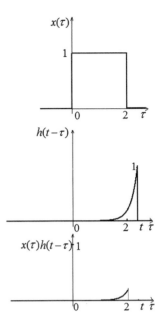

For $\tau > 0$, $h(t - \tau)$ is zero and the product of two functions is still zero. The integral of the product is then zero. So $y(t) = 0$ for $t < 0$.

For $0 < t < 2$ (Fig. 7.20)

For $\tau < 0$, the product of the two functions is again zero since $x(\tau) = 0$.

In the interval $0 < \tau < 2$, the product of the two functions is not null when $0 < \tau < t$. For $t < \tau$ the product is again zero.

Therefore $y(t) = \int_0^t e^{-a(t-\tau)} d\tau = e^{-at} \int_0^t e^{a\tau} d\tau = e^{-at} \frac{e^{a\tau}}{a}\big|_0^t = \frac{1}{a}(1 - e^{-at})$.

For $2 < t$ (Fig. 7.21)

For $\tau < 0$, the product of the two functions is still zero since $x(\tau) = 0$.

In the interval $0 < \tau < 2$ the product of the two functions is different from 0. When $2 < \tau$, the product becomes zero again.

Then $y(t) = \int_0^2 e^{-a(t-\tau)} d\tau = e^{-at} \int_0^2 e^{a\tau} d\tau = e^{-at} \frac{e^{a\tau}}{a}\big|_0^2 = \frac{1}{a} e^{-at}(e^{2a} - 1)$

In summary $y(t) = \begin{vmatrix} 0 & \text{if} & t < 0 \\ (1 - e^{-at})/a & \text{if} & 0 < t < 2 \\ e^{-at}(e^{2a} - 1)/a & \text{if} & 2 < t \end{vmatrix}$

The reader will verify that the function $y(t)$ is continuous at the boundaries of the interval $\{0, 2\}$.

Example 2 Let $x(t)$ be the rectangular window given by $x(t) = U(t) - U(t - T)$.

We look to assess the auto convolution $y(t) = x(t) \otimes x(t)$.

First we represent in Fig. 7.22 the functions $x(\tau)$ and $x(-\tau)$:

We now have four cases values depending on the possible values of t

If $t < 0$ (Fig. 7.23)

Fig. 7.22 Function $x(\tau)$ and its time reversed

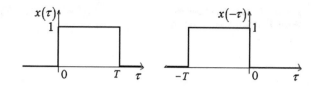

Fig. 7.23 Two functions to multiply (*top*); Their product (*bottom*)

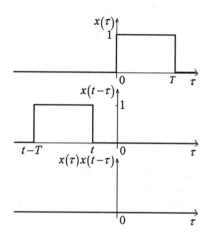

Fig. 7.24 Two functions to multiply (*top*); Their product (*bottom*)

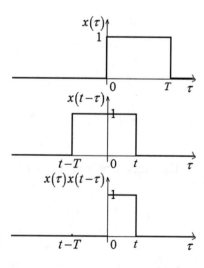

The product of $x(\tau)$ and $x(t-\tau)$ is zero $\forall \tau$ then $y(t) = 0$.

If $0 < t < T$ (Fig. 7.24)

The product of $x(\tau)$ and $x(t-\tau)$ is zero for $\tau < 0$ and for $\tau > t$. It is equal to 1 whenever $0 < \tau < t$. Thus $y(t) = \int_0^t d\tau = t$.

If $T < t < 2T$ (Fig. 7.25)

Fig. 7.25 Two functions to
multiply (*top*); Their product
(*bottom*)

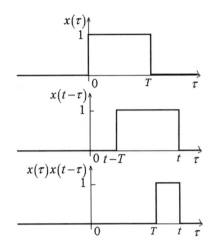

Fig. 7.26 Two functions to
multiply (*top*); Their product
(*bottom*)

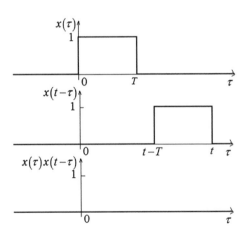

The product of $x(\tau)$ and $x(t - \tau)$ is zero for $\tau < t$ and for $\tau > T$. It is 1 for
$t < \tau < T$.

$$y(t) = \int_{t-T}^{T} d\tau = 2T - t.$$

If $2T < t$ (Fig. 7.26)
The product of $x(\tau)$ and $x(t - \tau)$ is always zero and whatever τ; $y(t) = 0$

In summary $y(t) = \begin{vmatrix} 0 & \text{if} & t < 0 \\ t & \text{if} & 0 < t < T \\ 2T - t & \text{if} & T < t < 2T \\ 0 & \text{if} & 2T < t \end{vmatrix}$

The function $y(t)$ is triangular with a base $2T$ (Fig. 7.27).

Fig. 7.27 Triangular window

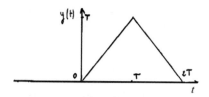

It can be seen in the preceding examples that the convolution of two causal functions is causal and that the autoconvolution of a rectangular window function is triangular.

Summary
We have applied theorems on Fourier transform and convolution for the calculations of Fourier transforms of common windows met in signal analysis. Rectangular, triangular, Hanning, Gaussian, are considered. Calculations of Fourier transforms were carried out in detail and give a good example of the use of integration of complex functions. We have related the decrease of the magnitude of Fourier transform of a window at infinite frequencies to the smoothness of its junction to zero at the edges of the window. Several examples of convolution were given. The chapter ends with a table of Fourier transforms of common functions.

Exercises

I. Gibbs phenomenon. Let the input signal in a filter be a rectangular pulse $x(t)$ equal to 1 between $-\tau/2$ and $\tau/2$ and null elsewhere. The filter is an ideal lowpass with $H(\omega) = Ke^{-j\omega t_0}$ when $-\omega_c < \omega < \omega_c$, and null elsewhere.

1. Give $X(\omega)$ and $Y(\omega)$ the Fourier transforms of the input and output signals $x(t)$ and $y(t)$.
2. Show that $y(t)$ is the difference of two sine integral functions $Si(u)$, where $Si(u) = \int_0^u \frac{\sin x}{x} dx$.
3. Use a Matlab simulation to obtain the results for $y(t)$ in Fig. 7.28 in the cases $K = 1$, $\tau = 2$, $t_0 = 2$, $\omega_c = 10$, $\omega_c = 20$.

Solution

1. $X(\omega) = \tau \, \text{sinc}\left(\frac{\omega\tau}{2}\right)$; $Y(\omega) = K\tau \, \text{sinc}\left(\frac{\omega\tau}{2}\right)e^{-j\omega t_0}$ for $-\omega_c < \omega < \omega_c$ and null elsewhere.
2. $y(t) = \frac{K\tau}{2\pi} \int_{-\omega_c}^{\omega_c} \text{sinc}(\omega\tau/2)e^{j\omega(t-t_0)} d\omega = \frac{K\tau}{\pi} \int_0^{\omega_c} \frac{\sin(\omega\tau/2)}{\omega\tau/2} \cos \omega(t - t_0)d\omega$. (The integral of the odd function $\frac{\sin(\omega\tau/2)}{\omega\tau/2} \sin \omega(t - t_0)$ over a symmetric interval is zero).

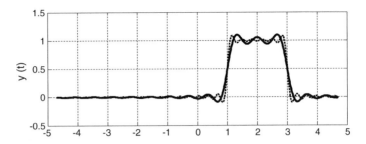

Fig. 7.28 Gibbs phenomenon after low-pass filtering

$$y(t) = \frac{K\tau}{2\pi} \int_0^{\omega_c} \frac{\sin\omega(t - t_0 + \tau/2)}{\omega\tau/2} - \frac{\sin\omega(t - t_0 - \tau/2)}{\omega\tau/2} d\omega.$$

$$y(t) = \frac{K}{\pi} \int_0^{\omega_c(t-t_0+\tau/2)} \frac{\sin x}{x} dx - \frac{K}{\pi} \int_0^{\omega_c(t-t_0-\tau/2)} \frac{\sin x}{x} dx.$$

$$y(t) = \frac{K}{\pi} \{ \mathrm{Si}(\omega_c(t - t_0 + \tau/2)) - \mathrm{Si}(\omega_c(t - t_0 - \tau/2)) \}.$$

3. $\omega_c = 10$ plain line; $\omega_c = 20$ dashed line.

II. Consider the function $f(t) = e^{-\frac{t^2}{2\sigma^2}}$ (Gaussian signal), where σ is a constant having the dimension of a time. Referring to the definitions in Chap. 6, show that the squares of spreads in time and frequency of this function are

$$\sigma_t^2 = \frac{1}{E} \int_{-\infty}^{\infty} t^2 |f(t)|^2 dt = \frac{\sigma^2}{2} \text{ et } \sigma_\omega^2 = \frac{1}{2\pi E} \int_{-\infty}^{\infty} \omega^2 |F(\omega)|^2 d\omega = \frac{1}{2\sigma^2}.$$

Show that the Heisenberg-Gabor relation for Gaussian signals is verified.

III. Consider the function $\psi(t) = e^{-\frac{t^2}{2\sigma^2}} e^{i\omega_0 t}$ (complex Morlet wavelet), where ω_0 and σ are constants. Calculate its Fourier transform $\Psi(\omega)$.

Solution: The FT of the product of functions is given by the convolution of their FT.

We have

$$\mathcal{F}\left(e^{-\frac{t^2}{2\sigma^2}}\right) = e^{-\frac{\sigma^2\omega^2}{2}} \sqrt{2\pi\sigma^2}; \ \mathcal{F}\left(e^{i\omega_0 t}\right) = 2\pi\delta(\omega - \omega_0).$$

Thus

$$\Psi(\omega) = \frac{1}{2\pi} \left(e^{-\frac{\sigma^2\omega^2}{2}} \sqrt{2\pi\sigma^2} \otimes 2\pi\delta(\omega - \omega_0) \right) = \sqrt{2\pi\sigma^2} e^{-\frac{\sigma^2(\omega-\omega_0)^2}{2}}.$$

The Heisenberg-Gabor relation follows naturally.

IV. Consider the function $h(t) = 1$ for $0 \le t \le 2$ and zero elsewhere.

Let $x(t) = \delta(t-3) + e^{-0.5t}(U(t) - U(t-5))$. Calculate the convolution $y(t) = x(t) \otimes h(t)$.

IV. In the table of Fourier transforms at the end of this chapter, we see that the FT of

$$e^{-at}\cos(\omega_0 t)U(t) \text{ is } \frac{a+j\omega}{(a+j\omega)^2 + \omega_0^2} \text{ and the FT of } e^{-at}\sin(\omega_0 t)U(t) \text{ is } \frac{\omega_0}{(a+j\omega)^2 + \omega_0^2}.$$

Comment on the asymptotic behavior of these two FT at high frequencies deriving on the discontinuities of these functions in the time domain.

Table of Fourier transforms

Time	Frequency
$\delta(t)$	1
1	$2\pi\delta(\omega)$
$f(t)$	$F(\omega)$
$f(t-t_0)$	$F(\omega)e^{-j\omega t_0}$
$f(t) \otimes g(t)$	$F(\omega)G(\omega)$
$f(t)g(t)$	$\frac{1}{2\pi}F(\omega) \otimes G(\omega)$
$e^{j\omega_0 t}$	$2\pi\delta(\omega - \omega_0)$
$\cos(\omega_0 t)$	$\pi(\delta(\omega - \omega_0) + \delta(\omega + \omega_0))$
$\sin(\omega_0 t)$	$-j\pi(\delta(\omega - \omega_0) - \delta(\omega + \omega_0))$
$\Pi_T(t) = \begin{vmatrix} 1 & \|t\| < \frac{T}{2} \\ 0 & \text{elsewhere} \end{vmatrix}$	$T \sin c\left(\frac{\omega T}{2}\right)$
$\Lambda_{2T}(t) = \Pi_T(t) \otimes \Pi_T(t)$	$T^2 \sin c^2\left(\frac{\omega T}{2}\right)$
$a > 0 \quad e^{-at}U(t)$	$\frac{1}{a+j\omega}$
$te^{-at}U(t)$	$\frac{1}{(a+j\omega)^2}$
$e^{-a\|t\|}U(t)$	$\frac{2a}{a^2+\omega^2}$
$\text{Re}(\alpha) \ge 0 \quad e^{-\alpha t^2}$	$\sqrt{\frac{\pi}{\alpha}}e^{-\frac{\omega^2}{4\alpha}}$
Pseudo function $Pf\left(\frac{1}{t}\right)$	$-j\pi\text{sgn}(\omega)$ (see chap. 9)
$\frac{1}{2}\delta(t) + \frac{j}{2\pi}Pf\left(\frac{1}{t}\right)$	$U(\omega)$
$U(t)$	$\pi\delta(\omega) + \frac{1}{j}Pf\left(\frac{1}{\omega}\right)$
$\text{sgn}(t)$	$\frac{2}{j}Pf\left(\frac{1}{\omega}\right)$
$e^{-at}\cos(\omega_0 t)U(t)$	$\frac{a+j\omega}{(a+j\omega)^2 + \omega_0^2}$
$e^{-at}\sin(\omega_0 t)U(t)$	$\frac{\omega_0}{(a+j\omega)^2 + \omega_0^2}$
$\text{Re}(v) > -1 \quad t^v U(t)$	$(j\omega)^{-v-1}\Gamma(v+1)$
$t^{-\frac{1}{2}}U(t)$	$(j\omega)^{-\frac{1}{2}}\sqrt{\pi}$
$\text{Re}(v) > -1 \quad t^v e^{at}U(t)$	$(j\omega - a)^{-v-1}\Gamma(v+1)$

(continued)

(continued)

Time	Frequency
$f(t - t_0)$	$F(\omega)e^{-j\omega t_0}$
$f(t) \otimes g(t)$	$F(\omega)G(\omega)$
$f(t)g(t)$	$\frac{1}{2\pi}F(\omega) \otimes G(\omega)$
$f(t)e^{j\omega_0 t}$	$F(\omega - \omega_0)$
$\Pi_T(t)\cos \omega_0 t$	$\frac{T}{2}\left\{\frac{\sin(\omega - \omega_0)\frac{T}{2}}{(\omega - \omega_0)\frac{T}{2}} + \frac{\sin(\omega + \omega_0)\frac{T}{2}}{(\omega + \omega_0)\frac{T}{2}}\right\}$
$\sum\limits_{n=-\infty}^{\infty} \delta(t - nT_0)$	$\frac{2\pi}{T_0}\sum\limits_{n=-\infty}^{\infty} \delta\left(\omega - n\frac{2\pi}{T_0}\right)$
$\delta'(t)$	$j\omega$
$\delta^{(n)}(t)$	$(j\omega)^n$

Chapter 8
Impulse Response of LTI Systems

It was shown in Chap. 5 that the impulse response of a LTI system is the inverse Fourier transform of the frequency response. It is given by the integral

$$h(t) = \frac{1}{2\pi} PV \int_{-\infty}^{+\infty} H(\omega)e^{j\omega t}d\omega. \tag{8.1}$$

The immediately apparent difficulty in the calculation of $h(t)$ is that the function $H(\omega)$ is in the general case a complex function of ω. The integral cannot generally be evaluated simply by the methods of integration of real functions. The integration is then performed in the complex plane by integration over a closed contour. The principles of the analysis and integration of a complex function are presented in Appendix A1. We use as examples the calculations of the impulse responses of first- and second-order systems.

It appears the important result that the causality of a stable physical system is implied by the position of the poles of the transfer function in the half complex plane with negative real parts.

The complex variable s encountered in the definition of transfer functions $H(s)$ was written as $s = \sigma + j\omega$. As $H(\omega) = H(s)|_{\sigma=0}$, relationship (8.1) becomes

$$h(t) = \frac{1}{2\pi j} PV \int_{-j\infty}^{+j\infty} H(s)e^{st}ds. \tag{8.2}$$

The integral is evaluated on the vertical axis on which $\sigma = j\omega$. Since one has to calculate the principal value of an integral, we can make the boundaries tend symmetrically to infinity. The residue method is generally used to calculate the integral.

© Springer International Publishing Switzerland 2016
F. Cohen Tenoudji, *Analog and Digital Signal Analysis*,
Modern Acoustics and Signal Processing, DOI 10.1007/978-3-319-42382-1_8

To begin this chapter, the impulse response $h(t)$ is calculated for simple examples to highlight the causal relationship between the stability of a system and the situation of the poles of the system transfer function in the complex plane.

8.1 Impulse Response of a First-Order Filter

We have seen that the frequency response of the RC circuit was

$$H(\omega) = \frac{1}{1 + jRC\omega} \tag{8.3}$$

The impulse response $h(t)$ is given by the inverse Fourier transformation (as discussed above, the principal value notation will be no longer mentioned hereinafter)

$$h(t) = \frac{1}{2\pi} \int_{-\infty}^{+\infty} \frac{1}{1 + jRC\omega} e^{j\omega t} d\omega \tag{8.4}$$

or, according to (8.2)

$$h(t) = \frac{1}{2\pi j} \int_{-j\infty}^{+j\infty} \frac{1}{1 + RCs} e^{st} ds \tag{8.5}$$

The integral is calculated by the residue method. For this, first of all we define a closed contour Γ which is composed of the vertical axis and a half great circle C in the left or half planes with radius tending toward infinity.

According to the residue theorem (see Appendix A1), the integral over a closed contour of $H(s)e^{st}$ is equal to the sum of the residues (labeled Res_i) inside the contour Γ

$$\int_{-j\infty}^{+j\infty} H(s)\, e^{st} ds + \int_C H(s)\, e^{st} ds = 2\pi j \sum_i \mathrm{Res}_i. \tag{8.6}$$

For the first-order filter

$$H(s) = \frac{-s_0}{s - s_0} \tag{8.7}$$

This function has a simple pole at s_0 where

$$s_0 = -\frac{1}{RC}.\tag{8.8}$$

One must distinguish two cases

- If $t > 0$, we can apply Jordan's lemma on C (see Appendix 1) if the exponential modulus of e^{pt} is bounded. For this to be so, in the case where t is positive, it is necessary that σ, the real part of s, is negative. Thus we close the contour of the half circle to the left (Fig. 8.1). According to Jordan's lemma, the integral tends to be zero as $R \to \infty$.

$$\int_C H(s)e^{st}ds = 0.\tag{8.9}$$

The contour encompasses the simple pole s_0. There is a residue of the integral for this contour called Bromwich contour then

$$\int_{-j\infty}^{+j\infty} H(s)e^{st}ds = 2\pi j(\text{Res})_{s=s_0} = 2\pi j \lim_{s->s_0}(s - s_0)H(s)e^{st} = -2\pi j s_0 e^{s_0 t}$$

$$\tag{8.10}$$

Finally if $t > 0$

$$h(t) = -s_0 e^{s_0 t} = \frac{1}{RC}e^{-\frac{1}{RC}t}.\tag{8.11}$$

Fig. 8.1 Integration
Bromwich contour for the
case $t > 0$

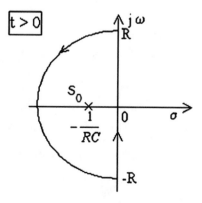

Fig. 8.2 Integration
Bromwich contour for the
case $t<0$

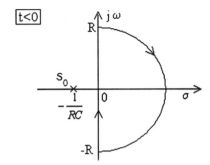

- Case $t<0$: To apply Jordan's lemma on C it is necessary for σ to be positive, so
  that the exponential e^{st} remains bounded when s follows the half circle. We
  therefore close the contour by the half circle on the vertical axis right (Fig. 8.2)
  and the integral tends to zero when $R\to\infty$. The contour surrounds no pole;
  there is no residue.

  So, according to Cauchy theorem

$$\int_{-j\infty}^{+j\infty} H(s)e^{st}ds + \int_C H(s)e^{st}ds = 0. \tag{8.12}$$

As the integral over the semicircle $\int_C H(s)e^{st}ds = 0$ vanishes, by the application of
Jordan's lemma, we get

$$\int_{-j\infty}^{+j\infty} H(s)e^{st}ds = 0, \tag{8.13}$$

or, for $t<0$,

$$h(t) = 0. \tag{8.14}$$

In summary, for any value of t we have

$$h(t) = \frac{1}{RC}e^{-\frac{t}{RC}}U(t), \tag{8.15}$$

where $U(t)$ is the Heaviside function, equal to zero for $t<0$ and to 1 for $t>0$
(Fig. 8.3).

Fig. 8.3 Impulse response

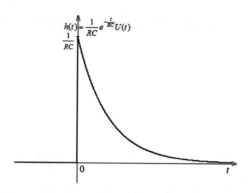

To complete, we now check as an exercise, that the solution $h(t)$ is correct by referring the previous result in the original differential equation and using the properties of the derivative of a discontinuous function.

The initial equation resulting from Ohm's law is

$$R\frac{dq(t)}{dt} + \frac{q}{C} = \delta(t), \tag{8.16}$$

with

$$h(t) = \frac{q(t)}{C} = \frac{1}{RC}e^{-\frac{t}{RC}}U(t). \tag{8.17}$$

The electric charge is

$$q(t) = \frac{1}{R}e^{-\frac{t}{RC}}U(t). \tag{8.18}$$

Its derivative is

$$\frac{dq(t)}{dt} = -\frac{1}{R^2C}e^{-\frac{t}{RC}}U(t) + \frac{1}{R}e^{-\frac{t}{RC}}\delta(t). \tag{8.19}$$

The second term of the second member $\frac{1}{R}e^{-\frac{t}{RC}}\delta(t)$ is equal to $\frac{1}{R}\delta(t)$ as the exponential is 1 for $t = 0$.

Indeed, by acting the distribution function $\frac{1}{R}e^{-\frac{t}{RC}}\delta(t)$ on a probe function $\varphi(t)$ in an integral

$$\int_{-\infty}^{+\infty} \frac{1}{R}e^{-\frac{t}{RC}}\delta(t)\varphi(t)dt = \frac{1}{R}\varphi(0). \tag{8.20}$$

The result is the same as for the distribution $\frac{1}{R}\delta(t)$

$$\int_{-\infty}^{+\infty} \frac{1}{R}\delta(t)\varphi(t)\mathrm{d}t = \frac{1}{R}\varphi(0).\tag{8.21}$$

So

$$\frac{\mathrm{d}q(t)}{\mathrm{d}t} = -\frac{1}{R^2C}e^{-\frac{t}{RC}}U(t) + \frac{1}{R}\delta(t).\tag{8.22}$$

We replace this derivative in the differential Eq. (8.16) and $e(t) = \delta(t)$ is found back in the second member

$$R\frac{\mathrm{d}q(t)}{\mathrm{d}t} + \frac{q}{C} = -\frac{1}{RC}e^{-\frac{t}{RC}}U(t) + \delta(t) + \frac{1}{RC}e^{-\frac{t}{RC}}U(t) = \delta(t).\tag{8.23}$$

To complete this study of the first-order system we check now that $H(\omega) = \frac{1}{1+jRC\omega}$ is the Fourier transform of $h(t) = \frac{1}{RC}e^{-\frac{t}{RC}}U(t)$.

$$\int_{-\infty}^{+\infty} \frac{1}{RC}e^{-\frac{t}{RC}}U(t)e^{-j\omega t}\mathrm{d}t = \frac{1}{RC}\int_{0}^{+\infty} e^{-\left(\frac{1}{RC}+j\omega\right)t}\mathrm{d}t = \frac{1}{RC}\left[\frac{e^{-\left(\frac{1}{RC}+j\omega\right)t}}{-\left(\frac{1}{RC}+j\omega\right)}\right]_{0}^{+\infty},\tag{8.24}$$

and then,

$$H(\omega) = \frac{1}{RC}\left(\frac{1}{\frac{1}{RC}+j\omega}\right) = \frac{1}{1+j\omega RC}.\tag{8.25}$$

8.2   Impulse Response of a Second Order Filter

Recall that the impulse response of a LTI filter is given by $h(t) = \frac{1}{2\pi j}\int_{-j\infty}^{+j\infty} H(s)\,e^{st}\mathrm{d}s$.
    With here

$$H(s) = \frac{1}{LCs^2 + RCs + 1} = \frac{1}{LC}\frac{1}{(s-s_1)(s-s_2)}.\tag{8.26}$$

Case where the roots the roots s_1 and s_2 are different

We can develop the rational fraction into simple elements. We write

$$\frac{1}{(s - s_1)(s - s_2)} = \frac{A}{(s - s_1)} + \frac{B}{(s - s_2)}. \tag{8.27}$$

We search A and B identifying the terms resulting from using the same denominator.

$$\frac{A(s - s_2) + B(s - s_1)}{(s - s_1)(s - s_2)} = \frac{1}{(s - s_1)(s - s_2)}, \tag{8.28}$$

yet $(A + B)s - As_2 - Bs_1 = 1$. This relationship should be verified for any s. It comes $B = -A$ and $-As_2 + As_1 = 1$.

Then $A = \frac{1}{(s_1 - s_2)}$ and $B = -\frac{1}{(s_1 - s_2)}$, and

$$H(s) = \frac{1}{LC}\left(\frac{1}{(s_1 - s_2)(s - s_1)} - \frac{1}{(s_1 - s_2)(s - s_2)}\right). \tag{8.29}$$

We meet a situation of two first-order systems in parallel. The impulse response is obtained from the result for the impulse response of the first-order filter.

$$h(t) = \frac{1}{LC}\frac{1}{(s_1 - s_2)}\left(e^{s_1 t} - e^{s_2 t}\right) U(t). \tag{8.30}$$

The poles s_1 and s_2 are the roots of the polynomial $LCs^2 + RCs + 1 = 0$ in the denominator. These roots depend on the value of the discriminant of the polynomial $\Delta = R^2 C^2 - 4LC$.

- If $\Delta > 0$,

$$s_1 - s_2 = \sqrt{\frac{R^2}{L^2} - \frac{4}{LC}}. \tag{8.31}$$

The impulse response will be in this case

$$h(t) = \frac{1}{\sqrt{R^2 C^2 - 4LC}} e^{-\frac{R}{2L}t}\left(e^{\sqrt{\frac{R^2}{4L^2} - \frac{1}{LC}}t} - e^{-\sqrt{\frac{R^2}{4L^2} - \frac{1}{LC}}t}\right) U(t). \tag{8.32}$$

This impulse response is constituted by the difference between two decreasing exponential.

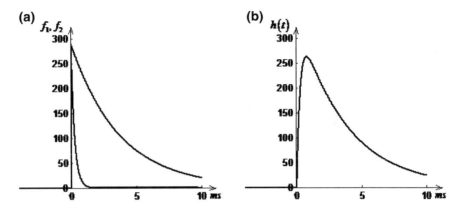

Fig. 8.4 a Functions appearing in (8.32); **b** Impulse response

Figure 8.4a shows the two functions appearing in (8.32). $h(t)$ shown in Fig. 8.4b is their difference. ($R = 40\ \Omega$, $L = 10^{-2}$ H, $C = 10^{-4}$ F.).

- If $\Delta < 0$

$$s_1 - s_2 = j\sqrt{\frac{4}{LC} - \frac{R^2}{L^2}}, \tag{8.33}$$

$$h(t) = \frac{1}{j2LC\sqrt{\frac{1}{LC} - \frac{R^2}{4L^2}}} e^{-\frac{R}{2L}t}\left(e^{j\sqrt{\frac{1}{LC} - \frac{R^2}{4L^2}}t} - e^{-j\sqrt{\frac{1}{LC} - \frac{R^2}{4L^2}}t}\right)U(t). \tag{8.34}$$

Finally

$$h(t) = \frac{1}{LC}e^{-\frac{R}{2L}t}\frac{\sin \omega_0 t}{\omega_0}U(t), \tag{8.35}$$

with

$$\omega_0 = \sqrt{\frac{1}{LC} - \frac{R^2}{4L^2}}. \tag{8.36}$$

The impulse response is shown in Fig. 8.5 for the values $R = 5\ \Omega$, $L = 10^{-2}$ H, $C = 10^{-4}$ F.

Fig. 8.5 Impulse response in a case where $\Delta < 0$

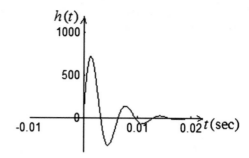

Case where the roots s_1 and s_2 are equal

This is the case if the discriminant $\Delta = 0$. The denominator has a real double root

$$s_1 = s_2 = -\frac{R}{2L}. \tag{8.37}$$

The transfer function is

$$H(s) = \frac{1}{LC}\frac{1}{(s - s_1)^2}. \tag{8.38}$$

To calculate $h(t)$ in this case one must calculate the residue at double pole s_1.

Reminder of the residue theorem: Let the function $F(s)$ with a pole of order n in $s = a$

The residue from the integration of $F(s)$ along a closed contour surrounding the pole is

$$\text{Res}_{\text{in } s=a} = \frac{1}{(n-1)!}\frac{d^{n-1}}{ds^{n-1}}[(s-a)^n F(s)]_{s=a}. \tag{8.39}$$

In the present case $\text{Res}_{\text{in } s=s_1} = \frac{d}{ds}\left[(s-s_1)^2 H(s)e^{st}\right]_{s=s_1} = \frac{1}{LC}\left[\frac{d}{ds}e^{st}\right]_{s=s_1}$.

Therefore $\text{Res}_{\text{in } s=s_1} = \frac{1}{LC}te^{s_1 t}$.

In this last case the impulse response is

$$h(t) = \frac{1}{LC}te^{-\frac{R}{2L}t}U(t) \tag{8.40}$$

The impulse response is shown for this critical damping in Fig. 8.6 for values $R = 20\ \Omega$,

$$L = 10^{-2}\ \text{H},\ C = 10^{-4}\ \text{F}.$$

Fig. 8.6 Impulse response in the case of critical damping

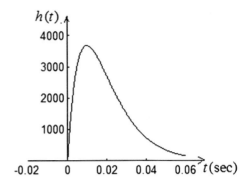

To complete this study, we check in one case that the function $h(t)$ is the solution of the differential equation of the second-order filter where the second member is $e(t) = \delta(t)$

We choose the case where $h(t)$ has the expression $h(t) = \frac{1}{LC} e^{-\frac{R}{2L}t} \frac{\sin \omega_0 t}{\omega_0} U(t)$

$$\frac{dh(t)}{dt} = \frac{d}{dt}\left(\frac{1}{LC} e^{-\frac{R}{2L}t} \frac{\sin \omega_0 t}{\omega_0}\right) U(t) + \frac{1}{LC} e^{-\frac{R}{2L}t} \frac{\sin \omega_0 t}{\omega_0} \delta(t) \qquad (8.41)$$

The second term vanishes since the function multiplying $\delta(t)$ is zero for $t = 0$.

$$\frac{dh(t)}{dt} = -\left(\frac{R}{2L^2C} e^{-\frac{R}{2L}t} \frac{\sin \omega_0 t}{\omega_0}\right) U(t) + \left(\frac{1}{LC} e^{-\frac{R}{2L}t} \cos \omega_0 t\right) U(t). \qquad (8.42)$$

Also

$$\frac{d^2h(t)}{dt^2} = \left(\frac{R^2}{4L^3C} e^{-\frac{R}{2L}t} \frac{\sin \omega_0 t}{\omega_0}\right) U(t) - \left(\frac{R}{2L^2C} e^{-\frac{R}{2L}t} \cos \omega_0 t\right) U(t) - \left(\frac{R}{2L^2C} e^{-\frac{R}{2L}t} \frac{\sin \omega_0 t}{\omega_0}\right) \delta(t)$$
$$- \left(\frac{R}{2L^2C} e^{-\frac{R}{2L}t} \cos \omega_0 t\right) U(t) - \left(\omega_0 \frac{1}{LC} e^{-\frac{R}{2L}t} \sin \omega_0 t\right) U(t) + \left(\frac{1}{LC} e^{-\frac{R}{2L}t} \cos \omega_0 t\right) \delta(t). \qquad (8.43)$$

Simplifying

$$\frac{d^2h(t)}{dt^2} = \left(\frac{R^2}{4L^3C} e^{-\frac{R}{2L}t} \frac{\sin \omega_0 t}{\omega_0}\right) U(t) - \left(\frac{R}{L^2C} e^{-\frac{R}{2L}t} \cos \omega_0 t\right) U(t)$$
$$- \left(\omega_0 \frac{1}{LC} e^{-\frac{R}{2L}t} \sin \omega_0 t\right) U(t) + \frac{1}{LC} \delta(t). \qquad (8.44)$$

Now we verify that $LC\frac{d^2h}{dt^2} + RC\frac{dh}{dt} + h(t) = \delta(t)$

$$\left(\frac{R^2}{4L^2}e^{-\frac{R}{2L}t}\frac{\sin \omega_0 t}{\omega_0}\right)U(t) - \left(\frac{R}{L}e^{-\frac{R}{2L}t}\cos \omega_0 t\right)U(t) - \left(\omega_0 e^{-\frac{R}{2L}t}\sin \omega_0 t\right)U(t) + \delta(t)$$

$$-\left(\frac{R^2}{2L^2}e^{-\frac{R}{2L}t}\frac{\sin \omega_0 t}{\omega_0}\right)U(t) + \left(\frac{R}{L}e^{-\frac{R}{2L}t}\cos \omega_0 t\right)U(t) + \left(\frac{1}{LC}e^{-\frac{R}{2L}t}\frac{\sin \omega_0 t}{\omega_0}\right)U(t)$$

$$= e^{-\frac{R}{2L}t}\frac{\sin \omega_0 t}{\omega_0}U(t)\left(-\frac{R^2}{4L^2} - \omega_0^2 + \frac{1}{LC}\right) + \delta(t) = \delta(t). \qquad (8.45)$$

we have written $\omega_0^2 = \frac{1}{LC} - \frac{R^2}{4L^2}$.

This calculation proves that the impulse response satisfies generalized Ohm's law when the second member is $e(t) = \delta(t)$.

Example We search the impulse response of the current for the electric circuit RC of the first order.

Ohm's law is written with variable $q(t)$

$$R\frac{dq}{dt} + \frac{q}{C} = e(t). \qquad (8.46)$$

Using the current in the circuit as a variable, Ohm's law is $Ri(t) + \frac{1}{C}\int i(t)dt = e(t)$

Calculation of the current transfer function: We note $e(t) = e^{st}$

We are looking for $i(t)$ in the form $i(t) = Ae^{st}$. Then $\int i(t)dt = \frac{1}{s}Ae^{st}$.

Using this expression in Ohm's law $RAe^{st} + \frac{1}{Cs}Ae^{st} = e^{st}$.

This equation is verified for all t if $A\left(R + \frac{1}{Cs}\right) = 1$, or when $A = \frac{1}{R + \frac{1}{Cs}} = \frac{sC}{1 + RCs}$.

So, the current transfer function is

$$H(s) = A = \frac{sC}{1 + RCs}. \qquad (8.47)$$

This function has a zero in $s = 0$ and a pole $s_0 = -\frac{1}{RC}$. $H(s)$ is decomposed into simple elements as: $H(s) = D + \frac{E(s)}{F(s)} = \frac{DF(s) + E(s)}{F(s)} = \frac{D(1 + RCs) + E(s)}{1 + RCs} = \frac{sC}{1 + RCs}$.

By identifying the coefficients of the powers of s it comes $D = \frac{1}{R}$ and $E = -\frac{1}{R}$.

Then

$$H(s) = \frac{1}{R} - \frac{1}{R(RCs + 1)}. \qquad (8.48)$$

Calculating $h(t)$ by integration given in (8.2) $h(t) = \frac{1}{2\pi j}\int_{-j\infty}^{j\infty}\left(\frac{1}{R} - \frac{1}{RCs + 1}\right)e^{st}ds.$

On the integration vertical axis we have $s = j\omega$, then

$$h(t) = \frac{1}{R}\frac{1}{2\pi} \int\limits_{-\infty}^{\infty} e^{j\omega t}d\omega - \frac{1}{2\pi j} \int\limits_{-j\infty}^{j\infty} \frac{1}{R}\frac{1}{RCs+1}e^{st}ds, \qquad (8.49)$$

Using a result of first-order filter, we finally get

$$h(t) = \frac{1}{R}\delta(t) - \frac{1}{R^2C}e^{-\frac{t}{RC}}U(t). \qquad (8.50)$$

The reader will verify as an exercise that this result maybe found by a time derivation of the impulse response in tension across the capacitor.

Summary
We have carried out in this chapter the calculation of the impulse responses of two important circuit cases in electricity. They are calculated by performing the inverse transform calculations of the response function $H(\omega)$ by integration in the complex plane on a closed contour and using the residue theorem. We have shown that the causality of a stable physical system is implied by the position of the poles of the transfer function in the half complex plane with negative real parts.

Next chapter on Laplace transform generalizes these results and studies on examples the properties of causality and stability.

Exercise
This exercise comes as a following of Exercise 1, Chap. 2.

(a) Give the expression for calculating the impulse response $h(t)$ from the frequency response function $H(\omega)$.
(b) By integration in the complex plane, calculate $h(t)$. Justify the causal nature of this function. Give the graph of $h(t)$.

Chapter 9
Laplace Transform

The direct and inverse Laplace transforms are defined in this chapter. As specified in formulas 9.1–9.4 of this chapter, the Fourier transformation appears as a special case of the Laplace transform for $s = j\omega$. In Chap. 2 we used the transfer function properties defined in the $(\sigma, j\omega)$ plane to explain the variations of the frequency response of a system which is a function of the single variable ω. The inversion formula for the Fourier transform recalled here (9.2) shows that the function $F(s)$ is the two-sided Laplace transform (integration from $-\infty$ to $+\infty$) of the function $f(t)$. Laplace transform has played a very important role in electrical engineering in the study of electronic systems responses, causal by essence. It was oriented primarily for the treatment of causal signals, zero for negative time. Historically the one-sided form of the Laplace transform was used. The transfer function of an electrical circuit, written in the form of a rational fraction was decomposed into simple elements. For canonical form of input signals, it was possible to calculate the output signal in the Laplace domain as products of simpler functions. Simple rules treated the boundary conditions at time $t = 0$.

With the use of distributions which allow generalizing all functions, the Fourier transform calculations and the development of digital computation, the Laplace transform is less dominant today, especially in theoretical calculations. An important goal of this chapter is to provide an understanding of the domain of definition of the Laplace transform and its association with causality and stability of a system.

The special case of a marginally stable system is also discussed.

9.1 Direct and Inverse Transforms

It has been shown in chapter 8 that calculation of an inverse Fourier transform of the general form

© Springer International Publishing Switzerland 2016
F. Cohen Tenoudji, *Analog and Digital Signal Analysis*,
Modern Acoustics and Signal Processing, DOI 10.1007/978-3-319-42382-1_9

$$f(t) = \frac{1}{2\pi} PV \int_{-\infty}^{+\infty} F(\omega)e^{j\omega t}d\omega \tag{9.1}$$

had been performed in the complex plane using the integral

$$f(t) = \frac{1}{2\pi j} PV \int_{-j\infty}^{+j\infty} F(s)e^{st}ds. \tag{9.2}$$

The function $F(s)$ is the two-sided Laplace transform of the function $f(t)$, given by the integral

$$F(s) = \int_{-\infty}^{\infty} f(t)\,e^{-st}dt, \tag{9.3}$$

where s is any complex number.

By posing $s = \sigma + j\omega$ as was done previously, we see that in the case where $\sigma = 0$, the value of the Laplace transform on the vertical axis where $\sigma = 0$ is given by

$$F(s)|_{\sigma=0} = \int_{-\infty}^{\infty} f(t)\,e^{-st}dt. \tag{9.4}$$

We recognize the right side as the Fourier transform of the function $f(t)$.

Thus, the Fourier transform $F(\omega)$ of a function appears to be a special case of the two-sided Laplace transform of that function. The term two-sided means that, both boundaries in the time integral tend to infinity. In the one-sided Laplace transform, the lower boundary of the integral is $t = 0$. In reason of its proximity to the Fourier transform, the two-sided Laplace transform is preferred in this course.

In summary, the two-sided Laplace transform is

$$\text{Direct transform}: \quad F(s) = \int_{-\infty}^{\infty} f(t)\,e^{-st}dt. \tag{9.5}$$

$$\text{Inverse transform}: \quad f(t) = \frac{1}{2\pi j} PV \int_{-j\infty}^{+j\infty} f(t)\,e^{-st}ds. \tag{9.6}$$

Of the foregoing, it is seen that the transfer function of an LTI system is the Laplace transform of its impulse response.

While the Fourier transform is defined on the $s = j\omega$ axis, the Laplace transform is defined in the plane s. This is a two-dimensional function as the FT is one-dimensional. By the fact it contains much more information on the function $f(t)$. In particular, the function $F(s)$ reveals the crucial importance played by its poles and zeros.

The integral (9.5) defining the Laplace transform converges only for certain values of s. The function $F(s)$ is not defined in some parts of the complex s plane.

9.1.1 Study of Convergence with an Example

Let the function

$$f(t) = e^{-\alpha t} U(t), \tag{9.7}$$

where α is a real number for any sign and $U(t)$ is the Heaviside function.
Then:

$$F(s) = \int_{\infty}^{\infty} e^{-\alpha t} e^{-st} dt = \frac{e^{-(s+\alpha)t}\big|_{\infty} - 1}{-(s+\alpha)}, \tag{9.8}$$

with $s = \sigma + j\omega$. The exponential in the numerator tends to a finite limit when t tends to infinity only if the exponential modulus $\left|e^{-(\sigma + j\omega + \alpha)t}\right| = \left|e^{-(\sigma + \alpha)t}\right|$ tends to zero as t tends to infinity. It is necessary that $\sigma + \alpha > 0$, equivalently

$$\sigma > -\alpha. \tag{9.9}$$

In this case the Laplace transform is

$$F(s) = \frac{1}{(s+\alpha)}. \tag{9.10}$$

Since the condition $\sigma > -\alpha$ must be verified to ensure convergence, the function $F(s)$ is defined only in the half s plane, to the right of the vertical $\sigma = -\alpha$.

It is interesting to detail the study of the convergence which depends upon the sign of α

- If $\alpha > 0$ (Fig. 9.1),
  $f(t) = e^{-\alpha t} U(t)$ is a decreasing function of t tending to zero as t tends to infinity.

Fig. 9.1 $f(t)$ in the case $\alpha > 0$

$$f(t) = e^{-\alpha t} U(t)$$
$$\alpha > 0$$

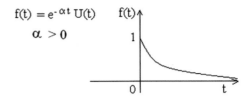

Fig. 9.2 Domain of definition of $F(s)$ when $\alpha > 0$

$$F(s) = \frac{1}{s + \alpha}$$

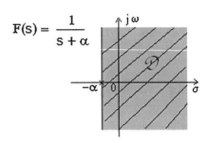

Fig. 9.3 $f(t)$ in the case $\alpha < 0$

$$f(t) = e^{-\alpha t} U(t)$$
$$\alpha < 0$$

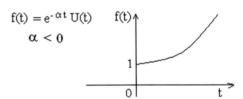

Fig. 9.4 Domain of definition of $F(s)$ when $\alpha < 0$

$$F(s) = \frac{1}{s + \alpha}$$

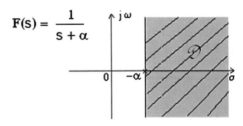

The vertical line $\sigma = -\alpha$ lies in the part of the complex plane where the real part of s is negative (Fig. 9.2).

In particular, $F(s)$ exists for $s = j\omega$. $F(\omega) = F(s)|_{s=j\omega} = \int_{-\infty}^{\infty} f(t) e^{-j\omega t} dt$.

We see that in this case the Fourier transform of the function $f(t)$ is defined.

- If $\alpha < 0$, $f(t) = e^{-\alpha t} U(t)$ is an increasing function of t tending to infinity as t tends to infinity (Fig. 9.3).

The vertical line $\sigma = -\alpha$ belongs to the right of the imaginary axis $\sigma = 0$. $F(s)$ is defined on the right of the vertical (Fig. 9.4). The Fourier transform of this signal does not exist. A system that would have this exponentially increasing impulse response is described as unstable.

Fig. 9.5 $f(t) = ^{-\alpha|t|}$ when
$\alpha > 0$

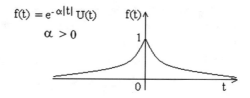

Fig. 9.6 Domain of
definition of $F(s)$ when $\alpha > 0$

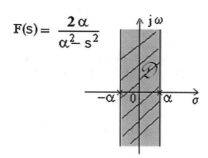

9.1.2   Another Example

Calculate the Laplace transform of $f(t) = ^{-\alpha|t|}$ assuming α is real positive (Fig. 9.5).

$$F(s) = \int_{-\infty}^{\infty} e^{-\alpha|t|}e^{-st}dt = \int_{-\infty}^{0} e^{\alpha t}e^{-st}dt + \int_{0}^{\infty} e^{-\alpha t}e^{-st}dt \tag{9.11}$$

The first integral converges for $\sigma < \alpha$. The second converges when $\sigma > -\alpha$. Then

$$F(s) = -\frac{1}{s-\alpha} + \frac{1}{s+\alpha} = \frac{2\alpha}{\alpha^2 - s^2}. \tag{9.12}$$

$F(s)$ will be defined for all s belonging to the vertical strip bounded by the two lines $-\alpha$ and $+\alpha$ (Fig. 9.6).The imaginary axis $\sigma = 0$ belonging to this area of convergence, the Fourier transform of $f(t)$ exists.

9.2   Stability of a System and Laplace Transform

As seen in the case of the first and second-order systems, impulse response decays exponentially in time. This is the usual case for physical systems.

The domain of definition in the Laplace plane of transfer function $H(s)$ of causal systems is a half plane on the right of a vertical line. When causal systems are damped, this vertical is located in the part of the plane where $\sigma < 0$.

This shows that for these systems, the frequency response $H(\omega)$ cannot be infinite for any value of the angular frequency ω. This ensures that for any input signal in the form $e^{j\omega t}$ in the system, the output signal $H(\omega)\,e^{j\omega t}$ has finite amplitude. It is said that such a system is stable. It can be shown that

A necessary and sufficient condition for a LTI system to be stable is that its impulse response $h(t)$ is such that $\int_{-\infty}^{\infty} |h(t)|\,dt < \infty$.

It follows from the foregoing that the poles of the transfer function of a causal and stable system have a strictly negative real part.

9.2.1   Marginal Stability

A special case of stability is encountered when the singularity of the transfer function of a system lies on the imaginary axis. This is the case when the frequency response is

$$H(\omega) = \frac{1}{j(\omega - \omega_0)}. \tag{9.13}$$

Let us determine its impulse response. It is given by

$$h(t) = \frac{1}{2\pi} PV \int_{-\infty}^{+\infty} H(\omega) e^{j\omega t} d\omega = \frac{1}{2\pi} PV \int_{-\infty}^{+\infty} \frac{1}{j(\omega - \omega_0)} e^{j\omega t} d\omega.$$

We note $\omega - \omega_0 = \Omega$, then $d\omega = d\Omega$. It comes

$$h(t) = \frac{1}{2\pi} PV \int_{-\infty}^{+\infty} \frac{1}{j\Omega} e^{j(\Omega + \omega_0)t} d\Omega = e^{j\omega_0 t} \frac{1}{2\pi} PV \int_{-\infty}^{+\infty} \frac{1}{j\Omega} e^{j(\Omega t)} d\Omega. \tag{9.14}$$

We look for a causal filter. We determine $h(t)$ using the artifice of adding a constant $\alpha > 0$ to the denominator and make α tend toward 0. Noting $s = j\Omega$, we have

$$h(t) = e^{j\omega_0 t} \lim_{\alpha \to 0} \left(\frac{1}{2\pi j} PV \int_{-j\infty}^{+j\infty} \frac{1}{\alpha + s} e^{st} ds \right). \tag{9.15}$$

Using results (9.10) and (9.7), we have

$$h(t) = e^{j\omega_0 t} \lim_{\alpha \to 0} (e^{-\alpha t} U(t)) = e^{j\omega_0 t} U(t). \tag{9.16}$$

The amplitude of $h(t)$ does not blow out to infinity for large t, nor decreases to 0. The filter is said marginally stable.

9.2.2 Minimum-Phase Filter

A minimum-phase filter is defined as a causal and stable filter whose inverse is causal and stable. It follows from this definition that the poles and zeros of this filter lie on the left of the imaginary axis in the Laplace plane.

We admit here the following property: from all filters with the same modulus of the frequency response, the minimum-phase filter is the one whose impulse response is the earliest. (This property is linked to the evolution of phase with frequency of this filter).

9.3 Applications of Laplace Transform

Many physical, electrical systems (generalized Ohm's law) or mechanical (fundamental laws of dynamics) satisfy the following general equation:

$$\frac{d^m y(t)}{dt^m} + a_1 \frac{d^{m-1} y(t)}{dt^{m-1}} + a_2 \frac{d^{m-2} y(t)}{dt^{m-2}} + \ldots + a_m y(t)$$
$$= b_0 \frac{d^r x(t)}{dt^r} + b_1 \frac{d^{r-1} x(t)}{dt^{r-1}} + \ldots + b_r x(t). \tag{9.17}$$

Note that in this type of equation, the functions and their derivatives appear only with a power 1. If it is not the case, the differential equation is not linear and the following development does not apply.

As seen in Chaps. 1 and 2, in this equation where appears only linear combinations of derivatives of input and output functions $x(t)$ and $y(t)$, the study of the system relies on its transfer function, i.e., on the system response $y(t)$:

$$y(t) = H(s)e^{st} \tag{9.18}$$

to the input $x(t)$ of the form

$$x(t) = e^{st}. \tag{9.19}$$

The r order derivative of $x(t)$ that appears in the Eq. (9.17) is in this case

$$\frac{\mathrm{d}^r x(t)}{\mathrm{d}t^r} = s^r \mathrm{e}^{st}. \tag{9.20}$$

Similarly, the m order derivative of $y(t)$ is

$$\frac{\mathrm{d}^m y(t)}{\mathrm{d}t^m} = s^m H(s) \mathrm{e}^{st}. \tag{9.21}$$

After substituting in Eq. (9.17) and by simplifying by e^{st}, we obtain the following expression for the transfer function $H(s)$ of the system

$$H(s) = \frac{b_0 s^r + b_1 s^{r-1} + \ldots + b_r}{s^m + a_1 s^{m-1} + a_2 s^{m-2} + \ldots + a_m}. \tag{9.22}$$

We obtain the impulse response of the system in developing $H(s)$ in simple elements and use of the following rule, which is a direct result of the calculation by residues as was used in Chap. 8

$$t^n \mathrm{e}^{s_i t} U(t) \overset{TL}{\leftrightarrow} \frac{n!}{(s - s_i)^{n+1}} \tag{9.23}$$

For example, if $n = 0$,

$$\mathrm{e}^{s_i t} U(t) \overset{TL}{\leftrightarrow} \frac{1}{s - s_i}, \tag{9.24}$$

if $n = 1$,

$$t\, \mathrm{e}^{s_i t} U(t) \overset{TL}{\leftrightarrow} \frac{1}{(s - s_i)^2}. \tag{9.25}$$

Example Consider the transfer function of a causal system of the form

$$H(s) = \frac{1}{s(s+1)^2}. \tag{9.26}$$

Decomposed into simple elements, $H(s)$ is written

$$H(s) = \frac{1}{s} - \frac{1}{(s+1)^2} - \frac{1}{(s+1)}. \tag{9.27}$$

The filter impulse response is then

$$h(t) = (1 - t e^{-t} - e^{-t}) U(t). \tag{9.28}$$

We note the following important results The domain of definition of the transfer function of a <u>causal</u> system is the half plane s on the right side of the singularity whose real part is the highest.

We recall that a physical system (causal) is <u>stable</u> if all the poles of its transfer function have a negative real part.

A stable physical system (causal) has a frequency response.

9.3.1 Response of a System to Any Input Signal

We know that the response of an LTI system with impulse response $h(t)$ to an input signal $x(t)$ is $y(t) = x(t) \otimes h(t)$. For causal signals and systems (to ensure the convergence of time integrals) or decreasing fast enough at infinity, it is possible to write $y(s) = x(s) H(s)$.

To obtain the time response $y(t)$, we will seek the images of time functions $x(t)$ and $h(t)$, carry out their product to have $y(s)$, and then go back into the time domain to obtain $y(t)$.

Summary
After giving the formulas of the two-sided Laplace transform, we studied on examples the convergence of the integral giving the transform, and determined the domain of definition of the Laplace transform function in the complex s plane. Marginally stable systems have been discussed. We have shown that the differential equations with constant coefficients encountered in electronic circuits Ohm's law take the form of a rational function in the Laplace domain. Formulas for the transition from time domain to Laplace plane are given in a table at the end of the chapter.

Exercise
Consider the marginally stable filter whose frequency response function is given by (9.13).

Let the input signal of this filter be $x(t) = e^{j\omega_0 t} U(t)$. Show that the output is

$$y(t) = t e^{j\omega_0 t} U(t).$$

Solution: A first manner is to use results (9.24) and (9.25).

A second manner is to calculate directly the convolution $y(t) = x(t) \otimes h(t) = e^{j\omega_0 t} U(t) \otimes e^{j\omega_0 t} U(t)$. $y(t) = \int\limits_{-\infty}^{+\infty} e^{j\omega_0 \tau} U(\tau) e^{j\omega_0(t-\tau)} U(t-\tau) \mathrm{d}\tau.$

$y(t) = e^{j\omega_0 t} \int\limits_{-\infty}^{\infty} U(\tau) U(t-\tau) \mathrm{d}\tau = t e^{j\omega_0 t} U(t)$. This result will be used in a further chapter when the Goertzel filter algorithm is detailed.

Tableof Laplace transforms

Function in time	Laplace transform
$\delta(t)$	1
$U(t)$	$\frac{1}{s}$
$t U(t)$	$\frac{1}{s^2}$
$t^n U(t)$	$\frac{n!}{s^{n+1}}$
$a > 0; \quad e^{-at} U(t)$	$\frac{1}{s+a}$
$t e^{-at} U(t)$	$\frac{1}{(s+a)^2}$
$t^n e^{-at} U(t)$	$\frac{n!}{(s+a)^{n+1}}$
$\cos(\omega_0 t) U(t)$	$\frac{s}{s^2 + \omega_0^2}$
$\sin(\omega_0 t) U(t)$	$\frac{\omega_0}{s^2 + \omega_0^2}$
$e^{-at} \cos(\omega_0 t) U(t)$	$\frac{s+a}{(s+a)^2 + \omega_0^2}$
$e^{-at} \sin(\omega_0 t) U(t)$	$\frac{\omega_0}{(s+a)^2 + \omega_0^2}$
$f(t)$	$F(s)$
$f(t - t_0)$	$F(s) e^{-st_0}$
$f(t) \otimes g(t)$	$F(s) G(s)$

Chapter 10
Analog Filters

Analog filters play an important role in signal processing. We deal here with some important, common cases. The analysis will be limited to the case of low-pass filters. We consider first, three classical all-poles filters: Butterworth, Chebyshev, and Bessel. We discuss their performances with regard to the situation of the poles of their transfer functions in the Laplace plane. It appears that the Butterworth filter has the flattest frequency response in the passband. The Chebyshev filter has the shortest transition region between the passband and the attenuated band. The Bessel filter will be used when minimum deformation of the signal through the filter is searched for. The band-pass or high-pass filters are deduced by moving the poles of the transfer function of the low-pass filters in the Laplace plane. The chapter ends with a comparison of the frequency responses of filters from each class.

10.1 Delay of a Signal Crossing a Low-Pass Filter

This paragraph is intended to define the delay of a signal passing through a filter. The reasoning is based on the frequency behavior of low-pass filters near zero frequency.

Consider a low-pass filter with frequency response $H(\omega)$. First of all we seek the approximate shape of the frequency response of the filter in the low-frequency limit. We perform a Taylor expansion of $H(\omega)$ in the vicinity of the zero frequency.

The module $|H(\omega)|$ is an even function. Its value approximated at zero order is

$$|H(\omega)| \cong H(0). \tag{10.1}$$

© Springer International Publishing Switzerland 2016
F. Cohen Tenoudji, *Analog and Digital Signal Analysis*,
Modern Acoustics and Signal Processing, DOI 10.1007/978-3-319-42382-1_10

The argument of $H(\omega)$ (phase shift taken by the signal $e^{j\omega t}$ when crossing the filter) developed to order 1 is written as

$$\varphi(\omega) \cong \varphi(0) + \omega \frac{\mathrm{d}\varphi}{\mathrm{d}\omega}. \qquad (10.2)$$

As $\varphi(\omega)$ is an odd function, continuous in $\omega = 0$, it is zero at $\omega = 0$. Therefore in the vicinity of the zero frequency

$$\varphi(\omega) \cong \omega \frac{\mathrm{d}\varphi}{\mathrm{d}\omega}. \qquad (10.3)$$

$\frac{\mathrm{d}\varphi}{\mathrm{d}\omega}$ has the dimensions of a time. We have seen in the examples of filters of the first and second order that in the vicinity of $\omega = 0$, the phase is a decreasing function of ω.

This leads us to write

$$\left.\frac{\mathrm{d}\varphi}{\mathrm{d}\omega}\right|_{\omega=0} = -\tau, \qquad (10.4)$$

where τ represents a positive time.

As will be explained in a following chapter, this delay is that taken by the envelope of a signal formed by the superposition of harmonic signals in a frequency band when crossing a band-pass filter. It is called the group delay. The phase delay will also be defined and calculated from the phase shift taken by an exponential monochromatic signal. In general, phase and group delays are functions of frequency.

$$\text{Group delay}: \ \tau_g(\omega) = -\frac{\mathrm{d}\varphi(\omega)}{\mathrm{d}\omega}. \qquad (10.5)$$

$$\text{Phase delay}: \ \tau_\varphi(\omega) = -\frac{\varphi(\omega)}{\omega}. \qquad (10.6)$$

Using relations from (10.1) to (10.4), it is thus possible to write in the vicinity of the zero frequency

$$H(\omega) \cong H(0)e^{-j\omega\tau} \qquad (10.7)$$

Filtering of a Low-Frequency Signal

Now we study the effect of the previous filter on a signal $x(t)$ presented to its input. It is assumed now that the spectral range of $x(t)$ does not extend beyond the area of validity of the LF approximation detailed above ($x(t)$ is a signal with very low frequency). In that case one may use the approximation

$$H(\omega) \cong H(0)e^{-j\omega\tau}. \tag{10.8}$$

The output signal $y(t)$ is then

$$y(t) = \frac{1}{2\pi} \int_{-\infty}^{+\infty} Y(\omega)e^{j\omega t}d\omega = \frac{1}{2\pi} \int_{-\infty}^{+\infty} X(\omega)H(\omega)e^{j\omega t}d\omega$$

$$\cong \frac{1}{2\pi} \int_{-\infty}^{+\infty} X(\omega)H(0)e^{-j\omega\tau}e^{j\omega t}d\omega. \tag{10.9}$$

And so

$$y(t) \cong H(0)\frac{1}{2\pi} \int_{-\infty}^{+\infty} X(\omega)e^{j\omega(t-\tau)}d\omega = H(0)x(t-\tau). \tag{10.10}$$

In summary

$$y(t) \cong H(0)x(t-\tau). \tag{10.11}$$

The filter output signal appears (within a multiplicative constant) identical to the delayed input signal by τ.

We treat as exercises at the end of this chapter some cases of distortions appearing when the filter frequency response departs slightly from (10.8).

10.2 Butterworth Filters

A low-pass Butterworth filter is defined from the square modulus of its frequency response

$$A(\omega) = H(\omega)H^*(\omega) = \frac{1}{1+(\omega^2)^n}. \tag{10.12}$$

ω is the normalized frequency $\frac{\omega}{\omega_c}$. n is an integer which defines the filter's order.

Figure 10.1 shows an example of the gain modulus $G(\omega) = \sqrt{A(\omega)}$ when $n = 6$

It is noted that the gain is flat at low frequency, equal to $\frac{1}{\sqrt{2}}$ at $\omega = 1$ (this corresponds to -3 dB attenuation at the cut-off frequency) and that the transition from passing to attenuating bands is quite gradual. The bigger n is the flatter and will gain when $|\omega| < 1$.

Fig. 10.1 Gain magnitude of
a Butterworth filter when
$n = 6$

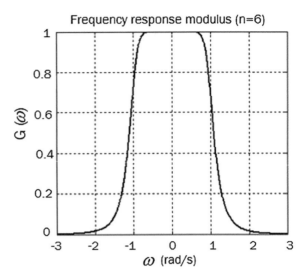

It is assumed that the coefficients of the various powers of s appearing in the transfer function are real, which implies that the poles are real or come in pairs of complex conjugates (arithmetic theorem).

We note $H(\omega)$ the frequency response of the filter that we look for. Let us detail the simple cases where $n = 1$ and 2.

If $n = 1$, $A(\omega) = H(\omega)H^*(\omega) = \frac{1}{1+\omega^2}$; $H(s)$ has one pole s_1.

$$H(\omega)H^*(\omega) = \frac{1}{j\omega - s_1}\frac{1}{-j\omega - s_1^*} = \frac{1}{\omega^2 + j\omega(s_1 - s_1^*) + s_1 s_1^*}.$$

We necessarily have $s_1 - s_1^* = 0$ and $s_1 s_1^* = 1$. It results that s_1 is real and $s_1 = -1$, as the causality of the filter requires a pole with a negative real part.

If $n = 2$

$$A(\omega) = H(\omega)H^*(\omega) = \frac{1}{1 + \omega^4}. \tag{10.13}$$

$H(s)$ has two poles s_1 and s_2. $H(s) = \frac{1}{s-s_1}\frac{1}{s-s_2}$;

$$H(\omega) = \frac{1}{j\omega - s_1}\frac{1}{j\omega - s_2} = \frac{1}{-\omega^2 - j\omega(s_1 + s_2) + s_1 s_2}.$$

To satisfy the canonical form of a second-order filter seen in Chap. 2, it is necessary that $s_2 = s_1^*$ and that $s_1 s_2$ is real. In addition to meet the form (10.13) it is necessary that $s_1 s_2 = 1$.

To ensure causality it is necessary that $\mathrm{Re}\,(s_1) = \mathrm{Re}(s_2) < 0$. It becomes

$$s_1 = -\frac{\sqrt{2}}{2} + j\frac{\sqrt{2}}{2}; \quad s_2 = -\frac{\sqrt{2}}{2} - j\frac{\sqrt{2}}{2}.$$

Let us study now the general cases

- **If n is even**

$A(s)$ will have the form

$$A(s) = \frac{1}{1 + s^{2n}}. \tag{10.14}$$

Let us search first the poles of $A(s)$. The roots of the denominator (roots of -1) are determined by solving the equation

$$s^{2n} = -1. \tag{10.15}$$

We search solutions in the form

$$s = e^{j\theta}. \tag{10.16}$$

It becomes

$$\theta_k = (2k+1)\frac{\pi}{2n} \quad \text{with} \quad k = 0, 1, 2, \ldots, 2n - 1. \tag{10.17}$$

It is noted that there are $2n$ roots on the unit circle. Poles of $A(s)$ come in groups of 4 (Fig. 10.2).

To satisfy the causality condition, the poles of the $H(s)$ are selected as the poles of $A(s)$ with negative real values.

- **If n is odd**

$A(s)$ will have the form

$$A(s) = \frac{1}{1 - s^{2n}}.$$

The poles of $A(s)$ are determined by solving the equation $s^{2n} = 1$.

We search solutions in the form $s = e^{j\theta}$. It becomes $\theta_k = \frac{k\pi}{n}$ with $k = 0, 1, 2, \ldots, 2n - 1$. At least two real poles will exist (Fig. 10.3).

Compared to other filters, Butterworth filters have the property that among all filters their **frequency response curve is the flattest** at zero frequency. On the other hand, the transition zone from passing to attenuating will be less steep than for a Chebyshev filter of the same order.

Fig. 10.2 Symmetry of $A(s)$ poles (n even)

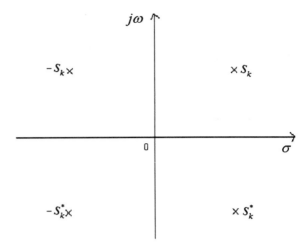

Fig. 10.3 Symmetry of $A(s)$ poles (n odd)

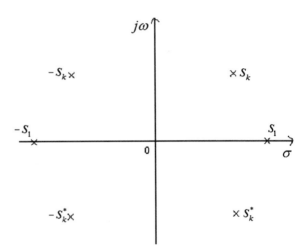

Exercise

Determining a third-order Butterworth low-pass filter with a -3 dB cut-off frequency $f_c = 1$ MHz . We pose $\omega_c = 2\pi f_c = 2\pi \times 10^6$ rad/s.

We have

$$A(\omega) = H(\omega)H^*(\omega) = \frac{1}{1 + \left(\dfrac{\omega}{\omega_c}\right)^6} = \frac{\omega_c^6}{\omega_c^6 + \omega^6}.$$

Since $A(\omega) = A(s)|_{s=j\omega}$, it becomes $A(s) = \dfrac{\omega_c^6}{\omega_c^6 - s^6} = -\dfrac{\omega_c^6}{s^6 - \omega_c^6}$.

All six poles of $A(s)$ are located on a circle of radius ω_c

$$s_k = \omega_c e^{j\theta_k} \quad \text{with} \quad \theta_k = k\frac{\pi}{3} \quad \text{and} \quad k = 0, 1, 2, \ldots, 5.$$

To build the transfer function $H(s)$ of the system assumed causal, we select poles whose real parts are negative

$$s_2 = \omega_c e^{j\frac{2\pi}{3}} = \omega_c\left(-\frac{1}{2} + j\frac{\sqrt{3}}{2}\right), \quad s_3 = \omega_c e^{j\pi} = -\omega_c \quad \text{and}$$

$$s_4 = \omega_c e^{j\frac{4\pi}{3}} = \omega_c\left(-\frac{1}{2} - j\frac{\sqrt{3}}{2}\right).$$

Note that the poles s_2 and s_4 are complex conjugates.

$$H(s) = \frac{\omega_c^3}{(s-s_2)(s-s_3)(s-s_4)} = \frac{\omega_c}{(s-s_3)}\frac{\omega_c^2}{(s-s_2)(s-s_2^*)} = H_1(s)H_2(s).$$

$H(s)$ is taken in the form of the product of a first-order and of a second-order transfer functions (in cascade).

The impulse response of the first-order system is

$$h_1(t) = \omega_c e^{s_3 t} U(t).$$

That of the second-order filter is

$$h_2(t) = \omega_c^2 \frac{1}{(s_2 - s_4)}(e^{s_2 t} - e^{s_4 t})U(t).$$

The product of the transfer functions corresponds to a convolution product of the impulse responses in the time domain. The frequency response is

$$h(t) = h_1(t) \otimes h_2(t).$$

It is necessary to calculate convolutions of the type $s(t) = e^{s_3 t} U(t) \otimes e^{s_2 t} U(t)$. Each convolution being causal, we get for example for $t > 0$:

$$s(t) = \int_{-\infty}^{\infty} e^{s_3(t-\tau)} U(t-\tau) e^{s_2 \tau} U(\tau) d\tau = \int_0^t e^{s_3(t-\tau)} e^{s_2 \tau} d\tau = e^{s_3 t} \int_0^t e^{(s_2 - s_3)\tau} d\tau$$

$$= e^{s_3 t} \frac{e^{(s_2 - s_3)\tau}}{(s_2 - s_3)}\bigg|_0^t = \frac{1}{(s_2 - s_3)}(e^{s_2 t} - e^{s_3 t}).$$

Therefore

$$h(t) = \frac{\omega_c^3}{(s_2 - s_4)} \left(\frac{1}{(s_2 - s_3)} (e^{s_2 t} - e^{s_3 t}) - \frac{1}{(s_4 - s_3)} (e^{s_4 t} - e^{s_3 t}) \right) U(t).$$

We can verify that this impulse response is real.
The frequency response is

$$H(\omega) = \frac{\omega_c}{(j\omega + \omega_c)} \frac{\omega_c^2}{\left(-\omega^2 + j\omega\omega_c + \omega_c^2 \right)}.$$

For the first order, we use a *RC* circuit, $RC = \frac{1}{\omega_c} = 1.59 \times 10^{-7}$ s, with for example $R = 10^3 \Omega$, $C = 1.59 \times 10^{-10}$ F. For the second order, we use a series *RLC* filter

$$LC = \frac{1}{\omega_c^2} = 2.53 \times 10^{-14}.$$

$RC = \frac{1}{\omega_c} = 1.59 \times 10^{-7}$ s, for example $R = 10^3 \Omega$, $C = 1.59 \times 10^{-10}$ F, $L = 1.59 \times 10^{-4}$ H.

10.3   Chebyshev Filters

Type 1 Chebyshev filter that is studied here shows an oscillation of the gain in the passband and has none in the stop band. It is defined from the square modulus of the frequency response

$$A(\omega) = H(\omega)H^*(\omega) = \frac{1}{1 + \varepsilon^2 T_n^2(\omega)}. \tag{10.18}$$

ε is a parameter related to the oscillation in the passband, $T_n(x)$ is a Chebyshev polynomial defined by

$$T_n(x) = \left| \begin{array}{ll} \cos\left(n \text{Arccos}(x)\right) & \text{if } |x| \leq 1 \\ \cosh\left(n \text{Arccosh}(x)\right) & \text{if } |x| > 1 \end{array} \right. . \tag{10.19}$$

$T_n^2(x)$ oscillates between 0 and 1 when $|x| \leq 1$. $A(\omega)$ oscillates between 1 and $\frac{1}{1+\varepsilon^2}$ in that interval.

We note that the Chebyshev filter has the distinctive feature of having a gain with a constant amplitude oscillation in passband.

It will be assumed here that the poles of the transmittance (transfer function) are simple and given by

$$\sigma_k = -\sinh a \sin\left(\frac{2k-1}{2n}\pi\right),\qquad(10.20)$$

and

$$\omega_k = \cosh a \cos\left(\frac{2k-1}{2n}\pi\right),\qquad(10.21)$$

with

$$a = \frac{1}{n}\text{Arcsinh}\left(\frac{1}{\varepsilon}\right)\quad\text{and}\quad k = 1, 2\ldots, n.\qquad(10.22)$$

The poles are located on an ellipse (Chebyshev ellipse) such that

$$\frac{\sigma_k^2}{\sinh^2 a} + \frac{\omega_k^2}{\cosh^2 a} = 1.\qquad(10.23)$$

Numerical Application

If $n = 6$ and $\varepsilon = 0.2$, it becomes $\cosh a = 1.0752$ and $\sinh a = 0.395$.

The equation of the ellipse is

$$\frac{x^2}{0.156} + \frac{y^2}{1.156} = 1.$$

The ellipse is elongated along the axis y. The length of the semi-major axis is 1.075. That of half the minor axis is 0.395 (Fig. 10.4). The poles of the transmittance are closer to the vertical axis than in the case of the Butterworth filter to which the poles are located on a circle.

This is this proximity to the vertical axis which causes oscillations on the gain in the passband of the Chebyshev filter that are not observed for the Butterworth filter (Fig. 10.5). It is also this proximity that ensures the rapid transition of the pass-band to the stopband of the transmittance of the Chebyshev filter.

By various numerical tests, we find that for a given order n, the oscillation in the passband in the case $\varepsilon = 0.1$ is less pronounced than for $\varepsilon = 0.5$, as seen above on the properties of $A(\omega)$ (the poles are furthest from the vertical axis), which is an advantage. But the disadvantage is that the passband—stopband transition is slower.

Fig. 10.4 Transmittance
poles of a Chebyshev filter;
$n = 6$; $\varepsilon = 0.2$

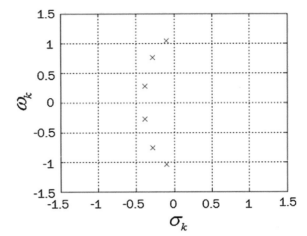

Fig. 10.5 Frequency gain of
a Chebyshev filter; $n = 6$;
$\varepsilon = 0.2$

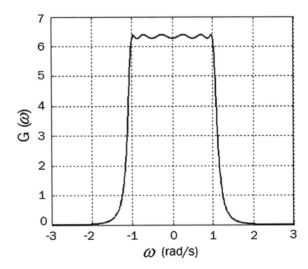

10.4   Bessel Filters

The disadvantage of both previous filters types, especially the Chebyshev filter, is
that the group delay of a component signal spectrum depends on the mean frequency of the signal. This may result in a distortion of the signal or variable delay
of a pulse with frequency in the filter. In the design of the Bessel filter, the
steadiness of the group delay is preferred giving less importance to the filter
selectivity. The idea beneath the reasoning of construction of this filter is the
following. A filter with the transfer function $H(s) = e^{-s\tau}$, and thus the frequency
response $H(\omega) = e^{-j\omega\tau}$, introduce phase and group delays $\tau_g = \tau_\varphi = \tau$ independent
of frequency.

To simplify the analysis it is noted in the following calculation $s\tau = s$. We cannot find an exact expression of $H(s)$ above in the form of a rational fraction. The series expansion of $e^{-s} = \frac{1}{e^s} = \frac{1}{1+s+\frac{s^2}{2}+\dots}$ limited to any order gives insufficient results when s is not very small. We choose the following approximation method; we write

$$e^{-s} = \frac{1}{\cosh s + \sinh s}. \tag{10.24}$$

The development of coth s in continuous fraction is used

$$\coth s = \frac{1}{s} + \frac{1}{\frac{3}{s} + \frac{1}{\frac{5}{s}+\frac{1}{\frac{7}{s}+\dots}}}. \tag{10.25}$$

Limiting the development to order n, an approximate expression of coth s is obtained.

At order 3 we get

$$\coth s \cong \frac{1}{s} + \frac{1}{\frac{3}{s} + \frac{1}{\frac{5}{s}}} = \frac{6s^2 + 15}{s^3 + 15s} \cong \frac{\cosh s}{\sinh s}. \tag{10.26}$$

We identify the numerators and denominators, and we take

$$\cosh s \cong 6s^2 + 15; \quad \sinh s \cong s^3 + 15s. \tag{10.27}$$

These expressions are replaced in the expression of $H(s)$.

At order 3

$$H(s) = \frac{1}{\mathrm{ch}s + \mathrm{sh}s} \cong \frac{1}{s^3 + 6s^2 + 15s + 15}. \tag{10.28}$$

The polynomial in the denominator is a third-order Bessel polynomial.

Limiting the development to first orders, we obtain the following polynomials:

$P_1 = s + 1$

$P_2 = s^2 + 3s + 3.$

$P_3 = s^3 + 6s^2 + 15s + 15.$

$P_4 = s^4 + 10s^3 + 45s^2 + 105s + 105.$

$P_5 = s^5 + 15s^4 + 105s^3 + 420s^2 + 945s + 945.$

$P_6 = s^6 + 21s^5 + 210s^4 + 1260s^3 + 4725s^2 + 10,395s + 10,395.$

$P_7 = s^7 + 28s^6 + 378s^5 + 3150s^4 + 17,325s^3 + 62,370s^2 + 135,135s + 135135.$

$P_8 = s^8 + 36s^7 + 630s^6 + 6930s^5 + 51,975s^4 + 270,270s^3 + 945,945s^2 + 2,027,025s + 2,027,025.$

We obtain the higher order polynomials by recurrence

$$P_n = (2n - 1)P_{n-1} + s^2 P_{n-2}. \tag{10.29}$$

10.5   Comparison of the Different Filters Responses

We choose to do the comparison for eighth order filters. The different behavior of frequency responses is explained by the respective positions of poles which are shown in Fig. 10.6. The poles of the Chebyshev filter (represented by x signs) are close to the imaginary axis. This explains the undulations of the gain that is sensitive to resonance represented by each pole. In the case of Chebyshev filter, the angle between the vector $\overrightarrow{MP_1}$ joining the point M on the ordinate axis $j\omega$ to the highest pole on the figure varies greatly when M passes the ordinate of this pole. This causes a rapid phase change and therefore an important group delay at the edge of the passband. Conversely, the highest frequency Bessel filter pole is relatively distant from the imaginary axis, inducing a slow variation of the phase. The pole with highest frequency of the Butterworth filter is at an intermediate distance from the imaginary axis compared to the two other filter types discussed here inducing an intermediate behavior.

Figure 10.7 shows the modules, phases, and group delays of the three types of filters studied here.

Fig. 10.6 Positions of the poles for the three 8th order filters

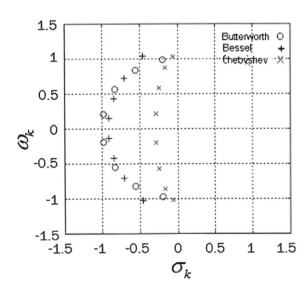

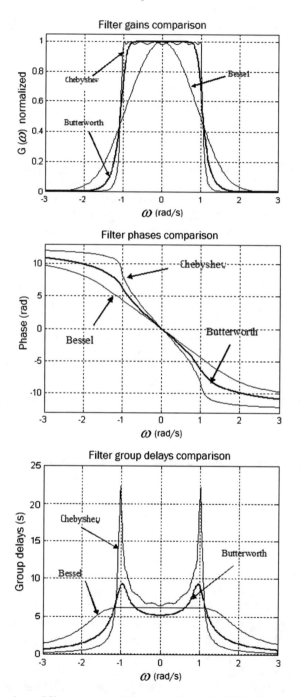

Fig. 10.7 Comparison of filter responses: Gains (*top*); Phases (*middle*); Group delays (*bottom*)

The Bessel filter gain modulus is a bell curve. This filter is poorly selective. The regular variation of the phase induces a constant group delay in the passband (main advantage of the filter).

The gain of the Butterworth filter is flat in the passband (main advantage of the filter), but the variation in the transition band is not very quick (compared to that of Chebyshev filter shown at the bottom of the figure). The pronounced variation of the phase in the vicinities of $|x = 1|$ induces a maximum group delay in these neighborhoods (remember that the group delay is given by the derivative of the phase with respect to ω).

The gain of the Chebyshev filter has oscillations (barely visible in the figure) in the pass bandwidth. The transition, passband to stopband, is very pronounced (main advantage of this filter). The strong variation of the phase in the vicinities of $|x = 1|$ induces a very important group delay in these regions. The oscillations of the group delay are apparent. They are caused by the rippling of the phase for each frequency close to the imaginary part of a pole.

Summary

In this chapter we have studied three classes of analog filters which play an important role in signal processing. Their transfer functions have rational fractions forms. We showed that their filtering properties are explained by the localization of the poles of the transfer functions in the Laplace plane. It appears that the Butterworth filter has the flattest frequency response in the passband. The Chebyshev filter has the shortest transition region between the passband and the attenuated band. The Bessel filter will be used when minimum deformation of the signal through the filter is searched for. Amplitude and phase distortions of signals are briefly discussed in the exercises section.

Exercises

I. A **Butterworth** filter is defined by the square modulus of its frequency response

$$A(\omega) = \frac{1}{1 + \left(\frac{\omega}{\omega_c}\right)^6}.$$

1. Determine the poles of the fraction $A(s)$ and give the expression of the realizable filter frequency response $H(\omega)$ with $A(\omega)$ being its square modulus. What is the filter's order?
2. Draw the graph of the filter gain according to frequency. What are in decibel the gains of the filter at the cut-off frequency f_c ? at $2f_c$?
3. What is the impulse response of this filter?
4. Remind the comparative advantages of Butterworth and Chebyshev filters. (Base the discussion on the pole positions in the Laplace plane).

II. **Chebyshev** filter: The square modulus of the frequency response of type 1 Chebyshev filter is given by $A(\omega) = H(\omega)H^*(\omega) = \frac{1}{1+\varepsilon^2 T_n^2(\omega)}$. ε is a parameter and $T_n(x)$ is a Chebyshev polynomial, oscillating between 0 and 1 when $|x| \leq 1$. Assume that $n = 12$ and $\varepsilon = 0.25$. We use $\text{Arcsinh}(\frac{1}{\varepsilon}) = 2.0947$.

1. Show that the poles of $A(s)$ are located on an ellipse which we determine as the main axes.
2. It is desired that the filter is physically realizable, i.e., causal. How to make the selection of $A(s)$ poles to build the transfer function $H(s)$? Give the expression of $H(s)$, then that of $H(\omega)$. Specify the nature of the filter.
3. Draw the shape of the square modulus of the gain $A(\omega)$ based on the properties of $T_n(x)$. Give an approximate value of the angular frequencies for which $A(\omega)$ has relative maxima.

III. **Bessel** filter: The transfer function of a third-order Bessel filter has the form

$$H(s) = \frac{1}{s^3 + 6s^2 + 15s + 15}.$$

The poles of $H(s)$ are $s_1 = -1.8389 + j1.7544$; $s_2 = -2.3222$; $s_3 = -1.8389 - j1.7544$.
The function $H(s)$ is defined for $\text{Re}(s) > -1.8389$.

1. Can we say that the filter is causal?
   Why can it be said that this filter has a frequency response? Give this frequency response.
2. This filter can result from cascading of two filters with real impulse responses. Give the transfer functions of these filters.
3. Derive the impulse response of the filter of order 3.
   Represent graphically the variations of the modulus and phase of the filter's frequency response.

IV. The squared modulus of the frequency response of an analog filter is denoted $A(\omega)$

$$A(\omega) = H(\omega)\, H^*(\omega).$$

1. Knowing that $A(s)$ is a rational function whose 12 poles are

$$-0.04 \pm j, -0.025 \pm 1.02j, -0.025 \pm 0.98j, 0.04 \pm j,$$
$$0.025 \pm 1.02j, 0.025 \pm 0.98j,$$

   locate these poles in the Laplace plane.

Based on a geometric argument, give the appearance of the frequency-gain modulus and specify the nature of the filter.

2. Build the frequency response of the physically realizable filter. Indicate the principle of the practical realization of this filter.

V. Amplitude response distortion.
The frequency response of an ideal low-pass filter is $H(\omega) = K\,e^{-j\omega t_0}$. We consider a filter with gain amplitude varying slightly in the band-pass. $|H(\omega)| = K(1 + a\cos(\omega\tau/2))$ for $-\frac{2\pi}{\tau} < \omega < \frac{2\pi}{\tau}$ and 0 elsewhere. a is supposed to be small. Evaluate the filter output $y(t)$ for an input $x(t)$.
Solution

$$Y(\omega) = H(\omega)X(\omega) = K\,e^{-j\omega t_0}\left(1 + a\cos(\omega\tau/2)\right)X(\omega).$$

$$Y(\omega) = K\,e^{-j\omega t_0}X(\omega) + K\,e^{-j\omega t_0}\frac{a}{2}e^{j\omega\tau/2}X(\omega) + K\,e^{-j\omega t_0}\frac{a}{2}e^{-j\omega\tau/2}X(\omega).$$

Using the time-delay property

$$y(t) = Kx(t - t_0) + K\frac{a}{2}x(t - t_0 + \tau/2) + K\frac{a}{2}x(t - t_0 - \tau/2).$$

$y(t) = y_0(t) + \frac{a}{2}y_0(t + \tau/2) + \frac{a}{2}y_0(t - \tau/2)$. The output signal is composed of the output in the absence of distortion and of two small echoes surrounding the main component.

VI. First-order phase distortion.
The frequency response of a non ideal low-pass filter is supposed to be $H(\omega) = K\,e^{-j(\omega t_0 - b\sin\omega\tau/2)}$ for $-\frac{2\pi}{\tau} < \omega < \frac{2\pi}{\tau}$ and 0 elsewhere. b is supposed to be small. Evaluate the filter output $y(t)$ for an input $x(t)$.
Solution

$$Y(\omega) = H(\omega)X(\omega) = K\,e^{-j(\omega t_0 - b\sin\omega\tau/2)}X(\omega).$$

We use the following development of the periodic exponential in first kind Bessel functions

$$e^{jb\sin\omega\tau/2} = \sum_{n=-\infty}^{\infty} J_n(b)e^{jn\omega\tau/2}.$$

$J_n(x)$ is given by the polynomial expansion

$$J_n(x) = \sum_{k=0}^{\infty} \frac{(-1)^k(x/2)^{n+2k}}{k!(n+k)!}.$$

For small x, limiting the development at first order we may write

$$J_1(-x) \cong -\frac{x}{2}, J_0(x) \cong 1, J_1(x) \cong \frac{x}{2}, J_n(x) \cong 0 \quad \text{for} \quad n > 1.$$

Then

$$H(\omega) \cong K\,e^{-j\omega t_0}\left(1 + \frac{b}{2}e^{j\omega\tau/2} - \frac{b}{2}e^{-j\omega\tau/2}\right).$$

$y(t) = y_0(t) + \frac{b}{2}y_0(t + \tau/2) - \frac{b}{2}y_0(t - \tau/2)$. In the output signal, two echoes with opposite amplitude surround the main signal component.

Chapter 11
Causal Signals—Analytic Signals

It was shown in Chap. 8 that the impulse response of a physical system is zero for negative time. This follows from the principle of causality: the output of the filter cannot precede the signal that created it, in this case, the Dirac distribution which is zero for negative time. The effect cannot precede the cause. The physical system, which satisfies the principle of causality, is said to be causal. By extension, the impulse response is said to be causal. More generally, we will call causal any function that is null for negative time. The general properties of these functions are discussed here starting from the properties of the Fourier transform of the Heaviside function. In the first paragraph, the Fourier transform of the pseudo-function $1/t$ is carried out, by integration in the complex plane, as a preliminary calculation that leads to the FT of the Heaviside function. We then show that the real and imaginary parts of the Fourier transform of a causal system are related by integration relationship formulas called the Hilbert transform. Analytic signals are defined as having a zero FT at negative frequencies. This notion brings an efficient tool to study several signal modulations and band-pass filtering.

11.1 Fourier Transform of the Pseudo-Function $\frac{1}{t}$

We desire to calculate the Fourier transform of the function $\frac{1}{t}$ defined by the integral

$$\int\limits_{-\infty}^{+\infty} \frac{1}{t} e^{-j\omega t} dt. \tag{11.1}$$

This integral does not converge because of the singularity in $t = 0$ of the integrand and its behavior at infinity. Strictly speaking, the function $\frac{1}{t}$ has no Fourier transform.

© Springer International Publishing Switzerland 2016
F. Cohen Tenoudji, *Analog and Digital Signal Analysis*,
Modern Acoustics and Signal Processing, DOI 10.1007/978-3-319-42382-1_11

We can only define the Cauchy principal value of the integral as

$$I = PV \int_{-\infty}^{+\infty} \frac{1}{t} e^{-j\omega t} dt = \lim_{\substack{\varepsilon \to 0 \\ R \to \infty}} \left\{ \int_{-R}^{-\varepsilon} \frac{1}{t} e^{-j\omega t} dt + \int_{\varepsilon}^{R} \frac{1}{t} e^{-j\omega t} dt \right\}. \qquad (11.2)$$

Then we say that I is the Fourier transform of the pseudo-function $Pf\left(\frac{1}{t}\right)$.

The principal value has an interest since it allows calculating the Fourier transform of the Heaviside function, as seen in the following.

Integration is used in the complex plane on the closed contour Γ compound of the real axis and the two semicircles of radii ε and R (Fig. 11.1)

The function being holomorph inside the contour Γ, the integral over this contour is zero (Cauchy theorem)

$$\underset{(1)}{\int_{-R}^{-\varepsilon} \frac{1}{x} e^{-j\omega x} dx} + \underset{(2)}{\int_{\alpha} \frac{e^{-j\omega z}}{z} dz} + \underset{(3)}{\int_{\varepsilon}^{R} \frac{1}{x} e^{-j\omega x} dx} + \underset{(4)}{\int_{C} \frac{e^{-j\omega z}}{z} dz} = 0. \qquad (11.3)$$

According to the definition of I we have

$$I = \lim_{\substack{\varepsilon \to 0 \\ R \to \infty}} \{(1) + (3)\}. \qquad (11.4)$$

$\lim_{R \to \infty} (4)$ will be zero only if the exponential modulus appearing in the integral can be bounded (Jordan's lemma). This modulus is

$$\left| e^{-j\omega z} \right| = \left| e^{-j\omega(x+jy)} \right| = e^{\omega y} = e^{\omega R \sin \theta}. \qquad (11.5)$$

We see that this modulus is bounded if the condition $\omega y < 0$ is satisfied.

If $\omega < 0$, the Γ contour is closed by the upper semicircle, so on C, one has $\omega y < 0$.

We have on the semicircle C

$$\int_{C} \frac{e^{-j\omega z}}{z} dz = 0. \qquad (11.6)$$

Fig. 11.1 Integration
contour Γ

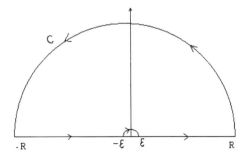

Calculation of integral (2) $\int_{\alpha} \frac{e^{-j\omega z}}{z} dz$

The integral over the semicircle can be set by the angle θ between the radius locating the point on the semicircle and the horizontal.

We can write $z = \varepsilon e^{j\theta}$, and on the circle $dz = \varepsilon e^{j\theta} j \, d\theta$,

$$\int_{\alpha}^{0} \frac{e^{jz\omega}}{z} dz = \int_{\pi}^{0} \frac{e^{j\omega\varepsilon e^{i\theta}}}{\varepsilon e^{j\theta}} e^{j\theta} j\varepsilon d\theta = j \int_{\pi}^{0} e^{j\varepsilon e^{i\theta}} d\theta, \tag{11.7}$$

where, as ε is small,

$$j \int_{\pi}^{0} e^{j\varepsilon e^{i\theta}} d\theta \cong j \int_{\pi}^{0} d\theta = -j\pi. \tag{11.8}$$

On the small semicircle

$$\lim_{\varepsilon \to 0}(2) = \int_{\alpha} \frac{e^{-j\omega z}}{z} dz = -j\pi, \tag{11.9}$$

then

$$I = PV \int_{-\infty}^{+\infty} \frac{1}{t} e^{-j\omega t} dt = j\pi \quad \text{if } \omega < 0. \tag{11.10}$$

If $\omega > 0$, we seek to have $\omega y < 0$ in the integral over C. We close the contour from below to have $y < 0$. Note that the small semicircle is browsed in the opposite direction of the previous case,

$$\int_{\alpha} \frac{e^{-j\omega z}}{z} dz = j\pi. \tag{11.11}$$

It results

$$I = PV \int_{-\infty}^{+\infty} \frac{1}{t} e^{-j\omega t} dt = -j\pi \quad \text{if } \omega > 0. \tag{11.12}$$

In summary

$$I = PV \int_{-\infty}^{+\infty} \frac{1}{t} e^{-j\omega t} dt = -j\pi \, \mathrm{sgn}\,(\omega). \qquad (11.13)$$

With the sign function of ω with the values

$$\mathrm{sgn}\,(\omega) = \begin{vmatrix} -1 & \text{if } \omega < 0 \\ 1 & \text{if } \omega > 0 \end{vmatrix}. \qquad (11.14)$$

We have found that $\frac{i}{\pi} Pf\left(\frac{1}{t}\right)$ has the Fourier transform $\mathrm{sgn}\,(\omega)$.

Now we deduce from the previous result the function whose Fourier transform is $U(\omega)$ Heaviside function in the frequency domain ω. First we see that

$$U(\omega) = \frac{1}{2}(1 + \mathrm{sgn}\,(\omega)). \qquad (11.15)$$

Since $\int_{-\infty}^{+\infty} \delta(t) e^{-j\omega t} dt = 1$, it can be said that 1 is the Fourier transform of $\delta(t)$. Therefore $\frac{1}{2}$ is the Fourier transform of $\frac{1}{2}\delta(t)$.

$$\text{Finally}: \quad \frac{1}{2}\delta(t) + \frac{j}{2\pi} Pf\left(\frac{1}{t}\right) \text{ has the FT } U(\omega). \qquad (11.16)$$

Similarly, we calculate the Fourier Transform of the Heaviside function of time $U(t)$. We can write

$$PV \int_{-\infty}^{+\infty} \frac{1}{\omega} e^{j\omega t} d\omega = j\pi \, \mathrm{sgn}\,(t). \qquad (11.17)$$

The sign change in the exponential induced the change of sign in the second member.

Therefore

$$\frac{1}{2\pi} PV \int_{-\infty}^{+\infty} \frac{1}{\omega} e^{j\omega t} d\omega = \frac{j}{2} \mathrm{sgn}\,(t). \qquad (11.18)$$

Thus, the inverse FT of $Pf\left(\frac{1}{\omega}\right)$ is $\frac{j}{2}\mathrm{sgn}(t)$.

So FT of $\mathrm{sgn}(t)$ is $\frac{2}{j} Pf\left(\frac{1}{\omega}\right)$.

As

$$U(t) = \frac{1}{2}(1 + \mathrm{sgn}\,(t)),$$ (11.19)

its Fourier transform is $\frac{1}{2}\left(2\pi\,\delta(\omega) + \frac{2}{j}Pf\left(\frac{1}{\omega}\right)\right)$. Finally

$$\boxed{U(t)\xrightarrow{\text{Fourier transf}}\pi\,\delta(\omega) + \frac{1}{j}Pf(\frac{1}{\omega}).}$$ (11.20)

11.2 Fourier Transform of a Causal Signal; Hilbert Transform

Let $x(t)$ be a causal signal, i.e. a signal $x(t)$ null for $t < 0$.
  We can write

$$x(t) = x(t)\,U(t) \quad \text{(where } U(t) \text{ is the Heaviside function).}$$ (11.21)

What are the properties of the Fourier transform of $x(t)$?
We recall the formula for the Fourier transform of a product of two functions

$$\int\limits_{-\infty}^{+\infty} x(t)\,U(t)\mathrm{e}^{-j\omega t}\mathrm{d}t = \frac{1}{2\pi}\int\limits_{-\infty}^{+\infty} X(\omega')\,V(\omega - \omega')\mathrm{d}\omega' = \frac{1}{2\pi}X(\omega)\otimes V(\omega),$$ (11.22)

where FT of $U(t)$ was noted $V(\omega)$.

$$V(\omega) = \pi\,\delta(\omega) + \frac{1}{j}Pf\left(\frac{1}{\omega}\right).$$ (11.23)

Taking the Fourier transforms of both sides of the Eq. (11.21),

$$X(\omega) = \frac{1}{2\pi}X(\omega)\otimes V(\omega).$$ (11.24)

We note $X(\omega) = A(\omega) + jB(\omega)$.
It becomes

$$A(\omega) + jB(\omega) = \frac{1}{2\pi}\left((A(\omega) + jB(\omega))\otimes\left(\pi\,\delta(\omega) + \frac{1}{j}Pf\left(\frac{1}{\omega}\right)\right)\right).$$ (11.25)

As $\delta(\omega)$ is the neutral element of the convolution, it becomes

$$\frac{A}{2}(\omega) + \frac{j}{2}B(\omega) = A(\omega) \otimes \frac{1}{2\pi j}Pf\left(\frac{1}{\omega}\right) + B(\omega) \otimes \frac{1}{2\pi}Pf\left(\frac{1}{\omega}\right). \qquad (11.26)$$

Equating the real and imaginary parts of the two members, we obtain the relationship between the real and the imaginary part of the FT of a causal signal

$$\boxed{A(\omega) = \frac{1}{\pi}B(\omega) \otimes Pf\left(\frac{1}{\omega}\right) \quad \text{or} \quad A(\omega) = \frac{1}{\pi}PV \int_{-\infty}^{\infty} B(\omega')\frac{1}{\omega - \omega'}d\omega'.} \qquad (11.27)$$

$$\boxed{B(\omega) = -\frac{1}{\pi}A(\omega) \otimes Pf(\frac{1}{\omega}) \quad \text{and} \quad B(\omega) = -\frac{1}{\pi}PV \int_{-\infty}^{\infty} A(\omega')\frac{1}{\omega - \omega'}d\omega'.}$$

$$(11.28)$$

It is thus seen that the real and imaginary parts of the Fourier transform of a causal signal are not independent. The integral in the second member of Eq. (11.27) is called a Hilbert integral. It is said that the real and imaginary parts $A(\omega)$ and $B(\omega)$ are Hilbert transforms of each other.

A physical system is always causal. The impulse response of a linear physical system is in consequence a causal function. The real and imaginary parts of the frequency response of these systems are related by the relationship demonstrated above. These relations are known in electromagnetism as the Kramers-Kronig relations, connecting the real and imaginary parts of the dielectric constant of a propagation medium that acts as a linear filter.

Relationship between the modulus and phase of the frequency response of a minimum phase filter

A minimum phase filter is such that all the poles and zeros of its transfer function $H(s)$ are located left of the imaginary axis.

The filter frequency response is noted $H(\omega)$. Using the magnitude and phase of $H(\omega)$

$H(\omega) = |H(\omega)|e^{j\varphi(\omega)}$. Taking the logarithm: $\log H(\omega) = \log|H(\omega)| + j\varphi(\omega)$.

All poles and zeros of $H(s)$ being to the left of the imaginary axis, the function $\log|H(s)|$ remains finite for $\text{Re}(s) > 0$, which means that $\log H(s)$ has all its zeros and poles to the left of the imaginary axis. Then, this function is the Laplace transform of a causal function. This property leads, as has been shown previously, that the real and imaginary parts of $\log H(\omega)$ are Hilbert transforms of each other. So we can write

$$\varphi(\omega) = -\frac{1}{\pi}PV \int_{-\infty}^{\infty} \frac{\log|H(\omega')|}{\omega - \omega'}d\omega'. \qquad (11.29)$$

This formula is very appealing because it implies that if we measure the amplitude versus the frequency of a phenomenon, we can deduce the phase variation law. However, this relationship is rarely used in practice because we must be able to experimentally measure the amplitude across the whole frequency axis, from minus infinity to plus infinity, which is rarely possible in practice.

However, this formula allows the evaluation of asymptotic behaviors of the module or phase in some cases.

Application example

What is the causal signal whose Fourier transform real part is a rectangular function?

We denote $\Pi_{2\Omega}(\omega)$ this function equal to 1 for $|\omega| < \Omega$ and zero elsewhere (Fig. 11.2)

$$A(\omega) = \Pi_{2\Omega}(\omega). \tag{11.30}$$

$$B(\omega) = -\frac{1}{\pi} PV \int_{-\Omega}^{+\Omega} \frac{1}{\omega - \omega'} \, d\omega'. \tag{11.31}$$

The result is (Fig. 11.3)

$$B(\omega) = \frac{1}{\pi} \text{Log} \left| \frac{\omega - \Omega}{\omega + \Omega} \right|. \tag{11.32}$$

The demonstration is instructive. For example, we study the case where the singularity is within the interval of integration

$$-\Omega < \omega < \Omega.$$

Then

$$PV \int_{-\Omega}^{+\Omega} \frac{d\omega'}{\omega - \omega'} = \lim_{\varepsilon \to 0} \left(\underbrace{\int_{-\Omega}^{\omega-\varepsilon} \frac{d\omega'}{\omega - \omega'}}_{(1)} + \underbrace{\int_{\omega+\varepsilon}^{\Omega} \frac{d\omega'}{\omega - \omega'}}_{(2)} \right) \tag{11.33}$$

Fig. 11.2 Rectangular window in frequency domain

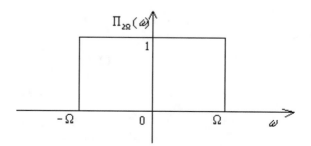

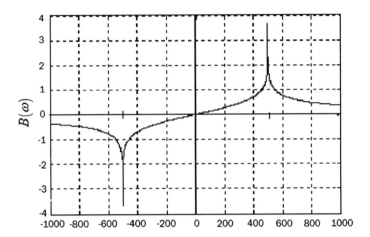

Fig. 11.3 Imaginary part $B(\omega)$ of analytic signal

Note that the principal value was used by taking ε common to lower and upper bounds.

In the first integral $\omega - \omega' > 0$; we set $X = \omega - \omega'$; $dX = -d\omega'$.

$$(1) = \int\limits_{\omega+\Omega}^{\varepsilon} -\frac{dX}{X} = \int\limits_{\varepsilon}^{\omega+\Omega} \frac{dX}{X} = \mathrm{Log}\,\frac{\omega+\Omega}{\varepsilon} \tag{11.34}$$

Note that when $\varepsilon \to 0$ this term tends to infinity.

In the second integral $\omega - \omega' < 0$; we set $X = \omega' - \omega$; $dX = d\omega'$.

$$(2) = \int\limits_{\varepsilon}^{\Omega-\omega} -\frac{dX}{X} = -\mathrm{Log}\,\frac{\Omega-\omega}{\varepsilon} \tag{11.35}$$

Again, this expression tends to infinity when $\varepsilon \to 0$, but the sum of (1) and (2) is limited.

There is compensation between the two diverging values

$$(1) + (2) = \mathrm{Log}\,\frac{\omega+\Omega}{\Omega-\omega}. \tag{11.36}$$

Thus

$$B(\omega) = -\frac{1}{\pi}\mathrm{Log}\frac{\omega+\Omega}{\Omega-\omega} = \frac{1}{\pi}\mathrm{Log}\frac{\Omega-\omega}{\omega+\Omega}. \tag{11.37}$$

One could look at the other cases for the situation of ω to find the complete formula (11.32) which contains absolute values.

Determination of $x(t)$

$$X(\omega) = A(\omega) + jB(\omega) = A(\omega) \otimes \left[\delta(\omega) - \frac{j}{\pi}Pf\left(\frac{1}{\omega}\right)\right] \tag{11.38}$$

$$X(\omega) = A(\omega) \otimes \frac{1}{\pi}\left[\pi\delta(\omega) + \frac{1}{j}Pf\left(\frac{1}{\omega}\right)\right] = A(\omega) \otimes \frac{1}{\pi}\mathcal{F}(U(t)) \tag{11.39}$$

Noting $a(t)$ the inverse FT of $A(\omega)$,

$$X(\omega) = \frac{1}{\pi}\mathcal{F}(a(t)) \otimes \mathcal{F}(U(t)) \quad \text{or} \quad X(\omega) = 2\mathcal{F}(a(t)\,U(t)).$$

and finally

$$x(t) = 2a(t)U(t) \tag{11.40}$$

It remains to calculate $a(t)$

$$a(t) = \frac{1}{2\pi}\int_{-\Omega}^{\Omega} e^{j\omega t}d\omega = \frac{1}{2\pi}\frac{e^{j\Omega t} - e^{-j\Omega t}}{jt} = \frac{1}{\pi t}\sin\Omega t \tag{11.41}$$

we deduce the signal

$$x(t) : \; x(t) = \frac{2}{\pi t}U(t)\sin(\Omega t). \tag{11.42}$$

11.3   Paley-Wiener Theorem

This theorem states a necessary and sufficient condition for causality of a signal from the modulus of its Fourier transform. Suffice it to state

Theorem *A necessary and sufficient condition for $x(t)$ be causal is that the following integral converges.*

$$\int\limits_{-\infty}^{+\infty} \frac{|\text{Log}|X(\omega)||}{1+\omega^2} \, d\omega < \infty. \tag{11.43}$$

An immediate application is that no physically realizable filters (i.e., causal) can have a transfer function with a modulus equal to zero over a whole frequency interval. This is an important result in signal analysis.

The Paley-Wiener integral can admit singularities at isolated points ($|\text{Log}(0)| = \infty$) *but not on a whole segment.*

We can deduce that we cannot find a physical filter *that completely eliminates a frequency band. A low-pass filter with the rectangular response* $\Pi_{2\Omega}(\omega) = 1$ *for* $-\Omega < \omega < \Omega$ *and zero elsewhere, cannot be physically realized.*

11.4   Analytic Signal

This is by definition a signal whose spectral amplitude is zero for negative frequencies. We can then write

$$X(\omega) = X(\omega)\, U(\omega) \tag{11.44}$$

In the time domain we have

$$x(t) = x(t) \otimes \mathcal{F}^{-1}(U(\omega)). \tag{11.45}$$

$$x(t) = x(t) \otimes \left(\frac{1}{2}\delta(t) + \frac{j}{2\pi} Pf\left(\frac{1}{t}\right) \right). \tag{11.46}$$

$x(t)$ is necessarily complex. We note

$$x(t) = x_1(t) + jx_2(t). \tag{11.47}$$

It becomes

$$x_1(t) + jx_2(t) = \frac{1}{2}x_1(t) + j\frac{1}{2}x_2(t) + (x_1(t) + jx_2(t)) \otimes \frac{j}{2\pi} Pf\left(\frac{1}{t}\right), \tag{11.48}$$

$$x_1(t) + jx_2(t) = \frac{j}{\pi}x_1(t) \otimes Pf\left(\frac{1}{t}\right) - \frac{1}{\pi}x_2(t) \otimes Pf\left(\frac{1}{t}\right). \tag{11.49}$$

By identifying the real and imaginary parts of two members

$$x_1(t) = -\frac{1}{\pi} x_2(t) \otimes Pf\left(\frac{1}{t}\right) \qquad (11.50)$$

$$x_2(t) = \frac{1}{\pi} x_1(t) \otimes Pf\left(\frac{1}{t}\right). \qquad (11.51)$$

Concept of instantaneous frequency of a signal

Let $x_1(t)$ a real signal and $x(t)$ the analytic signal associated with $x_1(t)$.

We call instantaneous frequency of the signal $x_1(t)$ the quantity

$$f_0 = \frac{1}{2\pi} \frac{d}{dt} \text{Arg}(x(t)).$$

Important properties

1.
$$X(\omega) = \mathcal{F}(x_1(t) + jx_2(t)), \qquad (11.52)$$

$$X(\omega) = \mathcal{F}(x_1(t)) + j\mathcal{F}\left(\frac{1}{\pi} x_1(t) \otimes Pf\left(\frac{1}{t}\right)\right), \qquad (11.53)$$

$$X(\omega) = \mathcal{F}\left(x_1(t) \otimes \left(\delta(t) + \frac{j}{\pi} Pf\left(\frac{1}{t}\right)\right)\right), \quad \text{or} \quad X(\omega) = 2X_1(\omega) U(\omega).$$
$$\qquad (11.54)$$

So $X(\omega)$ is obtained by taking twice the spectrum of $x_1(t)$ for positive frequencies.

2. The analytic signal can be rewritten

$$x(t) = x_1(t) + jx_2(t).$$

Its FT is written

$$X(\omega) = X_1(\omega) + jX_2(\omega) = 2X_1(\omega) U(\omega). \qquad (11.55)$$

So we must have for negative frequencies $X_1(\omega) + jX_2(\omega) = 0$, so

$$X_2(\omega) = jX_1(\omega) \quad \text{if } \omega < 0. \qquad (11.56)$$

For positive frequencies

$$X_1(\omega) + jX_2(\omega) = 2X_1(\omega),$$

so

$$X_2(\omega) = -jX_1(\omega) \quad \text{if } \omega > 0. \tag{11.57}$$

In summary

$$X_2(\omega) = -j \operatorname{sgn}(\omega)X_1(\omega). \tag{11.58}$$

Application

We calculate the analytic signal whose real part is $x_1(t) = \cos \omega_0 t$ (See Fig. 11.4).

$$X_1(\omega) = \int\limits_{-\infty}^{+\infty} \cos(\omega_0 t)e^{-j\omega t} \, dt = \frac{1}{2} \int\limits_{-\infty}^{+\infty} e^{-j(\omega - \omega_0)t} \, dt + \frac{1}{2} \int\limits_{-\infty}^{+\infty} e^{-j(\omega + \omega_0)t} \, dt,$$
$$\tag{11.59}$$

$$X_1(\omega) = \pi \, \delta(\omega - \omega_0) + \pi \, \delta(\omega + \omega_0). \tag{11.60}$$

The analytic signal is then, according to (11.54) (Fig. 11.5)

$$X(\omega) = 2\pi\delta(\omega - \omega_0). \tag{11.61}$$

Fig. 11.4 Real part $X_1(\omega)$ of analytic signal

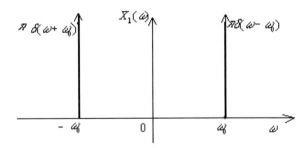

Fig. 11.5 Analytic signal $X(\omega)$

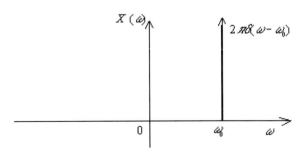

Taking the inverse FT, the analytic signal in the time domain is

$$x(t) = e^{j\omega_0 t}. \tag{11.62}$$

$x_1(t)$ is the real part of $x(t)$

$$x_1(t) = \cos \omega_0 t, \tag{11.63}$$

$x_2(t)$ is the imaginary part of $x(t)$

$$x_2(t) = \sin \omega_0 t. \tag{11.64}$$

The signal $x_2(t)$ is in phase quadrature with $x_1(t)$.

Exercise
Check that the FTs of $x_1(t) = \cos \omega_0 t$ and of $x_2(t) = \sin \omega_0 t$ satisfy the relationships (11.50) and (11.51).

Note

$$\sqrt{x_1^2(t) + x_2^2(t)} = 1. \tag{11.65}$$

The analytic signal modulus is constant and gives the amplitude of the cosine. Since

$$\mathrm{Arg}\left(\frac{x_2(t)}{x_1(t)}\right) = \omega_0 t, \tag{11.66}$$

the analytic signal argument is used to calculate the instantaneous phase.
By a derivative with respect to time we obtain the instantaneous frequency

$$\omega_0 = \frac{\mathrm{d}}{\mathrm{d}t}\mathrm{Arg}\left(\frac{x_2(t)}{x_1(t)}\right), \tag{11.67}$$

which is also in this case

$$\omega_0 = \frac{1}{t}\mathrm{Arg}\left(\frac{x_2(t)}{x_1(t)}\right). \tag{11.68}$$

Signal with slowly varying frequency Let the signal whose frequency varies with time

$$x_1(t) = \cos(\omega_0(t)\, t). \tag{11.69}$$

If the angular frequency slowly changes over time, we may write

$$x(t) \cong e^{j\omega_0(t)t}, \tag{11.70}$$

whereas previously

$$\omega_0(t) = \frac{1}{t}\mathrm{Arg}\left(\frac{x_2(t)}{x_1(t)}\right). \tag{11.71}$$

It is thus possible to follow the evolution of the signal frequency with time.

11.4.1 Instantaneous Frequency of a Chirp

Consider the signal

$$x_1(t) = \cos\left(\omega_0 t + \frac{\beta}{2}t^2\right). \tag{11.72}$$

The associated analytic signal is

$$x(t) = e^{j\left(\omega_0 t + \frac{\beta}{2}t^2\right)}. \tag{11.73}$$

$$\mathrm{Arg}(x(t)) = \left(\omega_0 t + \frac{\beta}{2}t^2\right). \tag{11.74}$$

The instantaneous frequency is

$$f(t) = \frac{1}{2\pi}(\omega_0 + \beta t). \tag{11.75}$$

In this chirp signal, the frequency increases linearly with time.

In Fig. 11.6 are shown the real and imaginary parts of the analytic signal of a chirp used in practice. The third figure shows the spectral amplitude and the fourth shows the instantaneous frequency of the signal calculated using formula (11.67).

11.4.2 Double-Sideband (DSB) Signal Modulation

Consider the signal $x_1(t)$ resulting from the product of an amplitude $a(t)$ by a carrier signal

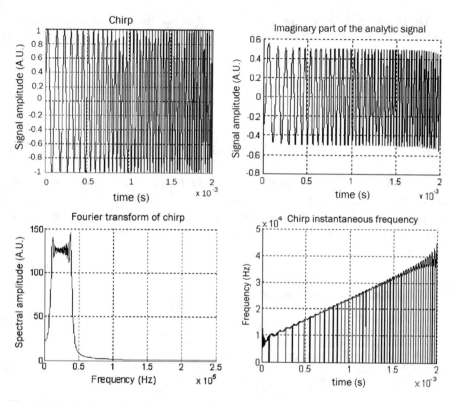

Fig. 11.6 **Left column** *Top* Chirp; *Bottom* Its Fourier transform; **Right column** *Top* Imaginary part of analytic signal; *Bottom* Instantaneous frequency

$$\cos \omega_0 t : \ x_1(t) = a(t) \cos \omega_0 t. \tag{11.76}$$

We assume that the variation with time of $a(t)$ is slow compared to that of $\cos \omega_0 t$. More precisely, it is assumed that the maximum frequency σ present in the spectrum of $a(t)$ is less than $\frac{\omega_0}{2}$: $\sigma < \frac{\omega_0}{2}$. The spectrum of $a(t)$ shown on Fig. 11.7 represents this condition

Fig. 11.7 Frequency limitation of spectrum $A(\omega)$

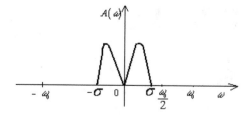

The Fourier transform formula for a product of two functions of time is used to calculate $X_1(\omega)$:

$$X_1(\omega) = \frac{1}{2\pi} A(\omega) \otimes [\pi\delta(\omega - \omega_0) + \pi\delta(\omega + \omega_0)], \qquad (11.77)$$

therefore (see Fig. 11.8)

$$X_1(\omega) = \frac{1}{2} A(\omega - \omega_0) + \frac{1}{2} A(\omega + \omega_0). \qquad (11.78)$$

We seek to determine the analytic signal of which $x_1(t)$ is the real part. As the maximum frequency in $A(\omega)$ is less than $\frac{\omega_0}{2}$, the function $A(\omega - \omega_0)$ is zero for negative frequencies and the function $A(\omega + \omega_0)$ is zero for positive frequencies (see Fig. 11.8). In this case, the analytic signal is (Fig. 11.9)

$$X(\omega) = A(\omega - \omega_0). \qquad (11.79)$$

In the time domain, the analytic signal is

$$x(t) = a(t)\, e^{j\omega_0 t} \qquad (11.80)$$

By expanding the exponential we have

$$x(t) = a(t)\, (\cos \omega_0 t + j \sin \omega_0 t). \qquad (11.81)$$

Fig. 11.8 Real part $X_1(\omega)$ of analytic signal

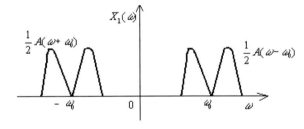

Fig. 11.9 Analytic signal $X(\omega)$ in DSB modulation

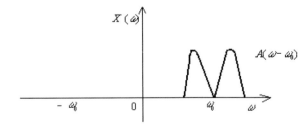

So:

$$x_1(t) = a(t)\cos\omega_0 t \quad\text{and}\quad x_2(t) = a(t)\sin\omega_0 t. \tag{11.82}$$

It follows that one can obtain the modulus of the modulation signal

$$|a(t)| = \sqrt{x_1^2(t) + x_2^2(t)}; \tag{11.83}$$

The absolute value of the envelope is obtained by taking the modulus of the analytic signal. The pulsation of the carrier is obtained by

$$\text{Arg}\left(\frac{x_2(t)}{x_1(t)}\right) = \omega_0 t; \quad \omega_0 = \frac{d}{dt}\text{Arg}\left(\frac{x_2(t)}{x_1(t)}\right). \tag{11.84}$$

Remark We will see in the exercises on amplitude modulation at the end of this chapter that it is possible to retrieve the modulation function and not only its magnitude by adding a constant to the modulation function.

11.4.3 Single-Sideband Signal Modulation (SSB)

We return to the previous example of amplitude modulation. Let $a(t)$ a real signal whose spectrum is limited to the low frequencies. $A(\omega) = 0$ for $|\omega| > \omega_0$. Let $x_1(t)$ be the signal $x_1(t) = a(t)\cos\omega_0 t$. As noted above, we have (Fig. 11.10)

$$X_1(\omega) = \frac{1}{2}A(\omega - \omega_0) + \frac{1}{2}A(\omega + \omega_0). \tag{11.85}$$

The analytic signal is (Fig. 11.11):

$$X(\omega) = 2X_1(\omega)U(\omega) = A(\omega - \omega_0). \tag{11.86}$$

Fig. 11.10 Amplitude modulated spectrum $X_1(\omega)$

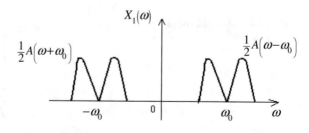

Fig. 11.11 Analytic
amplitude modulated
spectrum $X(\omega)$

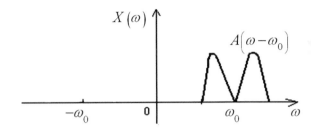

Calculating the inverse Fourier transform of $X(\omega)$ we obtain

$$x(t) = x_1(t) + jx_2(t) = a(t)e^{j\omega_0 t}, \tag{11.87}$$

so:

$$x_2(t) = a(t)\sin\omega_0 t.$$

$a(t)$ being real, it has the property

$$A(-\omega) = A^*(\omega).$$

It follows that $|X(\omega)|$ is symmetrical with respect to ω_0.

It is seen that the positive frequency spectrum contains all the information on $a(t)$.

Let us note (see Fig. 11.12)

$$Z(\omega) = 2A(\omega)U(\omega). \tag{11.88}$$

$z(t) = a(t) + j\tilde{a}(t)$, where $\tilde{a}(t)$ is the Hilbert transform of $a(t)$.

One can choose to limit the size of the frequency band during transmission, by transmitting only the signal $s(t)$ which is determined by the following relationships (Fig. 11.13):

$$Z_S(\omega) = Z(\omega - \omega_0).$$

Fig. 11.12 Single side band
analytic $Z(\omega)$

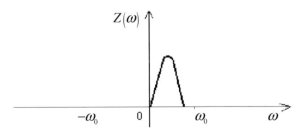

Fig. 11.13 Single side band modulated spectrum $Z_S(\omega)$

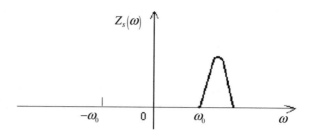

We have

$$z_S(t) = z(t)e^{j\omega_0 t} = (a(t) + j\tilde{a}(t))e^{j\omega_0 t}. \tag{11.89}$$

$s(t)$, the real part of $z_S(t)$, is written using the assumption that $a(t)$ is real

$$s(t) = a(t)\cos\omega_0 t - \tilde{a}(t)\sin\omega_0 t. \tag{11.90}$$

In reception, we can recover $a(t)$ from $s(t)$. Indeed multiplying $s(t)$ by $2\cos\omega_0 t$, we get

$$\begin{aligned}
2s(t)\cos\omega_0 t &= 2a(t)\cos^2\omega_0 t - 2\tilde{a}(t)\cos\omega_0 t\sin\omega_0 t \\
&= a(t) + a(t)\cos 2\omega_0 t - \tilde{a}(t)\sin 2\omega_0 t.
\end{aligned} \tag{11.91}$$

By low-pass filtering, it is possible to recover $a(t)$ at reception.

A delicate problem encountered in practice is that one does not always know a priori the frequency ω_0 used in the generation of the signal by modulation. It may be necessary to 'find' it at the reception. Furthermore a shift in frequency between the transmission and reception is accompanied by a distortion of the received signal (in SSB radio transmission, deformation of the voice is known as 'Donald Duck voice').

11.4.4 Band-pass Filtering of Amplitude Modulated Signal

Let $f(t)$ be a real signal as input to a system with the real impulse response $h(t)$. The output $g(t)$ of the system is given by $g(t) = f(t) \otimes h(t)$.

The different analytic signals in frequency domain are noted $Z_i(\omega)$, with

$$\begin{aligned}
Z_f(\omega) &= 2U(\omega)F(\omega); \quad Z_h(\omega) = 2U(\omega)H(\omega); \\
Z_g(\omega) &= 2U(\omega)G(\omega).
\end{aligned} \tag{11.92}$$

As $G(\omega) = H(\omega)F(\omega)$, we may write:

$$Z_g(\omega) = Z_h(\omega)F(\omega) = Z_f(\omega)H(\omega) = \frac{1}{2}Z_h(\omega)Z_f(\omega). \qquad (11.93)$$

Noting $z_g(t)$, $z_f(t)$ and $z_h(t)$, the temporal analytic signals corresponding respectively to the signals $g(t), f(t)$, and $h(t)$, we have the relations

$$z_g(t) = z_h(t) \otimes f(t) = z_f(t) \otimes h(t) = \frac{1}{2}z_h(t) \otimes z_f(t). \qquad (11.94)$$

Thus $g(t)$ can be obtained by one of the preceding convolutions.

For example, we see that if $f(t)$ is given by $f(t) = a(t)\cos \omega_0 t$ and since $h(t)$ is real, the filter output will be

$$g(t) = \mathrm{Re}\big(a(t)e^{j\omega_0 t}\big) \otimes h(t). \qquad (11.95)$$

It is assumed that the filter with impulse response $h(t)$ is band-pass, with frequency response $H(\omega)$, with modulus $|H(\omega)|$ and phase $\varphi(\omega)$. An example of modulus and phase for $H(\omega)$ is given in Fig. 11.14

We build the corresponding low-pass filter $H_b(\omega)$ defined by (Fig. 11.15):

$$H_b(\omega) = Z_h(\omega + \omega_0)e^{-j\varphi_0}, \quad \text{where} \quad \varphi_0 = \varphi(\omega_0). \qquad (11.96)$$

It is assumed in the following that $H(\omega)$ is a symmetric filter in the sense that $H_b(-\omega) = H_b^*(\omega)$. In this case, we remark that

$$Z_h(\omega) = H_b(\omega - \omega_0)e^{j\varphi_0}, \qquad (11.97)$$

and therefore $z_h(t) = h_b(t)e^{j(\omega_0 t + \varphi_0)}$, where $h_b(t)$ is the impulse response of the low-pass filter. As $h_b(t)$ is real, we have, taking the real part of $z_h(t)$:

$$h(t) = h_b(t)\cos(\omega_0 t + \varphi_0). \qquad (11.98)$$

Fig. 11.14 Band-pass filter response $H(\omega)$

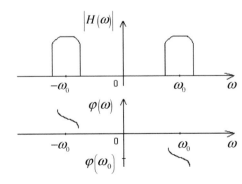

Fig. 11.15 Low-pass filter
response $H_b(\omega)$

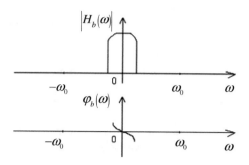

We now assume that the system input is the modulated signal $f(t)$ given by

$$f(t) = a(t) \cos \omega_0 t.$$

Then

$$Z_f(\omega) = A(\omega - \omega_0), \quad Z_h(\omega) = H_b(\omega - \omega_0)e^{j\varphi_0}.$$

We have

$$Z_g(\omega) = \frac{1}{2}Z_h(\omega)Z_f(\omega) = \frac{1}{2}A(\omega - \omega_0)H_b(\omega - \omega_0)^{j\varphi_0}$$
$$= \frac{1}{2}G_b(\omega - \omega_0)^{j\varphi_0},$$

and therefore

$$g(t) = \frac{1}{2}\mathrm{Re}\left(g_b(t)e^{j(\omega_0 t + \varphi_0)}\right). \tag{11.99}$$

We have written $g_b(t) = a(t) \otimes h_b(t)$.

The response of the system is a modulated signal whose envelope is the response
of the LF equivalent system filtering the envelope $a(t)$ of the input signal (within a
½ factor).

11.5   Phase and Group Time Delays

General property

We consider a low-pass filter and we assume that the spectrum of the input signal is
limited to very low frequencies. To estimate the filter response to such a signal, the
gain amplitude can be approximated by a constant and the phase by a linear law

$$|H(\omega)| \cong H(0) \quad \text{and} \quad \varphi(\omega) = -\omega\tau.$$

We then see that the system acts as a pure time shift filter.

In the case of the low-pass filter $H_b(\omega)$, the output is written in these conditions

$$g_b(t) = H_b(0)a(t - \tau).$$

In the case of a frequency response of the filter $H(\omega)$ operating in a frequency band around ω_0, we write

$$t_g(\omega_0) = -\left.\frac{d\varphi}{d\omega}\right|_{\omega=\omega_0} \quad \text{and} \quad t_p(\omega_0) = -\left.\frac{\varphi(\omega)}{\omega}\right|_{\omega=\omega_0} \tag{11.100}$$

It is now shown that t_g is the delay of the signal envelope (group delay) and t_p is the delay of one of the spectral components constituting the group (phase delay).

The phase introduced by the equivalent low-pass filter $H_b(\omega)$ described above has the form

$$\varphi_b(\omega) = \varphi(\omega + \omega_0) - \varphi_0(\omega).$$

We have

$$\varphi_b'(0) = \varphi'(\omega_0) = -t_g.$$

The derivative of the phase of the envelope has the value $-t_g$.

Furthermore

$$H_b(0) = Z_h(\omega_0)e^{-j\varphi_0}.$$

The response of the equivalent low-pass filter has the form $g_b(t) \cong 2|H(\omega_0)|a(t - t_g)$; it is real by hypothesis.

Since we have

$$g(t) = \frac{1}{2}\text{Re}\left(g_b(t)e^{j(\omega_0 t + \varphi_0)}\right),$$

it follows

$$g(t) \cong |H(\omega_0)|a(t - t_g)\cos(\omega_0 t + \varphi_0), \tag{11.101}$$

and therefore:

$$g(t) \cong |H(\omega_0)|a(t - t_g)\,\cos\omega_0(t - t_p).$$

The output signal consists of the envelope delayed by the group delay t_g, modulating a cosine shifted in phase by $\varphi_0 = -\omega_0 t_p$. t_p is the phase delay introduced by the filter.

11.6   Decomposition of a Voice Signal by a Filter Bank

The following example is an illustration of the modulation techniques used in the processing of voice signals. As explained above, the ear is sensitive only to the frequency content of an audible signal. A series of treatments in the frequency domain is acceptable if it renders the original spectrum, even without respect to the temporal shape of the signals. This is what allows the use of filter banks in speech processing. Nowadays, recording and transmission of voice signals are mainly digital. These actions are preceded by the conversion of analog signals into digital signals in a first step. This is detailed in the second part of the book devoted to digital signals. It is shown that the quality of the conversion depends on the number of quantization of the analog/digital converter levels. The greater the number of levels, the higher is the quality of the coding. But this accuracy requires a high number of quantization bits. In many applications, it is desired that the coding is done economically by reducing the number of bits while maintaining a sufficient quality during playback.

This is the case when the information transfer is limited by the throughput of the transmission channel. One is then led to divide the audio signal frequency band into subbands, to digitize, encode, and process signals contained in each of the subbands, and to reconstruct the audio signal at the end of the processing chain. A typical example is the following treatment:

Let us note $s(t)$ an original voice signal. The frequency band characteristic of the signal is divided into subbands by a filter bank (4 in Fig. 11.16).

We denote $s_n(t)$ the result of filtering of the original signal by the nth band-pass filter in $\{f_n, f_n + B_n\}$ whose impulse response is noted $h_n(t)$.

We thus have

$$s_n(t) = h_n(t) \otimes s(t). \tag{11.102}$$

Fig. 11.16 Example of filter bank

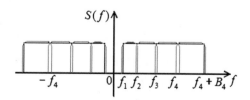

Fig. 11.17 a nth filter
response; **b** modulation
spectrum; **c** baseband
spectrum

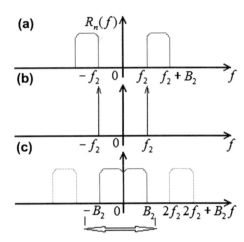

To bring this signal in the baseband it is multiplied by $\cos \omega_n t$. We note $r_n(t)$ this product

$$r_n(t) = s_n(t) \cos \omega_n t.$$

Figure 11.17 shows for the band $\{f_2, f_2 + B_2\}$, in a) the spectrum of $r_n(t)$, in b) and that of $\cos \omega_n t$ and in c) that of their product. A low-pass filtering allows to keep only the central portion of the spectrum $\{-B_2, B_2\}$. The signal from the low-pass filter is converted digitally.

It has been shown that the quantization noise was less impeding in this subband coding, and that one could use converters with a reduced number of bits. One gains in volume of information and in transfer speed. The signal is transmitted in digital form.

Fig. 11.18 a Baseband
spectrum; **b** modulation
spectrum; **c** band
reconstructed spectrum

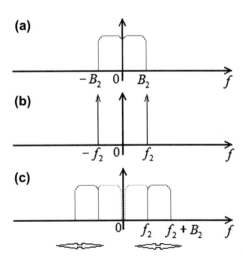

Results are presented in Fig. 11.18. At signal reception, each component in the baseband (spectrum Fig. 11.18a) for the subband 2 in the figure), is multiplied by $\cos \omega_n t$ (spectrum given in Fig. 11.18b). Pass-band filtering finally returns the subband (Fig. 11.18c). The final signal is reconstructed by adding the components of the different bands.

Summary

We have studied in this chapter the properties of causal functions which are null for negative time and analytic functions which are null for negative frequency. The general properties of these functions have been derived from the properties of the Fourier transform of the Heaviside function. That Fourier transform has been evaluated from the FT of the pseudo-function $1/t$. The real and imaginary parts of the FT of a causal system have been shown to be related by integration relationship formulas called the Hilbert transform. The properties of analytic signals have been used to study several types of signal modulations and band-pass filtering. The group and phase delays of the output signals of band-pass filters have been introduced. Frequency modulation is met in an exercise.

Exercises

I. Amplitude modulation.

A signal $f_1(t)$ comprises a carrier with angular frequency ω_0 modulated in amplitude by the signal $1 + \cos \Omega t; f_1(t) = (1 + \cos \Omega t) \cos \omega_0 t$.

1. Represent the appearance of $f_1(t)$ when $\Omega \ll \omega_0$.
2. What is the spectrum (representation in the Fourier domain) of the signal $f_1(t)$? Graph the spectrum assuming that $\Omega < \omega_0$. Give the bandwidth of the signal $f_1(t)$.
3. What is the analytic signal $z(t)$ whose $f_1(t)$ is the real part? Calculate the modulus of $z(t)$. Compare this result graphically with the representation of question 1.

II. Amplitude modulation.

Consider the signal $f(t) = (1 + a(t)) \cos \omega_0 t$. It is assumed that $a(t)$ is a slowly varying function with magnitude less than 1 and whose spectrum $A(\omega)$ is limited to the interval $\{-\omega_{max}, \omega_{max}\}$ with $2\omega_{max} < \omega_0$.

1. Draw the shape of the function $f(t)$.
2. Give the expression of $F(\omega)$, the Fourier transform of $f(t)$. Represent the shape of the spectrum $F(\omega)$.
3. Calculate the analytic signal $z_f(t)$ associated with $f(t)$.
4. To demodulate the signal $f(t)$ we multiply it by $\cos \omega_0 t$. We note $g(t)$ this product.
   What is the Fourier Transform of $g(t)$? How to retrieve $a(t)$ by filtering?

Solution

1. To draw $f(t)$ we choose as an example: $a(t) = 0.4\cos\omega_1 t$, with $f_1 = 5$ kHz and $f_0 = 127$ kHz. Shape of $f(t)$

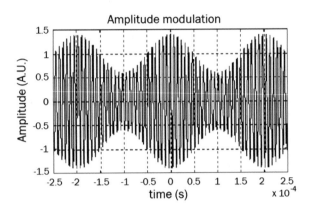

2. $F(\omega) = \frac{1}{2\pi}(2\pi\delta(\omega) + A(\omega)) \otimes \pi(\delta(\omega - \omega_0) + \delta(\omega + \omega_0))$.

To illustrate this we take a spectrum of $A(\omega)$ of the form

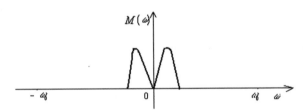

$$F(\omega) = \pi(\delta(\omega - \omega_0) + \delta(\omega + \omega_0)) + \frac{1}{2}(A(\omega - \omega_0) + A(\omega + \omega_0)).$$

Appearance of the spectrum $F(\omega)$:

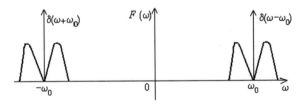

3. $Z_f(\omega) = 2\pi\delta(\omega - \omega_0) + A(\omega - \omega_0)$, $z_f(t) = (1 + a(t))e^{j\omega_0 t}$.

4. $G(\omega) = \frac{1}{2\pi}F(\omega) \otimes \pi(\delta(\omega - \omega_0) + \delta(\omega + \omega_0)) = \frac{1}{2}(F(\omega - \omega_0) + F(\omega + \omega_0))$.

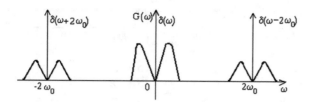

A low-pass filtering will extract $1 + a(t)$. $a(t)$ is extracted by filtering off the DC component. Note that this method only applies if $a(t)$ has no DC component, as in the case shown in figure representing $A(\omega)$. Speech or music signals have this property.

III. Frequency modulation:

This exercise exposes the principle of phase modulation of a carrier with frequency ω_0 by the signal $A_0 \sin \Omega t$. We assume that the angular frequency of the modulation signal verifies $\Omega \ll \omega_0$. A is the modulation index. We note $f_2(t)$ this phase modulated signal:

$$f_2(t) = \cos(\omega_0 t + A \sin \Omega t).$$

We accept here that the periodic function $e^{jA \sin \Omega t}$ has a development in Fourier series of the form $e^{jA \sin \Omega t} = \sum_{n=-\infty}^{\infty} J_n(A)e^{jn\Omega t}$, where the functions $J_n(A)$ are the Fourier series coefficients. They are the Bessel functions of the first kind of order n having the following properties:

a. J_n functions are damped oscillating functions like sine or cosine with a fairly low damping ($J_n(A)$ behaves as $\frac{1}{\sqrt{A}}$ for A large).
b. $J_n(-A) = (-1)^n J_n(A) = J_{-n}(A)$,
c. $J_n(A)$ becomes small for $n > A$.

Refer to the following table for some numerical values of $J_n(z)$.

1. Give the spectral representation of the signal $f_2(t)$.
2. Numerical application (Broadcasting radio in FM band 87.5–108 MHz): The frequency of the highest audio signal for radio transmission in the FM band is $f = 15$ kHz. A second baseband is used to encode stereo signals from $f = 26$ kHz to $f = 53$ kHz. A 3 kHz band is used around $f = 50$ kHz to code various information. The congestion standard in FM band is that the frequency excursion granted to an FM station must not exceed ± 75 kHz. It is assumed in this example that the carrier frequency is $f_0 = 100$ MHz and that the maximum frequency of the signal to be transmitted is 75 kHz. Based on the following table of Bessel functions, what is the maximum modulation index A if the tolerance for the amplitude of the spurious lines is one percent maximum of the amplitude of the main frequency line?

A	$J_0(A)$	$J_1(A)$	$J_2(A)$	$J_3(A)$	$J_4(A)$
0	1.0000	0.0000	0.0000	0.0000	0.0000
0.1	0.9975	0.0499	0.0012	0.0000	0.0000
0.2	0.9900	0.0995	0.0050	0.0002	0.0000
0.3	0.9776	0.1483	0.0112	0.0006	0.0000
0.4	0.9604	0.1960	0.0197	0.0013	0.0001
0.5	0.9385	0.2423	0.0306	0.0026	0.0002
0.6	0.9120	0.2867	0.0437	0.0044	0.0003
0.7	0.8812	0.3290	0.0588	0.0069	0.0006
0.8	0.8463	0.3688	0.0758	0.0102	0.0010
0.9	0.8075	0.4059	0.0946	0.0144	0.0016
1	0.7652	0.4401	0.1149	0.0196	0.0025
1.1	0.7196	0.4709	0.1366	0.0257	0.0036
1.2	0.6711	0.4983	0.1593	0.0329	0.0050
1.3	0.6201	0.5220	0.1830	0.0411	0.0068
1.4	0.5669	0.5419	0.2074	0.0505	0.0091
1.5	0.5118	0.5579	0.2321	0.0610	0.0118
1.6	0.4554	0.5699	0.2570	0.0725	0.0150

Solution

1.
$$f_2(t) = \frac{1}{2}\left(e^{j(\omega_0 t + A \sin \Omega t)} + e^{-j(\omega_0 t + A \sin \Omega t)}\right).$$

First we consider the exponential $e^{j(\omega_0 t + A \sin \Omega t)} = e^{j\omega_0 t} e^{jA \sin \Omega t} = \sum_{n=-\infty}^{\infty} J_n(A) e^{j(\omega_0 + n\Omega)t}$.

The spectrum of $f_2(t)$ which is a periodic function is a line spectrum. The amplitude of these lines are $J_n(A)$. It is noted that, since $J_{-n}(A) = (-1)^n J_n(A)$, the spectrum modulus is symmetrical about the center frequency ω_0. If the modulation index A is small, the property c. causes the number of lines around the carrier frequency to be low.

It is assumed that significant amplitudes lines are limited to two side lines. We can write

$$e^{j(\omega_0 t + A \sin \Omega t)} = J_0(A) e^{j\omega_0 t} + J_1(A) e^{j(\omega_0 + \Omega)t} + J_{-1}(A) e^{j(\omega_0 - \Omega)t} + J_2(A) e^{j(\omega_0 + 2\Omega)t} + J_{-2}(A) e^{j(\omega_0 - 2\Omega)t}$$
$$= e^{j\omega_0 t}(J_0(A) + j2J_1(A) \sin \Omega t + 2J_2(A) \cos 2\Omega t).$$

. Therefore

$$f_2(t) = \frac{1}{2}\left(e^{j\omega_0 t}(J_0(A) + j2J_1(A)\sin\Omega t + 2J_2(A)\cos 2\Omega t)\right)$$
$$+ \frac{1}{2}\left(e^{-j\omega_0 t}(J_0(A) - j2J_1(A)\sin\Omega t + 2J_2(A)\cos 2\Omega t)\right).$$
$$f_2(t) = J_0(A)\cos\omega_0 t - 2J_1(A)\sin\omega_0 t\sin\Omega t + 2J_2(A)\cos\omega_0 t\cos 2\Omega t.$$

In the above expression of $f_2(t)$, the spectrum is limited to the main line at frequency ω_0, to two lines at $\omega_0 \pm \Omega$ and two lines at $\omega_0 \pm 2\Omega$.
In the numerical example, these lines correspond to the frequency of the carrier at $f_0 = 100$ MHz and to frequencies $f_{1,-1} = 100$ MHz ± 75 kHz and $f_{2,-2} = 100$ MHz ± 150 kHz.

2. The frequencies of the second sidebands exceed the recommended frequency deviation. By limiting the modulation index to $A = 0.2$ the amplitude of the second sidebands remains limited to 1 % of the amplitude of the main frequency.
   In practice, the frequency difference between two FM radio stations is at minimum 400 kHz. This allows an excursion of ± 200 kHz, avoiding the embarrassment of overlapping second sidebands. Maintaining a low index of modulation, one can in principle transmit modulation signals whose frequency can reach 200 kHz.
   Finally, notice that the first sideband is in quadrature ($\frac{\pi}{2}$ phase shift) with the carrier, unlike what happens with the amplitude modulation where the phase difference is zero.

Chapter 12
Time–Frequency Analysis

Fourier analysis is not relevant to describe a signal when some of its properties change over time. This is the case, for example, for the chirp signal that we studied previously whose instantaneous frequency varies with time. Acoustically, the ear perceives for this type of signal an increase (or decrease) in the tone, while a simple Fourier transform of the signal does not provide easily interpretable information on the evolution of the "apparent frequency" of the signal over time. Although the frequency concept has been defined for a periodic signal, we continue to talk of frequency for this kind of signal. For example, for a chirp, we say that the frequency increases (or decreases) with time. The term frequency being inaccurate we rather speak of instantaneous frequency.

Simple Fourier analysis is unable to provide easily usable variables that are capable of describing the evolution with time of these signals' characteristics. These signals are called nonstationary and their analysis is called time–frequency analysis. Advances in the analysis of these signals have been important in recent years and allowed, among other things, to arrive at treatment techniques such as signal compression (MP3 audio, or JPEG video), signal detection in noisy environment, or restoring old recordings on 78 rpm discs.

This chapter reviews various methods for analyzing nonstationary signals. Multiplication of the signal by a sliding window leads to short-time Fourier analysis and spectrogram. In the Wigner–Ville distribution, the time reversed signal plays the role of a sliding window analyzer. The inconvenience of the preceding methods is that the width of the window of analysis is kept fixed. The analysis cannot be optimal both for a fast varying part and a slowly varying part of one signal. The continuous wavelet transform (CWT) principle is to explore the signal with a window whose width takes successively all possible values. We explain the theoretical basis of this method. Several wavelets are presented: Morlet, Mexican hat and Shannon wavelets. Later in this book, after developing the rules for calculations on digital signals, Chap. 19 is a continuation of this chapter for digital signals.

© Springer International Publishing Switzerland 2016   207
F. Cohen Tenoudji, *Analog and Digital Signal Analysis*,
Modern Acoustics and Signal Processing, DOI 10.1007/978-3-319-42382-1_12

12.1   Short-Time Fourier Transform (STFT) and Spectrogram

Let $x(t)$ be a signal and $w(t)$ a time window. The following quantity is called the short-time Fourier transform (STFT) or sliding Fourier transform:

$$\text{STFT}(t, \omega) = \int\limits_{-\infty}^{+\infty} x(s)w^*(s - t)e^{-j\omega s}ds. \tag{12.1}$$

This quantity is interpreted as follows: Within the integral, the time signal $x(s)$ is multiplied by the sliding window $w(s - t)$ (by its complex conjugate in the general case of complex signals) whose role is the selection of some part of the signal in a neighborhood of the instant t, then the Fourier transform of the product is calculated. $w(s)$ is a window centered at time $s = 0$ whose width is empirically chosen to discriminate as well as possible the evolutions of the signal on the time axis. Commonly used windows are Hanning or Gaussian (also called Gabor window).

The STFT is also called Gabor transform.

Spectrogram

The following quantity is called Spectrogram

$$S(t, \omega) = \left| \int\limits_{-\infty}^{+\infty} x(s)w^*(s - t)e^{-j\omega s}ds \right|^2. \tag{12.2}$$

It is the squared modulus of the STFT.

It is shown in the following that we also have the relationship

$$S(t, \omega) = \left(\frac{1}{2\pi} \right)^2 \left| \int\limits_{-\infty}^{+\infty} X(\omega')W^*(\omega' - \omega)e^{j\omega' t}d\omega' \right|^2. \tag{12.3}$$

To demonstrate this result, we first calculate the Gabor transform that we note as I for convenience

$$I = \int\limits_{-\infty}^{+\infty} x(s)w^*(s - t)e^{-j\omega s}ds, \tag{12.4}$$

$x(s)$ and $w^*(s - t)$ are then expressed from their Fourier transform within I

$$x(s) = \frac{1}{2\pi} \int\limits_{-\infty}^{+\infty} X(\omega')e^{j\omega' s}d\omega' \quad \text{and} \quad w^*(s - t) = \frac{1}{2\pi} \int\limits_{-\infty}^{+\infty} W^*(\omega'')e^{-j\omega''(s-t)}d\omega''$$

$$I = \left(\frac{1}{2\pi}\right)^2 \int\limits_{-\infty}^{+\infty} \int\limits_{-\infty}^{+\infty} \int\limits_{-\infty}^{+\infty} X(\omega')e^{j\omega's}d\omega' W^*(\omega'')e^{-j\omega''(s-t)}d\omega''e^{-j\omega s}ds. \quad (12.5)$$

We flip the order of integration to start by integrating upon s

$$I = \left(\frac{1}{2\pi}\right)^2 \int\limits_{-\infty}^{+\infty} \int\limits_{-\infty}^{+\infty} X(\omega')W^*(\omega'')d\omega'd\omega''e^{j\omega''t} \int_{-\infty}^{+\infty} e^{j(\omega'-\omega''-\omega)s}ds, \quad (12.6)$$

$$I = \frac{1}{2\pi} \int\limits_{-\infty}^{+\infty} \int\limits_{-\infty}^{+\infty} X(\omega')W^*(\omega'')e^{j\omega''t}\delta(\omega' - \omega'' - \omega)d\omega'd\omega''.$$

We then integrate upon ω'' for example

$$I = \frac{1}{2\pi} \int\limits_{-\infty}^{+\infty} X(\omega')W^*(\omega' - \omega)e^{j(\omega'-\omega)t}d\omega'. \quad (12.7)$$

$$S(t,\omega) = II^* = \left(\frac{1}{2\pi}\right)^2 \left| \int\limits_{-\infty}^{+\infty} X(\omega')W^*(\omega' - \omega)e^{j(\omega'-\omega)t}d\omega' \right|^2. \quad (12.8)$$

The term $e^{-j\omega t}$ disappears in the modulus computation,

$$S(t,\omega) = \left(\frac{1}{2\pi}\right)^2 \left| \int\limits_{-\infty}^{+\infty} X(\omega')W^*(\omega' - \omega)e^{j\omega't}d\omega' \right|^2, \quad \text{Q.E.D.} \quad (12.9)$$

It can be interpreted as follows: Let us first imagine that the spectrum $W(\omega)$ of the window lies predominantly in a neighborhood of zero frequency (spectrum of a low-pass filter). The bandwidth $W(\omega)$ is assumed to be small compared with that of $X(\omega)$.

As shown in Eq. (12.9), $S(t,\omega)$ is obtained by the multiplication of the signal spectrum $X(\omega')$ by the spectrum of the window $W(\omega')$ translated in frequency followed by an inverse FT. This is looking at time t, the contribution to the signal of the frequency band selected by $W^*(\omega' - \omega)$.

In practice, a spectrogram is obtained by tracking over time the changes in the outputs of a filter bank. This analysis was common when signals treatments were only made by analog ways.

A weakness of the concept expressed in formula (12.3) is that the bandwidth of the translated filter $W(\omega' - \omega)$ is the same for all frequency ranges. If this width may be sufficient to adequately assess the spectral amplitude in a certain frequency band, it may be quite inadequate to assess the spectral amplitude of the signal when the frequency lies in other bands.

One could imagine for overcoming this drawback to use a filter bank having the same Q-factor rather than same spectral width.

One can also express this idea in the time domain by saying that if a large time window is used to assess correctly the amplitude at low frequencies (to have a good resolution at low frequencies given the Heisenberg–Gabor uncertainty principle), this width is too large to account for rapid changes in high frequencies from one moment to the other within the time window. Rather, we would like a shorter time window to analyze the high frequencies. We are led to the concept of multiscale analysis.

Since the rise of digital computers where the possibilities of computing in the time domain are more important, the spectrogram analysis in the frequency domain has been supplanted by the calculation in the time domain.

Example of Spectrogram

Consider the following signal composed of a linearly increasing frequency chirp, to which is added a small sinusoidal component

$$x(t) = \sin\left(\omega_0 t + \frac{\beta}{2} t^2\right) + 0.3 \sin(2\pi f_1 t),$$

with $\omega_0 = 2 \times 10^3$ rad/s; $\beta = 6 \times 10^5$; $f_1 = 1.6 \times 10^5$ Hz.

Figure 12.1 shows the signal $x(t)$ (a) and the magnitude $|X(f)|$ of its Fourier transform (b)

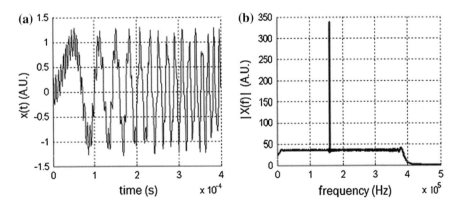

Fig. 12.1 Chirp signal and its FT: **a** Signal $x(t)$; **b** Its FT $|X(f)|$

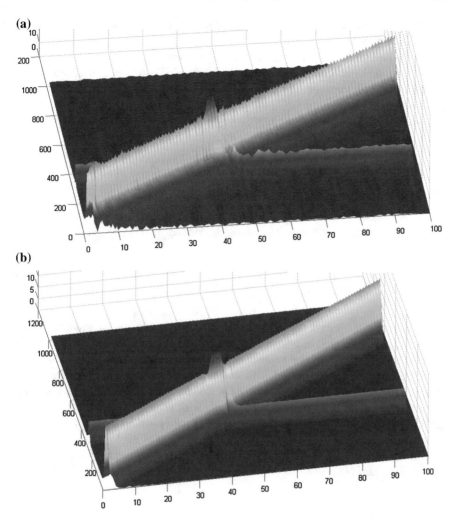

Fig. 12.2 a Spectrogram of $x(t)$ with rect. window; **b** Spectrogram of $x(t)$ with Gabor window

While the increase of the signal frequency is apparent on the temporal shape, the spectral amplitude shown in Fig. 12.1b bears no indication of the evolution in frequency.

Figure 12.2 represents in pseudo-3D two spectrograms calculated with windows of different shapes.

The spectrogram on the left is calculated by taking the FT after a multiplication by a rectangular moving window with a 100 μs width. The second one is obtained using a Gabor window (Gaussian window) with spread $\sigma_t = 31.6$ μs.

The horizontal axis from left to right is the time axis, the second horizontal axis is the frequency axis.

The permanent nature of the sine with frequency $f_1 = 1.6 \times 10^5$ Hz and the linear increase over time of the chirp are clearly visible on both spectrograms.

The peak occurs when the frequency of the chirp lies in the vicinity of the sinusoid frequency.

The smoother shape of the spectrogram given by the Gabor window is due to the apodization effect of a Gaussian window.

12.2   Wigner–Ville Distribution

The Wigner–Ville distribution of a signal is defined by the relation

$$W(t, \omega) = \int\limits_{-\infty}^{+\infty} x\left(t + \frac{\tau}{2}\right) x^*\left(t - \frac{\tau}{2}\right) e^{-j\omega\tau} d\tau. \tag{12.10}$$

Note that in the terms appearing in the product, the time τ integration variable appears with a plus sign in a term and with a minus sign in the second. The minus sign is characteristic of a time reversal of the signal (as is the case in a convolution).

If we compare this definition with that of the spectrogram, we note that in the integral, the time reversed signal plays the role of a sliding window analyzer. The parts with slow variations will select the parts with slow changes and fast changing parts select the parts with rapid changes, this property acts as a kind of window in the time domain that is appropriate to local variations of the signal.

The dimension of $W(t, \omega)$ is that of an energy, dependant on t and ω. The goal of this distribution is to give the energy of the signal at the frequency ω at a given time t.

Properties

$$\int\limits_{-\infty}^{+\infty} W(t, \omega) dt = |X(\omega)|^2. \tag{12.11}$$

Indeed, let us first write the expressions of the Fourier transforms of the time functions appearing in the right side

$$W(t, \omega) = \left(\frac{1}{2\pi}\right)^2 \int\limits_{-\infty}^{+\infty} e^{-j\omega\tau} d\tau \int\limits_{-\infty}^{+\infty} X(\omega') e^{j\omega'\left(t + \frac{\tau}{2}\right)} d\omega' \int\limits_{-\infty}^{+\infty} X^*(\omega'') e^{-j\omega''\left(t - \frac{\tau}{2}\right)} d\omega''. \tag{12.12}$$

We integrate this expression on t

$$\int_{-\infty}^{+\infty} W(t,\omega)\mathrm{d}t = \left(\frac{1}{2\pi}\right)^2 \int_{-\infty}^{+\infty} \mathrm{d}t \int_{-\infty}^{+\infty} e^{-j\omega\tau}\mathrm{d}\tau \int_{-\infty}^{+\infty} X(\omega')e^{j\omega'\left(t+\frac{\tau}{2}\right)}\mathrm{d}\omega' \int_{-\infty}^{+\infty} X^*(\omega'')e^{-j\omega''\left(t-\frac{\tau}{2}\right)}\mathrm{d}\omega''.$$

(12.13)

Switching the order of integrations, we first integrate over t

$$\int_{-\infty}^{+\infty} W(t,\omega)\mathrm{d}t = \left(\frac{1}{2\pi}\right)^2 \int_{-\infty}^{+\infty} e^{-j\omega\tau}\mathrm{d}\tau \int_{-\infty}^{+\infty} X(\omega')e^{j\frac{\omega'\tau}{2}}\mathrm{d}\omega' \int_{-\infty}^{+\infty} X^*(\omega'')e^{j\frac{\omega''\tau}{2}}\mathrm{d}\omega'' \int_{-\infty}^{+\infty} e^{j(\omega'-\omega'')t}\mathrm{d}t$$

$$\int_{-\infty}^{+\infty} W(t,\omega)\mathrm{d}t = \frac{1}{2\pi} \int_{-\infty}^{+\infty} e^{-j\omega\tau}\mathrm{d}\tau \int_{-\infty}^{+\infty} X(\omega')e^{j\frac{\omega'\tau}{2}}\mathrm{d}\omega' \int_{-\infty}^{+\infty} X^*(\omega'')e^{j\frac{\omega''\tau}{2}}\delta(\omega'-\omega'')\mathrm{d}\omega''.$$

We then integrate upon ω''

$$\int_{-\infty}^{+\infty} W(t,\omega)\mathrm{d}t = \frac{1}{2\pi} \int_{-\infty}^{+\infty} e^{-j\omega\tau}\mathrm{d}\tau \int_{-\infty}^{+\infty} X(\omega')X^*(\omega')e^{j\omega'\tau}\mathrm{d}\omega'.$$

We first do the integration on τ

$$\int_{-\infty}^{+\infty} W(t,\omega)\mathrm{d}t = \frac{1}{2\pi} \int_{-\infty}^{+\infty} X(\omega')X^*(\omega')\mathrm{d}\omega' \int_{-\infty}^{+\infty} e^{-j(\omega-\omega')\tau}\mathrm{d}\tau,$$

$$\int_{-\infty}^{+\infty} W(t,\omega)\mathrm{d}t = \int_{-\infty}^{+\infty} X(\omega')X^*(\omega')\delta(\omega-\omega')\mathrm{d}\omega' = |X(\omega)|^2.$$

(12.14)

Finally we obtain the relation (12.11)

$$\int_{-\infty}^{+\infty} W(t,\omega)\mathrm{d}t = |X(\omega)|^2$$

(12.15)

Similarly we have

$$\frac{1}{2\pi} \int_{-\infty}^{+\infty} W(t,\omega)\mathrm{d}\omega = |x(t)|^2.$$

(12.16)

Indeed

$$\frac{1}{2\pi} \int_{-\infty}^{+\infty} W(t,\omega)d\omega = \frac{1}{2\pi} \int_{-\infty}^{+\infty} \int_{-\infty}^{+\infty} x\left(t+\frac{\tau}{2}\right)x^*\left(t-\frac{\tau}{2}\right)e^{-j\omega\tau}dt\,d\omega. \quad (12.17)$$

Integrating firstly over ω,

$$\frac{1}{2\pi} \int_{-\infty}^{+\infty} W(t,\omega)d\omega = \frac{1}{2\pi} \int_{-\infty}^{+\infty} x\left(t+\frac{\tau}{2}\right)x^*\left(t-\frac{\tau}{2}\right)dt \int_{-\infty}^{+\infty} e^{-j\omega\tau}d\omega$$

$$= \frac{1}{2\pi} \int_{-\infty}^{+\infty} x\left(t+\frac{\tau}{2}\right)x^*\left(t-\frac{\tau}{2}\right)d\tau\,2\pi\,\delta(\tau). \quad (12.18)$$

Finally

$$\frac{1}{2\pi} \int_{-\infty}^{+\infty} W(t,\omega)\,d\omega = x(t)x^*(t) = |x(t)|^2. \quad (12.19)$$

We now show that the Wigner–Ville distribution of $x(t)$ can be expressed in the Fourier domain by the following expression:

$$W(t,\omega) = \int_{-\infty}^{+\infty} X\left(\omega+\frac{\omega'}{2}\right)X^*\left(\omega-\frac{\omega'}{2}\right)e^{j\omega't}d\omega'. \quad (12.20)$$

To do it, the following inverse FT terms are replaced in the definition (12.10)

$$x\left(t+\frac{\tau}{2}\right) = \frac{1}{2\pi} \int_{-\infty}^{+\infty} X(\omega')\,e^{j\omega'\left(t+\frac{\tau}{2}\right)}d\omega' \quad \text{and} \quad x^*\left(t-\frac{\tau}{2}\right) = \frac{1}{2\pi} \int_{-\infty}^{+} X^*(\omega'')e^{-j\omega''\left(t-\frac{\tau}{2}\right)}d\omega''.$$

$$W(t,\omega) = \left(\frac{1}{2\pi}\right)^2 \int_{-\infty}^{+\infty} X(\omega')\,e^{j\omega't}d\omega' \int_{-\infty}^{+\infty} X^*(\omega'')\,e^{-j\omega''t}d\omega'' \int_{-\infty}^{+\infty} e^{j\left(\frac{\omega'}{2}+\frac{\omega''}{2}-\omega\right)\tau}d\tau.$$

After integration on τ we get

$$W(t,\omega) = \frac{1}{2\pi} \int_{-\infty}^{+\infty} X(\omega')e^{j\omega't}d\omega' \int_{-\infty}^{+\infty} X^*(\omega'')\,e^{-j\omega''t}\delta\left(\frac{\omega'}{2}+\frac{\omega''}{2}-\omega\right)d\omega''.$$

After integration on ω'' $W(t,\omega) = \frac{1}{\pi} \int_{-\infty}^{+\infty} X(\omega')\,e^{j\omega't}X^*(2\omega-\omega')\,e^{-j(2\omega-\omega')t}\,d\omega'.$

Now the following change of variables is made: $\omega' = \omega - \frac{\omega''}{2}$; $d\omega' = -\frac{d\omega''}{2}$.
It becomes

$$W(t,\omega) = \frac{1}{2\pi} \int\limits_{-\infty}^{+\infty} X\left(\omega - \frac{\omega''}{2}\right) X^*\left(\omega + \frac{\omega''}{2}\right) e^{-j\omega''t} d\omega'',$$

which becomes after the last change of variables $\omega' = -\omega''$

$$W(t,\omega) = \frac{1}{2\pi} \int\limits_{-\infty}^{+\infty} X\left(\omega + \frac{\omega'}{2}\right) X^*\left(\omega - \frac{\omega'}{2}\right) e^{j\omega't} d\omega' \quad \text{Q.E.D.} \qquad (12.21)$$

The interest of this formula appears in the following examples.

Example 1 Calculus of the Wigner–Ville distribution of a signal whose frequency varies linearly over time

$$x(t) = e^{j\omega_0 t} e^{jat^2}. \qquad (12.22)$$

$$x\left(t + \frac{\tau}{2}\right) = e^{j\omega_0\left(t + \frac{\tau}{2}\right)} e^{ja\left(t + \frac{\tau}{2}\right)^2}; \quad x\left(t - \frac{\tau}{2}\right) = e^{j\omega_0\left(t - \frac{\tau}{2}\right)} e^{ja\left(t - \frac{\tau}{2}\right)^2};$$

$$x^*\left(t - \frac{\tau}{2}\right) = e^{-j\omega_0\left(t - \frac{\tau}{2}\right)} e^{-ja\left(t - \frac{\tau}{2}\right)^2}.$$

After calculation we get $x\left(t + \frac{\tau}{2}\right) x^*\left(t - \frac{\tau}{2}\right) = e^{j\omega_0\tau} e^{2jat\tau}$.

$$W(t,\omega) = \int\limits_{-\infty}^{+\infty} e^{j\omega_0\tau} e^{2jat\tau} e^{-j\omega\tau} d\tau = \int\limits_{-\infty}^{+\infty} e^{j(\omega_0 + 2at - \omega)\tau} d\tau.$$

$$W(t,\omega) = 2\pi\,\delta(\omega_0 + 2at - \omega). \qquad (12.23)$$

$W(t,\omega)$ is zero unless the condition $\omega = \omega_0 + 2at$ is satisfied.
   The distribution $W(t,\omega)$ is shown in Fig. 12.3 in the plane (t,ω)

Fig. 12.3 Wigner–Ville
distribution of $x(t)$

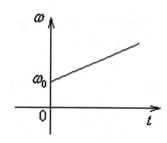

Example 2 Calculation of Wigner–Ville distribution for a signal whose frequency varies linearly with time and amplitude is modulated by a Gaussian shape

$$x(t) = e^{-\frac{(t-t_0)^2}{2\sigma^2}} e^{j\omega_0 t} e^{jat^2}. \tag{12.24}$$

$$x\left(t + \frac{\tau}{2}\right) = e^{-\frac{\left(t+\frac{\tau}{2}-t_0\right)^2}{2\sigma^2}} e^{j\omega_0\left(t+\frac{\tau}{2}\right)} e^{ja\left(t+\frac{\tau}{2}\right)^2};$$

$$x^*\left(t - \frac{\tau}{2}\right) = e^{-\frac{\left(t-\frac{\tau}{2}-t_0\right)^2}{2\sigma^2}} e^{-j\omega_0\left(t-\frac{\tau}{2}\right)} e^{-ja\left(t-\frac{\tau}{2}\right)^2}.$$

It becomes $x\left(t + \frac{\tau}{2}\right)x^*\left(t - \frac{\tau}{2}\right) = e^{-\frac{1}{\sigma^2}\left((t-t_0)^2 + \frac{\tau^2}{4}\right)} e^{j\omega_0\tau} e^{2ja\tau\tau}$.

$$W(t,\omega) = F\left(x\left(t + \frac{\tau}{2}\right)x^*\left(t - \frac{\tau}{2}\right)\right) = e^{-\frac{1}{\sigma^2}(t-t_0)^2} \frac{1}{2\pi} F\left(e^{-\frac{\tau^2}{4\sigma^2}}\right) \otimes F\left(e^{j\omega_0\tau} e^{2ja\tau\tau}\right).$$

As e^{-pt^2} has the FT $\sqrt{\frac{\pi}{p}}e^{-\frac{\omega^2}{4p}}$, $e^{-\frac{\tau^2}{4\sigma^2}}$ has the FT $\sqrt{\pi 4\sigma^2}e^{-\omega^2\sigma^2}$.
We then have

$$W(t,\omega) = 2\sigma\sqrt{\pi}e^{-\frac{1}{\sigma^2}(t-t_0)^2} e^{-\sigma^2(\omega-\omega_0-2at)^2}. \tag{12.25}$$

For a given t, the maximum of the distribution is such that $\frac{\partial W(t,\omega)}{\partial\omega} = 0$, then for

$$\omega = \omega_0 + 2at. \tag{12.26}$$

We expect to see in the time–frequency plane a track similar to that of Fig. 12.3. The maximum of the distribution in the time–frequency plane will occur at the point where the two partial derivatives are zero, for $\frac{\partial W(t,\omega)}{\partial\omega} = 0$ and for $\frac{\partial W(t,\omega)}{\partial t} = 0$. The first derivative has already been performed. To make zero the derivative with respect to time, we do the change of variables $u = t - t_0$. The factor in the exponential is written as

$$g(u) = -\frac{u^2}{\sigma^2} - \sigma^2(\omega - \omega_0 - 2au - 2at_0)^2.$$

Just impose $\frac{dg(u)}{du} = -\frac{2u}{\sigma^2} + 2a\sigma^2(\omega - \omega_0 - 2au - 2at_0) = 0$.
This gives

$$t - t_0 = \frac{\sigma^4 a(\omega - \omega_0 - 2at_0)}{1 + 2a^2\sigma^4}. \tag{12.27}$$

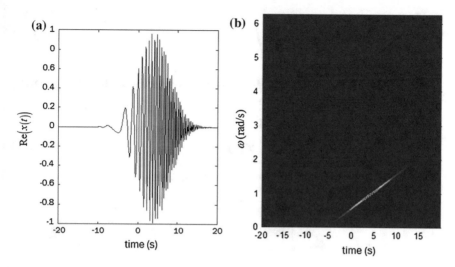

Fig. 12.4 a) chirp with a Gaussian envelope b) Wigner–Ville distribution of $x(t)$

Conditions (12.26) and (12.27) impose that the maximum of the distribution is reached for $t = t_0$.

We present in the following results of a numerical simulation which has been taken for $t_0 = 4$; $\sigma = 4$; $\omega_0 = 5.5$; and $a = 0.5$.

We recognize the Gaussian envelope of the signal on Fig. 12.4a. Figure 12.4b shows the linear track in the time–frequency plane of the higher amplitudes of the function. We numerically verify that the maximum of the function occurs for $t = t_0 = 4$.

It must be emphasised finally that for signals consisting of a superposition of signals, terms resulting from interference between the signals appear in the Wigner–Ville distribution. This results in difficulties in the interpretation of images for the detection of characteristics of the component signals. The wavelet analysis described in the following partially overcomes this problem.

12.3 Continuous Wavelet Transform

12.3.1 Examples of Wavelets

The aim of CWT is to decompose signals on a basis of functions providing good localization both in time and frequency domains. By assumption, the signal $x(t)$ to be decomposed are square-integrable ($\in L^2$). The wavelet $\psi(t)$ considered here is a continuous function in the time domain with a continuous FT. It should be also such that

$$\int_{-\infty}^{\infty} \psi(t)dt = 0. \tag{12.28}$$

This last condition is called the eligibility requirement for the function $\psi(t)$ to be a wavelet. We note $\Psi(\omega)$ the FT of $\psi(t)$.

The relationship (12.28) entails

$$\Psi(0) = 0. \tag{12.29}$$

The norm of the wavelet is assumed to be equal to 1

$$\sqrt{\left(\int_{-\infty}^{\infty} \psi(t)\psi^*(t)dt\right)} = 1. \tag{12.30}$$

In principle, a wavelet is localized around the origin with a narrow temporal spread.

A first example of wavelet is the "Mexican hat." It is the second derivative of the Gaussian function $f(t) = \frac{1}{\sqrt{2\pi\sigma^2}} e^{-\frac{t^2}{2\sigma^2}}$ with the sign changed: $\psi(t) = \frac{1}{\sqrt{2\pi\sigma^3}} e^{-\frac{t^2}{2\sigma^2}} \left(1 - \frac{t^2}{\sigma^2}\right)$.

The condition of eligibility is verified. It is represented in Fig. 12.5 for $\sigma = 0.1$.

A second example is the Morlet wavelet which is a monochromatic signal with frequency ω_0 modulated in amplitude by a Gaussian function $\psi(t) = e^{-\frac{t^2}{2\sigma^2}} e^{i\omega_0 t}$. Its FT is (see Chap. 7):

Fig. 12.5 Mexican hat

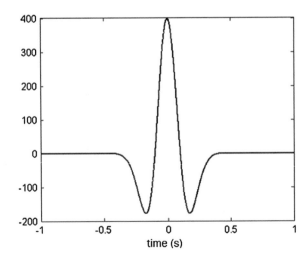

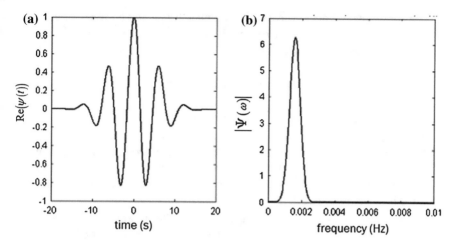

Fig. 12.6 **a** Morlet wavelet; **b** Spectral amplitude of Morlet wavelet

$$\Psi(\omega) = \sqrt{2\pi\sigma^2} e^{-\frac{\sigma^2(\omega-\omega_0)^2}{2}}.$$

The Gaussian shape of the envelope ensures the minimum of the product of time and frequency spreads (see Chap. 6). The eligibility condition is not satisfied by this function since $\Psi(0) \neq 0$, however, if $\omega_0 > \frac{5}{\sigma}$ the frequency peak of the Gaussian that is located in the vicinity of ω_0 is sufficiently far from the origin, and the decay controlled by $\frac{1}{\sigma}$ sufficient so that $\Psi(0) \simeq 0$. Figure 12.6 shows the real part of the Morlet wavelet and its spectral amplitude when $\omega_0 = 1$ and $\sigma = 5$. We see that $\Psi(0)$ is almost equal to zero ($\Psi(0) = 1.37 \times 10^{-4}$).

12.3.2 Decomposition and Reconstruction of a Signal with Wavelets

A wavelet basis consists of functions normalized to 1 $\psi_{a,b}(t) = \frac{1}{\sqrt{a}} \psi\left(\frac{t-b}{a}\right)$. These functions are obtained from the mother wavelet $\psi(t)$ by an expansion with the scale factor a (real > 0) and a time translation $b \in \mathbb{R}$.

By definition, the wavelet transform of $x(t)$ is

$$X_x(a, b) = \langle x(t), \psi_{a,b}(t) \rangle = \frac{1}{\sqrt{a}} \int_{-\infty}^{\infty} x(t) \psi^*\left(\frac{t-b}{a}\right) dt. \qquad (12.31)$$

This operation is referred as a **time-scale analysis**.

The above integral is the value of the correlation function between the functions $x(t)$ and $\psi_{a,b}(t)$ for a couple (a, b). We have already seen that a correlation is a convolution in which one function was reversed in time, so we can write

$$X_x(a, b) = x(b) \otimes \tilde{\psi}^*_{a,b}(b), \qquad (12.32)$$

where the time reversal of $\psi^*_{a,b}(b)$ was noted $\tilde{\psi}^*_{a,b}(b)$.

In interpreting the previous convolution as a filtering, one can say that $X_x(a, b)$ is the result of filtering $x(t)$ by a filter whose impulse response is an expanded version (if $a > 0$) or contracted (if $a < 0$) of the wavelet, reversed in time. Since the wavelet is such that the FT is such that $\Psi(0) = 0$, the filter appears as a band-pass filter. The bandwidth of the filter is determined by the scale factor a.

In this transformation, $|X_x(a, b)|^2$ appears as a cross power spectral density.

We will show in the following that in the case where the wavelet $\psi(t)$ is real, the function $x(t)$ can be found back from its transform defined in (12.31) by the following relationship:

$$x(t) = \frac{1}{C_\psi} \int\limits_0^\infty \int\limits_{-\infty}^\infty \frac{1}{a^2} X_x(a, b) \frac{1}{\sqrt{a}} \psi\left(\frac{t - b}{a}\right) db\, da \qquad (12.33)$$

where

$$C_\psi = \frac{1}{2} \int\limits_{-\infty}^\infty \frac{|\Psi(\omega)|^2}{|\omega|} d\omega. \qquad (12.34)$$

C_ψ is finite. This follows from condition (12.29) resulting in that the singularity in $\omega = 0$, caused by the term $\frac{1}{|\omega|}$ in the integral, is compensated by the zero of the numerator at this point.

Relation (12.33) also appears as a convolution product. $X_x(a, b)$, being itself given by a convolution, this relationship is then a double convolution.

The demonstration of the important result expressed by formula (12.33) is done here in the Fourier domain. It consists in showing that the FT of the two members of this relationship are equal. The FT of a convolution being equal to the product of FT, the demonstration can be written formally. It is preferred here to perform the calculation step by step in order to show how the technical difficulty presented by the presence of the scale factor a in the functions can be treated. The Fourier transform of $\tilde{\psi}^*_{a,b}(t)$ is calculated

$$\int\limits_{-\infty}^\infty \tilde{\psi}^*_{a,b}(t)e^{-j\omega t}dt = \int\limits_{-\infty}^\infty \psi^*_{a,b}(-t)e^{-j\omega t}dt = \int\limits_{-\infty}^\infty \psi^*_{a,b}(t)e^{j\omega t}dt = \Psi^*_{a,b}(\omega) \;(12.35)$$

$$\Psi_{a,b}(\omega) = \int\limits_{-\infty}^{\infty} \psi_{a,b}(t) e^{-j\omega t} dt = \frac{1}{\sqrt{a}} \int\limits_{-\infty}^{\infty} \psi\left(\frac{t-b}{a}\right) e^{-j\omega t} dt.$$

Writing $\frac{t-b}{a} = t'$ the previous integral becomes

$$\Psi_{a,b}(\omega) = \frac{1}{\sqrt{a}} \int\limits_{-\infty}^{\infty} \psi(t') e^{-j\omega(at'+b)} a\, dt' = \sqrt{a} e^{-j\omega b} \int\limits_{-\infty}^{\infty} \psi(t') e^{-j\omega a t'} dt'$$

$$= \sqrt{a} e^{-j\omega b} \Psi(a\omega). \tag{12.36}$$

Thus, the FT of $\tilde{\psi}^*_{a,b}(t)$ is $\sqrt{a} e^{j\omega b} \Psi^*(a\omega)$ because a and b are real.

We now show that the FT of the two members of Eq. (12.33) are equal, which necessarily causes the equality of these two terms. We first calculate the FT of the integral on b appearing in the right side. We set

$$I_1 = \int\limits_{-\infty}^{\infty} \int\limits_{-\infty}^{\infty} \frac{1}{a^2} X_x(a,b) \frac{1}{\sqrt{a}} \psi\left(\frac{t-b}{a}\right) e^{-j\omega t} db\, dt. \tag{12.37}$$

We write $\frac{t-b}{a} = u$; $t = au + b$; $dt = a\, du$.

$$I_1 = \int\limits_{-\infty}^{\infty} \int\limits_{-\infty}^{\infty} \frac{1}{a^2} X_x(a,b) \frac{1}{\sqrt{a}} \psi(u) e^{-j\omega a u} e^{-j\omega b} db\, a\, du.$$

$$I_1 = \int\limits_{-\infty}^{\infty} \frac{1}{a} \frac{1}{\sqrt{a}} X_x(a,b) e^{-j\omega b} \int\limits_{-\infty}^{\infty} \psi(u) e^{-j\omega a u} db\, du.$$

$$I_1 = \Psi(a\omega) \int\limits_{-\infty}^{\infty} \frac{1}{a} \frac{1}{\sqrt{a}} X_x(a,b) e^{-j\omega b} db.$$

$X_x(a,b)$ is replaced by the expression (12.31) which defines it

$$I_1 = \Psi(a\omega) \frac{1}{a^2} \int\limits_{-\infty}^{\infty} x(t') \int\limits_{-\infty}^{\infty} \psi^*\left(\frac{t'-b}{a}\right) e^{-j\omega b} db\, dt'.$$

The following integral on b is calculated

$$I_2 = \int\limits_{-\infty}^{\infty} \psi^*\left(\frac{t'-b}{a}\right) e^{-j\omega b} db. \tag{12.38}$$

We write $\frac{t'-b}{a} = u$; $b = t' - au$; $db = -adu$.

$$I_2 = \int_{-\infty}^{\infty} \psi^*(u)e^{-j\omega t'}e^{j\omega au}adu = ae^{-j\omega t'}\Psi^*(a\omega). \qquad (12.39)$$

Then

$$I_1 = \Psi(a\omega)\frac{1}{a^2}\int_{-\infty}^{\infty} x(t')ae^{-j\omega t'}\Psi^*(a\omega)dt' = \frac{1}{a}|\Psi(a\omega)|^2 X(\omega).$$

The FT of the two members of (12.33) leads to

$$X(\omega) = X(\omega)\frac{1}{C_\psi}\int_0^{\infty}\frac{1}{a}|\Psi(a\omega)|^2 da.$$

We write $a\omega = u$; $da = \frac{du}{\omega}$.

$$1 = \frac{1}{C_\psi}\int_0^{\infty}\frac{1}{u}|\Psi(u)|^2 du = \frac{1}{C_\psi}\frac{1}{2}\int_{-\infty}^{\infty}\frac{|\Psi(u)|^2}{|u|}du = 1,$$

from the definition of C_ψ in (12.34), and as $\psi(t)$ is real by assumption, the modulus of its FT is even.

It follows from the foregoing that the relationship (12.33) which expresses the reconstruction of $x(t)$ from its wavelet transform is demonstrated.

It is interesting to note that the convolution playing an essential role in the formulas of direct and inverse wavelet transforms, the calculations become simple products in the Fourier domain. Numerical calculations are carried out quickly and easily by using the fast Fourier transform which will be described later in this book in Chap. 16.

Finally, we note that in the continuous wavelet analysis, the parameters a and b vary within a continuum. In other words, to recover the function $x(t)$, it is necessary to use an infinite number of basis functions. We will see in Chap. 25 that it is possible to define a basis for the development of functions in L^2 made of wavelets limited in time. It will be technically necessary in that context to define a second function, the scaling function, which will be noted $\phi(t)$.

Figure 12.7 shows on the top a signal composed of two pulses with different spreads. On the figure below is shown its decomposition on Morlet wavelets basis calculated with MATLAB.

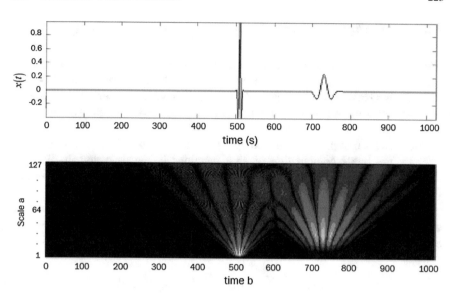

Fig. 12.7 Signal to analyze (*top*); Magnitude of decomposition coefficients $|X_{a,b}|$ (*bottom*)

Figure 12.8 shows on the top a signal composed of a sinusoid of fixed frequency mixed with a chirp of increasing frequency to which is superimposed a random signal. Below, the decomposition upon a Morlet wavelet basis clearly shows the three components of the signal.

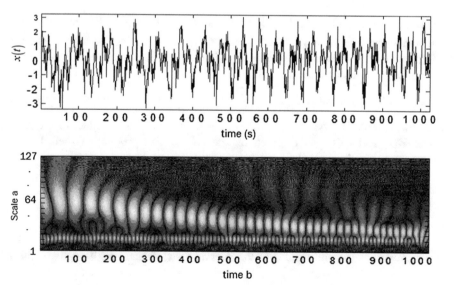

Fig. 12.8 Signal to analyze (*top*); Magnitude of decomposition coefficients $|X_{a,b}|$ (*bottom*)

12.3.3   Shannon Wavelet

We saw above in the example of the Morlet wavelet that a wavelet operates a band-pass filtering of the signal to be analyzed. The envelope of the frequency response is Gaussian in that case. We can try to have a more effective frequency selection by using rectangular windows in the frequency domain. The decomposition of the signal corresponds to a subband coding of the signal. The construction of this wavelet first passes by the recognition that we can build an ideal band-pass filter by the linear combination of sinc functions in the time domain. We first demonstrate that these functions may form an orthogonal basis for developing functions in the time domain.

Shannon–Whittaker sampling theorem: This theorem proved later in Chap. 26 stipulates that a function $f(t)$ whose spectrum is bounded on the interval $\left\{-\frac{1}{2T}, \frac{1}{2T}\right\}$ can be reconstructed from its sampled values with a step T.

$$f(t) = \sum_{n=-\infty}^{+\infty} f(nT) \frac{\sin\left(\frac{\pi}{T}(t-nT)\right)}{\frac{\pi}{T}(t-nT)}. \tag{12.40}$$

To make things easy, without loss of generality, we consider the particular case where $T = 1$. Let the space of functions whose spectrum is bounded on the interval $\left\{-\frac{1}{2}, \frac{1}{2}\right\}$ be noted V_0. The formula (12.40) becomes

$$f(t) = \sum_{n=-\infty}^{+\infty} f(n) \frac{\sin(\pi(t-n))}{\pi(t-n)}. \tag{12.41}$$

Let us write $\Theta(t) = \frac{\sin(\pi t)}{\pi t}$. The functions $\Theta(t-n) = \frac{\sin(\pi(t-n))}{\pi(t-n)}$ with $n \in \mathbb{Z}$ form an orthonormal basis in V_0. The function $\Theta(t)$ is the scaling function $\phi(t)$ associated with the Shannon wavelet from which the wavelet functions are built.

Let us now demonstrate the orthonormality of functions $\Theta(t-n)$ in the Fourier domain. We first calculate $P(\omega) = \int\limits_{-\infty}^{+\infty} \Theta(t)\Theta(t-n)e^{-j\omega t}dt$, then take the value of $P(\omega)$ in $\omega = 0$. The function $P(\omega)$, Fourier transform of the product of the functions $\Theta(t)$ and $\Theta(t-n)$ is given by the convolution product of their Fourier transforms

$$P(\omega) = \frac{1}{2\pi}\left[\mathcal{F}\left(\frac{\sin(\pi t)}{\pi t}\right) \otimes \mathcal{F}\left(\frac{\sin(\pi(t-n))}{\pi(t-n)}\right)\right]. \tag{12.42}$$

It is now recognized that $\frac{\sin(\pi t)}{\pi t}$ is the inverse FT of a rectangular window in the frequency domain, with value 1 in the interval $\{-\pi, \pi\}$ and zero elsewhere (check that)

$$\mathcal{F}\left(\frac{\sin(\pi t)}{\pi t}\right) = \Pi_{\{-\pi,\pi\}}(\omega).\tag{12.43}$$

Applying the shifting theorem, we get

$$\mathcal{F}\left(\frac{\sin(\pi(t-n))}{\pi(t-n)}\right) = \Pi_{\{-\pi,\pi\}}(\omega)e^{-jn\omega}.\tag{12.44}$$

Then $P(\omega) = \frac{1}{2\pi}\int\limits_{-\infty}^{+\infty} \Pi_{\{-\pi,\pi\}}(\omega - \omega')\Pi_{\{-\pi,\pi\}}(\omega')e^{-jn\omega'}d\omega'.$

$$P(\omega)|_{\omega=0} = \frac{1}{2\pi}\int\limits_{-\infty}^{+\infty} \Pi_{\{-\pi,\pi\}}(-\omega')\Pi_{\{-\pi,\pi\}}(\omega')e^{-jn\omega'}d\omega' = \frac{1}{2\pi}\int\limits_{-\pi}^{\pi} e^{-jn\omega'}d\omega',$$

$$P(\omega)|_{\omega=0} = \int\limits_{-\infty}^{+\infty} \Theta(t)\Theta(t-n)dt = 0 \quad \text{if} \quad n \neq 0,\tag{12.45}$$

and

$$P(\omega)|_{\omega=0} = \int\limits_{-\infty}^{+\infty} \Theta^2(t) = 1 \quad \text{if} \quad n = 0.\tag{12.46}$$

Note We have just shown that $\int\limits_{-\infty}^{+\infty} \frac{\sin^2(\pi t)}{\pi^2 t^2}dt = 1$. With a simple change of variable the value of the following classic integral is obtained $\int\limits_{-\infty}^{+\infty}\left(\frac{\sin x}{x}\right)^2 dx = \pi$.

The mother Shannon wavelet is defined as the difference of two sincs

$$\psi(t) = 2\text{sinc}(2\pi t) - \text{sinc}(\pi t) = \frac{\sin(2\pi t) - \sin(\pi t)}{\pi t}.\tag{12.47}$$

Figure 12.9 shows, in the time interval $\{-10, 10\}$, the two sincs contained in the definition (12.47), the mother Shannon wavelet, and below, their FT calculated numerically. The oscillations on the spectra are numerical artifacts. The Shannon wavelet acts as an ideal band-pass filter. As can be seen, the price to pay to have a very selective filtering is that the support of the wavelet is not compact.

We see in this example the band-pass character of the wavelet. The frequency bands of the wavelets derived from the mother wavelet by the scale factor a will scan the frequency axis and allow the analysis of a signal in different frequency bands.

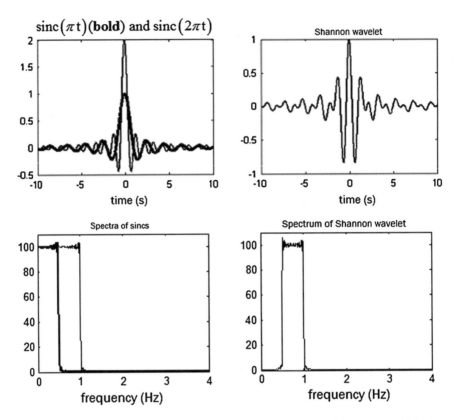

Fig. 12.9 Sincs functions and their spectra (*left*); Shannon wavelet and its spectrum (*right*)

Summary
This chapter is the presentation of methods developed to overcome the insufficiency of the Fourier transformation to describe signals whose properties vary in time. The sliding Fourier transform and its square modulus, the spectrogram, consist in multiplying the function by a sliding window before taking the FT for analyzing the signal locally. The window is moved successively along the whole time axis. The Wigner–Ville distribution is a related method where the signal itself is used as the analyzing function. These methods are however limited by the fact that the width of the analysis windows is fixed. The CWT allows the use of a window with variable width. It meets in an optimal way the detection of portions of the signal where the variation is fast as well as those where the signal varies slowly. The decomposition–reconstruction formulas of a signal on a wavelets basis have been demonstrated in the chapter. We will find in Chap. 19 the extension of wavelet analysis to digital signals.

Chapter 13
Notions on Digital Signals

Digital signals are sequences of numbers. The index in the sequence acts as the time. We say that time is discrete (or discontinuous). Digital signals may be purely synthetic (calculated algorithmically) or result from the conversion of analog signals by Analog/Digital converters (abbreviated ADC). Digital signals have nowadays become prominent, driven by continuing advances in microelectronics (Moore's Law, which has been verified for over thirty years, states that the ability to integrate electronic circuits doubles every 18 months).

Digital signals possess several advantages compared to analog signals: First, their treatment is more flexible. Processes on digital signals can be achieved that are impossible for analog signals. For example, one can easily change the parameters in a rule of calculation to improve filtering.

Second, the signal-to-noise ratio of digital signals can be large. For example, it can be maintained intact during propagation on a transmission channel unlike the situation for an analog signal that is always negatively affected during propagation. Indeed, the use of error-correcting codes allows finding back exactly the original digital signal at the output of the transmission channel.

In this chapter, we first give some idea of the analog to digital signal conversion and the error committed during this operation conditioned by the limited resolution of the converter. We show with a simple example the necessity of using a sampling frequency of the analog signal that is sufficiently high so that rapid variations of a signal can be correctly rendered in the digital signal. We also show with the simple example of digitizing a sine function how a frequency component higher than the sampling frequency can have the same digital image as a low-frequency signal caused by a stroboscopic effect (aliasing). These facts will be demonstrated mathematically in Chap. 19. We give at the end of this chapter, the expression of simple digital signals. We emphasize the fact that they appear as weighted sequences of Kronecker unit pulses.

© Springer International Publishing Switzerland 2016
F. Cohen Tenoudji, *Analog and Digital Signal Analysis*,
Modern Acoustics and Signal Processing, DOI 10.1007/978-3-319-42382-1_13

13.1   Analog to Digital Conversion

An Analog to Digital Converter (ADC) has two main characteristics:

- Its sampling frequency f_e which is the number of conversions per second.
- Its resolution (given by the number of possible levels at the converter output). While the values of the signal to be sampled are real numbers, their coding by the converter is in integer values.

The conversion operation requires the comparison of the value of the signal to different reference levels values. It requires a certain time which generally increases with the number of levels of comparison. It is conditioned by the speed of electronic circuits that are used. Increasing the accuracy of the conversion may require a change of technology and the decrease of the sampling frequency.

The quantization levels of a converter are generally uniformly distributed. However in some cases levels are used spaced by a logarithmic law. This is the case for some telephone connections (A law or μ law). Low levels are relatively close together allowing good rendering of small signals while high levels may still be digitized (approximately). This brings a good dynamic to the digitized signal.

The number of levels of a converter is generally a power of two: 2^M.

- For 8-bit converters, $M = 8$; There are $2^8 = 256$ levels.
- For 12-bit converters, $M = 12$; There are $2^{12} = 4096$ levels.

Several techniques are used to improve performance of conversion operations. They are based on the statistical properties of the signal to be sampled. It is rare to see the signal vary greatly from one sample to the next. The expected variation is small compared with the difference between the extreme values of the signal. Thus, a first method is to digitize the difference between the value of a sample and that of the previous sample whose value has been stored. Thus the amplitude range to be scanned is not required to cover the whole range of signal values. This principle can be extended to the use of values of several previous samples: One can make a prediction of the expected value for a sample from these earlier samples. The difference between the expected value and the true value is digitized (predictive coding). These converters are called Delta-Sigma (Sigma evokes the notion of sum and Delta the difference between the expected value and the value found). It has been shown that it is effective for the speech signals to decompose the signal by a bank of filters, to do a predictive coding conversion of each filtered component then digitally reconstruct the signal. A good reproduction of the signal is attained, even if the number of bits of the converter may seem insufficient a priori.

Scanning audible signals: The frequency of audible signals is less than 20 kHz. Today audible signals converters are used whose sampling frequencies range from 40 to 80 kHz (44.1 kHz in a digital compact disc). The sound is recorded digitally on a compact disc. When reading the record is processed and converted into an analog signal for listening (by a Digital to Analog Converter DAC). The ear is very sensitive to imperfections of the reproduced sound, so the precision of the converters must be high to get a good record: at least 16 bits or better, 24 bits to satisfy the ears of a musician in the case of a converter that does not use predictive coding.

The number of bits of a converter decreases as the frequency of the converter increases. Currently there are 16-bit digitizers to about 360 MHz. For 12-bit digitizers the sampling frequency rises to 3.5 GHz. Above this frequency, there are 8-bit digitizers (flash converters) up to frequencies of the order of 40 Gigahertz.

Example of accuracy of a converter

Take an 8-bit converter (256 levels) for sampling a bipolar signal in a ± 1 V range. We have the following correspondence between the levels and voltage:

Level		Voltage
0	$\rightarrow$	1 V
128	$\rightarrow$	0 V
255	$\rightarrow$	1 V–7.81 mV

Giving a precision $\frac{1}{128} = 7.81$ mV per bit.

The maximum error of the converter is 3.9 mV, that is to say half the least significant bit $\frac{1}{2}$ LSB.

The minimum relative error is

$$\frac{\text{error}}{\text{signal max}} = \frac{0.5}{128} = \frac{1}{256} = \frac{1}{2^M},$$

with a correspondence in decibels: $20 \log_{10} \frac{1}{256} = -48.16$ dB.

For 8-bit converted signals, the signal-to-noise ratio will not exceed 45 dB.

For a 16-bit converter we will have: $\frac{\text{error}}{\text{signal max}} = \frac{1}{2^{16}} = \frac{1}{65536} \Rightarrow -96.3$ dB. This is much better than for the 8-bit converter.

The presence of converter quantization levels therefore causes that the numerical values do not correspond exactly to the analog values. This error is equivalent to the superposition of a numerical error signal which fluctuates rapidly from one sample to another. The result is the emergence of broadband noise which is greater when the resolution of the converter is lower.

Experimentally, one can reduce the quantization error in two situations

1. If the sampling frequency of the analog–digital converter is high, much higher than the maximum frequency present in the signal to be digitized. This is the case, for example, when the signal supplied by a sensor has a maximum frequency of 5 MHz and the converter has a sampling frequency of 125 MHz. It is understood that since the signal varies relatively slowly, successive samples will land often on the same level of quantization. By smoothing the digital signal by a moving average digital filter, one softens the signal by creating signal values between the quantization levels. This operation is performed without unduly affecting the signal spectrum. This method brings about 2 quantization bits in the example.

2. Another method is possible in the case where the signal to be measured is certain, repetitive, with random noise superimposed on it. This situation is met in ultrasound echography for example where one can send the same impulsion repeatedly toward the target (the target must be fixed to make the method workable). Electronic noise present in the signal is usually sufficient to make the technique possible. The sum of several digitized signals is made using a computer. While the deterministic signal is affected by the same quantization error for each signal, the noise shifts randomly the noisy signal to different quantization levels, and so, at a given time, successive digitized signals will not land to the same level. Summing successive digitized signals, operates a statistical average of these values. Thus the average may have an intermediate value between two initial quantization levels. By averaging $N = 1024$ signals, we can expect to benefit from $\sqrt{N} = \sqrt{1024} = 32$ intermediate levels.

Of course, the noise must meet certain conditions; it should not be too low; its standard deviation should be at least of the order of 2–3 times the quantization interval. It should not be too large either to avoid that the value of the averaged signal would be far from that of the non-noisy signal. In this example, one can gain 3–4 quantization bits.

If natural electronic noise is too low, it is perfectly possible to consider adding an additional synthetic noise in order to make the operation possible.

In the rest of this course, we will not take into account the quantization error.

Notation of a digital signal f_e is the sampling frequency, the sampling step $T = \frac{1}{f_e}$.

The analog–digital conversion establishes the correspondence: $f(t) \rightarrow f(nT)$.
We write $f(t) \rightarrow f(nT) = f[nT] = f[n]$.
Note that for a digital signal, the time variable n is discrete. We note the time function $f[n]$ using square brackets around n, as the arrays in conventional programming languages such as assembly or C are noted.

13.2   Criterion for a Good Sampling in Time Domain

To achieve an acceptable sampling, it is necessary that the variations of the signal to be sampled are not too rapid between two sampling instants. More precisely:

Shannon Theorem *For the sampling of a signal to be correct in a spectral point of view, it is necessary that the spectral amplitude of the signal to be sampled are restricted to the frequency domain* $-\frac{f_e}{2}, \frac{f_e}{2}$.

If the Shannon condition is not met, aliasing *phenomenon occurs.*

Figure 13.1 shows qualitatively an example of digitalization of two signals with different frequency contents. On top, the samples follow well the variations of a

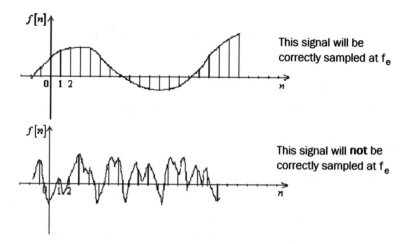

Fig. 13.1 Two sampled signals: correctly (*top*); incorrectly (*bottom*)

slowly varying signal. Using the same sample frequency, we see that the higher frequency signal is not sampled accurately. Some fast variations are not apparent on the samples.

This theorem will be proved in Chap. 16. It is preferred here to illustrate this condition on the simple example of a sinusoidal signal and bring up the problem caused by insufficient sampling frequency:

Let $f(t)$ the signal to digitize

$$f(t) = \sin \omega_0 t = \sin 2\pi f_0 t \tag{13.1}$$

At times nT we have:

$$f(nT) = \sin 2\pi f_0 nT \tag{13.2}$$

The digital signal will be

$$f[n] = \sin 2\pi f_0 nT \tag{13.3}$$

Since

$$T = \frac{1}{f_e}, \quad f[n] = \sin 2\pi \frac{f_0}{f_e} n \tag{13.4}$$

Let $f_2(t) = \sin 2\pi f_0' t$ the analog signal whose frequency exceeds the sampling frequency by the value f_0: $f_0' = f_e + f_0$.

$$f_2[n] = \sin 2\pi \frac{f'_0}{f_e} n = \sin\left(2\pi \frac{f_e}{f_e} n + 2\pi \frac{f_0}{f_e} n\right)$$
$$= \sin\left(2\pi \frac{f_0}{f_e} n + 2\pi n\right) = \sin 2\pi \frac{f_0}{f_e} n \qquad (13.5)$$

It is seen that the analog signals with frequencies f_0 and $f'_0 = f_e + f_0$ have the same numerical representation.

Similarly, if $f'_0 = f_e - f_0$,

$$f_2[n] = \sin 2\pi \frac{f'_0}{f_e} n = \sin\left(2\pi \frac{f_e}{f_e} n - 2\pi \frac{f_0}{f_e} n\right) = -\sin 2\pi \frac{f_0}{f_e} n \qquad (13.6)$$

The above two examples illustrate the aliasing.

This phenomenon is identical to the stroboscopic effect, well known in optics. In movies, where the pictures are presented with a time step of $T = 1/24$ s, we may see the wheels of a vehicle or the blades of a helicopter rotate very slowly or sometimes even backwards.

13.3 Simple Digital Signals

The basic digital signal is the **unit pulse** $\delta[n]$ (Kronecker function) defined by (Fig. 13.2):

$$\delta[n] = \begin{vmatrix} 0 & \text{if } n \neq 0 \\ 1 & \text{if } n = 0 \end{vmatrix}. \qquad (13.7)$$

For digital signals, this function plays the role of the Dirac distribution for continuous-time signals. We should not confuse this digital signal which is 1 at the origin with the Dirac distribution which is infinite at the origin.

Definition of the **step function** (Fig. 13.3):

$$U[n] = \begin{vmatrix} 0 & \text{for } n < 0 \\ 1 & \text{for } n \geq 0 \end{vmatrix}. \qquad (13.8)$$

Fig. 13.2 Kronecker unit pulse $\delta[n]$

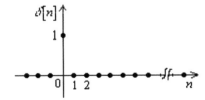

Fig. 13.3 Step function $U[n]$

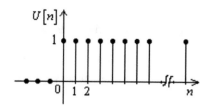

This function plays for digital signals the role of the Heaviside function for continuous-time signals.

Translation of the unit pulse
Translation to the right by one step

$$\delta[n-1] = \begin{vmatrix} 0 & \text{for } n \neq 1 \\ 1 & \text{for } n = 1 \end{vmatrix} \tag{13.9}$$

Translation by m steps

$$\delta[n-m] = \begin{vmatrix} 0 & \text{for } n \neq m \\ 1 & \text{for } n = m \end{vmatrix} \tag{13.10}$$

The translation is to the right as m positive, as is the case in Fig. 13.4:

Note Any numerical function can be considered as a linear combination of Kronecker pulses $\delta[n-m]$. So the step function can be written:

$$U[n] = \sum_{m=0}^{\infty} \delta[n-m]. \tag{13.11}$$

We will use in the sequel the following entry representing any function $f[n]$ as a linear combination of Kronecker functions

$$f[n] = \sum_{m=-\infty}^{\infty} f[m]\delta[n-m] \tag{13.12}$$

In this sum, factors $f[m]$ act as weights affecting each Kronecker pulse.

Fig. 13.4 Unit pulse translated by m

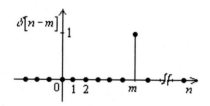

Summary

This chapter was an introduction to digital signals. We have detailed the two major characteristics of an analog to digital converter: the resolution of the quantization which depends on the number of levels of the converter evaluated in number of bits and the sampling frequency. On simple examples, we have shown that insufficient sampling frequency did not allow accounting for the rapid changes of the analog signal and we highlighted the problem of aliasing occurring in that case. We give the expression of simple digital signals and emphasize the fact that a digital signal appears as a succession of weighted unit Kronecker pulses.

Exercises

A digital rectangular window is a function with a constant value in a given time interval and zero elsewhere. Show that the function $\Pi[n] = U[n] - U[n-8]$ is a rectangular window. Represent this function. We can write $\Pi[n] = \sum_{m=0}^{7} \delta[n-m]$.

Solution:

I. The function $U[n]$ is zero for negative times and is 1 for positive times. The function $U[n-8]$ will be 1 starting at $n = 8$. Their difference will cancel for $n > 7$.

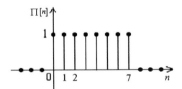

II. Graph the function $f[n] = (-1)^n U[n]$.

Answer:

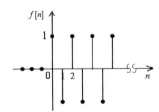

Chapter 14
Discrete Systems—Moving Average Systems

This first chapter on discrete systems is intended to show that simple rules of sums or differences in calculations on sequences of numbers can act as digital filters on these numbers, smoothing or enhancing certain spectral components of these signals. The digital filters have the decisive advantage of being easy to implement in a signal processing chain, easily modifiable, and able to vary over time to adapt to the evolutions of the signals to be processed (adaptive filtering, Kalman filtering).

This chapter is a mirror image of Chaps. 1 and 2 of this book that were dealing with analog signals. Drawing on an example of a smoothing filter (Moving Average, MA), the concepts of digital filter and linear time-invariant system are defined. We show that the function z^n is an eigenfunction of a time-invariant digital system as was the exponential e^{st} for analog systems. The impulse response, the transfer function and the frequency response are defined. The z-plane plays the role played by the Laplace plane for analog systems. A discrete convolution of the input signal by the impulse response provides the output signal. We study some examples of moving average filters and show how we can interpret geometrically the variation of the system's frequency response. The frequency response of a discrete system is inherently periodic in frequency. The advantages of the Moving Average filters are that they have a finite impulse response length. We see also that it is possible to create filters with zero phase shifts. A disadvantage of MA filters is that they are not very selective.

A discrete system associates to a sequence $f[n]$ considered as an input, another sequence $g[n]$ considered to be the output of the system.

$$\xrightarrow{f[n]} \boxed{\text{System}} \xrightarrow{g[n]}$$

Systems are commonly called filters.

© Springer International Publishing Switzerland 2016   235
F. Cohen Tenoudji, *Analog and Digital Signal Analysis*,
Modern Acoustics and Signal Processing, DOI 10.1007/978-3-319-42382-1_14

An example of system is

$$g[n] = \frac{1}{3}(f[n-1] + f[n] + f[n-1]). \tag{14.1}$$

This last system operates a smoothing of the signal $f[n]$. It provides a moving average of $f[n]$.

14.1   Linear, Time-Invariant Systems (LTI)

The properties that define a LTI system are
Linearity:
Let two arbitrary functions $f_1[n]$ and $f_2[n]$ enter the system:
Namely,

$$\xrightarrow{f_1[n]} \boxed{\text{System}} \xrightarrow{g_1[n]} \text{ and } \xrightarrow{f_2[n]} \boxed{\text{System}} \xrightarrow{g_2[n]}. \tag{14.2}$$

The system is linear if for any two constants a_1 and a_2, then

$$\xrightarrow{a_1 f_1[n] + a_2 f_2[n]} \boxed{\text{System}} \xrightarrow{a_1 g_1[n] + a_2 g_2[n]} \tag{14.3}$$

Time invariance:
Let

$$\xrightarrow{f[n]} \boxed{\text{System}} \xrightarrow{g[n]} \tag{14.4}$$

The system is translational invariant in time (we say also stationary) if its response to the delayed input is identical to the delayed initial response. That is to say, if and only if (iff)

$$\xrightarrow{f[n-m]} \boxed{\text{System}} \xrightarrow{g[n-m]} \tag{14.5}$$

14.2   Properties of LTI Systems

We call impulse response $h[n]$ the output function of the non-preprepared system for an input $\delta[n]$, pulse with unit amplitude at time $n = 0$.

$$\xrightarrow{\delta[n]} \boxed{\text{System}} \xrightarrow{h[n]} \tag{14.6}$$

Calculation of the output $g[n]$ when the input signal $f[n]$ has an arbitrary shape

It has been pointed out above that the function $f[n]$ can be regarded as a sequence of Kronecker functions $\delta[n]$ shifted

$$f[n] = \sum_{m=-\infty}^{+\infty} f[m]\, \delta[n-m]. \tag{14.7}$$

One can therefore consider that pulses $f[m]\,\delta[n-m]$ localized in $n = m$ successively enter the filter.

At the output of the LTI system, we will have, because of its properties of linearity and time invariance, the sum of the responses $h[n-m]$ to each of the pulses $\delta[n-m]$ that comprise the input signal.

We can then write

$$g[n] = \sum_{m=-\infty}^{+\infty} f[m]\, h[n-m]. \tag{14.8}$$

It can be shown by a simple change of variables that we also have

$$g[n] = \sum_{m=-\infty}^{+\infty} f[n-m]\, h[m]. \tag{14.9}$$

We say that the system output is the convolution (discrete) of the input signal with the impulse response. Symbolically

$$g[n] = f[n] \otimes h[n]. \tag{14.10}$$

It is easily shown that the convolution product is associative and distributive.

As an exercise, one will verify that the convolution of two causal signals is causal.

14.3   Notion of Transfer Function

Let z be an arbitrary complex number. The functions z^n are eigenfunctions of the operator describing the LTI filter. This means that there is a function $H(z)$ such that

$$\xrightarrow{\;f[n]=z^n\;} \boxed{\text{System}} \xrightarrow{\;g[n]=z^n H(z)\;} \tag{14.11}$$

Indeed, if $f[n] = z^n$, according to (14.8) we have $g[n] = \sum_{m=-\infty}^{+\infty} z^m\, h[n-m]$. Let $n - m = k$. We have $g[n] = \sum_{k=-\infty}^{+\infty} z^{n-k} h[k]$, or also

$$g[n] = z^n \sum_{k=-\infty}^{+\infty} h[k]z^{-k}. \tag{14.12}$$

The sum of the series appearing in the above equation is a function of z only which will be noted $H(z)$. This function is called the system transfer function. It is given by

$$H(z) = \sum_{k=-\infty}^{+\infty} h[k]z^{-k}. \tag{14.13}$$

We then have

$$g[n] = H(z)z^n \text{ if } f[n] = z^n \tag{14.14}$$

More generally, the z transform $F(z)$ of an arbitrary function $f[n]$, is defined by the relationship

$$F(z) = \sum_{n=-\infty}^{+\infty} z^{-n}f[n]. \tag{14.15}$$

We can see from that the filter transfer function is the z transform of its impulse response.

Theorem Consider a discrete LTI system. The z transform, $G(z)$ of the system output signal is the product of the z transform of the input signal $F(z)$ by the system transfer function $H(z)$: $G(z) = H(z)F(z)$.

Indeed $\xrightarrow{f[n]}$ $\boxed{\text{System}}$ $\xrightarrow{g[n] = \sum_{m=-\infty}^{+\infty} h[n-m]f[m]}$

The z transform of $g[n]$ is given by $G(z) = \sum_{n=-\infty}^{+\infty} g[n] z^{-n}$

$$G(z) = \sum_{n=-\infty}^{+\infty} \sum_{m=-\infty}^{+\infty} h[n-m]f[m]z^{-n} = \sum_{p=-\infty}^{+\infty} \sum_{m=-\infty}^{+\infty} h[p]f[m]z^{-m}z^{-p}.$$

We noted $n - m = p$. $z^{-n} = z^{-m}z^{-p}$.
The variables are separated, we can write

$$G(z) = \sum_{p=-\infty}^{+\infty} h[p] z^{-p} \sum_{m=-\infty}^{+\infty} f[m] z^{-m} = H(z) F(z). \tag{14.16}$$

Case where functions $f[n]$ and $h[n]$ have limited lengths N_1 and N_2

This is a case often encountered in practice. It can be reduced by simply changing the time origin to the case of two causal signals. The convolution product of two causal signals is causal. The sum of the durations N_1 and N_2 of $f[n]$ and $h[n]$ signals is a high bound to the signal duration of $g[n]$. The z transform of $g[n]$ may be written in this case

$$G(z) = \sum_{n=-\infty}^{+\infty} g[n]\, z^{-n} = \sum_{n=0}^{N_1+N_2} g[n]\, z^{-n} = \sum_{n=0}^{N_1+N_2} z^{-n} \sum_{m=0}^{n} f[m]\, h[n-m]. \quad (14.17)$$

The form of $G(z)$ is that of a polynomial in z^{-1}. The coefficients of the polynomial are terms of the convolution product of $f[n]$ and $h[n]$. The first coefficients of the polynomial are

$$f[0]h[0];\ f[0]h[1]+f[1]h[0];\ f[0]h[2]+f[1]h[1]+f[2]h[0];\ \text{etc}$$

The terms of the convolution product are obtained by the multiplication of the two polynomials in z^{-1}.

MATLAB uses this calculation method when calling the function conv(f, h).

14.4 Frequency Response of a LTI System

Consider the monochromatic input signal $f[n] = e^{j\omega nT}$. The digital signal obtained by sampling the analog signal $f(t) = e^{j\omega t}$ is recognized.

This signal has the form $f[n] = z^n$, with $z = e^{j\omega T}$.

According to (14.14) we necessarily have

$$g[n] = e^{jn\omega T} H(e^{j\omega T}). \quad (14.18)$$

$$\xrightarrow{e^{j\omega nT}} \boxed{\text{System}} \xrightarrow{H(e^{j\omega T})e^{j\omega nT}}$$

It is noted that the output signal is also monochromatic. It has the same angular frequency as the input signal. $H(e^{j\omega T})$ is the frequency response of the filter (also known as complex gain). It is the system transfer function evaluated on the unit circle.

$$H(e^{j\omega T}) = \sum_{n=-\infty}^{+\infty} h[n]\, e^{-j\omega nT}. \quad (14.19)$$

The system frequency response is the Fourier transform (in the sense of operations on digital signals) of the impulse response.

Important property The frequency response $H(e^{j\omega T})$ of a discrete filter is a periodic function of ω, with period $\omega_e = \frac{2\pi}{T}$. Indeed all the exponential $e^{-j\omega n T}$ contained in the sum are periodic functions of ω. Their periods are submultiples of ω_e. The period of the series sum is the greatest period, ω_e, appearing in the sum.

14.5 Moving Average (MA) Filters

These are filters whose impulse response is time-limited. They are also referred as finite impulse response (FIR) filters.

This impulse response is often written in the form $h[n] = \sum_{k=n_1}^{n_2} b[k]\, \delta[n-k]$, or $h[n] = \sum_{k=n_1}^{n_2} b_k\, \delta[n-k]$, with n_1 and n_2 both integers.

The general equation relating the input and output signals of these filters has the form

$$g[n] = \sum_{m=n_1}^{n_2} b_m f[n-m]. \tag{14.20}$$

To understand on an example the behavior of these filters, we return to the system previously met

$$g[n] = \frac{1}{3}[f[n-1] + f[n] + f[n+1]]. \tag{14.21}$$

Calculating the impulse response $h[n]$ (Fig. 14.1):

$$h[n] = \frac{1}{3}(\delta[n-1] + \delta[n] + \delta[n+1]). \tag{14.22}$$

It is observed on Fig. 14.1 that the impulse response of this filter is limited to 3 points. It is not causal, since it is not zero for $n < 0$.

Fig. 14.1 Impulse response

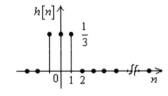

The transfer function of this filter function is given by

$$H(z) = \frac{1}{3}\left(z^{-1} + z^0 + z^{+1}\right) = \frac{1}{3}z^{-1}\left(z^2 + z + 1\right) = \frac{1}{3}z^{-1}(z - z_0)(z - z_0^*). \quad (14.23)$$

A z polynomial has always roots in the complex set $\mathbb{C}$. The number of roots is necessarily equal to the degree of the polynomial. In the present case, the two roots are complex conjugate because the polynomial coefficients are real.

Except for the $z = 0$ pole (which could be eliminated by translating the impulse response by one step to the right), the notable points are the zeros z_0 and z_0^*. This filter belongs to the moving average filters class (MA). These filters are also called all-zero filters.

Calculation of $H(z)$ zeros: The polynomial $P(z) = z^2 + z + 1$ has two roots

$$z_0, z_0^* = -\frac{b \pm \sqrt{\Delta}}{2} = -\frac{1}{2} \pm j\frac{\sqrt{3}}{2} = e^{\pm j\frac{2\pi}{3}}$$

14.6   Geometric Interpretation of Gain Variation with Frequency

Geometrically, the representative point M of the complex number $z = e^{j\omega T}$ lies on the circle of radius 1 (Fig. 14.2 a). Its position on the circle depends on the angular frequency ω. The frequency response module is $1/3$ times the product of the length

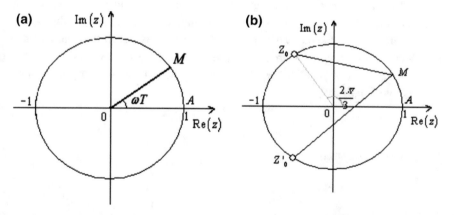

Fig. 14.2   a Representative point of a monochromatic signal; **b** segments controlling the gain

of two segments joining the point M to the two points Z_0 and Z_0' which correspond to the polynomial zeros (Fig. 14.2 b). The frequency response is

$$H(e^{j\omega T}) = \frac{1}{3}e^{-j\omega T}(e^{j\omega T} - z_0)(e^{j\omega T} - z_0^*). \tag{14.24}$$

Its modulus is $\left|H(e^{j\omega T})\right| = \frac{1}{3}\left|(e^{j\omega T} - z_0)\right|\left|(e^{j\omega T} - z_0^*)\right|.$

$$\left|H(e^{j\omega T})\right| = \frac{1}{3}\left|(e^{j\omega T} - z_0)\right|\left|(e^{j\omega T} - z_0^*)\right| = \frac{1}{3}MZ_0 \cdot MZ_0'. \tag{14.25}$$

When the frequency varies, the point M scans the unit circle. When this point approaches the point Z_0 (when ωT approaches $\frac{2\pi}{3}$) the length of segment MZ_0 decreases and becomes zero for M at point Z_0 (for $\omega_0 T = \frac{2\pi}{3}$, or $\omega_0 = \frac{2\pi}{T}\frac{1}{3} = \frac{\omega_e}{3}$). The gain will increase again when M scans the arc $Z_0\,Z_0'$ and cancels again for M in Z_0'. Thus one qualitatively explains the shape of the gain module function of ω shown in Fig. 14.3.

We could have directly calculated the frequency response $H(e^{j\omega T})$ as the FT of the impulse response

$$H(e^{j\omega T}) = \frac{1}{3}\left(e^{-j\omega T} + 1 + e^{j\omega T}\right) = \frac{1}{3}(1 + 2\cos \omega T). \tag{14.26}$$

Figure 14.3 shows that the filter gain is important at low frequencies. Although the gain is a periodic function and finds maximum values at frequencies multiple of $\frac{2\pi}{T}$, we still speak of a **low-pass filter** for the following reason. It is important to understand that the neighborhoods of $\omega_n = n\frac{2\pi}{T}$ are also low frequency neighborhoods. Indeed a monochromatic signal of the form $e^{jn\omega T}$ at these frequencies will have the form $e^{jn\frac{2\pi}{T}T} = e^{j2\pi n} = 1$ which is the epitome of a continuous signal.

In contrast, the frequency zone around $\omega = \omega_N = \frac{\pi}{T}$ is a high frequency area as the monochromatic signal with the form $e^{jn\omega T}$ will have the form $e^{jn\frac{\pi}{T}T} = e^{j\pi n} = (-1)^n$ for $\omega = \omega_N = \frac{\pi}{T}$, which signal changes sign at each instant being an example of a signal of high frequency.

Fig. 14.3 Gain magnitude

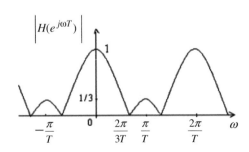

In conclusion, since the frequency response is a periodic function, we may restrict its analysis to a period, in the interval $\left\{-\frac{\pi}{T}, \frac{\pi}{T}\right\} = \left\{-\frac{\omega_e}{2}, \frac{\omega_e}{2}\right\}$, whose center being the low-frequency area, the vicinity of the edges being the high frequency area.

Example The above filter is not very selective. For a low-pass filter more selective, you can add two zeros located in the high-frequency area.

For $Z_1, \omega_1 T = \frac{5}{12} 2\pi$ (this corresponds to a 150° angle of 150° of OM with the horizontal axis). The second zero is chosen to be the complex conjugate of the previous to ensure the impulse response to be real. For $Z_1', \omega_1' T = -\frac{5}{12} 2\pi$.

$H(z)$ will have the form

$$H(z) = \left(z - e^{j\frac{2\pi}{3}}\right)\left(z - e^{-j\frac{2\pi}{3}}\right)\left(z - e^{j2\pi\frac{5}{12}}\right)\left(z - e^{-j2\pi\frac{5}{12}}\right). \tag{14.27}$$

In polynomial form

$$\begin{aligned}
H(z) &= (z - z_0)(z - z_0^*)(z - z_1)(z - z_1^*) \\
&= (z^2 - z(z_0 + z_0^*) + 1)(z^2 - z(z_1 + z_1^*) + 1),
\end{aligned} \tag{14.28}$$

$$H(z) = (z^2 - 2z \cos \omega_0 T + 1)(z^2 - 2z \cos \omega_1 T + 1), \tag{14.29}$$

ω_0 and ω_1 being the angular frequencies of the zeros z_0 and z_1 of the polynomial.

Because $\cos(150°) = -0.866$, we get

$$H(z) = (z^2 + z + 1)(z^2 + 1.732z + 1). \tag{14.30}$$

By developing (14.30)

$$H(z) = z^4 + 2.732z^3 + 3.732z^2 + 2.732z + 1. \tag{14.31}$$

If one wishes to impose a gain 1 at zero frequency ($z = 1$), it should be normalized by the factor

$$H(1) = 1 + 2.732 + 3.732 + 2.732 + 1 = 11.196. \tag{14.32}$$

As shown in Fig. 14.4, the gain modulus that is given by the expression $|H(e^{j\omega T})| = MZ_0 \cdot MZ_0' \cdot MZ_1 \cdot MZ_1'$ will be greater at lower frequencies (when the point M is in the right half plane) than at high frequencies (when the point M is located in the half plane to the left). The contrast passband stopband is more important in this case than in the first example because of the greater number of zeros in the left half plane.

Fig. 14.4 Segments
controlling the gain

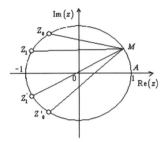

At Nyquist frequency: $\omega_N T = \pi$ and $z = -1$.

$$H(-1) = 1 + 2.732 + 3.732 - 2.732 + 1 = 0.268. \tag{14.33}$$

The attenuation at Nyquist frequency in decibels is $20 \log_{10} \frac{0.268}{11.196} = -32.4\,\text{dB}$.
Impulse response of this filter

$$h[n] = \delta[n+4] + 2.732\,\delta[n+3] + 3.732\,\delta[n+2] + 2.732\,\delta[n+1] + \delta[n]. \tag{14.34}$$

This filter is anti-causal. His transfer function has only positive powers of z.

We could have made the impulse response to be even by multiplying $H(z)$ by
z^{-2}, which amounts to shifting the time signal two steps to the right

$$H'(z) = z^{-2}(z^4 + 2.732\,z^3 \ldots) = z^2 + 2.732\,z + 3.732 + 2.732\,z^{-1} + z^{-2}, \tag{14.35}$$

$$h'[n] = \delta[n+2] + 2.732\,\delta[n+1] + 3.732\,\delta[n] + 2.732\,\delta[n-1] + \delta[n-2]. \tag{14.36}$$

Therefore

$$g'[n] = f[n-2] + 2.732f[n-1] + 3.732f[n] + 2.732f[n+1] + f[n+2]. \tag{14.37}$$

14.7 Properties of Moving Average (MA) Filters, also Called Finite Impulse Response (FIR)

Generally, a filter whose transfer function is a polynomial is called moving average
(MA) filter. This function has the general form

$$H(z) = \sum_{n=n_1}^{n=n_2} b[n]\,z^{-n}; \quad n_1 \text{ and } n_2 \text{ being two finite integers.} \tag{14.38}$$

Notable points of $H(z)$ are zeros of the polynomial. This function has no poles (except pole in $z = 0$ that can be removed by a time lag). For this reason this type of filter is also called all-zero filter.

The impulse response is limited to $N = n_2 - n_1 + 1$ points. We write

$$h[n] = \sum_{m=n_1}^{m=n_2} b[m] \, \delta[n - m]. \tag{14.39}$$

This time limitation justifies the name of finite response filter (FIR) for this type of filter.

The temporal limitation of the FIR brings benefits because the calculations are shorter.

The numerical calculation of the output signal at any instant n, can be carried out completely without the need to truncate the impulse response. In electronic circuits DSP signal processors (Digital Signal Processor), one can find a wired calculation circuit performing convolution and get the result in near-real time (with a time lag often not prohibitive).

These filters also have the advantage of allowing the realization of linear phase filters. In this case, the group and phase time delays are equal and do not depend upon frequency.

The latter property allows preserving the shape of the signal after passing through the filter.

In ultrasound echography for example, it brings an advantage for the discrimination of two close targets.

This type of filter has the major disadvantage of the slowness of the transition passband to block-band. We will use auto regressive (AR) filters which will be defined later if one wants to achieve a filter with a rapid transition from the frequency response between the passband and the attenuated band.

Linear phase filter

Let use consider a FIR filter whose impulse response is even.

$$h[n] = \ldots + a_2 \delta[n + 2] + a_1 \delta[n + 1] + a_0 \delta[n] + a_1 \delta[n - 1] + a_2 \delta[n - 2] + \ldots \tag{14.40}$$

The frequency response is

$$H(e^{j\omega T}) = 2 \left(\frac{a_0}{2} + a_1 \cos \omega T + a_2 \cos 2\omega T + \ldots \right). \tag{14.41}$$

The frequency response is real. The phase shift is always zero or π (this when the frequency response is negative).

By delaying the impulse response, for example to produce a causal filter, the phase shift is linear. Indeed if $h'[n] = h[n - m]$, $H'(e^{j\omega T}) = e^{-jm\omega T} H(e^{j\omega T})$.

The phase shift $\varphi = -m\omega T$ varies linearly with ω.

This operation is only possible with a finite impulse response filter. Causality generally required to an IIR filter prohibits most often the parity of the impulse response. In that case, the technique explained can be used in the following when a zero phase shift filter is desired.

Performing a zero phase filtering
We show now that it is possible to realize easily a zero phase filter using two consecutive filtering through the same filter and two time reversals. The supposedly real signal of limited duration is noted $f[n]$. The impulse response of the MA filter used is noted $h[n]$ and supposed to be real.

A first filtering of the signal $f[n]$ is performed in the filter. The output signal is $g[n] = f[n] \otimes h[n]$. This signal being the convolution of two real signals is real.

In the Fourier domain the following relation holds $G(e^{j\omega T}) = F(e^{j\omega T})H(e^{j\omega T})$.

Since $g[n]$ is real, FT of this signal reversed in time $g[-n]$ is $G^*(e^{j\omega T}) = F^*(e^{j\omega T})H^*(e^{j\omega T})$.

We denote $x[n]$ the result obtained by filtering $g[-n]$ with the same filter

$$x[n] = g[-n] \otimes h[n]$$

In the Fourier domain we have the relationship

$$X(e^{j\omega T}) = F^*(e^{j\omega T})H^*(e^{j\omega T})H(e^{j\omega T}) = F^*(e^{j\omega T})\left|H(e^{j\omega T})\right|^2.$$

If we time reverse $x[n]$ and note $y[n] = x[-n]$, we have in the Fourier domain

$$Y(e^{j\omega T}) = F(e^{j\omega T})\left|H(e^{j\omega T})\right|^2. \tag{14.42}$$

The output signal phase after this double filtering is identical to that of the input signal. The gain of the equivalent filter is $\left|H(e^{j\omega T})\right|^2$.

In MATLAB the filtfilt() function realizes this double filtering.

On this example, we can see the flexibility of numerical filtering operations on an operation impossible to perform (or at least very difficult) in analog signal processing.

14.8   Other Examples of All-Zero Filters (MA)

Low-Pass Filter: The moving average filter considered above, with impulse response given by (14.22), causes only an attenuation of 9.54 dB to a signal with the Nyquist frequency, relatively to continuous signal. At this frequency, a cosine signal changes sign at each value of n. (Quick change, so a high frequency character of this type of signal).

You may prefer a smoothing of the form described now which completely alleviates (obliterates) the Nyquist frequency component. The impulse response of this filter is

$$h[n] = \frac{1}{2}\delta[n+1] + \delta[n] + \frac{1}{2}\delta[n-1]. \qquad (14.43)$$

Its frequency response is

$$H(e^{j\omega T}) = \frac{1}{2}e^{j\omega T} + 1 + \frac{1}{2}e^{-j\omega T} = 1 + \cos \omega T. \qquad (14.44)$$

The shape of the gain with frequency is given in Fig. 14.5
The transfer function is

$$H(z) = \frac{z}{2} + 1 + \frac{z^{-1}}{2} = \frac{z^{-1}}{2}\left[z^2 + 2z + 1\right] = \frac{z^{-1}}{2}(z+1)^2. \qquad (14.45)$$

This function has a double root, in $z = -1$, or $z = e^{j\pi}$ (Fig. 14.6).

The zero of $H(z)$ in $z = -1$ makes it possible to completely attenuate the signal at the Nyquist frequency. We see from this example that the filter design is built by reasoning in terms of properties in the z plane. Since it was desired to eliminate the Nyquist frequency at the filter crossing, a zero of $H(z)$ was placed in $z = -1$.

Fig. 14.5 Gain magnitude

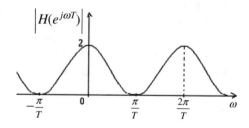

Fig. 14.6 Pole and zero locations

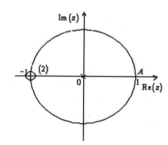

High-pass filter blocking the dc component

The following filter is often used to simulate deriving a digital signal as is done in analog processing. A zero (unique) is placed in $z = 1$ to obtain $H(e^{j\omega T})_{\omega=0} = 0$. The transfer function is

$$H(z) = z - 1. \tag{14.46}$$

A causal filter is preferred made by translating the impulse response by one time step. This amounts to multiply the transfer function by z^{-1}: $H(z) = 1 - z^{-1}$. ($H(1) = 0$ is verified).

The impulse response is

$$h[n] = \delta[n] - \delta[n - 1]. \tag{14.47}$$

The temporal filter equation is

$$g[n] = f[n] - f[n - 1]. \tag{14.48}$$

The frequency response is

$$H(e^{j\omega T}) = 1 - e^{-j\omega T}. \tag{14.49}$$

We can rewrite this function in the following form

$$H(e^{j\omega T}) = e^{-j\frac{\omega T}{2}}(e^{j\frac{\omega T}{2}} - e^{-j\frac{\omega T}{2}}) = e^{-j\frac{\omega T}{2}}2j\sin\frac{\omega T}{2}, \tag{14.50}$$

having a modulus: $|H(e^{j\omega T})| = 2|\sin\frac{\omega T}{2}|$.

We would have found qualitatively the shape of the variation of gain with frequency using geometric reasoning based on that $|H(e^{j\omega T})| = |1 - e^{-j\omega T}|$. This module is equal to the segment AM length, which increases when M moves away from the point A corresponding to zero frequency and reaches the maximum of 2 for $\omega = \frac{\pi}{T}$ (Fig. 14.7 a, b).

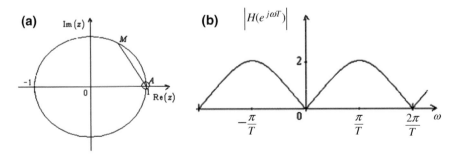

Fig. 14.7 a Segment AM controlling the gain; **b** gain magnitude

Low-frequency behavior: We are interested in the development of $H(e^{j\omega T})$ near $\omega = 0$ (For $\omega \ll \frac{\pi}{T}$).

$$H(e^{j\omega T}) = 1 - e^{-j\omega T} \cong 1 - (1 - j\omega T) = j\omega T \qquad (14.51)$$

The frequency response is proportional to $j\omega$. At low frequencies, this discrete filter has the same frequency response as the analog **derivative filter** (within a T constant factor).

Note: The frequency response of the analog derivative filter is given by $\frac{d}{dt} e^{j\omega t} = j\omega e^{j\omega t}$, giving $H_a(\omega) = j\omega$.

We note that the temporal filter equation $g[n] = \frac{1}{T}(f[n] - f[n-1])$ is a numerical approximation of the analog derivative filter.

The simulation is not as good for the digital signals of high frequency as the analog differentiator gain is π at Nyquist frequency when it is only 2 for the discrete filter.

In the following, another simulation of the derivation will be presented. The frequency response will be preserved, but at the cost of a less simple, and infinite duration, impulse response, thus less able to convolution calculation with DSP processors.

Summary

This chapter was the first treating discrete systems in this course. We have defined the concept of digital filters by using the example of a moving average filter. We have shown that for LTI systems, the function z^n is an eigenfunction of the system as was the exponential e^{st} for analog filters. The impulse response, the transfer function, and the frequency response have been defined. The z-plane plays the role played by Laplace s-plane for analog systems. We showed that the frequency response of a digital filter is periodic in frequency and made a parallel with the development in Fourier series studied in Chap. 3. We studied some examples of moving average filters and showed how we can interpret geometrically the variation of the frequency response function. The location of the zeros of the transfer function was shown to condition the frequency response.

Exercises

I. A filter has a single zero $z_1 = 1$ and a second single zero $z_2 = -1$. Give the expression of the transfer function of a filter having these zeros. Give its impulse and frequency responses, using a geometric argument to explain the variation of gain with frequency. Assuming a $F_e = 10^6$ Hz sampling frequency, what is the filter gain at frequency $f = 250\,\text{kHz}$?

Solution: We simply take $H(z) = (z - 1)(z + 1) = z^2 - 1$.

The impulse response is $h[n] = \delta[n+2] - \delta[n]$.

The frequency response is: $H(e^{j\omega T}) = e^{j2\omega T} - 1 = e^{j\omega T}\left(e^{j\omega T} - e^{-j\omega T}\right) = e^{j\omega T} 2j \sin \omega T$.

We see on the figure in the z-plane that the gain modulus $G = MA\ MB$ is zero for $\omega = 0$ and $\omega T = \pi$. It is maximum for $\omega T = \frac{\pi}{2}$, then for $f = \frac{F_e}{4} = 250\,\text{kHz}$, that is to say when the point M comes in P. The gain module is then $G = PA\ PB = 2$.

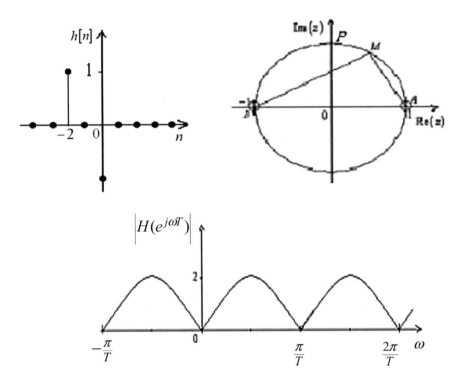

We would have found the gain analytically by writing: $G = \left|H(e^{j\omega T})\right| = 2|\sin \omega T|$. This last function cancels in $\omega = 0$ and $\omega = \frac{\pi}{T}$, and reaches its maximum for $\omega = \frac{\pi}{2T}$.

II. A filter transfer function is $H(z) = z^{-8} + 1$. What are the properties of this filter?

Solution: The zeros of the transfer function are the solutions of the equation $z^{-8} = -1$, or also $z^{8} = -1$. Written in trigonometric form: $\rho^{8} e^{j8\theta} = e^{j(\pi + 2k\pi)}$. The solutions $\rho_k e^{j\theta_k}$ are such that $\rho_k = 1$ and $\theta_k = \frac{\pi}{8} + k\frac{\pi}{4}$ (with $(k = 0, 1, \ldots 7)$). The impulse response is $h[n] = \delta[n] + \delta[n - 8]$.

The frequency response is: $H\left(e^{j\omega T}\right) = e^{-j8\omega T} + 1$. The frequency gain looks as follows:

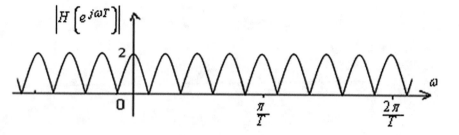

Chapter 15
Z-Transform

The z-transform plays for digital signals the role of the Laplace transform for analog signals. The circumference of the zero centered circle of radius 1 in the z-plane plays the role of the imaginary axis of frequencies in the Laplace plane. In this chapter, after having defined the z-transform and the Fourier transform of a numerical sequence, we specify the domain of convergence of the power series, that is to say, the domain of definition of the z-transform. It is shown that the domain of convergence of a causal sequence is the exterior of a disc centered at the origin with radius R_1 (conversely the interior for an anticausal sequence). In the case of a causal sequence, the Fourier transform exists if $R_1 < 1$. Using the properties of integration on a closed contour and the residue theorem in the complex plane, we demonstrate the inversion formula of the z-transform that allows the determination of the elements of a sequence provided one knows the z-transform. We show that the z-transform of a product of two series is given by a convolution formula in the frequency domain. Various properties of the z-transform are given in a table. Two interesting exercises on the z-transform of a time reversed function are given at the end of the chapter.

15.1 Definition

The z-transform of a sequence $f[n]$ is defined by the relation

$$F(z) = \sum_{n=-\infty}^{+\infty} f[n]z^{-n}. \tag{15.1}$$

This transformation is referred to as two sided since the boundaries of the sum extend to infinity on both sides.

© Springer International Publishing Switzerland 2016   253
F. Cohen Tenoudji, *Analog and Digital Signal Analysis*,
Modern Acoustics and Signal Processing, DOI 10.1007/978-3-319-42382-1_15

The Fourier transform of the digital signal is defined by

$$F(e^{j\omega T}) = \sum_{n=-\infty}^{+\infty} f[n]e^{-j\omega nT}. \tag{15.2}$$

It is the z-transform calculated for $z = e^{j\omega T}$ which belongs to the unit circle.

The function $F(z)$ appears as the sum of a power series in z. The values $f[n]$ act as the sequence of coefficients. $F(z)$ is a Laurent series.

$$F(z) = \ldots + f[-2]z^2 + f[-1]z^1 + f[0] + f[1]z^{-1} + f[2]z^{-2} + \ldots \tag{15.3}$$

_____

if the sequence $f[n]$ is anticausal

_____

if the sequence $f[n]$ is causal

For a given sequence $f[n]$, the convergence of the series, which implies the existence of $F(z)$, will be ensured only if z belongs to certain domains of the complex plane (for $|z|$ "sufficiently large" and/or $|z|$ "small enough").

More specifically

The convergence domain of a Laurent series is the intersection D of two discs centered at $z = 0$ with radii R_1 and R_2; $R_1 < |z| < R_2$ (Fig. 15.1).

Possibly, for a given sequence $f[n]$, one can have $R_1 = 0$ and $R_2 = \infty$.

For example, for a **causal** sequence, we have

$$F(z) = f[0] + f[1]z^{-1} + f[2]z^{-2} + \ldots. \tag{15.4}$$

It is clear that for $|z|$ large enough, the series giving $F(z)$ converges as long as $f[n]$ does not increase with n faster that $|z^n|$. The domain of convergence D is the outside of a disc with radius R_1. This domain $|z| > R_1$ is hatched in Fig. 15.2.

The distance from the origin $z = 0$ of the singularity of $F(z)$ the farthest from the origin is the series convergence radius. It follows that the circle with radius R_1 does not belong to the convergence domain.

The causal function $f[n]$ will have a FT or not depending whether or not $R_1 < 1$. In other words, FT exists if the circle $z = 1$ belongs to the convergence domain.

Fig. 15.1 Convergence
domain D in the general case

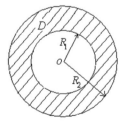

Fig. 15.2 Convergence
domain D of a causal
sequence

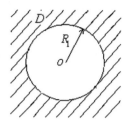

Example

The step function $U[n]$ has a z-transform for $|z| > 1$ but no Fourier transform.
Indeed its z-transform is

$$V(z) = \sum_{n=0}^{\infty} z^{-n}. \tag{15.5}$$

$V(z)$ is the sum of a geometric series with common ratio z^{-1} which converges if
$|z| > 1$.

$$V(z) = \sum_{n=0}^{\infty} z^{-n} = \frac{1}{1 - z^{-1}} = \frac{z}{z - 1} \text{ if } |z| > 1. \tag{15.6}$$

However, since the value $z = 1$ is a pole of $V(z)$, the convergence radius is
$R_1 = 1$.

Thus $U[n]$ has no FT.

For an **anticausal** sequence, the z-transform has the form

$$F(z) = \ldots f[-2]z^2 + f[-1]z^1. \tag{15.7}$$

This series converges for $|z| < R_2$ if the function is not growing too fast when
n tends towards minus infinity. The area of convergence is the hatched disc in
Fig. 15.3.

An **anticausal** function $f[n]$ has a FT if $R_2 > 1$.

Fig. 15.3 Convergence
domain D of an anticausal
sequence

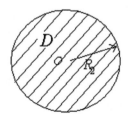

15.2   Inversion of z-Transform

The objective here is calculating $f[n]$ from its z-transform.

Beforehand, we prove the following result on integration of z^n in the complex plane on a circle centered at the origin:

$$\oint_C z^n \mathrm{d}z = \begin{vmatrix} 0 & \text{if } n \neq -1 \\ 2\pi \mathrm{j} & \text{if } n = -1 \end{vmatrix}. \tag{15.8}$$

The trigonometric form of z is used for this $z = \rho \mathrm{e}^{\mathrm{j}\theta}$. Then $z^n = \rho^n \mathrm{e}^{\mathrm{j}n\theta}$.

On the integrating circle centered in $z = 0$, we have $\mathrm{d}z = \rho \mathrm{e}^{\mathrm{j}\theta}\mathrm{j}\mathrm{d}\theta$ (Fig. 15.4).

$$\oint z^n \mathrm{d}z = \int_0^{2\pi} \rho^n \mathrm{e}^{\mathrm{j}n\theta} \rho \mathrm{e}^{\mathrm{j}\theta}\mathrm{j}\mathrm{d}\theta = \mathrm{j}\rho^{n+1} \int_0^{2\pi} \mathrm{e}^{\mathrm{j}(n+1)\theta}\mathrm{d}\theta.$$

If $n \neq -1$

$$I = \mathrm{j}\rho^{n+1} \frac{\left[\mathrm{e}^{\mathrm{j}(n+1)\theta}\right]_0^{2\pi}}{\mathrm{j}(n+1)} = 0. \tag{15.9}$$

If $n = -1$

$$I = \mathrm{j}\rho^0 \int_0^{2\pi} \mathrm{e}^0 \mathrm{d}\theta = \mathrm{j} \int_0^{2\pi} \mathrm{d}\theta = 2\pi \mathrm{j}. \tag{15.10}$$

We can calculate the values of $f[n]$ from integrals including $F(z)$.

We now show by way of example how $f[1]$ is obtained by the integration of $F(z)$ on a closed contour located in its domain of definition D and surrounding the origin.

Fig. 15.4 Integration contour

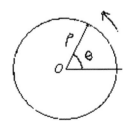

Assume that the conditions of term by term integration of the series are met

$$\oint_{C \in D} F(z)dz = \ldots \oint_C f[-1]zdz + \oint_C f[0]z^0dz + \oint_C f[1]z^{-1}dz + \ldots \quad (15.11)$$

By using the property demonstrated above, these integrals are zero except those where the term z^{-1} appears. We have therefore

$$\oint_{C \in D} F(z)dz = 2\pi j f[1]. \quad (15.12)$$

Likewise, after multiplication by z inside the integral

$$\oint_{C \in D} zF(z)dz = \ldots \oint_C f[-1]z^2dz + \oint_C f[0]z^1dz + \oint_C f[1]z^0dz + \oint_C f[2]z^{-1}dz + \ldots$$
$$= 2\pi j f[2]$$

These results may be extended to the general case. For any n we have

$$f[n] = \frac{1}{2\pi j} \oint_{C \in D} z^{n-1} F(z)dz. \quad (15.13)$$

Important note
The integration circle, centered at the origin, is obviously taken in the domain of definition of the function $F(z)$, otherwise the operation is meaningless.
   In practice, the residue theorem can be used for integration of $F(z) z^{n-1}$ using the residues at its poles included in the integration contour

$$f[n] = \sum_i \text{Residues}_i. \quad (15.14)$$

This development is valid when $F(z)$ is a rational fraction of z. This is usually the case in digital signal analysis.
   We recall here the formula given in Appendix 1 which calculates a residue. Let the function $f(z)$ having a pole of order n in $z = a$,

$$\text{Residue}\big|_{\text{in } z=a} = \frac{1}{(n-1)!} \frac{d^{n-\infty}}{dz^{n-\infty}} [(z-a)^n f(z)]_{z=a}. \quad (15.15)$$

In particular, if the circle $|z| = 1$ is located at the area of convergence, the Fourier transform does exist and the inversion formula takes a particular form on circle $z = 1$:

We note $z = e^{j\omega T}$; $dz = e^{j\omega T} jT d\omega$.

$$f[n] = \frac{1}{2\pi j} \int_0^{\frac{2\pi}{T}} e^{j(n-1)\omega T} F(e^{j\omega T}) jT e^{j\omega T} d\omega = \frac{T}{2\pi} \int_0^{\frac{2\pi}{T}} F(e^{j\omega T}) e^{jn\omega T} d\omega, \qquad (15.16)$$

or

$$f[n] = \frac{1}{\omega_e} \int_0^{\omega_e} F(e^{j\omega T}) e^{jn\omega T} d\omega. \qquad (15.17)$$

One would have found this formula considering that $F(e^{j\omega T})$ is a periodic function of ω with period $\omega_e = \frac{2\pi}{T}$. The values of the function $f[n]$ appear to be the coefficients of the Fourier series expansion of this periodic function.

15.3   z-Transform of the Product of Two Functions

Consider the function

$$y[n] = f[n]g[n]. \qquad (15.18)$$

We look to express the z-transform of $y[n]$ from those of $f[n]$ and $g[n]$.

$$Y(z) = \sum_{n=-\infty}^{+\infty} f[n] g[n] z^{-n}.$$

In the sum, we express $f[n]$ from its z-transform $F(z)$. The definition domain of $F(z)$ is noted D_f. $Y(z) = \frac{1}{2\pi j} \oint_{C \in D_f} z'^{-1} F(z') dz' \sum_{n=-\infty}^{+\infty} g[n] z^{-n} z'^n$.

$$Y(z) = \frac{1}{2\pi j} \oint_{C \in D_f} z'^{-1} F(z') dz' \sum_{n=-\infty}^{+\infty} g[n] \left(\frac{z}{z'}\right)^{-n}.$$

Noting D the intersection of definition domains of $F(z)$ and $G(z)$,

$$Y(z) = \frac{1}{2\pi j} \oint_{C \in D} F(z') G\left(\frac{z}{z'}\right) z'^{-1} dz' \qquad (15.19)$$

15.4   Properties of the z-Transform

Some important properties are summarized in the following table. Readers are invited to make their demonstrations as an exercise.

Table of z-transforms

	Sequence	Transform
	$\delta[n]$	1
	$\delta[n-m]$	z^{-m}
	$U[n]$	$\frac{z}{z-1}\,si\,|z|>1$
	$K^n U[n]$	$\frac{z}{z-K}\,si\,|z|>|K|$
	$f[n]$	$F(z)$
Delay by m steps	$f[n-m]$	$z^{-m}F(z)$
Conjugate	$f^*[n]$	$F^*(z^*)$
	$(-1)^n f[n]$	$F(-z)$
Time reversal		
• infinite signal duration	$f[-n]$	$F(z^{-1})$
• finite N signal duration	$f[-n]$	$z^{-N}F(z^{-1})$
Convolution	$f[n]\otimes g[n]$	$F(z)G(z)$
Multiplication by n	$nf[n]$	$-z\dfrac{\mathrm{d}F(z)}{\mathrm{d}z}$
$f[n]$ is symmetric conjugate	$f[n]=f^*[-n]$	$F(z)=F^*\left(\dfrac{1}{z^*}\right)$

15.5   Applications

A filter can be designed by selecting a priori its transfer function or frequency response. The impulse response of this filter is then obtained by the inversion formula.

As an example, we present the study of the digital equivalent of the analog derivative filter with conservation of the shape of the frequency response: The frequency response of the analog derivative filter is $H_a(\omega) = j\omega$. In the interval $-\frac{\omega_e}{2} < \omega < \frac{\omega_e}{2}$ the frequency response of the digital filter is selected to be $H(e^{j\omega T}) = j\omega T$. As $H(e^{j\omega T})$ is necessarily periodic with period $\omega_e = \frac{2\pi}{T}$, the frequency response over the entire frequency axis is obtained by repeating this pattern with a period ω_e (Fig. 15.5). The response of the analog filter is multiplied by T for obtaining a frequency response of the digital filter with the same dimension as the time signal, as it should be.

Fig. 15.5 Discrete
differentiator frequency
response

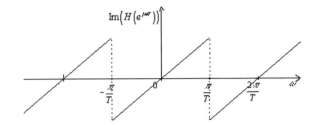

The impulse response is obtained by inverting the frequency response

$$h[n] = \frac{1}{\omega_e} \int_{-\frac{\omega_e}{2}}^{\frac{\omega_e}{2}} j\omega T e^{jn\omega T} d\omega = \frac{jT}{\omega_e} \int_{-\frac{\omega_e}{2}}^{\frac{\omega_e}{2}} \omega e^{jn\omega T} d\omega. \tag{15.20}$$

The integration is done by parts: We note

$$\omega = u; \ du = d\omega; \ e^{jn\omega T} d\omega = dv; \ v = \frac{e^{jn\omega T}}{jnT} \ (\text{if } n \neq 0).$$

$$h[n] = \frac{jT}{\omega_e} \left\{ \left[\frac{\omega e^{jn\omega T}}{jnT} \right]_{-\frac{\omega_e}{2}}^{\frac{\omega_e}{2}} - \int_{-\frac{\omega_e}{2}}^{\frac{\omega_e}{2}} \frac{e^{jn\omega T}}{jnT} d\omega \right\}$$

$$= \frac{jT}{\omega_e} \left\{ \frac{\frac{\omega_e}{2} e^{jn\frac{\omega_e}{2}T} + \frac{\omega_e}{2} e^{-jn\frac{\omega_e}{2}T}}{jnT} - \left[\frac{e^{jn\omega T}}{-n^2T^2} \right]_{-\frac{\omega_e}{2}}^{\frac{\omega_e}{2}} \right\}$$

$$h[n] = \frac{1}{n}\cos n\pi + \frac{jT}{\omega_e n^2 T^2}\left(e^{jn\frac{\omega_e}{2}T} - e^{-jn\frac{\omega_e}{2}T} \right); \ h[n] = \frac{1}{n}\cos n\pi + \underset{\underset{0}{\uparrow}}{\frac{j}{\omega_e n^2 T}} 2j\sin n\pi$$

The second term of the right side is always zero when $n \neq 0$.
Finally $h[n] = \frac{1}{n}\cos n\pi$ if $n \neq 0$;
For $n = 0$ the calculation is resumed at the beginning (Fig. 15.6):

$$h[0] = \frac{jT}{\omega_e} \int_{-\frac{\omega_e}{2}}^{\frac{\omega_e}{2}} \omega d\omega = \frac{jT}{\omega_e}\left[\frac{\omega^2}{2} \right]_{-\frac{\omega_e}{2}}^{\frac{\omega_e}{2}} = 0.$$

Fig. 15.6 Discrete differentiator impulse response

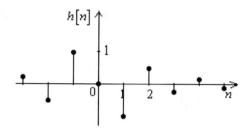

In summary

$$h[n] = (-1)^n \frac{1}{n} \text{ when } n \neq 0 \text{ and } h[n] = 0 \text{ for } n = 0. \qquad (15.21)$$

We verify that $h[n]$ is an odd function which gives a purely imaginary frequency response. We find again the slow decay of $h[n]$ as $\frac{1}{n}$. This function can be considered to consist of Fourier series expansion coefficients of a periodic function in frequency having discontinuities (finite step in $\frac{\omega_e}{2}$ in the first period).

Summary

This chapter provides the main properties of the z-transform: Convergence conditions of the z-power series and definition domains of z-transform, existing conditions of the FT. We demonstrated the z-transform and Fourier transform inversion formulas and made the connection with the formula giving the coefficients of the Fourier series encountered in Chap. 3. The main properties of the z-transform are given. Two exercises on time reversal of functions whose results will be used in following chapters are given at the end of the chapter.

Exercises

I. Calculate the z-transform of the signal $f[n] = (-1)^n$ for $n \geq 0$, and zero for $n < 0$, specifying its definition domain. (Answer: $F(z) = \frac{1}{1+z^{-1}}$ with $|z| > 1$).

II. Denoting $F(z)$ the z-transform of $f[n]$, connect the z-transform and Fourier transforms of $g[n] = (-1)^n f[n]$ and $f[n]$.
   [Answer: $G(z) = F(-z); G(e^{j\omega T}) = F(-e^{j\omega T}) = F(e^{j(\omega T + \pi)})$].

III. Let $f[n]$ the function defined on the support $\{0, N\}$. Its time reverse is noted $h[n]$.
   Link the z-transforms of $f[n]$ and $h[n]$ as well as their Fourier transforms. Show that a zero z_0 of the z-transform of $f[n]$ leads to a zero $\frac{1}{z_0}$ of its time reversed z-transform.
   Solution: $F(z) = \sum_{n=0}^{N} f[n] z^{-n}$; $H(z) = \sum_{n=0}^{N} f[N-n] z^{-n}$.

By a change of variable $p = N - n$, it comes

$$H(z) = \sum_{p=0}^{N} f[p]\, z^{-N+p} = z^{-N} \sum_{p=0}^{N} f[p]\, z^{p} = z^{-N} F\left(\frac{1}{z}\right).$$

$H(e^{j\omega T}) = e^{-jN\omega T} F(e^{-j\omega T}); \quad H(e^{j\omega T}) = e^{-jN\omega T} F^*(e^{j\omega T})$ if $f[n]$ is a real function.

If $F(z) = (z - z_0)$, then $H(z) = z^{-1}\left(\frac{1}{z} - z_0\right) = -z_0 z^{-2}\left(z - \frac{1}{z_0}\right)$.

IV. The function $f[n]$ is defined on the support $\{0, N\}$. Let the function $g[n] = (-1)^n \tilde{f}[n]$, where $\tilde{f}[n]$ is the time reversed of $f[n]$. Link the z-transforms of $f[n]$ and $g[n]$ as well as their Fourier transforms.

Solution: $G(z) = \sum_{n=0}^{N} (-1)^n f[N-n]\, z^{-n} = \sum_{p=0}^{N} f[p](-1)^{N-p} z^{-N+p} = z^{-N}(-1)^N F\left(-\frac{1}{z}\right).$

$$G(e^{j\omega T}) = e^{-jN\omega T}(-1)^N F^*(e^{j(\omega T + \pi)}).$$

V. Note $F(z)$ the z-transform of $f[n]$. What is the function of which $\frac{1}{2}(F(z) + F(-z))$ is the z-transform?

Answer: The function searched for is half the sum of the respective inverse z-transforms. According to the result of exercise 2, it is:
$\frac{1}{2}(f[n] + (-1)^n f[n]) = \begin{vmatrix} f[n] & \text{if } n \text{ even} \\ 0 & \text{if } n \text{ odd} \end{vmatrix}$.

Chapter 16
Fourier Transform of Digital Signals

This chapter presents the main properties of the Fourier transformation of digital signals. Having given the definition of the Fourier transform of a digital signal, we explain the Poisson summation formula which is the essential formula for Fourier analysis of periodic signals. Then, we demonstrate the Shannon theorem which sets the conditions for which sampling takes place without loss of information. We then show the Whittaker–Shannon theorem proving that an analog signal can be reconstructed from its samples if the sampling was done respecting the Shannon condition. We also demonstrate the Parseval energy theorem for sampled signals. Since computer analysis is necessarily performed on finite-length signals, it is interesting to look into the situation where initially infinite-length signals have their support truncated by the multiplication by a finite duration window keeping only the samples lying within a time interval, a rectangular window. After calculating the Fourier transform of a digital rectangular window, we are interested in the FT of a time-limited sine function. It is found that the abrupt truncation effected by the rectangular window causes oscillations in the spectrum (manifestation of the Gibbs phenomenon) which will spread in the spectral domain. We are led to consider more gradual selecting windows which greatly reduce the oscillation amplitudes (hence the name apodization window which they are given: from Latin, removing the foot). We show an example of how the multiplication of a finite duration signal with a Hanning window can be used to distinguish a small spectral component in a spectrum dominated by a frequency component of great amplitude.

Basically, a computer is only able to calculate a spectrum for a finite number of frequencies. By selecting these frequencies as uniformly distributed, it is possible to reduce the inverse Fourier transform operation involving an integral to a discrete summation. This comes at the price of a periodization in the time domain. For a signal with finite time duration N, a periodization with a period greater than N allows an exact recovery of the original time signal without superimposition. This operation wherein the direct and inverse Fourier transforms are discrete is called discrete Fourier transform. Cooley and Tukey have shown that by choosing the number of points of this transform to be a power of 2, it is possible to perform the calculation

© Springer International Publishing Switzerland 2016
F. Cohen Tenoudji, *Analog and Digital Signal Analysis*,
Modern Acoustics and Signal Processing, DOI 10.1007/978-3-319-42382-1_16

much faster by dichotomy. This calculation algorithm is known as the Fast Fourier Transform (FFT). It has revolutionized signal analysis techniques and allows using signal processor hardware specifically designated to perform spectrum calculations in near real time. We explain its principle in this chapter. The chapter ends with the presentation of the interpolation of a signal by adding zeros in the conjugate domain; it is followed by showing artifacts of Fourier analysis with a computer.

The Fourier transform $F(e^{j\omega T})$ of a numerical signal $f[n]$ resulting from the sampling of an analog signal at the frequency $f_e = \frac{1}{T}$ is defined by

$$F(e^{j\omega T}) = \sum_{n=-\infty}^{+\infty} f[n]e^{-j\omega n T}. \tag{16.1}$$

For this transform may exist, the series must converge. It is necessary that the function $f[n]$ decreases sufficiently rapidly at infinity. The function $F(e^{j\omega T})$ is periodic with period $\omega_e = 2\pi f_e = \frac{2\pi}{T}$. The function $f[n]$ is found back by the inversion formula demonstrated in Chap. 15:

$$f[n] = \frac{1}{\omega_e} \int_0^{\omega_e} F(e^{j\omega T})e^{jn\omega T} d\omega. \tag{16.2}$$

Shannon's theorem expresses the relationship between the Fourier transform of a function $f[n]$ obtained by digitizing an analog function $f(t)$ and the Fourier transform of this analog function. This relationship is based on the Poisson summation formula reviewed in the following section.

16.1   Poisson's Summation Formula

We want to evaluate the following sum:

$$Y(e^{j\Omega T}) = \sum_{n=-\infty}^{+\infty} e^{nj\Omega T}. \tag{16.3}$$

It is a Fourier series of the variable Ω. The function thus developed must be periodic in Ω of period $\frac{2\pi}{T} = \omega_e$. The Fourier coefficients are all equal to 1. The integrals must all give 1 for every n:

$$\frac{1}{\omega_e} \int_{-\frac{\omega_e}{2}}^{\frac{\omega_e}{2}} Y(e^{j\Omega T})e^{-jn\Omega T} d\Omega = 1 \quad \forall n. \tag{16.4}$$

Intuitively, we understand that the function $Y(e^{j\Omega T})$ must have a very particular form, since the result of the integration of its multiplication by functions $e^{-jn\Omega T}$ with oscillations varying with n has always the same result. We conclude that this function cannot be defined other than in $\Omega = 0$. It follows that we necessarily have $Y(e^{j\Omega T}) = \omega_e \delta(\Omega)$ in the interval $\left\{-\frac{\omega_e}{2}, \frac{\omega_e}{2}\right\}$. It is a Dirac distribution localized in $\Omega = 0$.

$Y(e^{j\Omega T})$ being a periodic function with period ω_e, and as the Dirac distributions do not overlap, we have on the entire frequency axis

$$Y(e^{j\Omega T}) = \omega_e \sum_{l=-\infty}^{+\infty} \delta(\Omega - l\omega_e). \tag{16.5}$$

Then we get the Poisson summation formula:

$$\sum_{n=-\infty}^{+\infty} e^{jn\Omega T} = \omega_e \sum_{l=-\infty}^{+\infty} \delta(\Omega - l\omega_e) \text{ with } \omega_e = \frac{2\pi}{T}. \tag{16.6}$$

This formula is fundamental to the study of periodic signals and sampled signals.

16.2   Shannon Aliasing Theorem

Let $f(t)$ be a function of the continuous-time (analog signal). Its Fourier transform is given by

$$F_a(\omega) = \int_{-\infty}^{+\infty} f(t) e^{-j\omega t} dt. \tag{16.7}$$

We sample $f(t)$ to create the sequence $f[n]$: $f[n] = f(nT)$.
The Fourier transform of the numerical signal is defined as

$$F(e^{j\omega T}) = \sum_{n=-\infty}^{+\infty} f[n] e^{-j\omega nT}. \tag{16.8}$$

The Fourier transform of the analog function $f(t)$ is noted as $F_a(\omega)$. The sampled values of $f(t)$ can be calculated by the inverse Fourier transform of the function $F_a(\omega)$:

$$f[n] = f(nT) = \frac{1}{2\pi} \int_{-\infty}^{+\infty} F_a(\omega') e^{j\omega' nT} d\omega'. \tag{16.9}$$

By replacing these expressions of $f[n]$ from integrals in the discrete sum, the numerical Fourier transform can be rewritten as

$$F\left(e^{j\omega T}\right) = \frac{1}{2\pi} \sum_{n=-\infty}^{+\infty} \int_{-\infty}^{+\infty} F_a(\omega')e^{j\omega'nT}e^{-j\omega nT}d\omega'. \qquad (16.10)$$

By swapping the order of summations, we have

$$F\left(e^{j\omega T}\right) = \frac{1}{2\pi} \int_{-\infty}^{+\infty} F_a(\omega')d\omega' \sum_{n=-\infty}^{+\infty} e^{j(\omega'-\omega)nT}. \qquad (16.11)$$

After applying the Poisson summation formula:

$$F\left(e^{j\omega T}\right) = \frac{\omega_e}{2\pi} \int_{-\infty}^{+\infty} F_a(\omega')d\omega' \sum_{l=-\infty}^{+\infty} \delta(\omega' - \omega - l\omega_e). \qquad (16.12)$$

We integrate over ω' and get

$$F\left(e^{j\omega T}\right) = \frac{1}{T} \sum_{l=-\infty}^{+\infty} F_a(\omega + l\omega_e). \qquad (16.13)$$

It appears that $F\left(e^{j\omega T}\right)$ is the sum of the analog Fourier transform and all its translations by a multiple of the sampling angular frequency ω_e. This is the formula that describes the folding of spectra.

We note the general property that the sampling of a function in a domain (here the time domain) is accompanied by a periodization in the conjugate domain (in this case, the Fourier domain).

2 Cases Are Possible

(a) The support of $F_a(\omega)$ is greater than ω_e. There will be overlap. In the case of Fig. 16.1, the support of $F_a(\omega)$ is limited to the interval $\{-\sigma, \sigma\}$ and the inequality $\omega_e < 2\sigma$ holds.

The information on $F_a(\omega)$ is lost in the summation.

(b) If the support of $F_a(\omega)$ is limited to interval $\{-\sigma, \sigma\}$ and $\omega_e > 2\sigma$, translated spectra do not overlap as shown in Fig. 16.2.

In this case of nonoverlapping, the information on $F_a(\omega)$ is not degraded by the summation. Between $-\frac{\omega_e}{2}$ and $\frac{\omega_e}{2}$, the numerical signal spectrum is identical (within the factor T) to the analog spectrum.

Fig. 16.1 Case of aliased
spectra overlapping

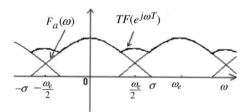

Fig. 16.2 Nonoverlapping
when Nyquist criterion is
verified

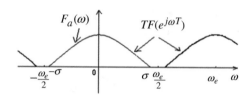

Shannon Condition

Thus, if the function $F_a(\omega)$ has a bounded support to 2σ and the sampling frequency is such that $\omega_e \geq 2\sigma$ the folding will not result in the superimposition of the motif $F_a(\omega)$ and its translated.

In this case $F(e^{j\omega T})$ and $F_a(\omega)$ will be identical in the first period of the Fourier transform of the numerical function.

This condition called **Shannon condition** requires that for a correct sampling of a signal (that is to say, so that the spectrum of the numerical signal represents exactly the one of the analog signal), it is necessary that the **sampling frequency is greater than twice the maximum frequency present in the spectrum**.

16.3   Sampling Theorem of Shannon–Whittaker

Let $f(t)$ be a continuous time function with Fourier transform $F_a(\omega)$

$$f(t) \overset{Fourier}{\longrightarrow} F_a(\omega) = \int\limits_{-\infty}^{+\infty} f(t)\, e^{-j\omega t} dt \qquad (16.14)$$

We seek a relationship between $f(t)$ and its sampled values $f(nT)$ used to define $F(e^{j\omega T})$. If the function $F_a(\omega)$ has a bound and that the Shannon condition $\omega_e > 2\sigma$ is met holds, we can write

$$f(t) = \frac{1}{2\pi} \int\limits_{-\infty}^{+\infty} F_a(\omega) e^{j\omega t} d\omega = \frac{1}{2\pi} \int\limits_{-\sigma}^{\sigma} F_a(\omega) e^{j\omega t} d\omega = \frac{1}{2\pi} \int\limits_{-\frac{\omega_e}{2}}^{\frac{\omega_e}{2}} F_a(\omega) e^{j\omega t} d\omega,$$

or using equality between $F_a(\omega)$ and $TF(e^{j\omega T})$ valid in the interval $\{-\frac{\omega_e}{2}, \frac{\omega_e}{2}\}$ and given by (16.13):

$$f(t) = \frac{T}{2\pi} \int_{-\frac{\omega_e}{2}}^{\frac{\omega_e}{2}} F(e^{j\omega T}) \, e^{j\omega t} d\omega. \tag{16.15}$$

Remember that $F(e^{j\omega T}) = \sum_{n=-\infty}^{+\infty} f(nT)e^{-jn\omega T}$. Copying this expression in (16.15), we write

$$f(t) = \frac{1}{\omega_e} \sum_{n=-\infty}^{+\infty} f(nT) \int_{-\frac{\omega_e}{2}}^{\frac{\omega_e}{2}} e^{j\omega(t-nT)} d\omega. \tag{16.16}$$

The integration on ω gives

$$\int_{-\frac{\omega_e}{2}}^{\frac{\omega_e}{2}} e^{j\omega(t-nT)} d\omega = \frac{e^{j\frac{\omega_e}{2}(t-nT)} - e^{-j\frac{\omega_e}{2}(t-nT)}}{j(t-nT)} = \frac{2\sin\left(\frac{\omega_e}{2}(t-nT)\right)}{t-nT}. \tag{16.17}$$

We multiply both terms of the fraction by $\frac{\omega_e}{2}$

$$f(t) = \frac{1}{\omega_e} \frac{\omega_e}{2} 2 \sum_{n=-\infty}^{+\infty} f(nT) \frac{\sin\frac{\omega_e}{2}(t-nT)}{\frac{\omega_e}{2}(t-nT)}, \tag{16.18}$$

$$f(t) = \sum_{n=-\infty}^{+\infty} f(nT) \sin c\left(\frac{\omega_e}{2}(t-nT)\right). \tag{16.19}$$

This results known as Shannon–Whittaker sampling theorem states that from samples $f(nT)$ taken at times nT, one can find back the value of function $f(t)$ for all t.

The information is not lost in the sampling operation. The cardinal sine functions act as interpolation functions.

It is interesting to analyze an example of the recovery of the function $f(t)$ from its samples $f[nT]$. In this example, assume that the Shannon condition is met and that the values of $f(nT)$ are nonzero only between $t = -T$ and $t = 7T$:

$f(nT) = 0.8; \ 1; \ 1.2; \ 1.6; \ 1.9; \ 2.2; \ 1.5; \ 0.7$. For $n = -1, 0, 1\ldots, 6$ (Fig. 16.3).

Note that each sin c has its zeros for time values equal to nT. The sum signal $f(t)$, in bold in Fig. 16.3, passes by all the sampled values $f[nT]$. The residual

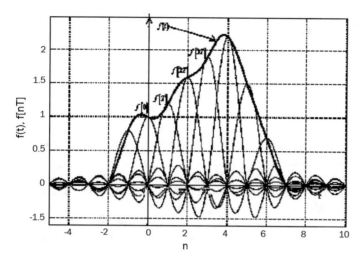

Fig. 16.3 Reconstruction of a function from its samples

oscillation that appears on $f(t)$ in this simulation is a manifestation of the Gibbs phenomenon: a limited frequency band signal often exhibits oscillations localized near the areas of rapid variation of the time signal.

16.4 Application of Poisson's Summation Formula: Fourier Transform of a Sine

Let $f[n] = \sin(n\omega_o T)$. Its Fourier transform is written as

$$F\left(e^{j\omega T}\right) = \sum_{n=-\infty}^{+\infty} \sin(n\omega_o T) e^{-jn\omega T} = \frac{1}{2j} \sum_{n=-\infty}^{+\infty} \left(e^{jn\omega_o T} - e^{-jn\omega_o T}\right) e^{-jn\omega T}. \quad (16.20)$$

$F\left(e^{j\omega T}\right)$ consists of two sums of exponential products which are calculated using the Poisson's summation formula:

$$\sum_{n=-\infty}^{+\infty} e^{jn(\omega_o-\omega)T} = \omega_e \sum_{n=-\infty}^{+\infty} \delta(\omega - \omega_o - n\omega_e). \quad (16.21)$$

Therefore,

$$F\left(e^{j\omega T}\right) = -\frac{j\omega_e}{2} \left(\sum_{n=-\infty}^{+\infty} \delta(\omega - \omega_o - n\omega_e) - \delta(\omega + \omega_o - n\omega_e) \right). \quad (16.22)$$

Fig. 16.4 Two Dirac combs
in the FT of a sine function

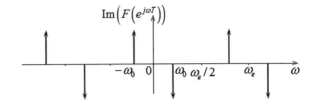

The spectrum of a digitized sine is thus constituted by two Dirac combs of period ω_e. (See Fig. 16.4).

The analysis of numerical signals is done in practice necessarily upon time-limited signals. Although we can theoretically calculate the Fourier transform of the sine function over the whole axis, arrays in computer calculation have necessarily finite length. Accordingly, numerical computation is only able to provide the Fourier transform of the sine function limited to a finite time interval.

16.5   Fourier Transform of a Product of Functions of Time

The multiplication of two signals in the time domain corresponds to their convolution in the frequency domain.

Note $y[n]$, the product of functions $f[n]$ and $w[n]$:

$$y[n] = f[n]w[n]. \tag{16.23}$$

$$\sum_{n=-\infty}^{+\infty} f[n]w[n]e^{-jn\omega T} = \sum_{n=-\infty}^{+\infty} f[n]\frac{1}{\omega_e}\int_0^{\omega_e} W\left(e^{j\omega'T}\right)e^{j\omega'nT}e^{-jn\omega T}d\omega'. \tag{16.24}$$

By swapping the orders of the integral and of summation:

$$\sum_{n=-\infty}^{+\infty} f[n]w[n]e^{-jn\omega T} = \frac{1}{\omega_e}\int_0^{\omega_e} W\left(e^{j\omega'T}\right)d\omega' \sum_{n=-\infty}^{+\infty} f[n]e^{-j(\omega-\omega')nT}. \tag{16.25}$$

Finally,

$$Y\left(e^{j\omega T}\right) = \frac{1}{\omega_e}\int_0^{\omega_e} W\left(e^{j\omega'T}\right)F\left(e^{j(\omega-\omega')T}\right)d\omega'. \tag{16.26}$$

This is a "convolution integral" in the frequency domain specific to the case of numerical signals.

We note that the Fourier transform of a product of two numerical signals is obtained from the "circular convolution" Fourier transforms of the two functions. The integration takes place over a period in the Fourier domain.

16.6   Parseval's Theorem

The energy E of a numerical signal is defined by

$$E = \sum_{n=-\infty}^{+\infty} |f[n]^2|. \tag{16.27}$$

Parseval theorem states that energy can also be calculated in the frequency domain. To demonstrate this, we treat the general case of a complex signal $f[n]$. To use the result of the preceding paragraph, we write

$$y[n] = f[n]f^*[n]. \tag{16.28}$$

Its Fourier transform is

$$\sum_{n=-\infty}^{+\infty} y[n]e^{-jn\omega T} = \sum_{n=-\infty}^{+\infty} f[n]f^*[n]e^{-jn\omega T} = \frac{1}{\omega_e} \int_0^{\omega_e} F\left(e^{j\omega'T}\right) F^*\left(e^{j(\omega-\omega')T}\right) d\omega'. \tag{16.29}$$

Taking $\omega = 0$ in the above relation, we get

$$E = \sum_{n=-\infty}^{+\infty} f[n]f^*[n] = \frac{1}{\omega_e} \int_0^{\omega_e} F(e^{j\omega'T})F^*\left(e^{-j\omega'T}\right) d\omega' = \frac{1}{\omega_e} \int_0^{\omega_e} \left|F\left(e^{j\omega'T}\right)\right|^2 d\omega'. \tag{16.30}$$

The latter relationship is the expression of the Parseval theorem for numerical signals:

$$\sum_{n=-\infty}^{+\infty} |f[n]|^2 = \frac{1}{\omega_e} \int_0^{\omega_e} \left|F(e^{j\omega T})\right|^2 d\omega. \tag{16.31}$$

If the signal $f[n]$ is real, Parseval theorem takes the form

$$\sum_{n=-\infty}^{+\infty} f^2[n] = \frac{1}{\omega_e} \int_0^{\omega_e} \left|F\left(e^{j\omega T}\right)\right|^2 d\omega. \tag{16.32}$$

16.7   Fourier Transform of a Rectangular Window

Let the function $w_r[n]$ be equal to 1 between 0 and $N - 1$ and zero elsewhere. Its Fourier transform is written as

$$W_r\left(e^{j\omega T}\right) = \sum_{n=0}^{N-1} e^{-j\omega nT} = \frac{1 - e^{-j\omega NT}}{1 - e^{-j\omega T}} = \frac{e^{-j\omega\frac{NT}{2}}}{e^{-j\frac{\omega T}{2}}} \frac{\sin N \frac{\omega T}{2}}{\sin \frac{\omega T}{2}}. \tag{16.33}$$

At zero frequency $W_r(\omega = 0) = N$. The first part of $W_r\left(e^{j\omega T}\right)$ is a phase term with modulus 1. In mathematics, the function $\frac{\sin N\frac{\omega T}{2}}{\sin \frac{\omega T}{2}}$ is called a Dirichlet function. In optics, it is called the grating function, because it is related to the amplitude of the diffracted wave by an optical grating. The maximum of the function is in $\omega = 0$, where the resolution of the indeterminate form gives the value N. The zeros of the Dirichlet function occur for frequencies such that $\sin N \frac{\omega T}{2} = 0$, thus for $N \frac{\omega_k T}{2} = k\pi$ with k integer, giving $\omega_k = k \frac{\omega_e}{N}$. There will be N zeros in the interval $\{0, \omega_e\}$.

This function is represented in linear scale on the left and in decibels on right in Fig. 16.5 in the case $N = 21$. It is verified that, except for the first peak, the zeros of the function are regularly distributed.

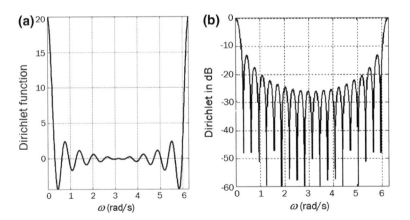

Fig. 16.5 Dirichlet function; **a** In linear scale; **b** In decibels

Fig. 16.6 FT of a rectangular window for N = 1000

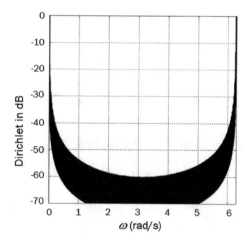

We now evaluate the amplitude the first secondary maximum when N is great. The oscillation of the sine in numerator is fast and thus the first zeros of the numerator are obtained for small values of the factor $\frac{\omega T}{2}$. So under these conditions we can use the following approximation $\sin \frac{\omega T}{2} \cong \frac{\omega T}{2}$ in the denominator. The first secondary maximum is obtained for $\sin N \frac{\omega_1 T}{2} = -1$, that is to say for $N \frac{\omega_1 T}{2} = \frac{3\pi}{2}$.
So

$$\left| W_r \left(e^{j\omega_1 T} \right) \right| \cong \frac{2}{\omega T} = \frac{2N}{3\pi}. \tag{16.34}$$

The relative amplitude of the first secondary maximum is $\frac{|W_r|_{1\text{stlobe}}}{|W_r|_{\text{maxi}}} = \frac{2}{3\pi}$. Expressed in $20 \log_{10} \frac{2}{3\pi} = -13.46\,\text{dB}$, as seen on Fig. 16.5b. This lobe has a significant relative importance.

The Dirichlet function plays for numerical signals a role similar to the $\sin c$ function for analog signals. As seen in Fig. 16.6, the height of the peaks of oscillations of the function slowly decreases with frequency. It is still about -60 dB at the Nyquist frequency in the case $N = 1000$.

16.8 Fourier Transform of a Sine Function Limited in Time

We want to calculate the Fourier transform of a sine limited in time. This limitation problem occurs generally when attempting to calculate numerically a Fourier transform, the signals then having of necessity a finite duration. Assume the signal length limited to N points.

Let us calculate

$$Y\left(e^{j\omega T}\right) = \sum_{n=0}^{N-1} \sin(n\omega_o T)e^{-jn\omega T}. \tag{16.35}$$

To bring out a Fourier transform which requires a summation from minus infinity to plus infinity on summation index n, we multiply $\sin(n\omega_o T)$ by the rectangular window $w_r[n]$ equal to $\begin{vmatrix} 1 & \text{from 0 to } N-1 \\ 0 & \text{elsewhere} \end{vmatrix}$. This allows extending the bounds of the summation to infinity. $Y\left(e^{j\omega T}\right)$ remains unchanged:

$$Y\left(e^{j\omega T}\right) = \sum_{n=-\infty}^{+\infty} w_r[n]\sin(n\omega_o T)e^{-jn\omega T}. \tag{16.36}$$

It appears the Fourier transform of a simple product whose result is the convolution of the Fourier transforms of these functions. As shown above, the Fourier transform of the rectangular window is

$$W_r\left(e^{j\omega T}\right) = e^{-j(N+1)\frac{\omega T}{2}}\frac{\sin N\frac{\omega T}{2}}{\sin\frac{\omega T}{2}}. \tag{16.37}$$

In computing the $Y\left(e^{j\omega T}\right)$, the Fourier transform of the sine being a Dirac comb, one will have to calculate convolution integrals of the form

$$I = \frac{1}{\omega_e}\omega_e \int_{0}^{\omega_e} \delta(\omega_o - \omega')e^{-j(N+1)(\omega-\omega')\frac{T}{2}}\frac{\sin\frac{N(\omega-\omega')T}{2}}{\sin\frac{(\omega-\omega')T}{2}}d\omega'. \tag{16.38}$$

$$I = e^{-j(N+1)(\omega-\omega_o)\frac{T}{2}}\frac{\sin\frac{N(\omega-\omega_o)T}{2}}{\sin\frac{(\omega-\omega_o)T}{2}}.$$

The zeros of this function noted ω_l are given by

$$N\left(\frac{\omega_l - \omega_o}{2}\right)T = l\pi.$$

So

$$(\omega_l - \omega_o) = \frac{2\pi}{T}\frac{l}{N}, \text{ or : } \omega_l = \omega_o + \frac{2\pi}{T}\frac{l}{N}. \tag{16.39}$$

Thus, in the Fourier transform of a sine limited in time, the Dirichlet functions

Fig. 16.7 Discrete FT of a sine at frequency ω_0

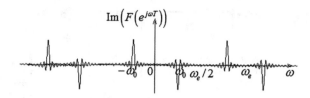

$$\sin \frac{N(\omega - n\omega_e - \omega_0)T}{2} \over \sin \frac{(\omega - n\omega_e - \omega_o)T}{2} \tag{16.40}$$

replace the Dirac distributions which appear in the FT of a continuous time sine. These functions are decreasing slowly as ω moves away from an angular frequency $n\omega_e - \omega_0$. (See Fig. 16.7).

This results in a spreading on the whole frequency axis of the Fourier transform of this truncated sine. This is particularly troublesome when attempting to identify small spectral components in a spectrum.

It is for this reason that a time-limited signal is often multiplied by a time window of a different shape whose Fourier transform spreading extends over a smaller interval of frequency. These windows are called apodization windows.

16.9 Apodization Windows

The general property of these windows is that the more the function is regular in the time domain (continuity of the function and its first derivatives at the edges of the window), the less the frequency spreading will be high.

For example, the Hann window which in its analog form is a continuous function and whose derivative is continuous, has a frequency spectrum more compact than that of the rectangular window having a discontinuity at its boundaries.

Hamming windows:

$$w_H[n] = \alpha + (1 - \alpha) \cos \frac{2\pi n}{N}, \tag{16.41}$$

where $-\left(\frac{N-1}{2}\right) \le n \le \frac{N-1}{2}$ and $0 < \alpha < 1$.

A special case of this window is the von Hann window (Hanning window) when $\alpha = 0.5$.

The advantage thereof is that the window function is connected continuously with zero values outside the central range.

The main lobe of the Fourier transform of this window is twice wider than that of the rectangular window, but the first side lobe has a much smaller amplitude (-31.4 dB instead of -13.6 dB).

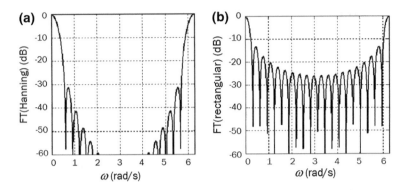

Fig. 16.8 FT transforms of two windows; **a** Hanning; **b** Rectangular

One can wonder about the interest of a Hamming window in case $\alpha \neq 0.5$. Indeed, in this case, the function is discontinuous at its edges, resulting in frequency spreading. The lobes of the Fourier transform in this window are less regular. One can take advantage of this irregularity, by a proper choice of α value, to lower the amplitude of a lobe to permit detection of a small signal at the frequency of the lobe.

Figure 16.8 shows on the same scale the moduli of the FT of a Hanning window (Fig. 16.8a) and a rectangular window (Fig. 16.8b) in the case $N = 21$.

Note the relatively rapid decay of the oscillations of the Hanning window. This rapid decrease is accompanied by an expansion by a factor of 2 of the main peak. The Hanning window is an example of the apodization windows which aims to reduce the amplitude of the oscillations peaks around the main peak.

Tukey window:

This window is also called edge cosine window. Its temporal form is a constant central plateau connected by half cosine cycles to zero at both ends. The decay of the oscillations in the Fourier domain is less rapid than that of the Hanning window but the signal energy is larger, which may be advantageous in the analysis of signals in the presence of random noise.

Blackman window:

$w[n] = a_0 - a_1 \cos\left(\frac{2\pi n}{N-1}\right) + a_2 \cos\left(\frac{4\pi n}{N-1}\right)$ with $a_0 = \frac{1-\alpha}{2}$; $a_1 = \frac{1}{2}$; $a_2 = \frac{\alpha}{2}$.

Typically for this window, $\alpha = 0.16$.

Kaiser window:

This window is considered excellent. The amplitude of the oscillation in frequency decreases more rapidly than the oscillation of the Hanning window.

$$w_K[n] = \frac{I_0\left(\beta\sqrt{1 - \left[\frac{2n}{(N-1)}\right]^2}\right)}{I_0(\beta)}. \tag{16.42}$$

I_0 is the modified Bessel function of zero order. β is a parameter varying in practice between 2 and 30. The greater is this parameter, the greater selection is effective. This window can give a -60 dB second lobe.

Practical calculation:

$$I_0(x) = 1 + \sum_{k=1}^{\infty} \left[\frac{(x/2)^k}{k!}\right]^2. \tag{16.43}$$

The summation can be limited to the first 5 terms for $x = 0.5$ and to 25 for $x = 19.0$.

In Matlab, the `kaiser()` function provides the numerical values of this window.

The rectangular window is the best for detecting a signal in noise. The width of the central lobe of the Fourier transform of the Blackman window is 1.9 that of Hanning 1.28.

The first Blackman secondary lobe is 7.4 dB below that of the Hanning window.

Energy ratios:

$$10 \log_{10}\left(\frac{\text{Hanning energy}}{\text{Blackman energy}}\right) = 0.9\,\text{dB};$$

$$\left(\frac{\text{Hanning energy}}{\text{Rectangular energy}}\right) = 0.375 \Rightarrow -4.26\,\text{dB}.$$

The following example illustrates the advantage of multiplying a signal by an apodization window. It is assumed that the signal $f[n]$ is limited to $N = 2048$ points and is composed by the sum of two sines with frequencies f_0 and f_1. We assume that the amplitude of the sine with frequency f_1 is relatively much lower (10^{-5}) than the first.

$f[n] = \sin(2\pi f_0 nT) + \varepsilon \sin(2\pi f_1 nT)$ with $\varepsilon = 10^{-5}$, $f_0 = f_e/50$, and $f_1 = 3.5 f_0$.

It is impossible to detect the presence of the small signal on the time display of the signal (Fig. 16.9a). Its Fourier transform is shown in Fig. 16.9b. The oscillations of the Fourier transform are not apparent because the number of points N of the FFT is the same as the length of the signal (this fact will be discussed in the following section on FFT). This frequency curve does not detect the presence of small signal, due to the spreading of the spectrum of the main sinus. This spreading is caused by the rectangular window effect due to the intrinsic limitation of the

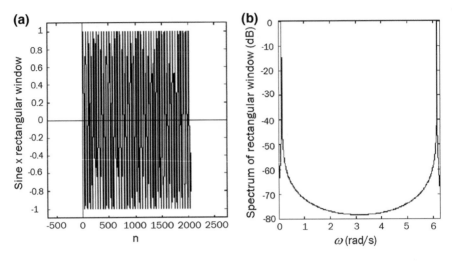

Fig. 16.9 Analysis of the sum of two sine functions (rectangular window); **a** In time domain; **b** In Fourier domain

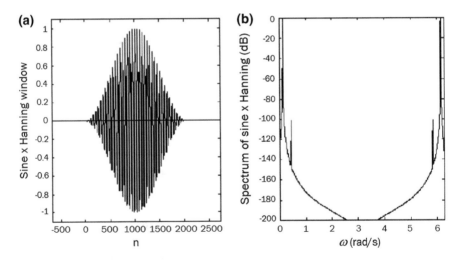

Fig. 16.10 Analysis of the sum of two sine functions (Hanning window); **a** In time domain; **b** In Fourier domain

signal length to 2048 points. This spreading completely invades the small sine spectrum area.

The result of the multiplication of the signal $f[n]$ by a Hanning window is shown below on the left. Its spectrum in decibels on his right shows clearly the component at frequency f_1. We further note the relative height of -100 dB between the peaks that can be evaluated in Fig. 16.10 corresponding to $\varepsilon = 10^{-5}$.

In summary, brutal truncation of the signal caused to the edges of the rectangular window has resulted in the frequency spreading. The gradual shift toward zero provided by the multiplication by the apodization window reduces spreading and allows detection of the presence of a small harmonic component and measurement of its amplitude. We see here again the interest of the logarithmic representation (dB) for the detection of small harmonic components.

16.10   Discrete Fourier Transform (DFT)

The Fourier transform of the numerical signal $f[n]$ is given by

$$F\left(e^{j\omega T}\right) = \sum_{n=-\infty}^{n=+\infty} f[n]e^{-jn\omega T}. \tag{16.44}$$

The inversion formula giving $f[n]$ is an integral over a period of the Fourier transform:

$$f[n] = \frac{1}{\omega_e} \int_0^{\omega_e} F\left(e^{j\omega T}\right)e^{jn\omega T}\,d\omega. \tag{16.45}$$

We place ourselves in the context of a numerical calculation of the integral where a discrete sum is used as an approximation of the integral. The result of this sum is only an approximation of $f[n]$ and is denoted differently, $f_p[n]$.

$$f_p[n] = \frac{1}{\omega_e} \sum_{k=0}^{N-1} F\left(e^{j\omega_k T}\right)e^{jn\omega_k T}\Delta\omega. \tag{16.46}$$

The interval $\{0, \omega_e\}$ has been divided into N intervals with width $\Delta\omega = \frac{\omega_e}{N}$. The samples of the function $F\left(e^{j\omega T}\right)$ at N points were noted $F\left(e^{j\omega_k T}\right)$, with angular frequencies

$$\omega_k = k\frac{\omega_e}{N} \quad \text{and } k = 0, 1, 2, \ldots, N-1. \tag{16.47}$$

We also note $F[k]$ these samples that are naturally given by:

$$F[k] = F\left(e^{j\omega_k T}\right) = \sum_{n=-\infty}^{+\infty} f[n]e^{-jn\omega_k T} \text{ with } \omega_k = k\frac{\omega_e}{N}. \tag{16.48}$$

It becomes

$$e^{jn\omega_k T} = e^{jnk\frac{\omega_e}{N}T} = e^{j2\pi\frac{nk}{N}} \text{ since } \omega_e T = 2\pi.$$

Therefore,

$$f_p[n] = \frac{1}{N}\sum_{k=0}^{N-1} F[k]e^{j2\pi\frac{nk}{N}}. \tag{16.49}$$

It is to be noted that because the exponential are periodic functions, $f_p[n]$ is also a periodic function of n, with period N while $f[n]$ was not a priori. This explains the subscript p used to note that function.

Now we establish the relationship between $f_p[n]$ and $f[n]$.

We replace $F[k] = F\left(e^{j\omega_k T}\right) = \sum\limits_{n=-\infty}^{+\infty} f[n']e^{-j2\pi\frac{n'k}{N}}$ in the expression giving $f_p[n]$.

It comes after swapping the order of summation:

$$f_p[n] = \frac{1}{N}\sum_{n'=-\infty}^{+\infty} f[n']\sum_{k=0}^{N-1} e^{-j2\pi\frac{(n'-n)k}{N}}. \tag{16.50}$$

It is recognized in the sum on k the sum of a geometric series.

$$S = \sum_{k=0}^{N-1} e^{-j\frac{2\pi k}{N}(n-n')} = \frac{1-e^{-j2\pi(n-n')}}{1-e^{-j\frac{2\pi}{N}(n-n')}}. \tag{16.51}$$

The numerator of this fraction is always zero since n and n' are integers.

The sum S will be zero unless the denominator is zero too. This is the case when $(n - n') = lN$.

In this case the sum is reassessed $S = \sum_{k=0}^{N-1} 1 = N$.

It becomes then

$$f_p[n] = \sum_{l=-\infty}^{+\infty} f[n+lN]. \tag{16.52}$$

We see that $f_p[n]$ is the sum of $f[n]$ and of all its translated by quantities lN. There is a temporal aliasing of function $f[n]$. $f_p[n]$ is a periodic function of period N. Sampling of the Fourier transform results in the time domain in a periodization of the signal and aliasing.

16.10.1   *Important Special Case: The DFT of a Bounded Support Function*

In the case where $f[n]$ is a function with support superiorly bounded to N points, there will be identity between $f[n]$ and $f_p[n]$ on the support of $f[n]$. In this case, we can write a pair of transforms:

$$F[k] = \sum_{n=0}^{N-1} f[n]\, e^{-j\frac{2\pi nk}{N}} \tag{16.53}$$

$$f[n] = \frac{1}{N} \sum_{k=0}^{N-1} F[k] e^{j\frac{2\pi nk}{N}}. \tag{16.54}$$

These two relationships define the Discrete Fourier Transform (DFT).

It is this transform which is numerically calculated by the Fast Fourier Transform algorithm (FFT) originally developed by Cooley and Tukey. In practice, the range of variation of the index k in the frequency domain is $\{0, N-1\}$ matching the frequency interval $\left\{0, (N-1)\frac{\omega_e}{N}\right\}$.

16.11   Fast Fourier Transform Algorithm (FFT)

FFT algorithms have allowed gaining a major factor in the calculation times of the discrete Fourier transform (a factor of 100 for signals with thousand points). This time saving made possible the real-time spectral analysis of signals. This contribution was decisive in the signal analysis by computer.

This very famous algorithm was proposed by Cooley and Tukey (1965). It is based on successive subdivisions of data to be analyzed in packages of two by decimation. It assumes that the number N of data is a power of 2: $N = 2^M$.

We now show how this time saving is possible in the calculation of the formula (16.53).

It is therefore assumed in the following that N is even. We set

$$W_N = e^{-j\frac{2\pi}{N}}. \tag{16.55}$$

Equation (16.53) is

$$F[k] = \sum_{n=0}^{N-1} f[n] W_N^{nk}. \tag{16.56}$$

We verify from (16.55) that

$$W_N^{(n+mN)(k+lN)} = W_N^{nk} \text{ for all } m, l \text{ integers.} \tag{16.57}$$

It is noted that the factor W_N^{nk} is periodic in k with period N, leading to a periodic $F[k]$ with the same period. So we need to just evaluate $F[k]$ in N points.

To estimate the gain in computing time provided by the FFT, we will assess the number of operations required to calculate the function $F[k]$. For each of the N values of k, if one uses the formula (16.56), N^2 complex multiplication of the function values $f[n]$ should be performed with the factors W_N^{nk}. We will calculate N^2 values of W_N^{nk} if we take N time as well as frequency points. For $N = 1000$, we must know 10^6 the values of W_N^{nk} which is considerable. The computing of a term is important W_N^{nk} because it is constituted by calculations of sine and cosine values that are relatively long. While these values can be set in advance in a table, one should always perform N^2 complex multiplications followed by N summations. When N is large, we can estimate to about N^2 the number of required operations.

Now we examine the principle of the Cooley Tukey FFT who managed to avoid N^2 operations to bring their total to approximately $N \log_2 N$. The algorithm proceeds by a decimation in the time domain. To expose its principle, we choose an even number N of points. The sum (16.56) is separated into a sum of two terms discriminating values of the function of even and odd ranks:

$$F[k] = \sum_{n=0}^{N/2-1} f[2n] W_N^{2nk} + \sum_{n=0}^{N/2-1} f[2n+1] W_N^{(2n+1)k}. \tag{16.58}$$

Note that

$$W_N^2 = e^{-j\frac{2\pi \cdot 2}{N}} = e^{-j\frac{2\pi}{N/2}} = W_{N/2}. \tag{16.59}$$

Noting $f_1[n] = f[2n]$ and $f_2[n] = f[2n+1]$, we can write

$$F[k] = \sum_{n=0}^{N/2-1} f_1[n] W_{N/2}^{nk} + W_N^k \sum_{n=0}^{N/2-1} f_2[n] W_{N/2}^{nk}, \tag{16.60}$$

or

$$F[k] = F_1[k] + W_N^k F_2[k], \tag{16.61}$$

where $F_1[k]$ and $F_2[k]$ appear as DFT calculated in $N/2$ points.

As we now show, the advantage of this decomposition is that one only needs to calculate half of $F[k]$ values, those on the interval $0 \le k \le \frac{N}{2} - 1$. We can deduce $F[k]$ for $\frac{N}{2} \le k \le N - 1$ from these first values.

Indeed, for values of k higher than $N/2$, we write $k' = k + \frac{N}{2}$, $0 \le k \le \frac{N}{2} - 1$.

We have

$$F[k'] = F_1[k'] + W_N^{k'} F_2[k'], \tag{16.62}$$

$$F_1[k'] = \sum_{n=0}^{N/2-1} f_1[n] e^{-j\frac{2\pi}{N/2}n(k+N/2)} = \sum_{n=0}^{N/2-1} f_1[n] e^{-j\frac{2\pi}{N/2}nk} e^{-j\frac{2\pi}{N/2}nN/2}$$

$$= \sum_{n=0}^{N/2-1} f_1[n] e^{-j\frac{2\pi}{N/2}nk}.$$

Therefore, $F_1[k'] = F_1[k]$. Similarly, we have

$$F_2[k'] = F_2[k]. \tag{16.63}$$

The factor $W_N^{k'}$ in the second term remains to be considered:

$$W_N^{k'} = W_N^{k+N/2} = e^{-j\frac{2\pi}{N}(k+N/2)} = e^{-j\frac{2\pi}{N}k} e^{-j\frac{2\pi}{N}(N/2)} = -e^{-j\frac{2\pi}{N}k} = -W_N^k.$$

We can summarize these results as follows:

$$F[k] = \begin{cases} F_1[k] + W_N^k F_2[k] & \text{for} \quad 0 \le k \le \frac{N}{2} - 1 \\ F_1\left[k - \frac{N}{2}\right] - W_N^{k-N/2} F_2\left[k - \frac{N}{2}\right] & \text{for} \quad \frac{N}{2} \le k \le N - 1 \end{cases}. \tag{16.64}$$

To fix ideas, we consider the case $N = 8$:
We have $F[0] = F_1[0] + W_N^{k=0} F_2[0]; F[1] = F_1[1] + W_N^{k=1} F_2[1];$

$$F[2] = F_1[2] + W_N^{k=2} F_2[2]; F[3] = F_1[3] + W_N^{k=3} F_2[3].$$

We deduce $F[4] = F_1[0] - W_N^{k=0} F_2[0]; F[5] = F_1[1] - W_N^{k=1} F_2[1];$

$$F[6] = F_1[2] - W_N^{k=2} F_2[2]; F[7] = F_1[3] - W_N^{k=3} F_2[3].$$

As an example, we see that between the calculations of $F[0]$ and $F[4]$, only the sign preceding $W_N^{k=0}$ has changed. We saved in the operation about 50 % of the computing time.

Decimation by a factor of 2 can be repeated several times (That is the reason why an initial number N to be a power of 2 is chosen) and lead to calculations involving only grouping two terms of the function $f[n]$. The calculation time decreases by about 50 % each time.

It may be shown that while the initial calculation required about N^2 operations, the number of operations required by Cooley Tukey algorithm is $N \log_2 N$.

For N = 1024, the gain in calculation time is a factor of about 100.
For N = 4048, the gain is about 300.
For N = 16,384, the gain is about 1000.

16.12   Matrix Form of DFT

According to (16.53), the DFT of $f[n]$ is written as $F[k] = \sum_{n=0}^{N-1} f[n]\, W_N^{nk}$, where $W_N = e^{-j\frac{2\pi}{N}}$ and $0 \le k \le N-1$. A matrix form can be used:

$$
\begin{pmatrix} F[0] \\ F[1] \\ F[2] \\ \cdots \\ F[N-1] \end{pmatrix}
=
\begin{pmatrix}
1 & 1 & 1 & \cdots & 1 \\
1 & W_N & W_N^2 & \cdots & W_N^{N-1} \\
1 & W_N^2 & W_N^4 & \cdots & W_N^{2(N-1)} \\
\cdots & \cdots & \cdots & \cdots & \cdots \\
1 & W_N^{N-1} & W_N^{2(N-1)} & \cdots & W_N^{(N-1)^2}
\end{pmatrix}
\begin{pmatrix} f[0] \\ f[1] \\ f[2] \\ \cdots \\ f[N-1] \end{pmatrix}.
$$

$$(16.65)$$

The Fourier matrix has remarkable properties. Its column vectors are orthogonal two by two. Conversely, we have

$$
\begin{pmatrix} f[0] \\ f[1] \\ f[2] \\ \cdots \\ f[N-1] \end{pmatrix}
=
\frac{1}{N}
\begin{pmatrix}
1 & 1 & 1 & \cdots & 1 \\
1 & W_N^* & W_N^{*2} & \cdots & W_N^{*N-1} \\
1 & W_N^{*2} & W_N^{*4} & \cdots & W_N^{*2(N-1)} \\
\cdots & \cdots & \cdots & \cdots & \cdots \\
1 & W_N^{*(N-1)} & W_N^{*2(N-1)} & \cdots & W_N^{*(N-1)^2}
\end{pmatrix}
\begin{pmatrix} F[0] \\ F[1] \\ F[2] \\ \cdots \\ F[N-1] \end{pmatrix}.
$$

$$(16.66)$$

Except for the factor $1/N$, the inverse matrix of the linear transformation is the conjugate of the original matrix.

This matrix form of DFT is used in many calculations.

16.13   Signal Interpolation by Zero Padding

Nowadays, Fourier analysis is performed mainly on digitized signals using the discrete Fourier (16.53) and (16.54)). The calculations are carried out using FFT algorithms. By construction, the DFT provides the value of the Fourier transform of the signal $F\left(e^{j\omega T}\right)$ in a finite number of points. It is assumed that the signal is sampled in accordance with Shannon's condition and that its length is finite and equal to points N. As shown above, the DFT $F[k] = \sum_{n=0}^{N-1} f[n] e^{-j\frac{2\pi nk}{N}}$ provides the value of the signal's FT at N frequency points $\omega_k = k\frac{\omega_e}{N}$ with $k \in \{0, N-1\}$.

It is shown below that the use of the DFT with addition of zeros (zero padding) is used to interpolate the signals in frequency or time.

Frequency interpolation: Assume that we add zeros N to the sequence $f[n]$ and that we calculate the DFT on $2N$ points: $F'[k] = \sum_{n=0}^{2N-1} f[n]e^{-j\frac{2\pi nk}{2N}}$.

For even values of k, we note $k = 2k'$, and we have:

$$F'[2k'] = \sum_{n=0}^{2N-1} f[n]e^{-j\frac{2\pi n 2k'}{2N}} = \sum_{n=0}^{N-1} f[n]e^{-j\frac{2\pi nk'}{N}}. \tag{16.67}$$

The upper boundary of the sum has been reduced to $N - 1$ because the function $f[n]$ is zero by hypothesis beyond this value. It is seen that for even values of k we find back the DFT values at N points. For odd values of k, one obtains new values of the Fourier transform $F(e^{j\omega T})$ of the signal $f[n]$. This is equivalent to an interpolation of the values given by the DFT on N points. Obviously, this operation can be conducted by filling the signal with a greater number of zeros so as to reach a desired frequency resolution.

Interpolation in time: Equivalently, one can interpolate in time. This has practical value when the sampling frequency is limited to a given value, but one desires to know closer values in time of the underlying analog signal . It is known that the DFT inversion formula is $f[n] = \frac{1}{N}\sum_{k=0}^{N-1} F[k]e^{j\frac{2\pi nk}{N}}$

The function $F[k]$ being periodic, with period N we can also write

$$f[n] = \frac{1}{N}\sum_{k=-\frac{N}{2}}^{\frac{N}{2}-1} F[k]e^{j\frac{2\pi nk}{N}}. \tag{16.68}$$

Let us complete the function $F[k]$ with zeros up to the boundaries $-N$ and N. This function is noted as $F'[k]$. Now we compute the inverse DFT on $2N$ points of $F'[k]$ by noting:

$$f'[n] = \frac{1}{2N}\sum_{k=-N}^{N-1} F'[k]e^{j\frac{2\pi nk}{2N}}. \tag{16.69}$$

For even values of n we note $n = 2n'$. We have

$$f'[2n'] = \frac{1}{2N}\sum_{k=-N}^{N-1} F'[k]e^{j\frac{2\pi 2n'k}{2N}} = \frac{1}{2N}\sum_{k=-N}^{N-1} F'[k]e^{j\frac{2\pi n'k}{N}}. \tag{16.70}$$

Since the function $F'[k]$ is zero outside the interval $\left\{-\frac{N}{2}, \frac{N}{2}\right\}$, we have

$$f'[2n'] = \frac{1}{2N} \sum_{k=-\frac{N}{2}}^{\frac{N}{2}-1} F'[k] e^{j\frac{2\pi n'k}{N}} = \frac{1}{2} f[n']. \tag{16.71}$$

We find for even values of the index n (within a factor of 2), the value of the function $f[n]$. For odd values of the index, interpolated values between two even values are obtained.

Note: In the above calculations negative indexes were used to facilitate the demonstration. However, the indexes of the functions used in the numerical calculations are positive or zero. As shown in the preceding figures, these indexes correspond to the frequency interval $\{0, \omega_e\}$. In the Fourier domain, adding zeros must be made in the central zone around $\frac{\omega_e}{2}$.

The interest of interpolation in the time domain appears in the following application: The standard sample rate for audio CDs is 44.1 kHz. According to the sampling theorem, it is possible in principle to reconstruct the audio signal $f(t)$ from the values $f[n] = f(nT)$ recorded on the CD, using the formula (16.19).

The difficulty is that it is impossible to electronically realize the sinc functions that appear in this formula.

The expedient to keep constant the value of the function during the time interval between samples ($f(t)$ is then approximated by a stairs function) is not acceptable from a point of view of the quality of the audio output (the ear is very sensitive to discontinuities at the edges of the intervals).

Even after a low-pass filtering of this step function, the audio quality of the reconstructed signal is insufficient. It is preferable to interpolate the signal numerically in time in a first step as previously described, so as to have samples closest in time, so as to create a step function with narrower steps before smoothing this function by low-pass filtering.

16.14 Artifacts of the Fourier Transform on a Computer

Almost anybody who has calculated and displayed the FFT of a sine function with a computer for the first time has been surprised that instead of observing a single frequency line, he observes a series of lines closely packed around the expected frequency. The phenomenon is normal and can be explained by the following:

The FFT algorithm generally used to calculate the Fourier transform samples the frequency range in N points evenly distributed in the frequency interval $0 \leftrightarrow \omega_e$. The frequency step is $\frac{\omega_e}{N} = \frac{2\pi}{T}\frac{1}{N}$. When calculating by a computer the Fourier transform of a sine function limited to N points, we have shown that it appeared in the frequency domain a Dirichlet function resulting from the FT of the rectangular window.

Fig. 16.11 Sampling of
spectrum when $\omega_0 \neq \omega_k$

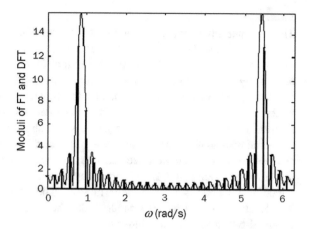

Fig. 16.12 Sampling of
spectrum when $\omega_0 = \omega_k$

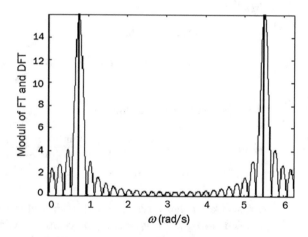

The step in the spectrum calculated by the method of Cooley and Tukey calculated in N points corresponds exactly to the interval between two zeros of the FT
of the window function.

Two situations may occur in practice:

(a) If ω_o falls somewhere between two values $\omega_k = k\frac{2\pi}{T}\frac{1}{N}$ and $\omega_{k+1} = (k+1)\frac{2\pi}{T}\frac{1}{N}$ the maximum spectrum DFT value does not reach $\frac{N}{2}$ and the other values fall somewhere on the secondary peaks of the Fourier transform, as shown in Fig. 16.11.

(b) If the sine frequency ω_o falls exactly on a frequency $\omega_k = k\frac{2\pi}{T}\frac{1}{N}$, a peak is then obtained in ω_o, but the zeros of the function $\sin\frac{N(\omega-\omega_o)T}{2}$ fall on the other multiples of $\frac{2\pi}{T}\frac{1}{N}$. And give 0 (Fig. 16.12). There is therefore a single peak in ω_o (with height $\frac{N}{2}$ due to the factor $\frac{1}{2}$ that appears in the exponential development of the sine).

Summary
This chapter gives the main properties of the Fourier transformation of digital signals. After having given the Poisson summation formula, we demonstrated the Shannon sampling theorem and the theorem of Shannon–Whittaker proving that an analog signal can be reconstructed from its samples if the sampling was done respecting the Shannon condition. For the treatment of limited duration signals, which are inherently multiplied by a rectangular window, we have calculated the Fourier transform of a rectangular window and showed that the multiplication of a time-limited signal by an apodization window allows a better quality of spectral analysis. We have defined the discrete Fourier transform and studied the FFT algorithm of Cooley and Tukey, operating on signals whose length is a power of 2, which can calculate spectra very rapidly. The chapter was completed by the interpolation method by zero padding and the peculiarities of the sampled spectra obtained by numerical calculations.

Exercises

I. Denote the causal Hanning window:

$$w_{\mathrm{H}}[n] = \begin{cases} \frac{1}{2}\left(1 - \cos\frac{2\pi n}{N-1}\right) & \text{for} \quad n = 0, 1, \ldots, N-1 \\ 0 & \text{elsewhere} \end{cases}.$$

Show that its Fourier transform is

$$W_{\mathrm{H}}\left(e^{j\omega T}\right) = e^{-j\omega T\frac{N-1}{2}}\left(\frac{1}{2}\frac{\sin(N\omega T/2)}{\sin(\omega T/2)} + \frac{1}{4}\frac{\sin(N(\omega T - \phi_0)/2)}{\sin((\omega T - \phi_0)/2)} + \frac{1}{4}\frac{\sin(N(\omega T + \phi_0)/2)}{\sin((\omega T + \phi_0)/2)}\right),$$

with $\phi_0 = \frac{2\pi}{N-1}$.

II. A function $f(t)$ has a spectrum $F_a(\omega)$ limited to the interval $\{-\omega_0, \omega_0\}$. It is sampled at frequency $\omega_e = 2\omega_0$, in accordance with Shannon's condition. We denote $F\left(e^{j\omega T}\right)$ the Fourier transform of the sampled function $f[n]$. Show that if the function $f(t)$ is sampled at frequency $\frac{\omega_e}{2}$, aliasing occurs, and that its Fourier transform in the range $\left\{-\frac{\omega_e}{2}, \frac{\omega_e}{2}\right\}$ is given by $F_2\left(e^{j\omega T}\right) =$
$\frac{1}{T}\left(F_a\left(e^{j\left(\omega + \frac{\omega_e}{2}\right)T}\right) + F_a\left(e^{j\omega T}\right) + F_a\left(e^{j\left(\omega - \frac{\omega_e}{2}\right)T}\right)\right)$, or

$$F_2\left(e^{j\omega T}\right) = \left(F\left(e^{j\left(\omega + \frac{\omega_e}{2}\right)T}\right) + F\left(e^{j\omega T}\right) + F\left(e^{j\left(\omega - \frac{\omega_e}{2}\right)T}\right)\right).$$

III. Quadrature mirror filters:
A numerical filter is defined by its temporal equation:

$$g_0[n] = f[n] + f[n-1].$$

1. Compute the impulse response $h_0[n]$ of this filter; its transfer function. Place the remarkable points of this function in the z-plane. Calculate the frequency response $H_0(e^{j\omega T})$. Graph the frequency gain. What is the character of this filter?

   A second filter is defined as follows: Its gain is mirror of the previous filter with respect to the angular frequency $\frac{\omega_e}{4}$. We have thus $|H_1(e^{j\omega T})| = |H_0(e^{j(\omega T - \pi)})|$. Make a drawing to give the aspect of $|H_1(e^{j\omega T})|$. Conclude that we can deduce the remarkable points of this filter from those of the first filter by symmetry. Calculate the transfer function, the frequency response $H_1(e^{j\omega T})$ and the impulse response $h_1[n]$ of this filter.

2. Prove the relationship $|H_0(e^{j\omega T})|^2 + |H_1(e^{j\omega T})|^2 = \text{Cte}$.

3. The scalar product of two signals is defined by $\langle s_1[n], s_2[n] \rangle = \sum\limits_{n=-\infty}^{\infty} s_1^*[n]s_2[n]$. Show that $h_0[n]$ and $h_1[n]$ are orthogonal by a calculation in the time domain. Find again the orthogonality by a calculation in the frequency domain.

4. Let the signal $f[n] = \begin{vmatrix} 1 & \text{if } 0 \le n \le 10 \\ 0 & \text{elsewhere} \end{vmatrix}$. Calculate the outputs $g_0[n]$ and $g_1[n]$ of the previous filters. Can we say that one of the two responses gives the slow part of the input and the other the details?

Chapter 17
Autoregressive Systems (AR)—ARMA Systems

We studied in Chap. 14 the digital Moving Average LTI systems. It was clear that those filters are simple to implement because their impulse response is finite. They allow one to filter out totally some chosen frequency components; however, they have the disadvantage of having a frequency response that varies slowly. That makes them not very selective and unsuitable for making band-pass filters. The autoregressive filters presented in this chapter do not have these disadvantages. They are digital equivalents of the analog filters presented in Chap. 10. They can be higly selective but their disadvantage is that their impulse response has an infinite length. This chapter begins with the presentation of the AR filters of the first and second order. We determine their impulse responses, study their stability, and calculate their transfer functions and their domain of definition in the z-plane. As before, the geometric interpretation of the frequency gain provides a thorough understanding of the filter's mode of operation and gives way to generalization toward ARMA filters (Autoregressive–Moving Average) of which several examples are studied. It is interesting to be able to use the many accumulated results in the literature on analog filters; various methods of passing from an analog filter to its digital equivalent are presented, but, in essence, the equivalence cannot be performed perfectly. We study the pros and cons of commonly used methods.

In the autoregressive type of filter, the time equation giving the value of the output signal $g[n]$ at time n contains terms representing the value of the output signal at earlier moments $g[n-1], g[n-2], \ldots, g[n-k]$. They are also referred as systems with feedback. The general form of the equation of a time autoregressive systems is:

$$g[n] = -\sum_{k=1}^{p} a_k g[n-k] + b_0 f[n]. \tag{17.1}$$

© Springer International Publishing Switzerland 2016
F. Cohen Tenoudji, *Analog and Digital Signal Analysis*,
Modern Acoustics and Signal Processing, DOI 10.1007/978-3-319-42382-1_17

17.1 Autoregressive First-Order System

This system is recursive and defined by the time equation:

$$g[n] = Kg[n-1] + f[n], \tag{17.2}$$

where K is a complex or real constant. This temporal equation alone does not define a system but two. An additional hypothesis is necessary to define a system: the causal or noncausal hypothesis.

17.1.1 Case of a Causal System

We now show that the hypothesis $h[-1] = 0$ is sufficient to ensure the causality of the filter.

In what follows, we calculate the impulse response by induction.

Let the input signal be $f[n] = \delta[n]$. In determining the impulse response, we consider separately the cases of negative and positive or zero times:

$$
\begin{array}{ll}
n \geq 0 & n < 0 \\
h[0] = Kh[-1] + 1 = 1, & h[-1] = 0 \text{ by assumption} \\
h[1] = K, & h[-2] = 0, \\
\cdots & \cdots \\
h[n] = K^n, & h[n] = 0.
\end{array}
$$

Therefore,

$$h[n] = K^n U[n]. \tag{17.3}$$

This function is called causal because its values are zero for negative time.

In the following, for ease of presentation, the constant K is assumed real. The results are easily generalized to the complex case.

Figure 17.1 shows that the impulse response is decreasing if $|K| < 1$, increasing $|K| > 1$, and alternate if $K < 0$.

For $|K| > 1$ the impulse response increases indefinitely in absolute terms, reflecting the instability of the system, since a finite input $f[n] = \delta[n]$ causes an output going to infinity.

Transfer function of first-order system

As seen before, this function is the z-transform of the impulse response:

$$H(z) = \sum_{n=-\infty}^{+\infty} K^n z^{-n} U[n] = \sum_{n=0}^{\infty} K^n z^{-n}. \tag{17.4}$$

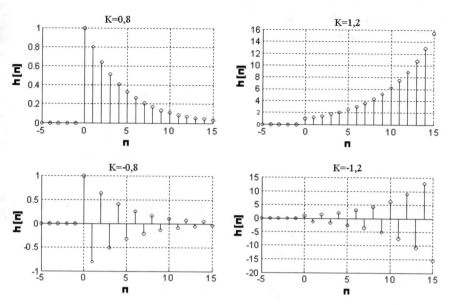

Fig. 17.1 Impulse response for the causal first-order system for different values of parameter K

It is the sum of a geometric series of common ratio Kz^{-1}.

It converges if the magnitude of the common ratio is less than 1 so $\left|Kz^{-1}\right| < 1$, or $|K| < |z|$. It becomes

$$H(z) = \frac{1}{1 - Kz^{-1}} = \frac{z}{z - K}. \tag{17.5}$$

This function has a singularity (a simple pole) in $z = K$.

We denote D the convergence domain of the series which will be the domain of definition of $H(z)$. It is the locus of points such that $|z| > |K|$. D is the outer area of a disc centered at the origin with radius $|K|$ (Fig. 17.2).

Fig. 17.2 Definition domain D of a causal first-order system

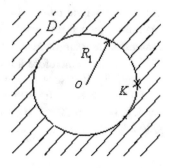

Fig. 17.3 Unit circle and
vector $\overrightarrow{PM}$ governing the gain

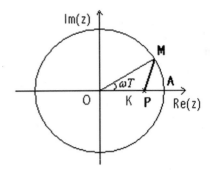

Note that if $|K| > 1$ the FT does not exist, because the circle $|z| = 1$ is not in the convergence domain. In this case, the system is not stable. As shown above, the impulse response increases indefinitely in absolute value.

If $|K| < 1$ the FT exists and is expressed by

$$H\left(e^{j\omega T}\right) = \frac{e^{j\omega T}}{e^{j\omega T} - K}. \tag{17.6}$$

Gain magnitude variation with frequency
Geometrically, we have (See Fig. 17.3)

$$\left|H\left(e^{j\omega T}\right)\right| = \frac{1}{\left|e^{j\omega T} - K\right|} = \frac{1}{PM}. \tag{17.7}$$

$$\text{For } \omega = 0, |H(1)| = \frac{1}{1 - K}. \quad \text{For } \omega T = \pi, |H(-1)| = \frac{1}{1 + K}. \tag{17.8}$$

The aspect of the gain magnitude is conditioned by the length of segment PM which connects the point M representative of the monochromatic signal $e^{jn\omega T}$ to the pole P with abscissa K on the real axis.

To illustrate the behavior of the gain with frequency, we take the case $K > 0$ as an example. When $\omega = 0$ the point M lies in A, the segment PM has a length $1 - K$, which is its minimum value. Its inverse $\frac{1}{PM}$ is maximum with value $\frac{1}{1-K}$. The maximum gain is in $\omega = 0$. When ω increases from 0, the segment PM length increases, the gain decreases to the minimum value $\frac{1}{1+K}$ obtained for $\omega = \frac{\pi}{T}$. For $K > 0$ we have therefore a low-pass filter.

Similarly, it is seen that in the case $K < 0$ the filter is high pass.

In Fig. 17.4 the frequency response of the magnitude's low-pass filter is on the left, on the right is the gain of the high-pass filter. The abscissa ωT ranges from 0 to 2π.

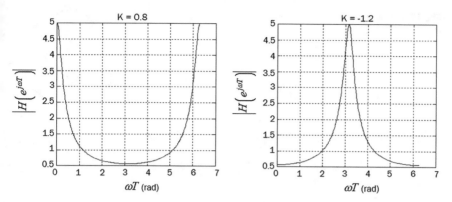

Fig. 17.4 Gain magnitude with positive (*left*) and negative (*right*) values of parameter K

17.1.2 Analysis of the Anticausal System

The hypothesis $h[1] = 0$ is sufficient to define it. It becomes

for $n \geq 0$

$h[1] = 0 = Kh[0] + 0$

$h[0] = Kh[-1] + 1 = 0 \Rightarrow h[-1] = -K^{-1}$

$\Rightarrow h[0] = 0$

. . .

$h[2] = Kh[1] + 0 = 0$

$h[n] = 0$

for $n < 0$

$h[-1] = Kh[-2] \Rightarrow h[-2] = -K^{-2}$

. . .

. . .

. . .

$h[n] = -K^n.$

In summary,

$$h[n] = -K^n U[-n - 1]. \tag{17.9}$$

Calculation of the anticausal filter transfer function $H(z)$

$$H(z) = - \sum_{n=-\infty}^{+\infty} z^{-n} K^n U[-n - 1] = - \sum_{n=-\infty}^{-1} K^n z^{-n}. \tag{17.10}$$

We write $m = -n - 1$,

$$H(z) = - \sum_{m=0}^{\infty} K^{-m-1} z^{m+1} = -\frac{z}{K} \sum_{m=0}^{\infty} \left(\frac{z}{K}\right)^m. \tag{17.11}$$

Fig. 17.5 Definition domain
D of an anticausal first-order
system

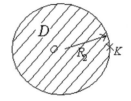

This geometric series converges if $|z| < |K|$, it becomes

$$H(z) = -\frac{z}{K}\frac{1}{1 - \frac{z}{K}} = -\frac{z}{K - z} = \frac{z}{z - K}.\tag{17.12}$$

The domain of definition of $H(z)$ is within the disc of radius $|K|$ (Fig. 17.5).
Note that if $|K| > 1$ the system possesses a frequency response.

It is noted that the expression of the transfer function of the anticausal filter
(17.12) is the same as that of the causal filter (17.5). What differentiates the two
transfer functions is their domain of definition in the z-plane. We see that it is
essential to specify the domain of definition of a system transfer function.

An instructive exercise is to recalculate the impulse response from the transfer
function. The calculation will be done by the residues method, taking the integrating circle radius of radius R surrounding the origin included in each case within
the definition domain: $|K| < R$ is taken to find the impulse response of the causal
filter and $R < |K|$ for the case of the anticausal filter.

Exercise

Calculation of the output of the causal filter of first order when the input signal is
the step function $f[n] = U[n]$

The system output is given by

$$g[n] = \sum_{m=-\infty}^{+\infty} K^m U[m] U[n - m] = \sum_{m=0}^{n} K^m \quad \text{if } n \geq 0,\tag{17.13}$$

and $g[n] = 0$ for $n < 0$. The output is also causal.

for $n = 0$: $g[0] = 1$
for $n = 1$: $g[1] = \frac{1-K^2}{1-K} = \frac{(1-K)(1+K)}{1-K} = 1 + K$
for any other n : $g[n] = \frac{1-K^{n+1}}{1-K}$
                 $g[n] = \frac{1-K^{n+1}}{1-K} = \frac{(1-K)}{1-K}(1 + K + \cdots + K^n) = 1 + K + \cdots + K^n.$

We have therefore

$$g[n] = (1 + K + \cdots + K^n)U[n].\tag{17.14}$$

We see directly that $g[n]$ has no FT since this function increases indefinitely with n.

Exercise

Repeat this calculation using the z-transform and residue theorem.

Show in first time that $G(z) = H(z) V(z) = \frac{1}{1-Kz^{-1}} \frac{1}{1-z^{-1}}$, then show in the case $|K| < 1$ that the definition domain of $G(z)$ is the exterior of the disc with radius 1 (that is, the intersection of definition domains of $H(z)$ and $V(z)$). Then calculate the first terms of function $g[n]$ after showing that this function is causal.

17.2 Autoregressive System (Recursive) of Second Order

This system is defined by the time equation:

$$g[n] = a_1 g[n-1] + a_2 g[n-2] + f[n], \tag{17.15}$$

where a_1 and a_2 are two complex constants in the general case. As in the case of first-order system, a hypothesis about the possible causality of the system is necessary to fully define the system.

17.2.1 Calculation of the System Transfer Function $H(z)$

As the calculation of the impulse response of this filter is more difficult than in the case of first-order AR system, we rather study the filter properties by addressing the problem by calculating first the transfer function. The system is linear, time invariant. Then, by definition of the transfer function, when the input is of the form $f[n] = z^n$, the output takes the form $g[n] = H(z)z^n$.

$$\xrightarrow{z^n} \boxed{\text{System}} \xrightarrow{H(z)z^n}$$

Using the translation property $g[n-m] = H(z)z^{n-m}$ and replacing it in temporal equation we get $H(z)z^n - a_1 H(z)z^{-1}z^n - a_2 H(z)z^{-2}z^n = z^n$, or simplifying by z^n:

$$H(z) = \frac{1}{1 - a_1 z^{-1} - a_2 z^{-2}} = \frac{z^2}{z^2 - a_1 z - a_2}. \tag{17.16}$$

Assume now that the system is causal. This assumption leads, as was shown in the previous chapter, that the definition domain of $H(z)$ is the outside of a disc centered at the origin and whose radius is the distance between the coordinate origin and the furthest singularity from the origin. We are led to identify the poles of $H(z)$. The second-degree polynomial at denominator always has two roots in $\mathbb{C}$.

For simplicity, we will restrict here the study to the case where both coefficients a_1 and a_2 are real (as will appear below, this ensures that the impulse response is real), then the roots are either real or complex conjugate.

The full discussion of the nature of the roots depending on the value of the discriminant of the second-degree polynomial is not done here. Is treated in the following only the most common case in practice where both roots are complex conjugate. We can write in this case

$$H(z) = \frac{z^2}{\left(z - z_p\right)\left(z - z_p^*\right)},\tag{17.17}$$

or also

$$H(z) = \frac{z^2}{z^2 - z\left(z_p + z_p^*\right) + z_p z_p^*}.\tag{17.18}$$

By identifying the coefficients of the powers of z one must have

$$a_1 = z_p + z_p^*; \quad a_2 = -z_p z_p^*.\tag{17.19}$$

We note $z_p = re^{j\omega_p T}$, ($\omega_p T$ is the argument of a pole).
We have therefore

$$z_p + z_p^* = 2r \cos \omega_p T; \quad z_p z_p^* = r^2.\tag{17.20}$$

Finally,

$$a_1 = 2r \cos \omega_p T \quad \text{and} \quad a_2 = -r^2.\tag{17.21}$$

Thus,

$$H(z) = \frac{z^2}{z^2 - 2zr \cos \omega_p T + r^2}.\tag{17.22}$$

If the frequency response exists, it is written as

$$H\left(e^{j\omega T}\right) = \frac{e^{2j\omega T}}{e^{2j\omega T} - 2e^{j\omega T} r \cos \omega_p T + r^2}.\tag{17.23}$$

Using the formula (17.17) in the case $z = re^{j\omega T}$, the magnitude of the frequency response is given by

$$\left|H\left(e^{j\omega T}\right)\right| = \frac{1}{\left|\left(e^{j\omega T} - re^{j\omega_p T}\right)\right|\left|\left(e^{j\omega T} - re^{-j\omega_p T}\right)\right|} \qquad (17.24)$$

We see that for the frequency response to exist, it is first necessary that the poles are not located on the unit circle. Moreover, the unit circle must belong to the definition domain of function $H(z)$.

In the following discussion, the filter is supposed causal and the frequency response defined, which requires that the poles' magnitudes are necessarily less than 1 ($r < 1$).

17.2.2   Geometric Interpretation of Variation of Frequency Gain Magnitude

According to (17.24), we can write

$$\left|H\left(e^{j\omega T}\right)\right| = \frac{1}{MP \cdot MP'}. \qquad (17.25)$$

Figure 17.6 shows the pole situation and the position of M on the unit circle for a given frequency.

The gain magnitude is the inverse of the product of the lengths of the segments MP and MP' connecting the point M to each of the poles. The gain will be great if one of the segments MP or MP' becomes small. There will be resonance when M approaches a pole (e.g., MP decreases when $\omega \rightarrow \omega_p$).

The sharpness of the resonance depends on the proximity of the pole and the circle radius of radius 1. If the pole is close to the circle, the resonance will be sharp, the amplitude at the resonance high, and the resonant frequency will be close to the frequency of the pole.

Fig. 17.6 Pole situation and vectors $\overrightarrow{PM}$ and $\overrightarrow{P'M}$ controlling the frequency gain

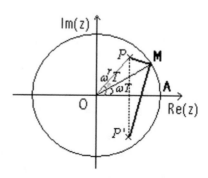

As shown in Fig. 17.7, the more distant the pole from the circle of radius 1, the less pronounced is the maximum, and the resonance frequency moves to lower frequencies. This shift toward the low frequencies is due to the growing influence of the second pole on the resonance of the first pole.

If the pole is near the circle, i.e., if r is close to 1, the resonance is sharp. It manifests for $\omega_r T \cong \pm\omega_p T$. It is assumed in the following that we are in this situation.

For M close to P: $MP \cong \sqrt{HM^2 + PH^2}$; $PH = 1 - r$, so $HM \cong \widehat{HM} = (\omega_p - \omega)T$ (see Fig. 17.8).

In the above relationship, the length of the segment HM was assimilated to the length of the arc $\widehat{HM}$. As seen in Fig. 17.8, one can write $MP' \cong PP' = 2r \sin \omega_p T$.

Fig. 17.7 Gain of a second-order AR filter for several pole magnitudes

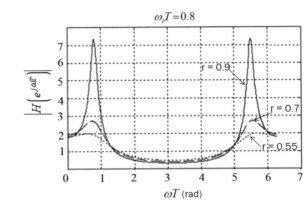

Fig. 17.8 Geometric situation in the case of sharp resonance

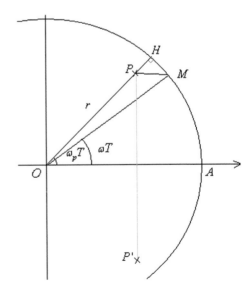

In the vicinity of the resonance frequency in the case of a sharp resonance, the expression of the gain takes the form

$$|H(e^{j\omega T})| = \frac{1}{\sqrt{(\omega_p - \omega)^2 T^2 + (1-r)^2}} \frac{1}{2r \sin \omega_p T} \qquad (17.26)$$

At resonance, for $\omega_r \approx \omega_p$, the gain is maximum. It is given by

$$|H|_{Max} = |H(e^{j\omega T})|_{\omega=\omega_p} = \frac{1}{(1-r)2r \sin \omega_p T} \qquad (17.27)$$

Calculation of the -3 bB bandwidth in the case of a sharp resonance

For two angular frequencies ω_1 and ω_2, the two terms under the square root in (17.26) are equal:

$$|(\omega_p - \omega_{1,2})T| = (1-r). \qquad (17.28)$$

Then,

$$|H(e^{j\omega_{1,2}T})| = \frac{1}{\sqrt{2(1-r)^2}2r \sin \omega_p T} = \frac{|H|_{Max}}{\sqrt{2}}. \qquad (17.29)$$

We choose $\omega_1 < \omega_p$. Equation (17.28) is writen as $(\omega_p - \omega_1)T = 1 - r$, or $\omega_1 = \omega_p - \frac{1-r}{T}$.

For the other angular frequency $\omega_2 > \omega_p$, we have $\omega_2 = \omega_p + \frac{1-r}{T}$.

The -3 dB bandwidth is then

$$\Delta\omega = \omega_2 - \omega_1 = 2\left(\frac{1-r}{T}\right) = \frac{2}{T}(1-r) = \frac{\omega_e}{\pi}(1-r). \qquad (17.30)$$

N.A. if $r = 0.999$, we get $\Delta\omega = \frac{\omega_e}{\pi}10^{-3}$.

Returning to frequency variable $f = \frac{\omega}{2\pi}$, the -3 dB bandwidth is

$$\Delta f = \frac{f_e}{\pi}(1-r) \qquad (17.31)$$

We also see that the sampling frequency being given, one can determine the pole magnitude r to achieve the desired -3 dB bandwidth.

N.A. Let $f_e = 20$ kHz. If one looks for a resonance frequency of $f_r = 1.2$ kHz and a bandwidth $\Delta f = 300$ Hz, we will take $f_p = f_r = 1.2$ kHz (hypothesis of sharp resonance that is justified a posteriori) and $r = 1 - \pi\frac{\Delta f}{f_e} = 0.9529$. Then,

$$\omega_p T = 2\pi f_p T = 2\pi \frac{f_p}{f_e} = 0.377.$$

We deduce the coefficients a_1 and a_2: $a_1 = 2r \cos \omega_p T = 2 \times 0.9529 \times \cos 0.377 = 1.772$;

$$a_2 = -r^2 = -0.9529^2 = -0.908.$$

The time equation of this filter is as follows: $g[n] = 1.772g[n-1] - 0.908g[n-2] + f[n]$.

17.2.3 Impulse Response of Second-Order System

It is assumed here that the filter is causal. We start from the expression of the transfer function given in (17.17).

After dividing the numerator and denominator by z^2, we write

$$H(z) = \frac{1}{\left(1 - \frac{z_p}{z}\right)\left(1 - \frac{z_p^*}{z}\right)}. \tag{17.32}$$

Since the filter is causal, the domain of definition D is the exterior of the disk of radius $r = |z_p|$, as was justified in the chapter on the z-transform. The division z^2 was permitted because the point $z = 0$ does not belong to the domain of definition of $H(z)$.

We recognize in $H(z)$ a product of first-order system transfer functions:

$$H_1(z) = \frac{1}{1 - \frac{z_p}{z}} \quad \text{and} \quad H_2(z) = \frac{1}{1 - \frac{z_p^*}{z}}. \tag{17.33}$$

We have therefore in the time domain a convolution:

$$h[n] = h_1[n] \otimes h_2[n]. \tag{17.34}$$

Assuming causal first-order systems,

$$h_1[n] = z_p^n U[n] \quad \text{and} \quad h_2[n] = z_p^{*n} U[n],$$

$$h[n] = \sum_{m=-\infty}^{+\infty} z_p^m U[m] z_p^{*n-m} U[n-m]. \tag{17.35}$$

$h[n]$ will be causal: $h[n] = \sum_{m=0}^{n} z_p^m z_p^{*n-m} U[n]$.

Fig. 17.9 Second AR
system; example of a resonant
impulse response

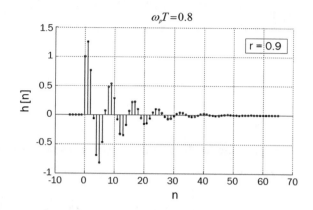

It is wise to use the trigonometric form: $z_P = re^{j\omega_p T}$,

$$h[n] = U[n] \sum_{m=0}^{n} r^m e^{j\omega_p m T} r^{n-m} e^{-j\omega_p(n-m)T}$$

$$= U[n] r^n e^{-j\omega_p n T} \sum_{m=0}^{n} e^{2j\omega_p m T} = U[n] r^n e^{-j\omega_p n T} \frac{1 - e^{j2\omega_p(n+1)T}}{1 - e^{2j\omega_p T}}.$$

Finally,

$$h[n] = r^n \frac{\sin \omega_p(n+1)T}{\sin \omega_p T} U[n]. \qquad (17.36)$$

Figure 17.9 shows an example of the impulse response of a resonant
second-order filter.

17.2.4   Functional Diagrams of the Digital System of Second Order

The output signal is given by the relation

$$g[n] = a_1 g[n-1] + a_2 g[n-2] + f[n]. \qquad (17.37)$$

1. **Basic chart:** (Fig. 17.10).

Fig. 17.10 Basic
representation

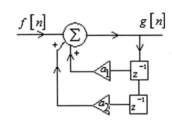

2. **Representation by two first-order systems in cascade:**

We have

$$G(z) = H(z)F(z) = \frac{z}{z - z_P}\frac{z}{z - z_P^*}F(z) = H_1(z)H_2(z)F(z). \qquad (17.38)$$

The filter can be realized by two filters of the first-order connected in cascade
(Fig. 17.11).

Their time equations are written as

$$\begin{aligned}
s[n] &= z_P s[n-1] + f[n], \\
g[n] &= z_P^* g[n-1] + s[n].
\end{aligned} \qquad (17.39)$$

3. **Realization by two first-order systems in parallel:**

$$H(z) = \frac{z^2}{(z - z_P)(z - z_P^*)}. \qquad (17.40)$$

$H(z)$ is decomposed into a sum of two simple elements (Fig. 17.12).

Fig. 17.11 Representation
by cascade of two first-order
systems

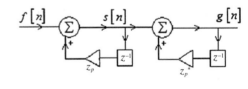

Fig. 17.12 Representation
by two first-order systems in
parallel

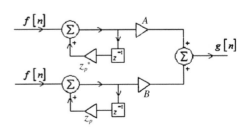

$$H(z) = \frac{Az}{z - z_P} + \frac{Bz}{z - z_P^*} = \frac{Az^2 - Azz_P^* + Bz^2 - Bzz_P}{D},$$

$$\Rightarrow A + B = 1; \quad Az_P^* + Bz_P = 0 \Rightarrow B = -\frac{Az_P^*}{z_P}. \tag{17.41}$$

$$A\left(1 - \frac{z_P^*}{z_P}\right) = 1; \quad A = \frac{1}{1 - \frac{z_P^*}{z_P}} = \frac{z_P}{z_P - z_P^*}; \quad B = \frac{z_P^*}{z_P^* - z_P}.$$

17.3   ARMA Filters

Generally, an ARMA filter (Autoregressive–Moving Average) is defined by the time equation:

$$g[n] + \sum_{k=1}^{p} a_k g[n-k] = \sum_{k=0}^{q} b_k f[n-k]. \tag{17.42}$$

Its transfer function has the form

$$H(z) = \frac{b_0 + b_1 z^{-1} + b_2 z^{-2} + \cdots + b_q z^{-q}}{1 + a_1 z^{-1} + \cdots + a_p z^{-p}} = b_0 \frac{\displaystyle\prod_{k=1}^{q} (1 - z_k z^{-1})}{\displaystyle\prod_{k=1}^{p} (1 - p_k z^{-1})}. \tag{17.43}$$

This is a rational fraction of polynomial functions in z.

By a choice of the position of zeros and poles of this polynomial in the complex plane, we can get the desired frequency response or approaching it closely. Optimization techniques for locating these remarkable points to approach the desired result have been developed.

In the design of ARMA filters, we also use the body of knowledge on analog filters for getting digital filters with similar properties using passing techniques from analog to digital, examples of which are presented below.

Example of an ARMA filter
Rejection filter (also called notch filter): The objective in the design of this type of filter is to obtain a frequency response as flat as possible, except in a narrow frequency band where the gain cancels at its center. The shape of the frequency response is that of a plank of wood in which we sawed a notch.

Principle: Place a zero of the transfer function on the unit circle to cancel the gain at frequency f_0 and also place a pole near the zero with the same argument (Fig. 17.13).

Fig. 17.13 Case of vicinity
of pole and zero

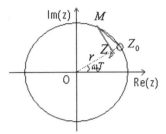

$$H(z) = \frac{(z - z_0)}{(z - z_p)} \quad \text{with } z_0 = e^{j\omega_0 T}, \quad z_p = re^{j\omega_0 T} \quad \text{and} \quad r \cong 1. \tag{17.44}$$

Using a geometric interpretation it is found that the frequency gain magnitude is given by the ratio of segments joining the point M representative of a monochromatic signal with frequency ω to the zero Z_0 and the pole Z_p of the transfer function:

$$\left| H\left(e^{j\omega T}\right) \right| = \frac{MZ_0}{MZ_p}. \tag{17.45}$$

This magnitude is nearly 1 for frequencies far from ω_0, as the distances of the point M to the neighboring points Z_0 and Z_p are close. However, the gain is zero for $\omega = \omega_0$ since the segment MZ_0 length is zero.

To obtain a real impulse response, it is necessary that the coefficients appearing in the expression of the transfer function are real, this is not the case in (17.44) for all frequencies.

This leads us to add complex conjugates of zero and pole in (17.46).

Thus, among the coefficients of the polynomials appear the sum and the product of complex conjugate numbers that are real quantities. The filter transfer function is then

$$H(z) = \frac{(z - z_0)(z - z_0^*)}{(z - z_p)(z - z_p^*)}. \tag{17.46}$$

In the following example, it was taken $f_e = 20$ kHz, $f_0 = 2$ kHz and $r = 0.98$.

Meanwhile in the interval $\left\{ -\frac{f_e}{2}, \frac{f_e}{2} \right\}$, the shape of the frequency response modulus is given in Fig. 17.14.

The reader is invited to explain, using a geometric argument, why in this example, the gain value at plateau is about 1.02 rather than 1.

This type of filter is used when we want to eliminate a parasitic frequency signal which sometimes overlaps with the signal (frequency of 60 Hz, for example, in highly amplified signals as is the case for electrocardiograms).

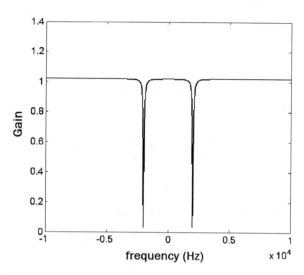

Fig. 17.14 Frequency response of a notch filter

Comb filter

It is a rejection filter for removing a fundamental frequency and its harmonics. We reproduce regularly along the unit circle the zero–pole configuration encountered in the previous example. To do this, the filter transfer function is

$$H(z) = \frac{1 - z^{-M}}{1 - A z^{-M}} = \frac{z^M - 1}{z^M - A}, \quad A \text{ is a real positive number close to 1.} \quad (17.47)$$

The zeros of $H(z)$ are $z_{0k} = e^{j\frac{2\pi k}{M}}$. Its poles are $z_{pk} = \sqrt[M]{A} e^{j\frac{2\pi k}{M}}$, with $k = 0, 1, 2, \ldots, M - 1$.

For a causal and stable filter, $A < 1$ is chosen and for definition domain of $H(z)$, the exterior of the circle of radius $\sqrt[M]{A}$. Note that zeros $z_{0k} = e^{j\frac{2\pi k}{M}}$ and poles $z_{pk} = \sqrt[M]{A} e^{j\frac{2\pi k}{M}}$ have same argument.

The frequency response is written as $H\left(e^{j\omega T}\right) = \frac{1 - e^{-jM\omega T}}{1 - A e^{-jM\omega T}}$.

N.A. If $M = 10$ and $A = 0.9$, poles' magnitude is $r = 0.9895$, a value very close to 1. We take $T = 1$. Figure 17.15 shows the remarkable points location (left) and the representation of the filter frequency gain magnitude (right).

Because of the shape of its gain, the filter is called comb filter.

The maximum gain is 1.053. The width of the notch depends on r, as shown below. The gain will be approximately $\frac{1}{\sqrt{2}}$ (−3 dB) if $\frac{ZM_1}{PM_1} = \frac{1}{\sqrt{2}}$.

$$\frac{1}{\sqrt{2}} = \frac{ZM_1}{\sqrt{ZM_1^2 + PM_1^2}} \Rightarrow ZM_1 = PZ = 1 - r.$$

Then $(\omega_1 - \omega_0)T \simeq 1 - r..$

Thus, the −3 dB bandwidth is given by $\Delta\omega = \omega_2 - \omega_1 = 2\frac{(1-r)}{T}$.

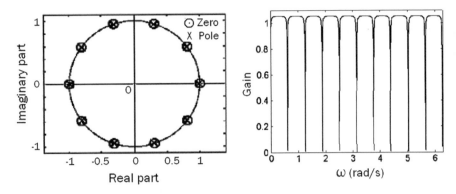

Fig. 17.15 Comb filter ($M = 10$); Pole–zeros situation (*left*); Frequency response (*right*)

As $r = \sqrt[10]{A} = 0.9895$, $\Delta\omega = 0.021$ if $T = 1$. The closer to 1 is r, the narrower will be the width of the rejected band.

From Eq. (17.47), we deduce that the time equation of the comb filter is as follows:

$$g[n] = Ag[n - M] + f[n] - f[n - M].\tag{17.48}$$

Goertzel algorithm

This algorithm provides a numerical efficient way of evaluating a spectral component at a chosen frequency ω_0. It is based on the property that the Fourier transform for a given frequency ω_0 of a digital signal limited in time to N points may be written as the output of a first-order AR system. Let us start with the expression of the DFT of a causal signal $f[n]$, assuming that $T = 1$. According to (16.53),

$$F[k] = \sum_{n=0}^{N-1} f[n]\, e^{-j\frac{2\pi nk}{N}}.\tag{17.49}$$

We note $W_N^k = e^{-j\frac{2\pi k}{N}}$. The last equation becomes

$$F[k] = \sum_{n=0}^{N-1} f[n]\, W_N^{kn} = \sum_{n=0}^{N-1} f[n]\, W_N^{-k(N-n)},\tag{17.50}$$

since $W_N^{-kN} = e^{j\frac{2\pi Nk}{N}} = 1$, k being an integer.

$F[k]$ appears to be the convolution of $f[n]$ and $W_N^{-nk}U[n]$ evaluated at $n = N$. The step function has been introduced to impose $F[k] = 0$ for $N < 0$.

In conclusion $F[k]$ is the output of a filter with an input $f[n]$. The transfer function $H(z)$ of this filter is the z-transform of $W_N^{-nk}U[n]$. $H(z) = \sum_{-\infty}^{\infty} W_N^{-nk}U[n]z^{-n} = \sum_{0}^{\infty} W_N^{-nk}z^{-n}$ The sum of this geometric series is

$$H(z) = \frac{1}{1 - W_N^{-k}z^{-1}} = \frac{1}{1 - e^{-j\frac{2\pi k}{N}}z^{-1}}.$$ (17.51)

The filter is a first-order AR with a pole on the unit circle. Extending this result for any frequency ω_0, we write

$$H(z) = \frac{1}{1 - e^{-j\omega_0 T}z^{-1}}.$$ (17.52)

The situation of the pole is analog to that met in Chap. 10 for a system marginally stable with a pole on the imaginary axis. The magnitude of the impulse response should increase as n, the time variable.

The advantage of this result is that when we need to calculate a spectral component amplitude at one (or a small number of frequencies) the numerical calculus is faster and much simpler to implement in hardware than a FFT.

A refinement is provided by the following algorithm which avoids the complex calculation implied by (17.52) and in consequence is faster.

It consists of a cascade of a second-order AR filter with its complex conjugate poles $z_{p_{1,2}}$ on the unit circle at frequency ω_0.

$$H_1(z) = \frac{1}{(1 - z^{-1}e^{j\omega_0 T})(1 - z^{-1}e^{-j\omega_0 T})}.$$ (17.53)

This filter is followed by a single-zero MA filter. The zero is located at one pole of the first filter to cancel out the effect of that pole:

$$H_2(z) = 1 - z^{-1}e^{j\omega_0 T}$$ (17.55)

The transfer function of the cascade is

$$H(z) = H_1(z)H_2(z) = \frac{1}{(1 - z^{-1}e^{-j\omega_0 T})}.$$ (17.56)

Let us look at the problem in the time domain. We note $f[n]$ the input signal and $w[n]$ the output of the first filter according to the time equation:

$$w[n] = 2\cos(\omega_0 T)\dot{w}[n-1] - w[n-2] + f[n].$$ (17.57)

$w[n]$ is the input of the second filter whose output is $y[n]$. We have

$$y[n] = w[n] - w[n-1]e^{j\omega_0 T}$$ (17.58)

The calculation is faster that the one resulting from (17.52) since (17.56) is a calculation with real numbers and (17.57) is a complex equation which is performed only once, for $n = N$.

Fig. 17.16 Filter outputs;
Top 690 Hz; *Bottom* 770 Hz

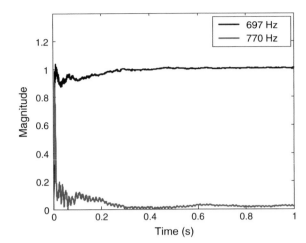

Example An example of application is the recognition in a noisy environment of frequencies used in the Dual-Tone Multi-Frequency signaling (DTMF) when punching a key in a touch key telephone.

The touch 1 generates $f_0 = 697\,\text{Hz}$ and $f_0' = 1209\,\text{Hz}$.

The touch 4 generates $f_1 = 770\,\text{Hz}$ and $f_1' = 1209\,\text{Hz}$.

Let us see on a simulation with Matlab in Fig. 17.16, how the discrimination between touch 1 and touch 4 is possible with the Goelter algorithm.

The sampling frequency is $F_e = 10\,\text{kHz}$; signals at $f_0 = 697\,\text{Hz}$ and $f_1 = 770\,\text{Hz}$ with amplitude 1; zero mean Gaussian noise with standard deviation is added to the signals. For the plot, the output of the filter has been divided by the index n in the sequence. The selection frequency of the filter is $f_0 = 697\,\text{Hz}$.

It is clear on 17.16 that after 100 ms (time inferior to the pressured key time) touch 1 with its frequency $f_0 = 697\,\text{Hz}$ is recognized by the value 1 of the output amplitude.

17.4   Transition from an Analog Filter to a Digital Filter

Several techniques are used to design digital filters with properties close to those of analog filters that are known for a long time. In the general case, no method is perfect in the sense that none provides simultaneous equal frequency responses and impulse responses of digital and analog filters.

17.4.1   *Correspondence by the Bilinear Transformation*

This first mapping rule, widely used, is obtained by a bilinear transformation in the complex plane allowing the passage of the transfer function $H_a(s)$ of the analog filter to that of the digital filter $H(z)$. s is the Laplace variable $s = \sigma + j\omega$. The bilinear transformation is

$$s = \frac{2}{T}\frac{1 - z^{-1}}{1 + z^{-1}}. \tag{17.62}$$

T is the sampling interval. The factor ensures $\frac{2}{T}$ that s has the dimension of the inverse of a time.

Discussion of this transformation: This mapping transforms the imaginary axis $j\omega$ of the Laplace plane in the unit circle of the z-plane. $s = 0$ (analog frequency zero) corresponds to the point $z = 1$ (digital zero frequency). $s = j\infty$ corresponds $z = -1$ (Nyquist frequency, the highest discrete frequency).

The imaginary axis in the Laplace plane (vertical axis of angular frequencies for which $s = j\omega$) is transformed in the unit circle in digital, as can be seen by expressing z as a function of s from (17.62):

$$z = \frac{1 + \frac{sT}{2}}{1 - \frac{sT}{2}}. \tag{17.63}$$

If $s = j\omega$ (s is on the imaginary axis, representative monochromatic signals), we see that z is the ratio of a complex number and its conjugate complex. This complex number is noted here in its trigonometric form $\rho e^{j\theta}$:

$$z = \frac{1 + j\frac{\omega T}{2}}{1 - j\frac{\omega T}{2}} = \frac{\rho e^{j\theta}}{\rho e^{-j\theta}} = e^{2j\theta} = e^{j\varphi} \quad \text{with } \varphi = 2\theta. \tag{17.64}$$

Thus, for all points of the pure imaginary axis $s = j\omega$ in the Laplace plane we have $|z| = 1$, which reflects the transformation of the imaginary axis for s into the circle of radius 1 for z.

Then,

$$\tan\theta = \frac{\omega T}{\frac{2}{1}} = \frac{\omega T}{2} \Rightarrow \text{Arg}(z) = 2\theta = 2\arctan\frac{\omega T}{2}. \tag{17.65}$$

Conversely, transformation of unit circle in the imaginary axis will be written as $s = j\omega = j\frac{2}{T}\tan\theta = j\frac{2}{T}\tan\frac{\varphi}{2}$, where φ is the argument of z.

Example of the digital filter corresponding to a first-order analog filter:

The analog filter transfer function has the form

$$H_a(s) = \frac{1}{s - s_1}. \tag{17.66}$$

Using the relation (17.62), that of the digital filter is then

$$H(z) = \frac{1}{\frac{2}{T}\left(\frac{1-z^{-1}}{1+z^{-1}}\right) - \frac{2}{T}\left(\frac{1-z_1^{-1}}{1+z_1^{-1}}\right)} = \frac{T}{4}\frac{(z+1)(z_1+1)}{z-z_1}. \tag{17.67}$$

We see in particular that the transmittance of the analog filter decreases to zero as the frequency approaches infinity, making the filter a low-pass filter. The transfer function of the corresponding digital filter has a zero at the Nyquist frequency $(z = -1)$, the highest frequency of the digital signal. This frequency corresponds to $\omega = \infty$ for the analog filter. The digital filter is also low pass.

N.A. Fig. 17.17 shows the frequency responses of a causal first-order analog filter and that of the digital filter obtained by bilinear transformation. We took $s_1 = -10 + j100$ and $T = 10^{-3}$. The pole s_1 was chosen close to the imaginary axis so as to ensure a sizable dynamic peak for a limited interval of ω. T was chosen to be relatively small to obtain a location of the pole z_1 in the first quadrant, in the low frequency area, so that the peak occurs at low frequency, far from the Nyquist frequency. It is observed that at high frequencies the digital gain falls below the analog gain due to the zero of the digital transfer function for $\frac{\omega_e}{2} = \frac{\pi}{T} = 3141.6$ rad/s.

17.4.2 Correspondence by Impulse Response Sampling

The digital filter impulse response is selected as a sampling of the frequency response of the analog filter $h[n] = h_a(nT)$. As seen earlier on the properties when sampling, the digital filter frequency response is given by the infinite sum of analog responses translated by multiple of ω_e:

Fig. 17.17 Frequency responses of analog and digital filters (bilinear transform)

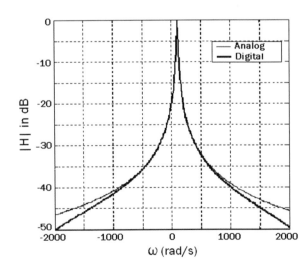

$$H\left(e^{j\omega T}\right) = \frac{1}{T} \sum_{n=-\infty}^{+\infty} H_a(\omega - n\omega_e) \tag{17.68}$$

Again we take for example, the first-order analog filter met before. The impulse response of the analog filter is $h_a(t) = -\frac{1}{s_1} e^{s_1 t} U(t)$, which is complex in this digital implementation since $p_1 = -10 + j100$ is complex. The filter gain is shown in Fig. 17.18.

At high frequencies, the value of the digital gain rises above that of the analog gain due to the summation operation of the translated spectra.

17.4.3 Correspondence by Frequency Response Sampling

The principle of this technique is to sample the frequency response of the analog filter and assign these values $H_a(\omega_k)$ (within the factor $1/T$) to the frequency response $H\left(e^{j\omega_k T}\right)$ of the digital filter. As essentially the function $H\left(e^{j\omega T}\right)$ is periodic, the operation is realistic only if the support of $H_a(\omega)$ is bounded or if the "forgotten" values are negligible. Digitally, the frequency domain is a continuum. It is necessary to evaluate the frequency response between two sampled values. The technique is based on the discrete Fourier transform. At first the digital impulse response is calculated by the inverse discrete Fourier transform of $H\left(e^{j\omega_k T}\right)$. This function has necessarily a finite duration since one cannot indefinitely extend the computations time. Moreover, we seek to minimize the inevitable folding error

Fig. 17.18 Frequency responses of analog and digital filters (impulse response sampling)

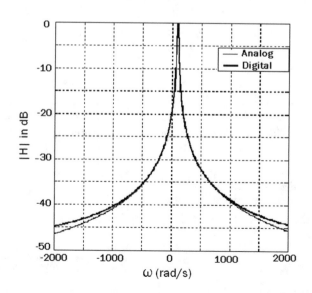

(here in the time domain) accompanying the discretization (here in the frequency domain) by taking a sufficiently tight sampling of the frequency response.

Once the digital impulse response is calculated, we can evaluate the frequency response for any frequency using the Fourier transform. The overall process provides a frequency response on a continuum which passes through the sampled points.

Example of building an ideal low-pass filter without phase:

By hypothesis, $H_a(\omega) = 1$ for $|\omega| < \Omega$ and is zero elsewhere. This function is sampled and we take $H(e^{j\omega_k T}) = 1$ for $|\omega| < \Omega$. The interval width is divided into intervals. To highlight the effects, the number of points used here is low $N = 16$. We take $H(e^{j\omega_k T}) = 1$ for integer values of k: $k \in [-3, 3]$ and 0 elsewhere.

$h[n]$ is constructed by inverse discrete Fourier transformation on N points. The obtained impulse response is limited to N points. In Fig. 17.19 the continuous line represents the frequency response of the low-pass filter and 16 samples of the function.

The impulse response with $N = 16$ points is given by $h[n] = \frac{1}{16} \sum_{k=-3}^{3} e^{j2\pi\frac{nk}{16}}$.

It is noted that $h[n]$ is an even function of n.

$h[0] = 0.4375$, $h[1] = 0.3142$, $h[2] = 0.0625$, $h[3] = -0.0965$, $h[4] = -0.0625$, $h[5] = 0.0417$, $h[6] = 0.0625$, $h[7] = -0.0124$, $h[8] = -0.0625$.

Using these values of $h[n]$, we can calculate $H(e^{j\omega T})$ $\forall \omega$ by Fourier transformation. We can obtain a sampling of this Fourier transform by performing the FFT calculation on any number of points M taken between 0 and f_e. Note that M can be taken large compared to the original number N that was used to limit the duration of the impulse response of the FIR filter to be N.

We may take $M = 512, 1024, 2048$, etc.

$$H(e^{j\omega_k T}) = \sum_{n=-8}^{8} h[n] e^{-jn\omega_k T} \quad \text{with } \omega_k = \frac{k}{M}\omega_e. \tag{17.69}$$

Figure 17.20 shows different frequency responses in the interval $\{0, \omega_e\}$. $f_e = 1\,\text{Hz}$.

Note that the function $H(e^{j\omega T})$ passes through the sampled values and its oscillations in the vicinities of the function $H_a(\omega)$ transitions are due to the Gibbs

Fig. 17.19 Sampling of a frequency response

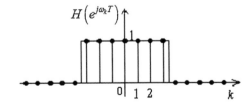

Fig. 17.20 Sampled
frequency response and
interpolated spectrum

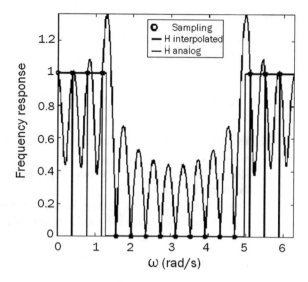

phenomenon in frequency. By sampling the analog frequency response more
tightly, we get a closer match of the original function, without nevertheless
removing oscillations caused by the discontinuities of the function $H_a(\omega)$.

Warning: We cannot ensure in the above example an even function $h[n]$ with an
even $N = 16$ number. In this case, it is necessary not to use a sample at the end of
the interval to ensure the periodicity of the inverse DFT on 16 points. It then
introduces a phase which is absent in the analog filter.

Summary
The autoregressive filters presented in this chapter are digital equivalents of the
analog filters met in Chap. 10. Their advantage is their high selectivity they allow.
Yet, their disadvantage is that their impulse response has an infinite length,
implying that, in principle, we should wait a time infinite to get the result of the
convolution of an input signal with the impulse response. This chapter began with
the presentation of the first and second-order AR filters. We have determined their
impulse responses, the conditions of stability, their transfer functions, and the
definition domain of these functions in the z-plane. As usual, the geometric interpretation of the frequency gain has provided a thorough understanding of the
operation of the filter and generalization to interpret the ARMA filters
(Autoregressive–Moving Average) of which several examples have been studied.
To use the many accumulated results in the literature on analog filters, various
methods of passing from an analog filter to its digital equivalent have been exposed.
In essence, the transition cannot be done accurately. We study the pluses and
minuses of commonly used methods.

Exercises

I. It is assumed that the signals are sampled at frequency $f_e = 50\,\text{MHz}$.
   A digital filter is defined by the position of the two poles of its transfer function:

$$p_0 = 0.97\frac{\sqrt{2}}{2}(1+i) \quad \text{and} \quad p_0^* = 0.97\frac{\sqrt{2}}{2}(1-i)$$

1. Give the expressions of the filter transfer function $H(z)$ and frequency response $H(e^{j\omega T})$.
2. Give the aspect of this frequency response magnitude justifying it by the situation of poles of $H(z)$.
3. What is the resonance frequency for a real input signal? Assuming that the resonance is sharp, give the -3 dB bandwidth.
4. Give the impulse response of the filter.

II. Consider the filter defined by its difference equation:

$$g[n] = 1/4(f[n-1] + 2f[n] + f[n+1]).$$

1. What is the frequency response of this filter? Represent the gain versus frequency. What is the nature of the filter? Give its -3 dB bandwidth.
2. What is the impulse response? Write the difference equation of a causal filter having the same gain (in magnitude) than the last. Compare its frequency response to that of the previous filter.
3. Calculate an all-zeros filter whose four zeros are on the unit circle in $\pm 2\pi/3$ and $\pm(2\pi/3 + \pi/6)$.
   Give the aspect of the frequency response. Can we speak of a low-pass filter? Give attenuation at Nyquist frequency.
   Give its impulse response assuming causality with nonzero response time at $n = 0$.

III. Fourier analysis of a digital signal:
   Let the digital sinusoidal signal be $s_0[n] = \sin(2\pi f_0 nT)$ (with $f_0 = 2\,\text{kHz}$ and $f_e = 20\,\text{kHz}$).

1. Calculate the Fourier transform $S_0(e^{j\omega T})$ of this signal. Represent this function.
2. We assume that only a portion $y[n]$ of the sine limited to $N = 2048$ points is available (from 0 to $N - 1$).

a. Write $y[n]$ in the form of the product of the sine function and a rectangular window.
b. Deduce the FT $Y(e^{j\omega T})$ as a convolution. Qualitatively represent the modulus of $Y(e^{j\omega T})$. Taking as reference the modulus at $f_0 = 2\,\text{kHz}$ give an upper bound to the modulus at 4 kHz.
c. Do we distinguish in the module's representation the presence of a second sinusoidal signal at frequency 4 kHz and amplitude $\varepsilon = 10^{-3}$ superimposed to the previous signal ($s[n] = s_0[n] + \varepsilon s_1[n]$)?
3. To improve the detectability of a small signal superimposed on the first, the signal $s[n]$ is multiplied by a Hanning window $w_H[n]$ consisting of one cycle of a sine signal equal to 0 in $n = 0$ and $n = N - 1$.
a. Give the expression of $w_H[n]$ and calculate its Fourier transform $W_H(e^{j\omega T})$. Represent approximately the latter function modulus.
b. Explain the reason for the increased detectability of the low spectral component when the signal $s[n] = s_0[n] + \varepsilon s_1[n]$ is multiplied by the Hanning window. Perform numerical evaluation.

IV. Digital Filtering: The transfer function of a digital filter is

$$H(z) = \frac{(z-e^{j\frac{\pi}{3}})(z-e^{-j\frac{\pi}{3}})}{(z-0.99e^{j\frac{\pi}{3}})(z-0.99e^{-j\frac{\pi}{3}})}.$$ Knowing   that   the   sampling   frequency   is

$f_e = 20\,\text{kHz}$, give the aspect of the frequency gain. Give the impulse response of this filter.

Note Assume that digital signals come from sampling at 20 kHz.

V. A digital signal $x[n]$ consists of an infinite succession of Kronecker pulses whose repetition frequency f_0 is 100 Hz.

A.

1. Give the expression of the signal $x[n]$ and represent graphically this signal.
2. What is the Fourier transform of this signal? Represent this FT.

B. The signal $x[n]$ is used as input to a first-order causal system with the parameter K.

3. Give the expression of the output signal $y[n]$.
4. On what condition upon K the system response to a Kronecker pulse does not exceed, at the entrance to the next excitement, 5 % of its initial value? Represent approximately the output signal in the case where K has the limit value (take three significant figures for K).
5. Give the expression of FT of the output signal. Represent this FT.

C. The signal $x[n]$ is now the input of a second-order system. One pole is noted $z_p = re^{j\omega_p T}$. Assume that the resonance is sharp.

6. Remind the expression of the impulse response of the second-order system. How to choose r to be in the limiting case of the Question B4 ($r = r_{5\%}$)?
7. Demonstrate the expression of the second-order system frequency response valid in the vicinity of the resonance frequency (strong). Demonstrate formula giving the -3 dB bandwidth.
8. Place the poles in the complex plane in the case where the resonance frequency is, in succession, 600 Hz, 1000 Hz, and 2400 Hz, and $r = r_{5\%}$. Calculate the -3 dB bandwidth.
9. The three systems of the previous question are placed in parallel, with the input signal $x[n]$ defined in A. It is desired that the spectral amplitudes of the output signal are such that the frequency components at 600 Hz and 1000 Hz have the same amplitude, with a -20 dB relative amplitude to that at 2400 Hz. Show that it is necessary to precede the filter by amplifiers whose gains are independent of frequency to readjust the output amplitudes to desired levels.
10. Give the expression of the output signal as a combination of convolutions.
11. The excitation signal has not an infinite duration but lasts 1/4 s. Describe in time and frequency the effect of this limitation.

NOTE: The problem models the digital synthesis of the vowel a (Attention a en anglais ne sonne pas come le a français). The three frequencies are the first three formants of a. The excitation Kronecker pulses are provided by the vocal cords which interrupt the output of air from the lungs in the form of very short pulses. This air enters the resonant cavities, larynx, mouth, whose adjustments provide the desired formants. The first two formants of the vowel i are approximately 200 and 2400 Hz. The frequency of such pulses for a female voice (of the order of 200 Hz) is twice that for a male voice.

VI. It is assumed that the digital signals originate from sampling with 10 kHz frequency.

Consider the digital filter defined by the equation: $g[n] = f[n] - 1.99858f[n-1] + f[n-2]$.

1. Determine the transfer function $H(z)$ of this filter. Show that it has two zeros on the unit circle. Represent the position of the zeros in the z-plane.
2. Conclude by a geometric argument the aspect of its frequency response after calculating the gain at zero frequency and 5 kHz.
   Can you use this filter to remove 60 Hz AC noise? Justify.
3. Give the impulse response of this filter.

VII. Let the digital filter defined by the temporal equation be

$$y[n] = \frac{1}{N} \sum_{l=-\left(\frac{N-1}{2}\right)}^{\frac{N-1}{2}} x[n+l], \text{ with } N \text{ odd.}$$

1. What is the impulse response of the filter? Represent it when $N = 9$.
2. Determine in this case the filter transfer function $H(z)$. Locate its zeros in the complex plane. From the position of these zeros, predict the shape of the filter frequency gain magnitude.
3. Give the expression of the frequency response $H(e^{j\omega T})$ (T is the sampling interval).

   Why could we predict that this function was real?

   What are the signal frequencies blocked by the filter in the case where the sampling frequency is $f_e = 1$ MHz. Accurately represent the frequency response.

VIII. Let the causal digital filter defined by the equation be

$$y[n] = 1.98\cos\left(\frac{\pi}{4}\right)y[n-1] - 0.99^2 y[n-2] + x[n] + x[n-1].$$

1. Calculate the system transfer function $H(z)$. Represent its remarkable points in the complex plane.
2. Show that the filter results from the cascade of two filters: a MA filter (Filter 1, $H_1(z)$) and an AR filter (Filter 2, $H_2(z)$). Give the functions $H_1(z)$ and $H_2(z)$ as well as their remarkable points. Discuss the problem in terms of stability and causality.
3. Calculate the frequency response of the MA filter and specify the nature of this filter. Represent its gain. Give the impulse response $h_1[n]$ of this filter.
   What is the response of the filter to the input $x[n] = U[n] - U[n-4]$?
4. Specify the character of the AR filter and give the appearance of its gain with frequency.
   Making the approximation valid if the pole is near the unit circle, give an approximate value of its resonant frequency and its bandwidth (in Hertz).
   Give the impulse response $h_2[n]$ of this filter and its aspect as a function of n.
5. Deduct from the above questions, the impulse response of the complete filter.

IX. Let the digital filter defined by the temporal equation be $g[n] = rg[n-1] + f[n] - f[n-1]$ with $r = 0.99$. The filter is supposed causal.

1. What is the transfer function of this filter? Place its notable points in the complex plane.
2. Give qualitatively using a geometric argument the shape of the frequency response (magnitude and phase) by specifying the values for frequencies $f_0 = 0$, $f_0 = 5$ kHz, and $f_0 = 10$ kHz. Compare the effects of this filter to those of the filter in question 1.
3. What is the impulse response of the filter?

X. Let the digital filter defined by the following time equation be

$$g[n] = -f[n] - f[n-1] + f[n-2] + f[n-3].$$

1. Give the impulse response $h[n]$ of the filter and represent this function. Is the filter causal?
2. Calculate the system transfer function $H(z)$. Having noticed that $z = 1$ is a root of $H(z) = 0$ determine the remarkable points of $H(z)$ and represent them in the z-plane.
3. Deduct from the position of these notable points of $H(z)$, the shape of the frequency response modulus $H(e^{j\omega T})$. What are the frequencies of the signals blocked by the filter? Calculate the expressions of the frequency response $H(e^{j\omega T})$ and of its modulus.
4. The input signal is now $f_0[n]$ which is the time reversal of $h[n]$: $f_0[n] = h[-n]$. Calculate the filter output signal $g_0[n]$. What is the Fourier transform of this signal?

Chapter 18
Minimum-Phase Systems—Deconvolution

In this chapter we introduce the notion of the minimum-phase system. We show with simple examples for two causal FIR systems having the same amplitude of the frequency gain, that a filter whose zeros are located within the unit circle will have a lower variation of phase with frequency. It follows that the impulse response of this filter is earlier. Since a minimum-phase causal filter has its zeros inside the unit circle, its inverse will be causal with its poles inside the unit circle, resulting in its stability. Deconvolving a signal is thus possible, that is to say, finding back the input signal of a filter by filtering the output signal of that filter. Then, we present the general problem of deconvolution with its frequency and time aspects. Deconvolution by the complex cepstrum method is introduced. It is illustrated with an example inspired from seismic measurements.

18.1 Minimum-Phase Systems

18.1.1 Notion of Minimum-Phase System

We begin this study by discussion of the phase shift generated by a very simple filter: an FIR for which the transfer function is limited to two terms:

$$H(z) = b_0 + b_1 z^{-1}. \tag{18.1}$$

This function is defined in the whole complex plane except at the point $z = 0$. We deduce that the filter is causal. It is stable, as the unit circle is within the domain of definition of $H(z)$.

The time equation of this filter is

$$g[n] = b_0 f[n] + b_1 f[n-1]. \tag{18.2}$$

© Springer International Publishing Switzerland 2016
F. Cohen Tenoudji, *Analog and Digital Signal Analysis,*
Modern Acoustics and Signal Processing, DOI 10.1007/978-3-319-42382-1_18

Its impulse response is

$$h[n] = b_0\, \delta[n] + b_1\, \delta[n-1]. \tag{18.3}$$

It consists of a pulse in time $n = 0$ followed by a second pulse at time $n = 1$. The zero of the transfer function is $z_0 = -\frac{b_1}{b_0}$. The relation (18.1) can be rewritten as

$$H(z) = b_0\left(1 - z_0\, z^{-1}\right). \tag{18.4}$$

For a reason which will appear later, we are interested in a filter whose transfer function $H_1(z)$ results by multiplying $H(z)$ by the following term:

$$\frac{z^{-1} - z_0^*}{1 - z_0\, z^{-1}} = (-z_0^*) \frac{1 - \left(1/z_0^*\right)z^{-1}}{1 - z_0\, z^{-1}}. \tag{18.5}$$

For $H_1(z)$, this term compensates the zero z_0 and adds the zero $1/z_0^*$ whose modulus is the inverse of z_0 modulus and with the same argument.

$$H_1(z) = b_0\left(1 - z_0\, z^{-1}\right)\frac{z^{-1} - z_0^*}{1 - z_0\, z^{-1}} = b_0\left(z^{-1} - z_0^*\right) = -b_0\, z_0^*\left(1 - \frac{z^{-1}}{z_0^*}\right). \tag{18.6}$$

The moduli of frequency responses of the two filters $\left|H\left(e^{j\omega T}\right)\right|$ and $\left|H_1\left(e^{j\omega T}\right)\right|$ are equal because the multiplier term modulus evaluated on the unit circle is 1:

$$\left|\frac{e^{-j\omega T} - z_0^*}{1 - z_0\, e^{-j\omega T}}\right| = \left|e^{-j\omega T}\right|\left|\frac{1 - z_0^* e^{j\omega T}}{1 - z_0\, e^{-j\omega T}}\right| = 1. \tag{18.7}$$

In the last fraction appeared the modulus of the ratio of a complex number and of its conjugate complex which is 1.

We can also say that the filter whose transfer function is the ratio given in (18.5) is an all-pass filter because the frequency response modulus is 1 at all frequencies.

Of course, the phase shifts generated by the two filters $H(z)$ and $H_1(z)$ are different.

The transfer function of the second filter is

$$H_1(z) = -b_0\, z_0^*\left(1 - \frac{z^{-1}}{z_0^*}\right) \text{ with } z_0^* = -\frac{b_1^*}{b_0^*}. \tag{18.8}$$

It is easier to continue the discussion with real coefficients b_0 and b_1 which ensure a real impulse response.

Since $z_0 = -\frac{b_1}{b_0}$ the zero of $H_1(z)$, $\frac{1}{z_0^*}$ is $\frac{1}{z_0^*} = \frac{1}{z_0} = -\frac{b_0}{b_1}$.

$H_1(z) = b_1\left(1 + \frac{b_0}{b_1}z^{-1}\right) = b_1 + b_0\, z^{-1}$. Hence $h_1[n] = b_1\, \delta[n] + b_0\, \delta[n-1]$.

The $H_1(z)$ filter impulse response is the time reverse of that of the original filter.

The z_0 position in the complex plane with respect to the unit circle depends on the value of the ratio $\frac{b_1}{b_0}$.

Assuming $|b_0| > |b_1|$, then $|z_0| < 1$. The zero of $H(z)$ lies within the unit circle while that of $H_1(z)$ is located outside this circle.

The effect of these zeros position on the phase of the frequency responses is now studied in an example.

By choosing $b_0 = 1$ and $b_1 = 0.5$, $z_0 = -0.5$, the impulse response $h[n]$ and the z_0 position obtained are shown in Fig. 18.1a, b.

The impulse response $h_1[n]$ and the zero position of the filter $H_1(z)$ are shown in Fig. 18.2a, b.

The gain magnitude is shown in Fig. 18.3a, equal for both filters (by construction) and in Fig. 18.3b the respective phases of frequency responses are plotted. It is noted that the two phases are negative for positive frequencies, as is the case for causal filters, and that the phase shift created by the second filter is always more negative than that of the first.

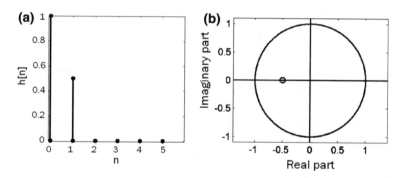

Fig. 18.1 a Impulse response; b zero of transfer function

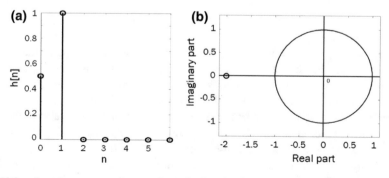

Fig. 18.2 a Impulse response; b zero of transfer function

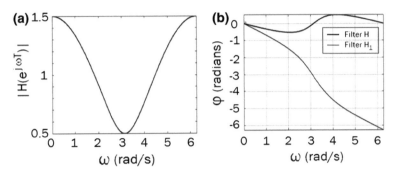

Fig. 18.3 a Gain of both filters; **b** Phases

By comparison it is said that the first filter is minimum phase.

We now show geometrically why this difference in phase variation is due to the fact that the zero of the transfer function is inside the unit circle for the first filter and is outside for the second filter.

The first filter frequency response is

$$H\left(e^{j\omega T}\right) = b_0 + b_1 e^{-j\omega T}. \tag{18.9}$$

We are still in the case $b_0 > b_1$. The argument of $H\left(e^{j\omega T}\right)$ is equal to the angle made by the vector sum of the vector b_0 and the vector $b_1 e^{-j\omega T}$ with the real axis. When ωT varies from 0 to 2π, the point M circulates counterclockwise on the unit circle, and the end of the vector sum sweeps the circle centered at b_0 with radius b_1 as shown in bold in Fig. 18.4a. It is noted that in the case of the left figure the phase will be contained in an interval within the range $\left\{-\frac{\pi}{2}, \frac{\pi}{2}\right\}$.

The frequency response of the second filter is

$$H_1\left(e^{j\omega T}\right) = b_1 + b_0 e^{-j\omega T}. \tag{18.10}$$

The end of the vector sum travels in a circle centered in b_1 and radius b_0 (Fig. 18.4b). As $b_0 > b_1$ this circle surrounds the origin, leading the phase decrease from 0 down to -2π.

In summary, we have shown on an example the general property that a filter whose zero lies inside the unit circle is causal, stable, and minimum phase. One whose zero is outside the unit circle is causal, stable but is not minimum phase.

For these filters with one zero, the impulse response is limited to two elements. If the magnitude of the first term of the response exceeds that of the second, the filter is minimum phase. Qualitatively we could say that energy comes faster (since the first term is the largest) from the minimum-phase filter (mpf).

The importance of the concept of minimum-phase causal filter is that this filter has a causal and stable inverse filter, that is to say that there is a causal and stable filter whose transfer function denoted here $H_{\text{inv}}(z)$ is such that $H_{\text{inv}}(z) = \frac{1}{H(z)}$.

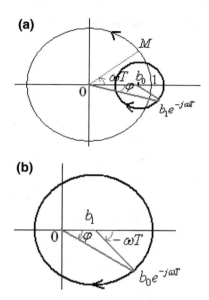

Fig. 18.4 **a** $H(e^{j\omega T}) = b_0 + b_1 e^{-j\omega T}$; **b** $H_1(e^{j\omega T}) = b_1 + b_0 e^{-j\omega T}$

If $|b_0| > |b_1|$, as has been seen above, the zero of the filter transfer function $H(z) = b_0(1 - z_0 z^{-1})$ is inside the unit circle. For its inverse filter $H_{\text{inv}}(z) = \frac{1}{b_0(1 - z_0 z^{-1})}$, the zero turned into a pole. As this pole is within the unit circle, the causal filter is stable. It is a first-order autoregressive system whose impulse response is

$$h_{\text{inv}}[n] = \frac{1}{b_0} \left(-\frac{b_1}{b_0}\right)^n U[n]. \qquad (18.11)$$

If the filter and its inverse are placed in cascade, the overall transfer function is

$$H(z) H_{\text{inv}}(z) = 1, \qquad (18.12)$$

resulting in the time domain

$$h[n] \otimes h_{\text{inv}}[n] = \delta[n]. \qquad (18.13)$$

By applying the inverse filter to the signal consisting of the impulse response, the impulse unit is obtained.

This principle is often used, particularly in the analysis of seismic signals, and it is then called deconvolution. This technique will be detailed hereinafter.

18.1.2 Properties of Minimum-Phase Systems

This analysis was developed on a filter with a single zero to clearly show the properties that will now be generalized. First, it is easily seen that the product of two minimum-phase transfer functions is minimum phase.

More generally, a minimum-phase polynomial is a polynomial in which all zeros are located within the unit circle. A minimum-phase system is a linear, time-invariant system over time, causal, whose transfer function is a rational function:$H(z) = \frac{B(z)}{A(z)}$ where $A(z)$ and $B(z)$ are minimum-phase polynomials.

Thus the poles and zeros of $H(z)$ are contained in the unit circle. As a result, the inverse of such a system whose transfer function is $H_{\text{inv}}(z) = \frac{1}{H(z)}$ is also a minimum-phase system (the poles and zeros are exchanged by the inversion).

A non-minimum phase system whose poles and zeros are not located on the unit circle can give passage to a minimum-phase system having an equal frequency response modulus by moving the poles and zeros lying outside the unit circle within this circle. Thus by multiplying the transfer function by terms of the type in (18.5), the modulus of the frequency response is unchanged.

We note $h[n]$ and $h_{\min}[n]$ the impulse responses of two filters having the same frequency response magnitude $|H(e^{j\omega T})|$, the second corresponding to the minimum-phase filter. It is shown mathematically that energy emerges faster from mpf. Thus, we have whatever n:

$$\sum_{k=0}^{n} |h[k]|^2 \leq \sum_{k=0}^{n} |h_{\min}[k]|^2. \tag{18.14}$$

The response of the mpf is earlier.

In the following example, we consider two FIRs having same frequency response. The zeros of the minimum-phase filter are shown in Fig. 18.5a. All zeros are inside the unit circle. In Fig. 18.5b, two conjugate complex zeros of the mpf

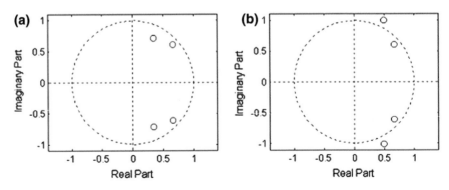

Fig. 18.5 Poles location of two filters with same gain. **a** Mpf; **b** Not mpf

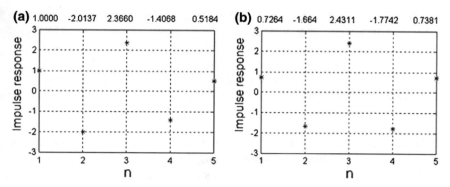

Fig. 18.6 Impulse responses of two filters with same gain. **a** Mpf; **b** Not mpf

have been displaced outside of the unit circle using the transformation rule of a zero z_0 in a zero $1/z_0^*$.

Figure 18.6 shows the two impulse responses (mpf in a) with their values at every moment written above the graph. We can verify that at every moment n, the property (18.14) is verified.

Zeros of the minimum-phase filter:

$z_0 = 0.9^*(\cos(\text{pi}/4.2) + i^* \sin(\text{pi}/4.2)); z_0^* = 0.9^*(\cos(\text{pi}/4.2) - i^*\sin(\text{pi}/4.2));$

$z_1 = 0.8^*(\cos(\text{pi}/2.8) + i^* \sin(\text{pi}/2.8)); z_1^* = 0.8^*(\cos(\text{pi}/2.8) - i^* \sin(\text{pi}/2.8));$

Zeros of the second filter:

$z_0 = 0.9^*(\cos(\text{pi}/4.2) + i^* \sin(\text{pi}/4.2)); z_0^* = 0.9^*(\cos(\text{pi}/4.2) - i^* \sin(\text{pi}/4.2));$

$z_1 = 1.12^*(\cos(\text{pi}/2.8) + i^*\sin(\text{pi}/2.8)); z_1^* = 1.12^*(\cos(\text{pi}/2.8) - i^*\sin(\text{pi}/2.8));$

18.2   Deconvolution

18.2.1   *Interest of Deconvolution*

When a signal reaches the observer, it has a history. It propagated over a communication medium and has undergone transformations. In general, during this process, noise is added on the original signal. It is often important to search to recover the original signal existing before these deformations. This research is an inverse problem: knowing the distortion of the transmission system properties, how, from the final distorted signal, can we find back the original signal? There are many situations in the case of the analysis of sounds. For example, can we find back the musical quality of orchestras that were recorded a century ago with imperfect recording means? Can we remove the creaking old recordings on disk phonograph?

This issue has generated a great deal of research. Numerous results have been obtained, either with linear techniques, or with nonlinear processes. The digital signal processing made possible unachievable results by analog processing.

As part of LTI systems study developed in this work, this objective is a deconvolution. In digital, the problem is expressed as follows: the received signal $y[n]$ resulting from the filtering of a signal $x[n]$ by an LTI filter with impulse response $h[n]$. We have

$$y[n] = x[n] \otimes h[n].$$

The question that arises is how to operate inverse filtering, that is to say find back the signal $x[n]$ from the measured signal $y[n]$?.

This problem is illustrated by the following diagram:

By modeling of physical problem, sometimes it is possible to have a good approximation of $h[n]$. How to retrieve $x[n]$ related to $y[n]$ by the convolution

$$y[n] = \sum_{m=-\infty}^{+\infty} x[m]h[n-m]. \tag{18.15}$$

18.2.2 Deconvolution Techniques

Deconvolution by complex spectral amplitudes division

Convolution (18.15) becomes a simple product in the frequency domain: $Y(e^{j\omega T}) = X(e^{j\omega T})H(e^{j\omega T})$. Since $X(e^{j\omega T}) = \frac{Y(e^{j\omega T})}{H(e^{j\omega T})} = Y(e^{j\omega T})H_{inv}(e^{j\omega T})$, with $H_{inv}(e^{j\omega T}) = \frac{1}{H(e^{j\omega T})}$, we can write

$$x[n] = \frac{1}{\omega_e} \int_0^{\omega_e} \frac{Y(e^{j\omega T})}{H(e^{j\omega T})} e^{j\omega n T} d\omega. \tag{18.16}$$

In the frequency domain, the deconvolution is illustrated by the scheme

Thus, by taking the ratio of the FT of the received signal $y[n]$ and that of the impulse response $h[n]$ the previously measured or otherwise determined, it is in principle possible to get $x[n]$ by an inverse Fourier transform.

Very attractive in principle, this technique is rarely used in practice for the following reasons: noise, however small, in signal $y[n]$ has spectral components in frequency ranges in which the size of $H(e^{j\omega T})$ which is in the denominator is low or even zero. Division by zero of these non-zero values of $Y(e^{j\omega T})$ because of the noise will give some very important values to the ratio. These values lead to an aberrant result of the estimation of $x[n]$.

We may in some cases use empirical methods which give fairly good results. Having found for example that a transmission channel weakens the high frequencies, we may enhance the high frequencies in reception to find satisfactory spectral amplitude. This spectrum recovery technique is widely used in analog or digital processing in restoring sound signals recorded a century ago.

When it is possible to calculate the impulse response $h_{inv}[n]$ of an inverse filter, the original signal $x[n]$ can be obtained by convolving the signal $y[n]$ therewith:

$$y[n] \otimes h_{inv}[n] = x[n] \otimes h[n] \otimes h_{inv}[n] = x[n].$$

Since by definition $h[n] \otimes h_{inv}[n] = \delta[n]$.

Inverse filtering deconvolution

When it is possible to calculate the impulse response $h_{inv}[n]$ of an inverse filter, the original signal $x[n]$ can be obtained by convolving the signal $y[n]$ therewith:

$$y[n] \otimes h_{inv}[n] = x[n] \otimes h[n] \otimes h_{inv}[n] = x[n],$$

Since by definition $h[n] \otimes h_{inv}[n] = \delta[n]$.

We saw earlier that a minimum-phase causal filter has a causal and stable inverse. It is then possible in that case to determine numerically $h_{inv}[n]$. This operation is illustrated in the following example. Figure 18.7a shows a FIR wavelet with 21 coefficients which can be considered as the impulse response of a minimum-phase filter as the all zeros of the filter transfer function are inside the unit circle (Fig. 18.7b): $h[n] = \sum\limits_{m=0}^{20} b_m \delta[n - m]$.

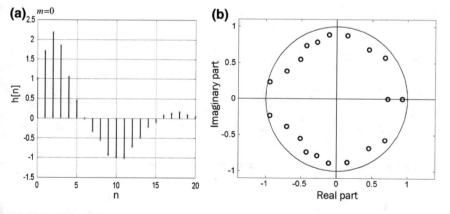

Fig. 18.7 FIR wavelet: **a** Wavelet; **b** zeros situation

The transfer function of the causal inverse filter is calculated by converting the zeros into poles. In principle, one can compute the impulse response $h_{inv}[n]$ of the inverse filter analytically. This function has an infinite duration since the inverse filter is AR. In practice, we proceed by numerically inverse DFT a sampling of function $H_{inv}(e^{j\omega T})$ on a number of points N. The resulting $h_{p\,inv}[n]$ is limited to N points. It imperfectly represents the impulse response $h_{inv}[n]$. If the number of points N is sufficiently large, the temporal aliasing will be low, and the function obtained will be close enough to the real response $h_{inv}[n]$ of the inverse filter. In Fig. 18.8a we see the calculated function $h_{p\,inv}[n]$ using an inverse FFT on $N = 64$ points of the initial wavelet and in Fig. 18.8b the result of the convolution $h[n] \otimes h_{p\,inv}[n]$. On the convolution product we recover the value 1 of the function $\delta[n]$ in $n = 0$, but a small parasitic signal appeared beyond $n = 63$ due to the limiting of $h_{p\,inv}[n]$ support to $N = 64$ points.

The case of a wavelet, which is the impulse response of a non-mpf filter having same frequency response module as above is now presented. The wavelet (not mpf) is shown in Fig. 18.9a. The impulse response of the causal inverse filter, evaluated on 64 points, is shown in Fig. 18.9b. We can see from the figure that this inverse causal filter is unstable. The deconvolution is impossible.

Deconvolution by the complex Cepstrum method

The deconvolution technique by applying the inverse filter developed in the preceding paragraph assumes that we a priori know the wavelet present in the composite signal. In this case, this method is very effective to pinpoint the arrival time and amplitude of the wavelet. It is able to separate the arrival times of two replicas of the wavelet very near even if they overlap within the composite signal.

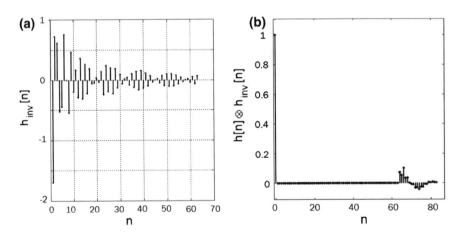

Fig. 18.8 a Causal filter inverse response; **b** Deconvolution result

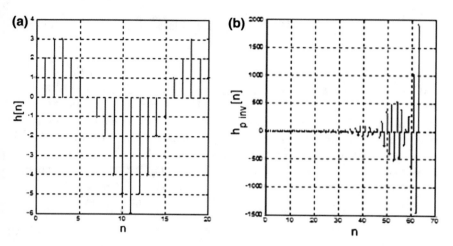

Fig. 18.9 a Non-mpf filter wavelet; **b** impulse response

In practice, however, the wavelet constituting a signal is generally not known a priori. This is the case, for example, in the processing of seismic signals, where even if the temporal shape of the excitation signal from an explosion generating a disturbance in the ground is known, the signal received by a remote sensor results in filtering the original signal by the propagation medium which arrives completely deformed.

The deconvolution method by the complex cepstrum does not require a priori knowledge of wavelet buried in the composite signal. The name of this technique comes from a pun based on the inversion of the word spectrum.

Consider a signal $x[n]$. Its z-transform (its complex spectrum) is noted $X(z)$. The time function called complex cepstrum $\hat{x}[n]$ is defined by

$$\hat{x}[n] = \frac{1}{2\pi j} \oint_C \log(X(z)) z^{n-1} dz. \qquad (18.17)$$

The first reason for the use of the logarithm is that the logarithm of a product is equal to the sum of the logarithms of the members of that product. It is worth noting here the notion of logarithm of a complex number. By showing the modules and arguments of $X(z)$, $X(z) = |X(z)|e^{j\mathrm{Arg}(X(z))}$, one can write

$$\log(X(z)) = \log(|X(z)|) + j\mathrm{Arg}(X(z)). \qquad (18.18)$$

Now it is assumed that $x[n]$ is the convolution of two functions $f[n]$ and $g[n]$: $x[n] = f[n] \otimes g[n]$. It follows that $X(z) = F(z)G(z)$, product of z-transforms of $f[n]$ and $g[n]$. Then

$$\log(X(z)) = \log(F(z)) + \log(G(z)). \tag{18.19}$$

It then comes, due to the linearity of the z-transform, $\hat{x}[n] = \hat{f}[n] + \hat{g}[n]$.
Complex cepstrum of $x[n]$ is the sum of the complex cepstra of $f[n]$ and $g[n]$.

If we can make a separation in the time domain of functions $\hat{f}[n]$ and $\hat{g}[n]$, the deconvolution is successful. We can get back $f[n]$ by first taking $\log(F(z))$, the logarithm of $f[n]$'s z-transform. Then, by taking the exponential of this result, we get $F(z)$ and finally eventually performing the inverse z-transform we obtain $f[n]$.

An example of a typical deconvolution in a seismic situation is now presented. It is assumed that the signal consists of a wavelet $f[n]$ and a replica (echo by a subterranean layer, for example) located at a later time $g[n] = af[n - n_0]$.
We assume that $|a| < 1$, which is natural for the case of an echo.So,

$$x[n] = f[n] + af[n - n_0]. \tag{18.20}$$

We can rewrite it as

$$x[n] = f[n] + af[n] \otimes \delta[n - n_0] = f[n] \otimes (\delta[n] + a\,\delta[n - n_0]).$$

Taking $z = e^{j\omega T}$, we can write

$$
\begin{aligned}
X\left(e^{j\omega T}\right) &= F\left(e^{j\omega T}\right) FT(\delta[n] + a\,\delta[n - n_0]), \\
X\left(e^{j\omega T}\right) &= F\left(e^{j\omega T}\right)\left(1 + a\,e^{-jn_0\omega T}\right).
\end{aligned}
\tag{18.21}
$$

$$\log\left(X\left(e^{j\omega T}\right)\right) = \log\left(F\left(e^{j\omega T}\right)\right) + \log\left(1 + a\,e^{-jn_0\omega T}\right). \tag{18.22}$$

Since it has been assumed that $|a| < 1$, one can use the development of $\log(1 + u)$ in the vicinity of $u = 0$ and write

$$\log\left(1 + a\,e^{-jn_0\omega T}\right) \simeq a\,e^{-jn_0\omega T} + a^2 e^{-2jn_0\omega T} + \ldots \tag{18.23}$$

The inverse FT of this infinite sum is

$$\hat{g}[n] = a\,\delta[n - n_0] + a^2\delta[n - 2n_0] + \ldots,$$

therefore

$$\hat{x}[n] = \hat{f}[n] + \hat{g}[n] = \hat{f}[n] + a\,\delta[n - n_0] + a^2\delta[n - 2n_0] + \ldots. \tag{18.24}$$

The cepstrum of $x[n]$ is composed of the sum of the cepstrum of the wavelet $f[n]$ and a series of δ functions located at the instants kn_0. It is then possible to identify first the delay n_0 of the replica with a study of the periodicity of these peaks. It is also possible to remove these peaks from $\hat{x}[n]$, e.g., by interpolating the values in kn_0 by the average of the values on the left and right of the abscissa. A good evaluation is then obtained for $\hat{f}[n]$. It remains only to take its FT $\log\big(F\big(e^{j\omega T}\big)\big)$, then perform the exponential of this function, and finally take the inverse FT for obtaining the wavelet $f[n]$.

A numerical simulation is presented in the following example. To create the composite signal $x[n]$, the wavelet $f[n]$ is the impulse response of a second-order filter $h[n] = a_1\,h[n-1] + a_2\,h[n-2] + \delta[n]$ with $a_1 = 1.65$ and $a_2 = -0.8$. The time of arrival of the replica is $n_0 = 10$. Its relative amplitude is $a = -0.7$. The sum of these two signals is performed: $x[n] = f[n] + af[n-n_0]$, which is represented in Fig. 18.10a. Figure 18.10b shows the cepstrum $\hat{x}[n]$. Note the peaks with periodicity 10. This allows finding the delay $n_0 = 10$ of the replica.

We remove the peaks by replacing the value of $\hat{x}[n]$ in these points with the average of the adjacent point values. We then have a good assessment of the cepstrum of the wavelet $\hat{f}[n]$ (Fig. 18.9a). The wavelet is derived in accordance with the end of the algorithm described above. The calculation result is shown in Fig. 18.11b. It overlaps very well with the original wavelet. Deconvolution succeeded in this case.

By subtracting $x[n]$ of the wavelet reconstructed, that is to say by calculating $x[n] - f[n]$, one can calculate the replica. It superimposes very well to the function $af[n-n_0]$, as shown in Fig. 18.12.

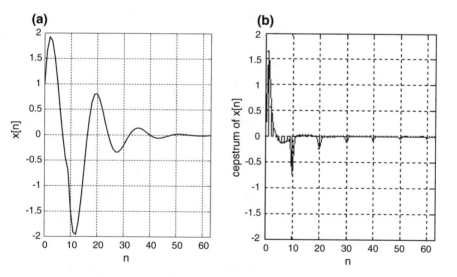

Fig. 18.10 **a** Sum of a signal and its replica; **b** Cepstrum of $x[n]$

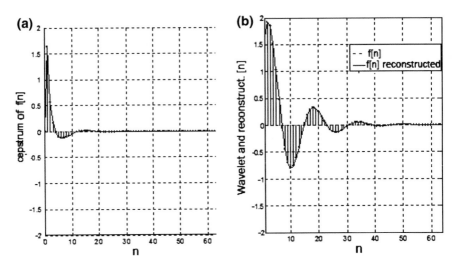

Fig. 18.11 **a** Cepstrum of $f[n]$; **b** Reconstructed wavelet

Fig. 18.12 Replica and its reconstruction

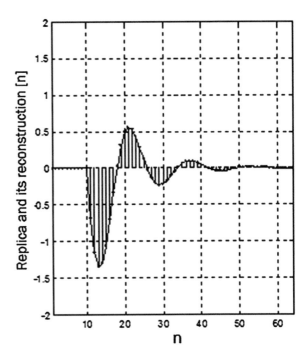

Summary
In this chapter we have first introduced the notion of minimum-phase system. We have shown with simple examples of two causal FIR systems having the same amplitude of the frequency gain, where filter whose zeros are located within the unit circle will have a lower variation of phase with frequency. It follows that the impulse response of this filter is earlier. Since a minimum-phase causal filter has its zeros inside the unit circle, its inverse is causal with its poles inside the unit circle and then causal. Deconvolving a signal is possible in some cases by finding back the input signal of a filter by filtering the output signal of that filter. We have presented the general problem of deconvolution with its frequency and time aspects. Deconvolution by the complex cepstrum method has been introduced. It was illustrated by an example inspired from seismic measurements.

Chapter 19
Wavelets; Multiresolution Analysis

This chapter follows Chap. 12 on time-frequency analysis. It has been shown how decomposition on a wavelet basis allows highlighting effectively changes with time of the properties of a signal. Signal processing is mainly done digitally today; wavelet bases with compact support have been searched which could be used by simple filtering operations. These bases must also allow reconstructing the signal accurately and easily from the decomposition coefficients. This treatment, which is called multi-resolution analysis, is remarkably effective in data compression, especially for image processing.

To begin this chapter, we return to the general problem. The amount of information exchanged and stored digitally today is enormous. To make these operations possible with transmission channels with physically intrinsic limited throughput and storage capacity, compression techniques that allow information made acceptable for audio or video were sought. Along with the steady increase in the speed of electronic components, intensive research has been conducted to develop new coding algorithms for data compression. A striking example of the results of this research is found when consulting an aerial view of a location on Internet. First appears the globe with few details. One may rotate the view to center approximately on the desired location, and then ask for magnification. The first image is blurred without much detail. This leaves the user some time to adjust the centering of the map on his place of interest. It would be very inefficient to convey all the details of a map (operation that takes a long time) for a card that is not properly centered. The operation continues with increasing magnifications. This shows the efficiency during the transmission of an image to transmit firstly a view without the details, and then transmit the detailed information when needed.

It is known that the details of a signal are contained in the high frequencies. Hence the idea to separate information contained in high and low frequencies. This is the principle of the use of filter banks. The progress of the analysis of these filter banks led to multiresolution analysis. It allows numerical separation of frequency bands recursively while still allowing the ability to reconstruct the signal without loss of information in a second step. The presentation of these concepts in this chapter

© Springer International Publishing Switzerland 2016
F. Cohen Tenoudji, *Analog and Digital Signal Analysis*,
Modern Acoustics and Signal Processing, DOI 10.1007/978-3-319-42382-1_19

begins with the principles of the dyadic decomposition-reconstruction of a signal. The initial Haar Transform mathematical developments are given as an example. Their understanding makes it easier to address the concepts of multi-resolution analysis. The space devoted to this problem being limited in this book, readers are invited to deepen the concepts briefly described here with the many books dedicated to wavelet analysis, especially the books of S. Mallat and I. Daubechies.

There is a relationship between filter banks and wavelets that we are aiming to put in evidence here.

19.1 Dyadic Decomposition-Reconstruction of a Digital Signal; Two Channels Filter Bank

Dyadic decomposition now exposed allows by a simple linear filtering operation the decomposition of a digital signal into two components. One component contains the low frequencies and the other the high frequencies. It is possible under certain conditions using a second filtering operation on the two components to fully recover the original signal.

The question raises: A signal $x[n]$ is filtered by a filter with impulse response $h_0[n]$. One sample over 2 of the output signal is set to zero (this operation is symbolized in the graph below by the symbol decimation by 2 and then re extension by 2). The result is filtered by the filter with impulse response $g_0[n]$. On a parallel branch the same operation is made with filters with impulse responses $h_1[n]$ and $g_1[n]$. The output signals are added (see Fig. 19.1).

We denote $y_0[n]$ the signal at the output of the first filter in the upper branch of the graph. Its z-transform is: $Y_0(z) = X(z)H_0(z)$. We note $\hat{y}_0[n]$ the signal resulting from zeroing a sample over 2 of $y_0[n]$ and $\hat{Y}_0(z)$ its z-transform.

$\hat{Y}_0(z)$ is related to $Y_0(z)$ by the relationship:

$$\hat{Y}_0(z) = \frac{1}{2}(Y_0(z) + Y_0(-z)). \tag{19.1}$$

Indeed, the development of $Y_0(z)$ is:

$$Y_0(z) = \ldots + y_0[-2]z^2 + y_0[-1]z^1 + y_0[0] + y_0[1]z^{-1}y_0[2]z^{-2} + \ldots$$

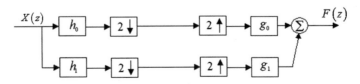

Fig. 19.1 Two channels filter bank decomposition-reconstruction

We have: $Y_0(-z) = \ldots + y_0[-2]z^2 - y_0[-1]z^1 + y_0[0] - y_0[1]z^{-1} + y_0[2]z^{-2} + \ldots$

Then $\frac{1}{2}(Y_0(z) + Y_0(-z)) = \ldots + y_0[-2]z^2 + y_0[0] + y_0[2]z^{-2} + \ldots$

We recognize the z-transform of the series $\hat{y}_0[n]$ in which all the terms of even ranks were canceled.

We note $\hat{Y}_0(z)$ and $\hat{Y}_1(z)$ respectively the z-transform of $\hat{y}_0[n]$ and $\hat{y}_1[n]$. The z-transform of the output signal is:

$$F(z) = G_0(z)\hat{Y}_0(z) + G_1(z)\hat{Y}_1(z). \tag{19.2}$$

Using the relation (19.1):

$$F(z) = \frac{1}{2}G_0(z)(Y_0(z) + Y_0(-z)) + \frac{1}{2}G_1(z)(Y_1(z) + Y_1(-z)). \text{ Or,}$$

$$F(z) = \frac{1}{2}G_0(z)(H_0(z)X(z) + H_0(-z)X(-z)) + \frac{1}{2}G_1(z)(H_1(z)X(z) + H_1(-z)X(-z))$$

Thus:

$$F(z) = \frac{1}{2}(H_0(z)G_0(z) + H_1(z)G_1(z))X(z) + \frac{1}{2}(H_0(-z)G_0(z) + H_1(-z)G_1(z))X(-z). \tag{19.3}$$

We want now $F(z)$ to be a filtering of $X(z)$ without aliasing. It is then necessary that the factor of $X(-z)$ is zero in (19.3). Thus, we necessarily have:

$$H_0(-z)G_0(z) + H_1(-z)G_1(z) = 0. \tag{19.4}$$

Then:

$$F(z) = \frac{1}{2}(H_0(z)G_0(z) + H_1(z)G_1(z))X(z). \tag{19.5}$$

The condition (19.4) may be verified by different functions combinations.

A particularly interesting case is met when the filters are such that the output $F(z)$ is equal to a delayed version of $X(z)$. In other words we are looking for a combination of filters that will allow a reconstruction of the signal $x[n]$ after its decomposition in two components for compression purpose, for example.

We seek these 4 filters $H_0(z)$, $H_1(z)$, $G_0(z)$ and $G_1(z)$ as causal.

Because of their causality, the filters necessarily generate a delay for the output signal. If we note $m > 0$ the delay of $x[n]$ by the crossing through the filter, we will have:

$$F(z) = X(z)z^{-m}. \tag{19.6}$$

A possible configuration is:

$$G_0(z) = z^k H_1(-z) \text{ and } G_1(z) = -z^k H_0(-z). \tag{19.7}$$

Another possibility is given by a change in sign:

$$G_0(z) = -z^k H_1(-z) \text{ and } G_1(z) = z^k H_0(-z), \tag{19.8}$$

where k may be any integer.

In the case (19.7),

$$F(z) = \frac{1}{2} z^k (H_0(z)H_1(-z) - H_1(z)H_0(-z))X(z). \tag{19.9}$$

Let us write $P(z) = H_0(z)H_1(-z)$. Taking (19.4), in account, relation (19.3) is written:

$$F(z) = \frac{1}{2}(P(z) - P(-z))z^k X(z).$$

In this case we should have

$$P(z) - P(-z) = 2 \text{ and } k = -m. \tag{19.10}$$

This equation appears as a sufficient condition for the quadruplet of filters allows the decomposition- reconstruction of the signal without loss of information.

In the case (19.8) with again $P(z) = H_0(z)H_1(-z)$ we get

$$P(z) - P(-z) = -2. \tag{19.11}$$

The following two paragraphs are simple examples of this decomposition. We will expose their limitations. Wavelet decomposition allows the determination of more efficient decompositions.

Haar transform

The decomposition by the Haar transform is a first example of the previous results.

The Haar transform of a pair of variable $\begin{pmatrix} x_1 \\ x_2 \end{pmatrix}$ in a pair $\begin{pmatrix} y_1 \\ y_2 \end{pmatrix}$ is defined

as: $\begin{pmatrix} y_1 \\ y_2 \end{pmatrix} = T\begin{pmatrix} x_1 \\ x_2 \end{pmatrix}$, wherein the matrix T is:

$$T = \frac{1}{\sqrt{2}}\begin{pmatrix} 1 & 1 \\ 1 & -1 \end{pmatrix}. \tag{19.12}$$

Inversely we have:

$$y_1 = \frac{1}{\sqrt{2}}x_1 + \frac{1}{\sqrt{2}}x_2 \text{ and } y_2 = \frac{1}{\sqrt{2}}x_1 - \frac{1}{\sqrt{2}}x_2. \tag{19.13}$$

The matrix T shown in (19.12) is symmetric. Its determinant has an absolute value of 1 and its column vectors are orthogonal. It follows that its inverse equals its transpose:

$$T^{-1} = T^T = \frac{1}{\sqrt{2}}\begin{pmatrix} 1 & 1 \\ 1 & -1 \end{pmatrix}. \tag{19.14}$$

Therefore

$$x_1 = \frac{1}{\sqrt{2}}y_1 + \frac{1}{\sqrt{2}}y_2 \text{ and } x_2 = \frac{1}{\sqrt{2}}y_1 - \frac{1}{\sqrt{2}}y_2. \tag{19.15}$$

We can apply this transformation in signal processing. Consider a sequence of values which constitutes a signal $x[n]$. Based on this transform, two data sequences $y_0[n]$ and $y_1[n]$ may be created by:

$$y_0[n] = \frac{1}{\sqrt{2}}x[n-1] + \frac{1}{\sqrt{2}}x[n] \text{ and } y_1[n] = \frac{1}{\sqrt{2}}x[n-1] - \frac{1}{\sqrt{2}}x[n]. \tag{19.16}$$

In practice, the signal $x[n]$ is causal. $y_0[n]$ and $y_1[n]$ are the output signals of two causal MA filters. The function $y_0[n]$ represents a smoothed version of the signal $x[n]$ and $y_1[n]$ is the derivation (numerical) of that signal. The respective impulse responses of these filters are:

$$h_0[n] = \frac{1}{\sqrt{2}}(\delta[n-1] + \delta[n]) \text{ and } h_1[n] = \frac{1}{\sqrt{2}}(\delta[n-1] - \delta[n]). \tag{19.17}$$

Their transfer functions are:

$$H_0(z) = \frac{1}{\sqrt{2}}(z^{-1}+1) \text{ and } H_1(z) = \frac{1}{\sqrt{2}}(z^{-1}-1). \tag{19.18}$$

The first filter is low-pass as its transfer function has a zero at $z = -1$. The second has a transmittance zero at $z = 1$, which makes it a high-pass filter.

We take $T = 1$ without loss of generality. The frequency responses are:

$$H_0(e^{j\omega}) = \frac{1}{\sqrt{2}}(e^{-j\omega}+1) \text{ and } H_1(e^{j\omega}) = \frac{1}{\sqrt{2}}(e^{-j\omega}-1). \tag{19.19}$$

Reconstruction of the signal from its components

We use the notations $x_1 = x[n-1]$ and $x_2 = x[n]$ with $y_1 = y_0[n]$ and $y_2 = y_1[n]$ in (19.15).

Relations (19.16) imply that to each signal value $x[n]$ correspond two values $y_0[n]$ and $y_1[n]$ that both contain information on the signal $x[n]$ at the instants n and $n-1$. Thus, according to the Eq. (19.15), we have:

$$x[n-1] = \frac{1}{\sqrt{2}} y_0[n] + \frac{1}{\sqrt{2}} y_1[n], \ \forall n, \qquad (19.20)$$

and

$$x[n] = \frac{1}{\sqrt{2}} y_0[n] - \frac{1}{\sqrt{2}} y_1[n], \ \forall n. \qquad (19.21)$$

If we keep only one value among 2 of $y_0[n]$ and $y_1[n]$ (decimation by 2 or, in other words, sub-sampling by 2), we can still go back to the function $x[n]$ by an inverse transformation using the matrix given in (19.14). There is no loss of information.

Since we used one value over 2 of $y_1[n]$ and $y_2[n]$, one can define the functions $\hat{y}_0[n]$ and $\hat{y}_1[n]$ such that $\hat{y}_0[n] = y_0[n]$ and $\hat{y}_1[n] = y_1[n]$ for even values of n and are zero for odd values.

It is then possible to reconstruct the signal $x[n]$ from the signals $\hat{y}_0[n]$ and $\hat{y}_1[n]$ writing for any n:

$$x[n] = \frac{1}{\sqrt{2}} (\hat{y}_0[n+1] + \hat{y}_0[n]) + \frac{1}{\sqrt{2}} (\hat{y}_1[n+1] - \hat{y}_1[n]), \ \forall n. \qquad (19.22)$$

$$x[n] = g_0[n] \otimes \hat{y}_0[n] + g_1[n] \otimes \hat{y}_1[n] \text{ or } X(z) = G_0(z)\hat{Y}_0(z) + G_1(z)\hat{Y}_1(z),$$

with

$$g_0[n] = \frac{1}{\sqrt{2}} (\delta[n+1] + \delta[n]) \text{ and } g_1[n] = \frac{1}{\sqrt{2}} (\delta[n+1] - \delta[n]).$$

With

$$G_0(z) = \frac{1}{\sqrt{2}} (z+1) \text{ and } G_1(z) = \frac{1}{\sqrt{2}} (z-1). \qquad (19.23)$$

Using (19.18), we may write:

$$G_0(z) = -zH_1(-z) \text{ and } G_1(z) = zH_0(-z).$$

We find that:

$$G_0(z)H_0(z) + G_1(z)H_1(z) = -zH_1(-z)H_0(z) + zH_0(-z)H_1(z) = -2z, \quad (19.24)$$

which is consistent with (19.10),
   and

$$G_0(z)H_0(-z) + G_1(z)H_1(-z) = 0. \tag{19.25}$$

The last expression (19.25) is the relation (19.4).

Filters with transfer functions $H_0(z)$ and $H_1(z)$ appear as decomposition filters (analysis filter bank) signal. Filters with transfer functions $G_0(z)$ and $G_1(z)$ are reconstruction filters (synthesis filters) of this signal.

Note The decomposition filters of the Haar transform $H_0(e^{j\omega})$ and $H_1(e^{j\omega})$ have an interesting property. We note on (19.19) that $H_1(e^{j\omega}) = -H_0\left(e^{j(\omega+\pi)}\right)$. Thus:

$$\left|H_1(e^{j\omega})\right| = \left|H_0\left(e^{j(\omega+\pi)}\right)\right|.$$

The following relationship stands:

$$\left|H_0(e^{j(\omega)})\right|^2 + \left|H_1(e^{j\omega})\right|^2 = \left|H_0(e^{j(\omega)})\right|^2 + \left|H_0\left(e^{j(\omega+\pi)}\right)\right|^2 = 2. \tag{19.26}$$

The gains amplitudes are represented in Fig. 19.2 in the case $0 \le \omega \le \pi$. These moduli are symmetrical with respect to $\omega = \frac{\pi}{2}$. These filters are called quadrature mirror filters.

This filter bank is simple and works very well. However, we note that the filters are not very selective, so it is not very effective in the separation of HF and LF. After passing through the LF filter, there are too much HFs remaining and vice versa in the HF filter. We are led to search for other, more efficient, filters.

Fig. 19.2 Spectral amplitudes of Haar filters for $\{0 \le \omega < \pi\}$

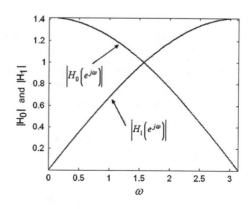

Daubechies wavelets are designed to solve this problem. They are detailed in Sect. 19.3 of this chapter.

LeGall-Tabatabai filter bank 5–3

We take as a second example the LeGall-Tabatabai filter bank 5–3 (It is used for image compression and is part of the JPEG2000 image coding standard). By assumption, the filters are linear-phase FIR, that is to say, their impulse responses do have a center of symmetry.

The role of high-pass filter is assigned to $H_1(e^{j\omega T})$. In the example discussed here, its impulse response is limited to three values. The high-pass property is achieved when $H_1(z)$ has a double zero in $z = 1$. This is the case for the causal filter with transfer function:

$$H_1(z) = \frac{1}{4}z^{-2}(z-1)^2 = \frac{1}{4} - \frac{1}{2}z^{-1} + \frac{1}{4}z^{-2}. \tag{19.27}$$

By hypothesis, (as the impulse response is limited to 5 elements), $H_0(z)$ has the form:

$$H_0(z) = a_2 + a_1 z^{-1} + a_0 z^{-2} + a_1 z^{-3} + a_2 z^{-4}. \tag{19.28}$$

One notices the symmetry of the coefficients which ensures linear phase shifts by the filters.

$$P(z) = H_0(z)H_1(-z) = \frac{a_2}{4} + \left(\frac{a_1}{4} + \frac{a_2}{2}\right)z^{-1} + \left(\frac{a_0}{4} + \frac{a_1}{2} + \frac{a_2}{4}\right)z^{-2} + \left(\frac{a_1}{4} + \frac{a_0}{2} + \frac{a_1}{4}\right)z^{-3}$$
$$+ \left(\frac{a_0}{4} + \frac{a_1}{2} + \frac{a_2}{4}\right)z^{-4} + \left(\frac{a_1}{4} + \frac{a_2}{2}\right)z^{-5} + \frac{a_2}{4}z^{-6}. \tag{19.29}$$

$$P(-z) = \frac{a_2}{4} - \left(\frac{a_1}{4} + \frac{a_2}{2}\right)z^{-1} + \left(\frac{a_0}{4} + \frac{a_1}{2} + \frac{a_2}{4}\right)z^{-2} - \left(\frac{a_1}{4} + \frac{a_0}{2} + \frac{a_1}{4}\right)z^{-3}$$
$$+ \left(\frac{a_0}{4} + \frac{a_1}{2} + \frac{a_2}{4}\right)z^{-4} - \left(\frac{a_1}{4} + \frac{a_2}{2}\right)z^{-5} + \frac{a_2}{4}z^{-6}.$$

$$P(z) - P(-z) = \left(\frac{a_1}{2} + a_2\right)z^{-1} + \left(\frac{a_1}{2} + a_0 + \frac{a_1}{2}\right)z^{-3} + \left(\frac{a_1}{2} + a_2\right)z^{-5}. \tag{19.30}$$

The center of symmetry of the output signal is located in $m = 3$. To satisfy the condition we must have: $P(z) - P(-z) = 2z^{-3}$.

It is therefore necessary that: $\frac{a_1}{2} + a_2 = 0$, thus: $a_2 = -\frac{a_1}{2}$.

Also it is necessary that $a_0 + a_1 = 2$.

If we take $a_0 = \frac{3}{2}$, then $a_1 = \frac{1}{2}$ and $a_2 = -\frac{1}{4}$.

Finally: $h_0 = \left(-\frac{1}{4}, \frac{1}{2}, \frac{3}{2}, \frac{1}{2}, -\frac{1}{4}\right)$ (low-pass) and $h_1 = \left(\frac{1}{4}, -\frac{1}{2}, \frac{1}{4}\right)$ (high-pass). These two expressions are the impulse responses of the analysis filters.

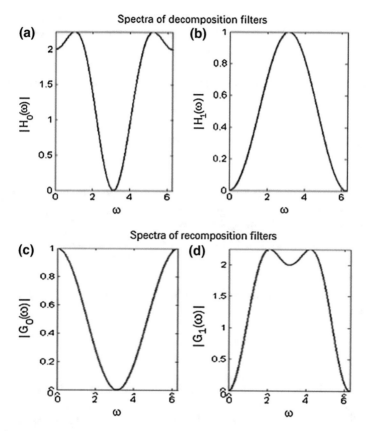

Fig. 19.3 Spectral amplitudes of LeGall-Tabatabai filters; Top, decomposition **a** Low pass; **b** High pass; Bottom, reconstruction **c** Low pass; **d** High pass

The impulse responses of the synthesis filters are deduced from relationships (19.7).

$$g_0 = \left(\frac{1}{4}, \frac{1}{2}, \frac{1}{4} \right) \text{ (low-pass) and } g_1 = \left(\frac{1}{4}, \frac{1}{2}, -\frac{3}{2}, \frac{1}{2}, \frac{1}{4} \right) \text{ (high-pass).}$$

The frequency responses of the LeGall-Tabatabai 5–3 filters are shown in Fig. 19.3.

In the previous example the filters were calculated through the determination of the coefficients of z^{-1} polynomials. This determination is made empirically, with no overall strategy for filtering. The discrete wavelet analysis which is developed in the next section provides a theoretical framework for the research of analysis filters and provides a more fruitful approach to the problem. We will find that the results of this research result as does the previous method, in the determination of the coefficients of FIR type filters.

As is shown in the following, relations (19.24) and (19.25) are also encountered in the decomposition of a signal on a basis of compactly supported wavelets (Daubechies wavelets, for example). It will be seen that the functions $h_0[n]$ and $h_1[n]$ encountered in (19.17) are used for construction of the scaling function and the Daubechies wavelet with two coefficients Db2 which is identical to the Haar wavelet.

19.2 Multiresolution Wavelet Analysis

Haar functions basis
In this presentation, the Haar wavelet is used to address simply the principles of discrete wavelet analysis. Although in practice the signal processing are done by numerical calculations, the functions under study are functions of continuous time t. The analysis of the properties of these treatments being done in the frequency domain, a difficulty which is encountered in the following lies in the coexistence of analog and digital Fourier transforms in the calculations.

Let us first define the **scaling function** $\phi(t)$ on the interval $[0, 1]$ over which it is equal to 1 (Fig. 19.4):

$$\phi(t) = \begin{vmatrix} 1 & \text{for } t \in [0, 1] \\ 0 & \text{elsewhere} \end{vmatrix}. \tag{19.31}$$

The time axis t is divided into contiguous intervals with widths equal to 1. The functions $\phi(t - k)$ will be equal to 1 in the intervals $[k, k + 1]$, $\forall k \in \mathbb{Z}$. A piecewise function $f_0(t)$ constant in each of the intervals $[k, k + 1]$ can be written as a linear combination of the functions $\phi(t - k)$:

$$f_0(t) = \sum_{k=-\infty}^{\infty} a_k \phi(t - k). \tag{19.32}$$

We denote V_0 the space generated by the set of functions $\phi(t - k)$.

Example Let a function $f_0(t)$ equal to 3 in the interval $[0, 1]$ and to -2 in the interval $[4, 5]$ and zero elsewhere. We get by identification in (19.32) $a_0 = 3$ et $a_4 = -2$. Other expansion coefficients are zero.

Fig. 19.4 Haar scaling function $\phi(t)$

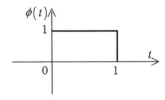

A more general method for determining the coefficients uses a scalar product. Here, the dot product of two real functions $f(t)$ and $g(t)$ is defined by:

$$<f(t), g(t)> = \int_{-\infty}^{\infty} f(t)g(t)dt. \tag{19.33}$$

The norm of $f(t)$ is: $\|f\| = \sqrt{<f(t), f(t)>} = \sqrt{\int_{-\infty}^{+\infty} f^2(t)dt}$.

The functions $f(t)$ and $g(t)$ must be square integrable ($f(t) \in L^2$).

It is readily apparent that since the intersection of the functions supports $\phi(t)$ and $\phi(t-k)$ is zero for $k \neq 0$, we have $<\phi(t), \phi(t-k)> = \int_{-\infty}^{\infty} \phi(t)\phi(t-k)dt = 0$ for $k \neq 0$.

The functions $\phi(t-k)$ form an orthogonal basis of the space V_0. The elements of this basis are normalized to 1 as

$$\|\phi(t-k)\|^2 = <\phi(t-k), \phi(t-k)> = \int_{-\infty}^{\infty} \phi^2(t-k)dt = 1. \tag{19.34}$$

The expansion coefficients a_k in the development (19.32) can be determined as being the projections of the function $f_0(t)$ on the functions of the basis:

$$<f_0(t), \phi(t-k)> = \int_{-\infty}^{\infty} \sum_{k'=-\infty}^{\infty} a_{k'} \phi(t-k')\phi(t-k)dt = a_k \tag{19.35}$$

By this method we find again in the previous example $a_0 = 3$ and $a_4 = -2$.

One can increase the resolution in the analysis of piecewise constant functions along the time axis by dividing by 2 the width of each interval. We are led to use for the representation the functions $\phi(2t)$ and $\phi(2t-1)$ which are compressed versions of $\phi(t)$ on the t-axis. The function $\phi(2t-1)$ is a delayed version of function $\phi(2t)$ by the delay $t = \frac{1}{2}$ (Fig. 19.5).

It is easily seen that we have $\phi(t) = c_0\phi(2t) + c_1\phi(2t-1)$, with $c_0 = 1$ and $c_1 = 1$.

Fig. 19.5 Haar scaling functions **a** $\phi(2t)$; **b** $\phi(2t-1)$

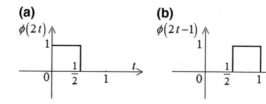

The functions $\phi(2t - k)$ generate the space V_1 of piecewise constant functions on intervals of width $\frac{1}{2}$. They form an orthogonal basis. It appears that a function belonging to the space V_0 also belongs to V_1. We therefore have $V_0 \subset V_1$.

We then notice that the function $\phi(t)$ is normalized to 1, but $\phi(2t)$ is not, as $\int_{-\infty}^{\infty} \phi^2(2t)dt = \frac{1}{2}$. To overcome this drawback, the orthonormal functions $\phi_{jk}(t)$ are defined:

$$\phi_{jk}(t) = 2^{j/2}\phi(2^j t - k). \tag{19.36}$$

In the case of a 2 compression factor, j is 1. It comes $\phi_{1k}(t) = 2^{1/2}\phi(2t - k)$.

Example The development of the function $f_1(t) \in V_1$ equal to -1 in the interval $\left[0, \frac{1}{2}\right]$ and 3 in the interval $\left]\frac{1}{2}, 1\right]$ takes the form:

$$f_1(t) = \sum_{k=-\infty}^{\infty} a_k\phi_{1k}(t) = a_0\phi_{10}(t) + a_1\phi_{11}(t) = a_0\sqrt{2}\phi(2t) + a_1\sqrt{2}\phi(2t - 1).$$

The coefficients a_0 and a_1 of the development of $f_1(t)$ are determined by projecting that function on the functions of the basis. We have:

$$<f_1(t), \phi_{10}(t)> \ = \sqrt{2}\int_0^1 f_1(t)\phi(2t)dt = -\frac{\sqrt{2}}{2} = a_0.$$

Similarly, $a_1 = \ <f_1(t), \phi_{11}(t)> \ = \ <f_1(t), \sqrt{2}\phi(2t - 1)>$. This leads to $a_1 = \frac{3}{\sqrt{2}}$.

Again, dividing by 2 the support of the basis functions, one can represent functions having 4 possibly different successive values in the range $[0, 1]$. A new basis of the space V_2 of all these functions is built on the contracted functions by a factor of 4 of the scaling function $\phi(t)$. It consists of functions:

$$\phi_{20}(t) = 2\phi(4t), \phi_{21}(t) = 2\phi(4t - 1), \phi_{22}(t) = 2\phi(4t - 2) \text{ and } \phi_{23}(t)$$
$$= 2\phi(4t - 3).$$

Approximation of any function

We now consider a function $f(t)$, a priori non-constant piecewise, square integrable: $\int_{-\infty}^{+\infty} |f(t)|^2 dt < \infty$, $(f(t) \in L^2)$. Assume that each interval $[k, k + 1]$ is divided into 2^j contiguous equal intervals. We note a_k the orthogonal projection of $f(t)$ on the function $2^{j/2}\phi(2^j t - k)$ which is equal to 2^j on the interval $\left[\frac{k}{2^j}, \frac{k+1}{2^j}\right]$.

Fig. 19.6 Haar wavelet $\psi(t)$

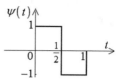

The sum $f_j(t) = 2^{j/2} \sum_{k=0}^{2^j-1} a_k \phi(2^j t - k)$ is an approximation with the resolution 2^{-j} of the function $f(t)$. The approximation will be better as the width of the interval 2^{-j} is smaller.

The mechanism by successive divisions by 2 of the intervals where the function is constant is clear. The production of bases of orthogonal functions is easy, but these bases, built solely on compressions of function $\phi(t)$, do not have much interest. Indeed, the knowledge of all the coefficients (2 coefficients in the example above) is required to estimate the function, even for a rough estimate.

An additional function $\psi(t)$ is introduced, the Haar **wavelet**, which will be used to build more efficient orthogonal bases. The Haar mother wavelet is defined as (Fig. 19.6):

$$\psi(t) = \begin{vmatrix} 1 & \text{for } t \in [0, 1/2] \\ -1 & \text{for } t \in]1, 1/2] \end{vmatrix} \text{ and } 0 \text{ elsewhere.} \qquad (19.37)$$

It is a wavelet as its integral is zero; that it is condensed in a restricted time. Furthermore its support is compact.

To consider a basis using this wavelet, let us return to the space V_1 of functions constant on intervals of widths $\frac{1}{2}$. An alternative basis for this space is constituted by the two functions $\phi(t)$ and $\psi(t)$. These functions are orthogonal and normalized to 1, as can easily be verified.

We now want the development on this basis of the function $f_1(t)$ met above, with value -1 in the interval $\left[0, \frac{1}{2}\right]$ and equals 3 in the interval $\left]\frac{1}{2}, 1\right]$.

Writing $f_1(t) = a_0 \phi(t) + b_0 \psi(t)$, we have:

$$<f_1(t), \phi(t)> \ = \ \int_{-\infty}^{\infty} f_1(t)\phi(t)dt = \int_0^1 f_1(t)\phi(t)dt = a_0 \int_0^1 \phi(t)\phi(t)dt = a_0.$$

This scalar product is $<f_1(t), \phi(t)> \ = \ \int_0^1 f_1(t)\phi(t)dt = -\int_0^{1/2} \phi(t)dt + 3\int_{1/2}^1 \phi(t)dt = 1,$

thus $a_0 = 1$ and $b_0 = \int_0^1 f_1(t)\psi(t)dt = -1 \int_0^{1/2} \psi(t)dt + 3 \int_{1/2}^1 \psi(t)dt = -2.$

Thus $f_1(t) = \phi(t) - 2\psi(t)$.

The function $\phi(t)$ is constant in the interval $[0, 1]$, the projection of $f_1(t)$ on $\phi(t)$ provides the average value of this function on the interval. The positive value $a_0 = 1$ reflects the fact that the function is more often positive than negative in this

Fig. 19.7 Haar functions;
a $\psi(2t)$; **b** $\psi(2t-1)$

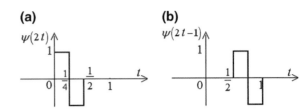

interval. The fact that $b_0 = -2 < 0$ informs us that either $f_1(t)$ is highly negative in the interval $\left[0, \frac{1}{2}\right]$, or greatly positive in the range $\left]\frac{1}{2}, 1\right]$. In summary, $\phi(t)$ gives an average value and $\psi(t)$ an unbalance.

Functions $\phi(t-k)$ span the space V_0. Functions $\psi(t-k)$ generate a space W_0 orthogonal noted V_0. The union of these spaces is the space V_1. Thus $V_1 = V_0 \oplus W_0$.

We now subdivide each interval of width 1 into 4 equal intervals, the function $f_2(t)$ belongs to the space noted V_2. One is led to seek a compound basis using functions $\phi(t-k)$ and several versions contracted or not of $\psi(t-k)$.

To make the mechanism readily apparent, we focus on the sub space of V_2 of functions null outside the interval $[0, 1]$. An orthonormal basis of this sub space is made of $\phi(t)$, $\psi(t)$, $\sqrt{2}\psi(2t)$ and $\sqrt{2}\psi(2t-1)$. These last two functions are shown in Fig. 19.7.

By noting W_1 the space generated by the two functions $\sqrt{2}\psi(2t)$ and $\sqrt{2}\psi(2t-1)$, one has $V_2 = V_1 \oplus W_1$ or equivalently $V_2 = V_0 \oplus W_0 \oplus W_1$.

Example Consider a piecewise constant function $f_3(t)$ defined upon the interval $[0, 1]$. This interval is divided into $2^3 = 8$ intervals $(j = 3)$. This function equals 6 within the interval $\left[0, \frac{1}{8}\right]$ and -2 in the interval $\left[\frac{3}{8}, \frac{4}{8}\right]$. We look for the development of this function on the basis consisting of the 8 orthogonal functions (the verification of orthogonality is left to the reader):

$$\phi(t), \psi(t), \psi(2t), \psi(2t-1), \psi(4t), \psi(4t-1), \psi(4t-2), \psi(4t-3).$$

The transition to a vector notation is useful. Each component of a vector is equal to the value of the function in the successive intervals. Thus it is written:

$$f_3(t) \rightarrow f_3 = \begin{pmatrix} 6 \\ 0 \\ 0 \\ 0 \\ -2 \\ 0 \\ 0 \\ 0 \end{pmatrix}$$

These column vectors taking up much space in the written page we prefer to use the following writing of a transposed row vector: $f_3 = (6 \quad 0 \quad 0 \quad 0 \quad -2 \quad 0 \quad 0 \quad 0)^T$.

In the calculations, the scalar product of $f_3(t)$ with the functions of a basis, for example $\psi(t)$, $<f_3(t), \psi(t)> = \int_0^1 f_3(t)\psi(t)dt$, is replaced by the scalar product of the vector f_3 with the vector $\psi = (1 \quad 1 \quad 1 \quad 1 \quad -1 \quad -1 \quad -1 \quad -1)^T$ after normalization to 1:

$$\psi = \frac{1}{2\sqrt{2}}(1 \quad 1 \quad 1 \quad 1 \quad -1 \quad -1 \quad -1 \quad -1)^T.$$

It comes:

$$<f_3(t), \psi(t)> = (6 \quad 0 \quad 0 \quad 0 \quad -2 \quad 0 \quad 0 \quad 0)\frac{1}{2\sqrt{2}}(1 \quad 1 \quad 1 \quad 1 \quad -1 \quad -1 \quad -1 \quad -1)^T = 2\sqrt{2}$$

The different standard basis vectors will be noted v_k:

$$v_1 = \phi_t = \frac{1}{2\sqrt{2}}(1 \quad 1 \quad 1 \quad 1 \quad 1 \quad 1 \quad 1 \quad 1)^T, v_2 = \psi_t = \frac{1}{2\sqrt{2}}(1 \quad 1 \quad 1 \quad 1 \quad -1 \quad -1 \quad -1 \quad -1)^T,$$

$$v_3 = \psi_{2t} = \frac{1}{2}(1 \quad 1 \quad -1 \quad -1 \quad 0 \quad 0 \quad 0 \quad 0)^T, v_4 = \psi_{2t-1} = \frac{1}{2}(0 \quad 0 \quad 0 \quad 0 \quad 1 \quad 1 \quad -1 \quad -1)^T,$$

$$v_5 = \psi_{4t} = \frac{1}{\sqrt{2}}(1 \quad -1 \quad 0 \quad 0 \quad 0 \quad 0 \quad 0 \quad 0)^T, v_6 = \psi_{4t-1} = \frac{1}{\sqrt{2}}(0 \quad 0 \quad 1 \quad -1 \quad 0 \quad 0 \quad 0 \quad 0)^T,$$

$$v_7 = \psi_{4t-2} = \frac{1}{\sqrt{2}}(0 \quad 0 \quad 0 \quad 0 \quad 1 \quad -1 \quad 0 \quad 0)^T, v_8 = \psi_{4t-3} = \frac{1}{\sqrt{2}}(0 \quad 0 \quad 0 \quad 0 \quad 0 \quad 0 \quad 1 \quad -1)^T.$$

The expansion coefficients of $f_3(t)$ are: $a_k = \{\sqrt{2}, 2\sqrt{2}, 3, -1, 3\sqrt{2}, 0, -\sqrt{2}, 0\}$.

These coefficients are the numerical wavelet transform of the original function. We verify that:

$$f_3(t) = \sum_{k=1}^{8} a_k v_k(t). \tag{19.38}$$

It is interesting to observe graphically the results of an approximation of $f_3(t)$ consisting of a sum of terms reduced to a number lower than 8. First of all we show in Fig. 19.8 the perfect reconstruction of $f_3(t)$ obtained by the sum (19.38).

Figure 19.9 shows an approximation $f_{3_1}(t)$ with the sum limited to its first 4 coefficients. We find that we have lost in resolution (in optics, it would be said that the image is blurred) but both pulses at 0 and $\frac{1}{2}$ are fairly well localized:

Comparison with a limitation in the Fourier domain

We want to test here the reconstruction of the signal when using only a part of the expansion coefficients of the discrete Fourier transform. The inversion of the DFT formula is denoted here in matrix form. It is:

Fig. 19.8 Function $f_3(t)$
under analysis and its perfect
reconstruction

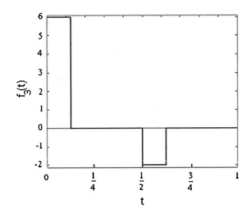

Fig. 19.9 Reconstruction
with a sum limited to first 4
elements

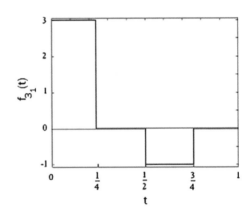

$$
\begin{pmatrix} f[0] \\ f[1] \\ f[2] \\ \cdots \\ f[7] \end{pmatrix} = \frac{1}{N} \begin{pmatrix} 1 & 1 & 1 & \cdots & 1 \\ 1 & W_8^* & W_8^{*2} & \cdots & W_8^{*7} \\ 1 & W_8^{*2} & W_8^{*4} & \cdots & W_8^{*14} \\ \cdots & \cdots & \cdots & \cdots & \cdots \\ 1 & W_8^{*7} & W_8^{*14} & \cdots & W_8^{*49} \end{pmatrix} \begin{pmatrix} F[0] \\ F[1] \\ F[2] \\ \cdots \\ F[7] \end{pmatrix}.
$$

Of course, the reconstruction is perfect when all 8 Fourier coefficients are used in
the inverse DFT. Reconstitution is then tested when we keep only 5 coefficients
corresponding to the lower frequencies by imposing $F[3] = F[4] = F[5] = 0$. The
result is reproduced in Fig. 19.10:

We see the resulting oscillation of the Gibbs phenomenon on all t axis.
Reconstruction is not as good as when using the wavelets.

The Haar functions $\phi(t)$ are discontinuous; they are ill-suited to serve as a basis
for approximations for smooth functions since reconstructions will necessarily
present discontinuities. Many compressed wavelets would be necessary to reduce

Fig. 19.10 Reconstruction
with a Fourier sum limited to
low frequencies

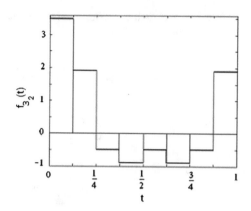

the effect of these discontinuities and achieve sufficient approximation. One is led to
seek other bases wavelets functions, continuous with compact support. Mallat and
Daubechies established the theoretical framework for determining that type of
wavelets. They calculated wavelets adapted to dyadic multiresolution (successive
divisions of the support by a factor of 2).

19.3   Daubechies Wavelets

Definition of the scaling function $\phi(t)$
The search for these wavelets begins by determining a scaling function with the
following properties:

The scaling function $\phi(t)$ must be real, causal, with a compact support ($\phi(t)$ is
zero outside a closed interval $[0, N]$ of the variable t). The upper bound N of this
interval is a positive integer. In practice, this integer will be small (less than a few
dozens). Its value will depend upon the desired level of resolution of the analysis.
A continuous function $\phi(t)$ on the entire time axis is sought, that implies that at
both ends of its support,

$$\phi(0) = \phi(N) = 0. \tag{19.39}$$

In the literature on wavelets, the scaling function is called the "father" function.
The desired scaling function $\phi(t)$ is normalized by hypothesis. It must verify:

$$\int_{-\infty}^{+\infty} \phi^2(t)dt = 1. \tag{19.40}$$

We impose to the function $\phi(t)$ and to its translated by an integer m to be orthonormal:

$$< \phi(t), \phi(t-m) > \ = \ \int_{-\infty}^{+\infty} \phi(t)\phi(t-m)\mathrm{d}t = \delta[m] \text{ with } m \in \mathbb{Z}. \qquad (19.41)$$

The key point of multiresolution analysis is that $\phi(t)$ has to be obtained by a linear combination of its compressed versions by a factor of 2 and translated by integer values:

$$\phi(t) = \sum_{k=0}^{N} c_k \phi(2t-k) \text{ with } k \in \mathbb{N}. \qquad (19.42)$$

It appears that the upper limit N of the sum index has the value of the right boundary of the support of $\phi(t)$.

The desired function $\phi(t)$ should be such that its compressed versions by a factor of 2 and translated by an integer k are orthogonal:

$$\langle \phi(2t), \phi(2t-k) \rangle = \int_{-\infty}^{+\infty} \phi(2t)\phi(2t-k)\mathrm{d}t = 0 \quad \text{if } k \neq 0. \qquad (19.43)$$

For $\forall k \in \mathbb{Z}$, the following property is verified:

$$\begin{aligned}
\|\phi(2t-k)\|^2 = \langle \phi(2t-k), \phi(2t-k) \rangle &= \int_{-\infty}^{+\infty} \phi^2(2t-k)\mathrm{d}t \\
&= \frac{1}{2}\int_{-\infty}^{+\infty} \phi^2(y)\mathrm{d}y = \frac{1}{2}.
\end{aligned} \qquad (19.44)$$

Given (19.42) the constraint (19.43) implies a relation between the coefficients c_k. Indeed:

$$\int_{-\infty}^{\infty} \phi(t)\phi(t-m)\mathrm{d}t = \int_{-\infty}^{\infty} \sum_{k=0}^{N} c_k \phi(2t-k) \sum_{k'=0}^{N} c_{k'} \phi(2t-2m-k')\mathrm{d}t.$$

The integrals of the various products are zero unless $k' = k - 2m$. Then:

$$\int_{-\infty}^{\infty} \phi(t)\phi(t-m)\mathrm{d}t = \frac{1}{2}\sum_{k=0}^{N} c_k c_{k-2m} = \delta[m].$$

The following relation between the coefficients must be verified:

$$\sum_{k=0}^{N} c_k c_{k-2m} = 2\delta[m].$$

(19.45)

In particular, when $m = 0$, $\sum_{k=0}^{N} c_k^2 = 2$.

Definition of the wavelet $\psi(t)$

The wavelet is defined as:

$$\psi(t) = \sum_{k=0}^{N} (-1)^k c_{N-k} \phi(2t - k)$$

(19.46)

The coefficients c_{N-k} are those of the development of $\phi(t)$ but taken in reverse order, starting with the end. The following will justify this choice for the coefficients of the development.

In the literature on wavelets, the function $\psi(t)$ is called "mother" wavelet.

By assumption, the wavelet $\psi(t)$ is orthogonal to the scale function:

$$\int_{-\infty}^{\infty} \phi(t)\psi(t)\mathrm{d}t = 0.$$

(19.47)

This implies an other relationship between the coefficients c_k. Replacing the functions in the last integral by their developments, it comes:

$$\int_{-\infty}^{\infty} \sum_{k=0}^{N} c_k \phi(2t - k) \sum_{k'=0}^{N} (-1)^{k'} c_{N-k'} \phi(2t - k')\mathrm{d}t = 0.$$

(19.48)

Due to the orthogonality of the functions $\phi(2t - k')$ and relation (19.43), we have:

$$\frac{1}{2}\sum_{k=0}^{N} (-1)^k c_k c_{N-k} = 0 \text{ and finally } \sum_{k=0}^{N} (-1)^k c_k c_{N-k} = 0.$$

(19.49)

For this cancellation to occur, it is necessary that the number of coefficients be even to avoid a central coefficient to be zero. That is to say that N must be odd. This is assumed in the following discussion.

For the Haar wavelet, which is within the scope of Daubechies wavelets (it is sometimes called Db2), although it is not continuous, we have $N = 1$, $c_0 = 1$, $c_1 = 1$.

A new constraint on the value of the coefficients c_k is provided by an other requirement imposed on the wavelet $\psi(t)$. We require that its integral is zero (see Chap. 12):

$$\int_{-\infty}^{\infty} \psi(t)\mathrm{d}t = 0, \tag{19.50}$$

then,

$$\sum_{k=0}^{N} (-1)^k c_{N-k} \int_{-\infty}^{\infty} \phi(2t - k)\mathrm{d}t = 0.$$

By a simple change of variables we see that $\int_{-\infty}^{\infty} \phi(2t - k)\mathrm{d}t = \frac{1}{2} \int_{-\infty}^{\infty} \phi(y)\mathrm{d}y = Cte.$
It follows that the following relation between the coefficients must be satisfied:

$$\sum_{k=0}^{N} (-1)^k c_{N-k} = 0. \tag{19.51}$$

This is a third connection between the coefficients c_k.
For the wavelet Db4 (N = 3) relation (19.51) writes $c_3 - c_2 + c_1 - c_0 = 0$.

Properties in the Fourier domain
The standard notation in the literature on wavelets for the FT of the scaling function $\phi(t)$ is used: $\hat{\phi}(\omega) = \int_{-\infty}^{\infty} \phi(t)\mathrm{e}^{-j\omega t}\mathrm{d}t.$

It follows that the FT of $\phi(2t)$ can be written:

$$F(\phi(2t)) = \int_{-\infty}^{\infty} \phi(2t)\mathrm{e}^{-j\omega t}\mathrm{d}t = \frac{1}{2} \int_{-\infty}^{\infty} \phi(y)\mathrm{e}^{-j\omega\frac{y}{2}}\mathrm{d}y = \frac{1}{2}\hat{\phi}\left(\frac{\omega}{2}\right). \tag{19.52}$$

By a simple change of variable, the shift theorem is expressed:

$$F(\phi(2t - k)) = \frac{1}{2} \int_{-\infty}^{\infty} \phi(y)\mathrm{e}^{-j\omega\frac{(y+k)}{2}}\mathrm{d}y = \frac{1}{2}\hat{\phi}\left(\frac{\omega}{2}\right)\mathrm{e}^{-j\omega\frac{k}{2}}.$$

By calculating the FT of the two members of the relationship (19.42), we have:

$$\hat{\phi}(\omega) = \int\limits_{-\infty}^{\infty} \sum_{k=0}^{N} c_k \phi(2t - k) e^{-j\omega t} dt = \frac{1}{2}\hat{\phi}\left(\frac{\omega}{2}\right) \sum_{k=0}^{N} c_k e^{-j\omega\frac{k}{2}} = P\left(\frac{\omega}{2}\right)\hat{\phi}\left(\frac{\omega}{2}\right)$$

$$(19.53)$$

where we write

$$P(\omega) = \frac{1}{2}\sum_{k=0}^{N} c_k e^{-j\omega k}. \qquad (19.54)$$

$P(\omega)$ is the Fourier transform of the function formed by the sequence of the coefficients c_k.

Note A difficulty using this notation appears here: $P(\omega)$ which is the Fourier transform of a numerical function will be noted in the same way than $\hat{\phi}(\omega)$ which is the Fourier transform of an analog function. Consistency with the rest of the book would ask that the notation $P(e^{j\omega})$ is used rather than $P(\omega)$ but the literature on wavelet has adopted $P(\omega)$. In the calculations, we must be careful to the difference in nature of $P(\omega)$ and $\hat{\phi}(\omega)$. The lack of hat in $P(\omega)$ recalls the difference. By abuse of notation, we write in a simplified manner:

$$P(\omega)\big|_{\omega=0} = P(e^{j\omega})\big|_{\omega=0} = P(0).$$

Based on the results of the Fourier transform of digital functions, we know that the function $P(\omega)$ is periodic, with period 2π. Imposing $\omega = 0$ in the relationship (19.53), it comes,

$$\hat{\phi}(0) = P(0)\hat{\phi}(0) \text{ which imposes } P(0) = 1. \qquad (19.55)$$

Thus using (19.54) with $\omega = 0$, we see that the expansion coefficients c_k must verify the sum rule:

$$\sum_{k=0}^{N} c_k = 2P(0) = 2. \qquad (19.56)$$

Iterating relation (19.53), we may write: $\hat{\phi}(\omega) = P\left(\frac{\omega}{2}\right)P\left(\frac{\omega}{4}\right)\hat{\phi}\left(\frac{\omega}{4}\right)$.
   Generalizing, we have:

$$\hat{\phi}(\omega) = \prod_{j=1}^{J} P\left(\frac{\omega}{2^j}\right)\hat{\phi}\left(\frac{\omega}{2^J}\right). \qquad (19.57)$$

If $\hat{\phi}(\omega)$ is continuous at $\omega = 0$, then $\lim\limits_{J\to\infty} \hat{\phi}\left(\frac{\omega}{2^J}\right) = \hat{\phi}(0)$.

It comes:$\hat{\phi}(\omega) = \prod\limits_{j=1}^{\infty} P\left(\frac{\omega}{2^j}\right)\hat{\phi}(0)$. In the following we show that $\hat{\phi}(0) = 1$.
Therefore

$$\hat{\phi}(\omega) = \prod_{j=1}^{\infty} P\left(\frac{\omega}{2^j}\right). \tag{19.58}$$

It is interesting to note here that the Fourier transform of the scale function is given by the Fourier transform of the sequence of coefficients c_k. It is useful for the following discussion to define the function $P(z)$, half of the z-transform of the sequence of coefficients c_k:

$$P(z) = \frac{1}{2}\sum_{k=0}^{N} c_k z^{-k}. \tag{19.59}$$

We have the property $P(\pi) = P(z)|_{\omega=\pi} = 0$. Indeed, according to (19.51) we have:

$$P(\pi) = \frac{1}{2}\sum_{k=0}^{N} c_k e^{-j\pi k} = \frac{1}{2}\sum_{k=0}^{N} (-1)^k c_k = 0. \tag{19.60}$$

Thus, the function $P(z)$ defined in (19.59) has a zero in the z plane at $z = -1$.

It is seen from Eq. (19.53) that this zero of $P(\omega)$ at $\omega = \pi$ leads to a zero $\hat{\phi}(\omega)$ for $\omega = 2\pi$.

Because it is periodic function of ω, $P(\omega)$ has zeros at $3\pi, 5\pi, \ldots$ which implies that $\hat{\phi}(\omega)$ has zeros at $6\pi, 10\pi, \ldots$.

Writing $\omega = 4\pi$ in (19.53), we have $\hat{\phi}(4\pi) = P(2\pi)\hat{\phi}(2\pi) = 2\hat{\phi}(2\pi) = 0$ from the foregoing. A generalization can be deduced:

$$\hat{\phi}(2l\pi) = 0 \text{ for } l \neq 0. \tag{19.61}$$

These properties are apparent in Fig. 19.11 which shows the variations of $|P(\omega)|$ and of $\left|\hat{\phi}(\omega)\right|$ with ω (the first function represented over a period) for Db4. $|P(\omega)|$ was calculated from the FFT of coefficients c_k whose values will be determined in the following. The function $\left|\hat{\phi}(\omega)\right|$ was calculated in an approximate way from the relationship (19.58), limiting the product to the first 6 terms.

The scalar product (19.41) appears as a correlation function that depends on the discrete variable m. It is noted $C[m]$. Noting $\tilde{\phi}(t)$ the time reversal of $\phi(t)$, we write:

$$C[m] = \int_{-\infty}^{\infty} \phi(t)\tilde{\phi}(m - t)\mathrm{d}t. \tag{19.62}$$

Fig. 19.11 Spectral amplitudes; **a** $|P(\omega)|$; **b** $\left|\hat{\phi}(\omega)\right|$

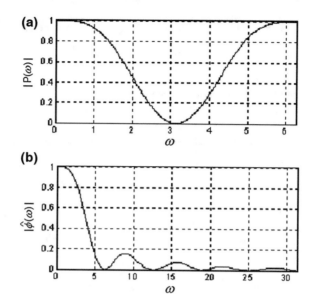

In the previous integral we recognize the value of the convolution product of $\phi(t)$ and $\tilde{\phi}(t)$ evaluated at $t = m$. $C[m]$ appears as a sampling of the correlation function $C(t)$ for integer values m of time t.

The FT of $C(t)$ is $\hat{C}(\omega) = \hat{\phi}(\omega)\hat{\tilde{\phi}}(\omega) = \hat{\phi}(\omega)\hat{\phi}^*(\omega) = \left|\hat{\phi}(\omega)\right|^2$.

According to the Shannon aliasing theorem, the discrete FT of $C[m]$ is the infinite sum $\sum_{l=-\infty}^{\infty} \hat{C}(\omega + l2\pi)$. This sum is 1 since the numerical FT of $C[m] = \delta[m]$ is 1. Thus, we arrive at the important result

$$\sum_{l=-\infty}^{\infty} \left|\hat{\phi}(\omega + l2\pi)\right|^2 = 1. \tag{19.63}$$

In particular, if we make $\omega = 0$ in the previous equation, we have

$$\sum_{l=-\infty}^{\infty} \left|\hat{\phi}(l2\pi)\right|^2 = 1. \tag{19.64}$$

The scaling function is assumed to have a compact support. It follows that from the time-frequency uncertainty relation, its spectrum necessarily extends to infinity. The relation (19.64) states that the sum of the values of the squared modulus of the FT of $\phi(t)$ at abscissas corresponding to all multiples of 2π equals 1.

Given (19.61), it follows that $\left|\hat{\phi}(0)\right|^2 = 1$. Then $\hat{\phi}(0) = \pm 1$. By convention the + sign is chosen. So we have: $\hat{\phi}(0) = \int_{-\infty}^{\infty} \phi(t)dt = 1$. We note that this is only a constraint of a scale factor on $\phi(t)$.

Fourier transform of the wavelet $\psi(t)$

To complete the definition of Daubechies wavelet Db4 which has 4 non-zero coefficients, one last condition is imposed on the wavelet $\psi(t)$. His first moment in t must be zero, which will allow a better analysis of very regular functions:

$$\int_{-\infty}^{\infty} t\psi(t)dt = 0. \tag{19.65}$$

We now show that this condition leads to the following relationship:

$$\sum_{k=0}^{N} (-1)^k k c_{N-k} = 0. \tag{19.66}$$

The proof is performed in the Fourier domain.
One notes:

$$\hat{\psi}(\omega) = \int_{-\infty}^{\infty} \psi(t)e^{-j\omega t}dt. \tag{19.67}$$

We first remark that condition (19.50) implies $\hat{\psi}(0) = 0$.

Differentiating (19.67) under the integral sign: $\frac{d\hat{\psi}(\omega)}{d\omega} = -j \int_{-\infty}^{\infty} \psi(t)e^{-j\omega t}dt.$ As a result

$$j\frac{d\hat{\psi}(\omega)}{d\omega}\bigg|_{\omega=0} = j\hat{\psi}'(0) = \int_{-\infty}^{\infty} t\psi(t)dt = 0. \tag{19.68}$$

Since $\hat{\psi}(0) = 0$ and $\hat{\psi}'(0) = 0$, taking the Taylor expansion of the function $\hat{\psi}(\omega)$ in the vicinity of $\omega = 0$, we see that $\hat{\psi}(\omega)$ varies as ω^2 if in the neighborhood of $\omega = 0$ $\hat{\psi}''(0) \neq 0$.

To continue the proof, analogously to the relationship (19.53), it is shown that:

$$\hat{\psi}(\omega) = G\left(\frac{\omega}{2}\right)\hat{\phi}\left(\frac{\omega}{2}\right), \tag{19.69}$$

where $G(\omega)$ is the discrete transform of the sequence of coefficients in (19.46):

$$G(\omega) = \frac{1}{2}\sum_{k=0}^{N}(-1)^k c_{N-k} e^{-j\omega k}.\qquad(19.70)$$

The relation (19.60) implies that

$$G(0) = 0.\qquad(19.71)$$

We derive the product (19.69): $\hat{\psi}'(\omega) = G'\left(\frac{\omega}{2}\right)\hat{\phi}\left(\frac{\omega}{2}\right) + G\left(\frac{\omega}{2}\right)\hat{\phi}'\left(\frac{\omega}{2}\right)$. We evaluate it at $\omega = 0$. $\hat{\psi}'(0) = G'(0)\hat{\phi}(0) + G(0)\hat{\phi}'(0) = G'(0)$ (We used Eq. (19.71) and the fact that $\hat{\phi}(0) = 1$).

Using the relation (19.68), we get

$$G'(0) = 0.\qquad(19.72)$$

Since $G'(\omega) = -\frac{j}{2}\sum_{k=0}^{N}(-1)^k k c_{N-k} e^{-j\omega k}$, it comes $G'(0) = -\frac{j}{2}\sum_{k=0}^{N}(-1)^k k c_{N-k} = 0$, and finally

$$\sum_{k=0}^{N}(-1)^k k c_{N-k} = 0.\qquad(19.73)$$

The fourth condition on the coefficients makes the ensemble of conditions sufficient to allow the determination of the coefficients of the wavelet Db4 Daubechies, for which $\phi(t)$ is defined on the interval $t \in [0, 3]$ and is continuous at its boundaries $\phi(0) = \phi(3) = 0$.

These coefficients are: $c_0 = \frac{1}{4}\left(1 + \sqrt{3}\right); c_1 = \frac{1}{4}\left(3 + \sqrt{3}\right); c_2 = \frac{1}{4}\left(3 - \sqrt{3}\right); c_3 = \frac{1}{4}\left(1 - \sqrt{3}\right)$.

The reader can verify that these values of the coefficients satisfy the relations previously encountered. This will allow him to review all of these conditions. In the following we give a preferred method for determining the value of these coefficients.

Mallat, Meyer theorem

It has been shown previously (19.63) that $\sum_{l=-\infty}^{\infty}\left|\hat{\phi}(\omega + l2\pi)\right|^2 = 1$.

Reporting in this relation $\hat{\phi}(\omega)$ as written in (19.53): $\hat{\phi}(\omega) = P\left(\frac{\omega}{2}\right)\hat{\phi}\left(\frac{\omega}{2}\right)$, we have:

$$\sum_{l=-\infty}^{\infty}\left|P\left(\frac{\omega}{2} + l\pi\right)\right|^2\left|\hat{\phi}\left(\frac{\omega}{2} + l\pi\right)\right|^2 = 1$$

or, as ω is arbitrary,

$$\sum_{l=-\infty}^{\infty} |P(\omega+l\pi)|^2 \left|\hat{\phi}(\omega+l\pi)\right|^2 = 1.$$

We separate the terms of even and odd ranks. We use the fact that $P(\omega)$ is periodic with period 2π. $P(\omega+2n\pi) = P(\omega); P(\omega+(2n+1)\pi) = P(\omega+\pi)$.

So we have $|P(\omega)|^2 \sum_{n=-\infty}^{\infty} \left|\hat{\phi}(\omega+2n\pi)\right|^2 + |P(\omega+\pi)|^2 \sum_{n=-\infty}^{\infty} \left|\hat{\phi}(\omega+(2n+1)\pi)\right|^2 = 1.$

The two sums being equal to 1 in the previous equation according to (19.63), we have finally:

$$|P(\omega)|^2 + |P(\omega+\pi)|^2 = 1. \tag{19.74}$$

Figure 19.12 shows $|P(\omega)|^2$, $|P(\omega+\pi)|^2$, and their sum, equal to 1 for Db4 Daubechies wavelet.

From the results of Chap. 15, we know that the numerical Fourier transform $G(\omega)$ of the wavelet coefficients is related to that of the coefficients of the development of $\phi(t)$ by

$$G(\omega) = e^{-jN\omega}(-1)^N P^*(\omega+\pi). \tag{19.75}$$

Then $|G(\omega)| = |P(\omega+\pi)|$.
Relation (19.74) also takes the form

$$|P(\omega)|^2 + |G(\omega)|^2 = 1. \tag{19.76}$$

From these results we may say that the digital filters consisting of the coefficients of the expansion of the scaling function and those developing the wavelet upon the functions $\phi(2t-k)$ are mirror filters.

Fig. 19.12 Spectral amplitudes $|P(\omega)|$ and $|P(\omega+\pi)|^2$ for $\{0 \le \omega < 2\pi\}$

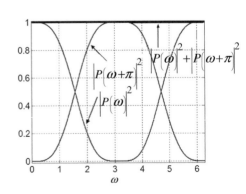

The first filter of frequency response $P(\omega)$ is low-pass, the second with frequency response $G(\omega)$ is high-pass.

The following property is verified:

$$P(\omega)G^*(\omega) + P(\omega+\pi)G^*(\omega+\pi) = 0. \tag{19.77}$$

Indeed, according to (19.75):

$$P(\omega)G^*(\omega) = P(\omega)e^{jN\omega}(-1)^N P(\omega+\pi). \tag{19.78}$$

and

$$P(\omega+\pi)G^*(\omega+\pi) = P(\omega+\pi)e^{jN(\omega+\pi)}(-1)^N P(\omega). \tag{19.79}$$

These last two relationships lead (19.77) since $e^{jN(\omega+\pi)} = -e^{jN\omega}$ as N is odd. We have also:

$$P(\omega)P^*(\omega+\pi) + G(\omega)G^*(\omega+\pi) = 0, \tag{19.80}$$

because $P(\omega)P^*(\omega+\pi) + e^{-jN\omega}(-1)^N P^*(\omega+\pi)e^{jN(\omega+\pi)}(-1)^N P(\omega) = 0$.

The wavelet is such that

$$\hat{\psi}(2\omega) = G(\omega)\hat{\phi}(\omega) \tag{19.81}$$

[according to (19.69)].

For Db4, the variation of $G(\omega)$ as ω^2 around $\omega = 0$ corresponds to a variation of $P(\omega)$ as $(\omega - \pi)^2$ around $\omega = \pi$.

Equivalently, we can say that the function $P(z)$ has a double zero at $z = -1$. For Db6, we require $\hat{\psi}''(0) = 0$. The first term of its Taylor expansion around $\omega = 0$ is in ω^3, the second moment is zero: $\int_{-\infty}^{\infty} t^2 \psi(t)dt = 0$. Then $P(z)$ has a triple zero at $z = -1$. We see that the number of zeros of $P(z)$ at $z = -1$ increase with the order of a Daubechies wavelet (See Byrne in Signal Processing—A mathematical approach, Peters Ltd Ed.).

Returning to Db4, we now express the consequence of the double zero of $P(z)$ en $z = -1$. $P(z)$ has necessarily the form $P(z) = (z+1)^2 Q(z)$.

In Fig. 19.13 we see the double zero in $z = -1$ and one zero in $z = 0.2679$ for the function $P(z)$ of Db4.

Posing $z = e^{j\omega}$, we have $P(\omega) = (e^{j\omega} + 1)^2 Q(\omega)$.

The development of $(e^{j\omega} + 1)^2$ is:

$$\left(e^{j\omega} + 1\right)^2 = e^{2j\omega} + 2e^{j\omega} + 1 = e^{j\omega}\left(e^{j\omega} + 2 + e^{-j\omega}\right) = 2e^{j\omega}(1 + \cos\omega)$$
$$= 4e^{j\omega}\cos\frac{\omega}{2},$$

Fig. 19.13 Zero-poles
locations of $P(z)$ function for
Db4

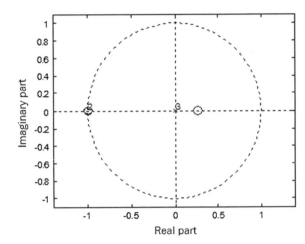

Thus:

$$|P(\omega)|^2 = 16\cos^2\frac{\omega}{2}|Q(\omega)|^2.$$

In the case where $|P(\omega)|^2 = \cos^2\frac{\omega}{2}$, (which is the case of Db2 and not of Db4), we have

$$|P(\omega+\pi)|^2 = \cos^2\frac{(\omega+\pi)}{2} = \sin^2\frac{\omega}{2},$$

and of course, we have:

$$|P(\omega)|^2 + |P(\omega+\pi)|^2 = \cos^2\frac{\omega}{2} + \sin^2\frac{\omega}{2} = 1 \qquad (19.82)$$

Daubechies Db4 wavelet

For the construction of Daubechies wavelets, we can use the following method, which is preferred. It consists in raising both sides of the above equation to an odd power $N = 2n-1$ with n a positive integer. If $n = 2$, then $N = 3$, we get Daubechies Db4.

$\left(\cos^2\frac{\omega}{2} + \sin^2\frac{\omega}{2}\right)^3 = 1$. The development of this expression shows the binomial coefficients C_N^k.

$$\left(\cos^2\frac{\omega}{2} + \sin^2\frac{\omega}{2}\right)^3 = \cos^6\frac{\omega}{2} + 3\cos^4\frac{\omega}{2}\sin^2\frac{\omega}{2} + 3\cos^2\frac{\omega}{2}\sin^4\frac{\omega}{2} + \sin^6\frac{\omega}{2}.$$

$|P(\omega)|^2 = \cos^6\frac{\omega}{2} + 3\cos^4\frac{\omega}{2}\sin^2\frac{\omega}{2} = \cos^4\frac{\omega}{2}\left(\cos^2\frac{\omega}{2} + 3\sin^2\frac{\omega}{2}\right)$ may be taken.

We still verify (19.74) as the second part of the binomial expansion is equal to $|P(\omega + \pi)|^2$. We must have $P(\omega) = \cos^2 \frac{\omega}{2} \left(\cos \frac{\omega}{2} + \sqrt{3} \sin \frac{\omega}{2} \right) e^{j\theta(\omega)}$.

Taking $e^{j\theta(\omega)} = e^{-j\frac{3\omega}{2}}$, a polynomial is obtained:

$$P(\omega) = \frac{1}{2} \left(c_0 + c_1 e^{-j\omega} + c_2 e^{-j2\omega} + c_3 e^{-j3\omega} \right).$$

The coefficients of all other powers of $e^{-j\omega}$ are zero.
Indeed:

$$\cos \frac{\omega}{2} = \frac{e^{j\frac{\omega}{2}} + e^{-j\frac{\omega}{2}}}{2} ; \cos^2 \frac{\omega}{2} = \frac{1}{4} \left(e^{j\omega} + e^{-j\omega} + 2 \right).$$

$$P(\omega) = \frac{1}{4} \left(e^{j\omega} + e^{-j\omega} + 2 \right) \left(\frac{e^{j\frac{\omega}{2}} + e^{-j\frac{\omega}{2}}}{2} + j\sqrt{3} \frac{e^{j\frac{\omega}{2}} - e^{-j\frac{\omega}{2}}}{2j} \right) e^{-j\frac{3\omega}{2}},$$

$$P(\omega) = \frac{1}{8} \left(e^{j\omega} + e^{-j\omega} + 2 \right) \left(e^{-j\omega} \left(1 + \sqrt{3} \right) + e^{-j2\omega} \left(1 - \sqrt{3} \right) \right).$$

The 4 coefficients c_k of Db4 are determined by identification:

$$c_0 = \frac{1}{4} \left(1 + \sqrt{3} \right); c_1 = \frac{1}{4} \left(3 + \sqrt{3} \right); c_2 = \frac{1}{4} \left(3 - \sqrt{3} \right); c_3 = \frac{1}{4} \left(1 - \sqrt{3} \right).$$

The problem of using the wavelet Db4 is resolved at this stage. This may seem surprising as we do not know yet neither the scaling function $\phi(t)$ nor the wavelet $\psi(t)$.

Firstly we recall that the support of the function $\phi(t)$ is the interval [0, 3] for Db4.

For Db4, $N = 3$, which implies that the support of the function $\phi(t)$ has a width 3. Arbitrarily the left boundary of the support of $\phi(t)$ is placed at $x = 0$.

The rapid decrease at infinity of $\hat{\phi}(\omega)$ accompanies the fact that the multiple zero of $P(\omega)$ in $\omega = \pi$ affects the decay of $\hat{\phi}(\omega)$. This is a consequence of the relation (19.58).

From the decay properties at infinity of FT function encountered in Chap. 7, it follows that the function does not present discontinuities across the support for Db4. It entrains that to allow continuity at the support boundaries for Db4, it is necessary that

$$\phi(0) = \phi(3) = 0. \tag{19.83}$$

In the following the Db4 wavelet is taken as an example to demonstrate the calculation of $\phi(t)$ for any rational abscissa. We show that we can calculate the values of $\phi(1)$ and $\phi(2)$ from the coefficients c_0 to c_3 and then we can deduce the values at other points by iteration.

Since $\phi(t) = \sum\limits_{k=0}^{3} c_k \phi(2t - k)$, and taking in account (19.83) we can write:

$$\phi(1) = \sum_{k=0}^{3} c_k \phi(2 - k) = c_0 \phi(2) + c_1 \phi(1).$$

We also have:

$$\phi(2) = \sum_{k=0}^{3} c_k \phi(4 - k) = c_2 \phi(2) + c_3 \phi(1).$$

These two equations can be grouped in the form of a linear system:

$$\begin{pmatrix} \phi(1) \\ \phi(2) \end{pmatrix} = \begin{pmatrix} c_1 & c_0 \\ c_3 & c_2 \end{pmatrix} \begin{pmatrix} \phi(1) \\ \phi(2) \end{pmatrix}. \tag{19.84}$$

$\begin{pmatrix} \phi(1) \\ \phi(2) \end{pmatrix}$ appears to be an eigenvector of the matrix $\begin{pmatrix} c_1 & c_0 \\ c_3 & c_2 \end{pmatrix}$ for the eigenvalue $\lambda = 1$. The numerical calculation of the coefficient matrix in (19.84) shows
that it actually has the eigenvalue $\lambda = 1$ with the corresponding eigenvector
$\begin{pmatrix} \phi(1) \\ \phi(2) \end{pmatrix} = a \begin{pmatrix} 0.9659 \\ -0.2588 \end{pmatrix}$, where a is any constant. As may be verified on the
final values of the function $\phi(t)$, we must have $a = \sqrt{2}$ to finally get
$\int_{-\infty}^{\infty} \phi(t)dt = 1$. It comes then: $\begin{pmatrix} \phi(1) \\ \phi(2) \end{pmatrix} = \sqrt{2} \begin{pmatrix} 0.9659 \\ -0.2588 \end{pmatrix} = \begin{pmatrix} 1.3660 \\ -0.3660 \end{pmatrix}$.

Having determined $\phi(1)$ and $\phi(2)$, we can deduce the values of $\phi(t)$ for the
half-integer abscissa by the recurrence relation. For example:

$$\phi(0.5) = \sum_{k=0}^{3} c_k \phi(1 - k) = c_0 \phi(1) + c_1 \phi(0) = c_0 \phi(1) = 0.9330.$$

Likewise:

$$\phi(1.5) = \sum_{k=0}^{3} c_k \phi(3 - k) = c_0 \phi(3) + c_1 \phi(2) + c_2 \phi(1) + c_3 \phi(0)$$
$$= c_1 \phi(2) + c_2 \phi(1) = 0.0012.$$

Continuing the process $\phi(0.25)$, $\phi(0.75)$ can be determined, and so on.

Note that we have access to the values of the scaling function $\phi(t)$ for values of
t located in a grid with as fine resolution as desired, but we do not know yet this
function.

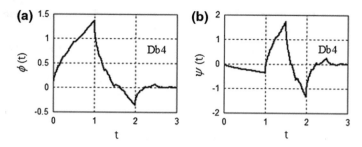

Fig. 19.14 Daubechies Db4 (N = 3); **a** Scaling function $\phi(t)$; **b** Wavelet $\psi(t)$

Figures 19.14 and 19.15 show the scaling function and the Daubechies wavelet for two values of the number N of coefficients.

Remember that if $N = 1$ you have Db2 which is identical to the Haar wavelet defined between 0 and 1:

$$\phi(0) = 0; \phi(0.5) = 1; \phi(1) = 1$$
$$\psi(0) = 0; \psi(0.5) = 1; \psi(1) = -1.$$

Scaling function and wavelet Daubechies wavelet Db32 ($N = 3$): (Fig. 19.14)
Scaling function and wavelet Daubechies wavelet Db32 ($N = 31$): (Fig. 19.15)

Decomposition and reconstruction of a function on a wavelet basis:
A function of the set V_0 can be decomposed as follows:

$$f(t) = \sum_k a_k \phi(t - k) = \sum_k b_k \phi\left(\frac{t}{2} - k\right) + \sum_k d_k \psi\left(\frac{t}{2} - k\right). \qquad (19.85)$$

Taking the FT of both sides of the previous equation and proceeding analogously to the demonstration of the (19.53), we obtain the following relationship in the Fourier domain:

$$\hat{f}(\omega) = a(\omega)\hat{\phi}(\omega) = 2b(2\omega)\hat{\phi}(2\omega) + 2d(2\omega)\hat{\psi}(2\omega). \qquad (19.86)$$

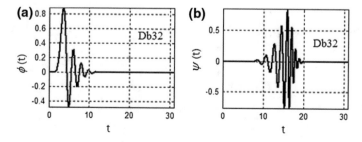

Fig. 19.15 Daubechies Db32 (N = 31); **a** Scaling function $\phi(t)$; **b** Wavelet $\psi(t)$

The Fourier transforms of the sequences of the coefficients a_k, b_k and d_k are noted $a(\omega)$, $b(\omega)$ and $d(\omega)$.

As $\hat{\phi}(2\omega) = P(\omega)\hat{\phi}(\omega)$ according to (19.53), it follows by using (19.81):

$$a(\omega)\hat{\phi}(\omega) = 2b(2\omega)P(\omega)\hat{\phi}(\omega) + 2d(2\omega)G(\omega)\hat{\phi}(\omega),$$

Then:

$$a(\omega) = 2b(2\omega)P(\omega) + 2d(2\omega)G(\omega). \tag{19.87}$$

We check now that we have:

$$\boxed{b(2\omega) = \tfrac{1}{2}(a(\omega)P^*(\omega) + a(\omega + \pi)P^*(\omega + \pi)),} \tag{19.88}$$

and

$$\boxed{d(2\omega) = \tfrac{1}{2}(a(\omega)G^*(\omega) + a(\omega + \pi)G^*(\omega + \pi)).} \tag{19.89}$$

To do this we replace in (19.87) $b(2\omega)$ and $d(2\omega)$ given by the above expressions:

$$a(\omega) = a(\omega)P^*(\omega)P(\omega) + a(\omega + \pi)P^*(\omega + \pi)P(\omega)$$
$$+ a(\omega)G^*(\omega)G(\omega) + a(\omega + \pi)G^*(\omega + \pi)G(\omega).$$
$$a(\omega) = a(\omega)(P^*(\omega)P(\omega) + G^*(\omega)G(\omega))$$
$$+ a(\omega + \pi)(P(\omega)P^*(\omega + \pi) + G(\omega)G^*(\omega + \pi)).$$

This relationship verified $\forall \omega$ validates relationships (19.88) and (19.89), taking into account the relationships (19.76) and (19.80).

In summary, formula (19.88) will be used to calculate the blurred version.

The formula (19.89) will be used to calculate signal details. The formula (19.87) will be used to reconstruct the signal from its two components.

Finally, we show how one can calculate the coefficients of the filters with frequency responses $b(\omega)$ and $d(\omega)$.

We note $\mathcal{F}_N(c_k)$ the numerical Fourier transform of the sequence of coefficients c_k.

For Db4, we have: $P(z) = \tfrac{1}{2}\sum_{k=0}^{3} c_k z^{-k}, P(\omega) = \tfrac{1}{2}\sum_{k=0}^{3} c_k e^{-j\omega k} = \tfrac{1}{2}\mathcal{F}_N(c_k);$

$$P^*(\omega) = \frac{1}{2}\sum_{k=0}^{3} c_k e^{j\omega k} = \frac{1}{2}\mathcal{F}_N^*(c_k)$$

We also have: $P(\omega + \pi) = \dfrac{1}{2}\sum\limits_{k=0}^{3} c_k e^{-j(\omega+\pi)k} = \dfrac{1}{2}\sum\limits_{k=0}^{3}(-1)^k c_k e^{-j\omega k} = \dfrac{1}{2}\mathcal{F}_N((-1)^k c_k).$

and $P^*(\omega + \pi) = \dfrac{1}{2}\mathcal{F}_N^*((-1)^k c_k).$

Then

$$b(2\omega) = \frac{1}{2}\left(\mathcal{F}_N(a[k])\frac{1}{2}\mathcal{F}_N^*(c[k]) + \mathcal{F}_N\left((-1)^k a[k]\right)\frac{1}{2}\mathcal{F}_N^*\left((-1)^k c[k]\right)\right).$$

We recognize in the first term the FT of the cross-correlation function of a [k] and c [k].

The second term is the FT of the cross-correlation function of $(-1)^k a[k]$ and $(-1)^k c[k]$.

To clarify, we apply this formula to Db4. It provides in the time domain:

$$b\left[\frac{k}{2}\right] = \frac{1}{4}(a[k] \otimes (c_3\delta[k+3] + c_2\delta[k+2] + c_1\delta[k+1] + c_0\delta[k]))$$

$$+ \frac{1}{4}\left((-1)^k a[k] \otimes (-c_3\delta[k+3] + c_2\delta[k+2] - c_1\delta[k+1] + c_0\delta[k])\right).$$

Note the reverse in the time domain to move from correlation to convolution. For even values of k, we get:

$$b\left[\frac{k}{2}\right] = a[k] \otimes \frac{1}{2}(c_3\delta[k+3] + c_2\delta[k+2] + c_1\delta[k+1] + c_0\delta[k]). \qquad (19.90)$$

For odd values of k,

$$b\left[\frac{k}{2}\right] = 0. \qquad (19.91)$$

We can write these relations in the form:

$$b[k] = a[k] \otimes h_{0D}[k], \qquad (19.92)$$

With

$$h_{0D}[k] = \frac{1}{2}(c_3\delta[k+3] + c_2\delta[k+2] + c_1\delta[k+1] + c_0\delta[k]) \text{ for } k \text{ even.} \qquad (19.93)$$

and $b[k] = 0$ for odd values of k.

$h_{0D}[k]$ is the impulse response of a low-pass non-causal filter. In practice, a causal filter created by a delay of three steps of this impulse response is used. We will write then:

$$h_{0D}[k] = \frac{1}{2}(c_3\delta[k] + c_2\delta[k-1] + c_1\delta[k-2] + c_0\delta[k-3]). \qquad (19.94)$$

We now look for the impulse response of the second decomposition filter given by the relation (19.89). The FT $G(\omega)$ is given by (19.70). For Db4, we have:

$$G(\omega) = \frac{1}{2}\sum_{k=0}^{3}(-1)^k c_{3-k}e^{-j\omega k} = \frac{1}{2}\mathcal{F}_{\mathcal{N}}((-1)^k \tilde{c}_k).$$

Then

$$G^*(\omega) = \frac{1}{2}\sum_{k=0}^{3}(-1)^k c_{3-k}e^{j\omega k} = \frac{1}{2}\mathcal{F}_{\mathcal{N}}^*((-1)^k \tilde{c}_k).$$

$$G(\omega+\pi) = \frac{1}{2}\sum_{k=0}^{3}(-1)^k c_{3-k}e^{-j(\omega+\pi)k} = \frac{1}{2}\sum_{k=0}^{3}c_{3-k}e^{-j\omega k} = \frac{1}{2}\mathcal{F}_{\mathcal{N}}(\tilde{c}_k).$$

$$G^*(\omega+\pi) = \frac{1}{2}\mathcal{F}_{\mathcal{N}}^*(\tilde{c}_k).$$

$$d(2\omega) = \frac{1}{2}\left(\mathcal{F}_{\mathcal{N}}(a[k])\frac{1}{2}\mathcal{F}_{\mathcal{N}}^*\left((-1)^k\tilde{c}[k]\right) + \mathcal{F}_{\mathcal{N}}\left((-1)^k a[k]\right)\frac{1}{2}\mathcal{F}_{\mathcal{N}}^*(\tilde{c}_k)\right).$$

In the time domain we have:

$$d\left[\frac{k}{2}\right] = \frac{1}{4}\left(a[k]\otimes(c_3\delta[k] - c_2\delta[k+1] + c_1\delta[k+2] - c_0\delta[k+3])\right)$$

$$+ \frac{1}{4}\left((-1)^k a[k]\otimes(c_3\delta[k] + c_2\delta[k+1] + c_1\delta[k+2] + c_0\delta[k+3])\right)$$

For even values of k, we get:

$$d\left[\frac{k}{2}\right] = a[k]\otimes\frac{1}{2}(c_3\,\delta[k] - c_2\,\delta[k+1] + c_1\,\delta[k+2] - c_0\,\delta[k+3]).$$

For odd values of k, $d[k] = 0$.

We can write this relationship as: $d[k] = a[k]\otimes h_{1D}[k]$, with $h_{1D}[k] = \frac{1}{2}(-c_0\,\delta[k+3] + c_1\,\delta[k+2] - c_2\,\delta[k+1] + c_3\,\delta[k])$ for k even.

and always $d[k] = 0$ for odd values of k.

$h_{1D}[k]$ is the impulse response of a non-causal high-pass filter. In practice, using a causal filter created with a delay of three steps of this impulse response. We then write:

$$h_{1D}[k] = \frac{1}{2}(-c_0\delta[k] + c_1\,\delta[k-1] - c_2\,\delta[k-2] + c_3\,\delta[k-3]). \qquad (19.95)$$

The filters reconstructing the signal from its LF and HF components have respective frequency responses $P(\omega)$ and $G(\omega)$, as the relationship (19.87) states.

Within a factor of 2, Decomposition and reconstruction filters satisfy relations (19.24) and (19.25).

This factor 2 is not important since it reflects a simple multiplication of the filters gains by a constant. We have here:

$$P(z)H_0(z) + G(z)H_1(z) = 1, \tag{19.96}$$

and

$$P(z)H_0(-z) + G(z)H_1(-z) = 0. \tag{19.97}$$

From the decomposition $P(\omega)$ and $G(\omega)$ we get the reconstruction filters:

$$h_{0R}[k] = c_0\delta[k] + c_1\delta[k-1] + c_2\delta[k-2] + c_3\delta[k-3].$$

$$h_{1R}[k] = c_3\delta[k] - c_2\delta[k-1] + c_1\,\delta[k-2] - c_0\delta[k-3].$$

We note the dissymmetry by the factor ½ between the decomposition and reconstruction filters responses. Matlab uses symmetric formulas imposing a common multiplying factor $\sqrt{2}$ to all impulse responses.

The impulse responses of the different filters given by Matlab for Db4 (Db4 is called db2 by Matlab) appear in Fig. 19.16:

Finally, the frequency responses of the analysis filters (decomposition) are given in Fig. 19.17:

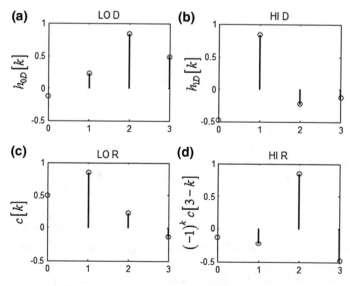

Fig. 19.16 Daubechies Db4 impulse responses; *Top* decomposition **a** Low pass; **b** High pass; *Bottom* reconstruction **c** Low pass; **d** High pass

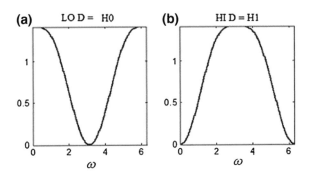

Fig. 19.17 Daubechies Db4 spectral amplitudes of decomposition filters; a Low pass; b High pass

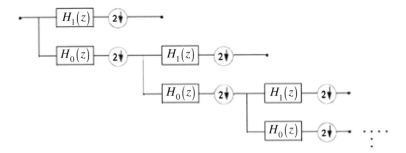

Fig. 19.18 Dyadic decomposition scheme

We encounter again the major role played by these relations in the dyadic multiresolution analysis.

The decomposition of a signal may proceed recursively: In the first step the blurred component is extracted (the approximation) of the signal [by filtering LF represented by $H_0(z)$ whose impulse response is given by (19.94)] and the component containing details [given by the HF filter represented by $H_1(z)$]. The operation is repeated and the blurred part is again decomposed in LF and HF components, etc., in a tree structure.

This multiresolution analysis can be represented by the diagram in Fig. 19.18:

The reconstruction is also recursive, starting from right to left, up into the tree structure, which is to render more and more details of the signal.

Summary
This chapter was dedicated to the analysis of a signal with multi-resolution. The dyadic decomposition-reconstruction scheme allows the separation of the frequency components in low frequencies (the shape) and high frequencies (the details) by two filters. The reconstruction may be performed with the use of the two associated filters. The decomposition may be performed recursively and, at any step, the signal

may be reconstructed exactly. After having exposed the principle of the dyadic decomposition, we have detailed the Haar transform and the LeGall-Tabatabai decomposition. Haar wavelet has been introduced but it is ill-suited to serve as a basis for approximations of smooth functions since reconstructions will necessarily present discontinuities. Mallat and Daubechies have established the theory and created wavelets which can be generated with the use of a small number of coefficients, that can be continuous with continuous derivatives up to a chosen order, and which allow the dyadic scheme. At a given order of continuity corresponds a Daubechies wavelet and the 2 decomposition-reconstruction filter pairs.

Exercises

I. Let the signal $x[n] = U[n] - U[n-10]$. Show using a picture that this signal is a digital rectangular window. This signal is filtered by two filters in parallel with transfer functions: $H_0(z) = (z^{-1} + 1)$ and $H_1(z) = (z^{-1} - 1)$. Calculate the two output signals $y_0[n]$ and $y_1[n]$. Show that the first filter provides a smoothing (a blurring) of the input signal and that the second detects transitions at the edge of the function $x[n]$ (in a two-dimensional filtering one would speak of contour detection in an image). Compare this result to the Haar transform encountered above.

II. Calculate numerically using Matlab the functions $P(\omega)$ and $G(\omega)$ for the Daubechies wavelet Db4 from the FFT of coefficients c_k. Derive approximate values of functions $\hat{\phi}(\omega)$ and $\hat{\psi}(\omega)$ from relations (19.58) and (19.69).

We will have limited the infinite products to the first six terms. Deduct by the inverse Fourier transform (ifft()function), the scaling functions and the wavelet Db4. Compare the results to those presented in the figures of these functions given in this chapter.

Chapter 20
Parametric Estimate—Modeling of Deterministic Signals—Linear Prediction

In this chapter our goal is the modeling of a digital signal in the time domain, i.e. we want to find a finite number of coefficients as small as possible, which allows the possibility to reconstruct exactly or approximately the signal with the use of these coefficients. We focus to model the signal as the impulse response of a LTI ARMA system. We only model causal signals.

In the general case, a finite number of coefficients do not allow to estimate a signal without making an error. The principle of the method is to minimize the error in the least squares sense. In the first section we show, using the frequency domain to demonstrate that property, that the equations derived from the least squares method are nonlinear, difficult, or impossible to solve. So we need to look at other methods necessarily less efficient in principle. We first study the Padé representation of the signal, which is accurate on a number of points equal to the number of coefficients chosen for the model, but whose estimate of the signal outside this range is very poor. The search for the coefficients reduces to solving a linear system of equations whose solutions are sought by matrix methods. One is led to release the accuracy constraints on the first points of the signal and seek to minimize the error on larger parts of the time axis. This is the principle of Prony's method and its improvement by the method of Shanks. All-pole modeling (AR) detailed then gives very good results in speech synthesis [it is known under the name LPC (Linear Predictive Coding)]. Techniques called correlation and covariance methods are used for time-limited signals.[1]

The preceding methods are useless in the cases where the properties of the systems involved in the signal production vary with time. Adaptive filtering has been developed for nonstationary signals analysis. The filter coefficients are reassessed as the signal evolves in time. This type of filtering is studied at the end of this chapter.

---

[1]To go further the reader is invited to refer to the excellent books of C.W. Therrien and M.H. Hayes listed in the bibliography at the end of this book.

© Springer International Publishing Switzerland 2016
F. Cohen Tenoudji, *Analog and Digital Signal Analysis,*
Modern Acoustics and Signal Processing, DOI 10.1007/978-3-319-42382-1_20

20.1 Least Square Method

Consider a digital deterministic signal $x[n]$. We desire to model this signal using a limited number of parameters, either to compress the information, or to interpolate and find absent data or to predict future values, yet unknown, of the signal. These techniques are widely used, like in speech processing or telecommunications, for example.

The estimator of the signal is noted $\hat{x}[n]$ and the error $e[n]$:

$$e[n] = x[n] - \hat{x}[n]. \tag{20.1}$$

We assume hereinafter that the signal $x[n]$ is zero for negative time $n < 0$. It is the same for the estimator $\hat{x}[n]$.

The choice of the estimator parameters is here performed by seeking to minimize the squared error over the whole time axis:

$$\varepsilon = \sum_{n=0}^{\infty} |e[n]|^2 = \sum_{n=0}^{\infty} |x[n] - \hat{x}[n]|^2. \tag{20.2}$$

The moduli express here the general treatment of complex signals.

Using the signals z transform, we get from (20.1):

$$E(z) = X(z) - \hat{X}(z). \tag{20.3}$$

We focus here on the choice often used where the estimator $\hat{X}(z)$ is sought in the form of a rational function (This technique is known as Linear Prediction Coding, LPC):

$$\hat{X}(z) = \frac{B_q(z)}{A_p(z)} = \frac{\sum_{k=0}^{q} b_q[k]z^{-k}}{1 + \sum_{k=1}^{p} a_p[k]z^{-k}}. \tag{20.4}$$

We look for a causal estimator $\hat{x}[n]$ (as is $x[n]$) and stable. The definition domain of $\hat{X}(z)$ therefore includes the unit circle. Thus, the Fourier transforms $X(e^{j\omega T})$ and $\hat{X}(e^{j\omega T})$ must exist.

We show now that the least squares method leads to the resolution, difficult, of nonlinear equations. Using Parseval theorem for the energy of the error we write:

$$\varepsilon = \sum_{n=0}^{\infty} |e[n]|^2 = \frac{1}{\omega_e} \int_0^{\omega_e} |E(e^{j\omega T})|^2 d\omega = \frac{1}{\omega_e} \int_0^{\omega_e} |X(e^{j\omega T}) - \hat{X}(e^{j\omega T})|^2 d\omega. \tag{20.5}$$

The error is minimum when the partial derivatives of ε with respect to the unknown parameters are zero:

$$\frac{\partial \varepsilon}{\partial a_p^*[k]} = 0 \quad \text{for } k = 1, 2, \ldots p,$$ (20.6a)

and

$$\frac{\partial \varepsilon}{\partial b_q^*[k]} = 0 \quad \text{for } k = 0, 1, 2, \ldots, q.$$ (20.6b)

Using the fact that $a_p^*[k]$ and $a_p[k]$ can be considered as independent variables, we have:

$$
\begin{aligned}
\frac{\partial \varepsilon}{\partial a_p^*[k]} &= \frac{1}{\omega_e} \int_0^{\omega_e} \frac{\partial}{\partial a_p^*[k]} \left[\left(X^*(e^{j\omega T}) - \hat{X}^*(e^{j\omega T}) \right) \left(X(e^{j\omega T}) - \hat{X}(e^{j\omega T}) \right) \right] d\omega \\
&= \frac{1}{\omega_e} \int_0^{\omega_e} -\frac{\partial}{\partial a_p^*[k]} \left[\hat{X}^*(e^{j\omega T}) \left(X(e^{j\omega T}) - \hat{X}(e^{j\omega T}) \right) \right] d\omega \\
&= \frac{1}{\omega_e} \int_0^{\omega_e} \left(X(e^{j\omega T}) - \frac{B_q(e^{j\omega T})}{A_p(e^{j\omega T})} \right) \frac{B_q^*(e^{j\omega T})}{\left(A_p^*(e^{j\omega T}) \right)^2} e^{jk\omega T} d\omega.
\end{aligned}
$$ (20.7)

Similarly:

$$
\begin{aligned}
\frac{\partial \varepsilon}{\partial b_q^*[k]} &= \frac{1}{\omega_e} \int_0^{\omega_e} -\frac{\partial}{\partial b_q^*[k]} \left[\left(X(e^{j\omega T}) - \frac{B_q(e^{j\omega T})}{A_p(e^{j\omega T})} \right) \frac{B_q^*(e^{j\omega T})}{A_p^*(e^{j\omega T})} \right] e\omega \\
&= -\frac{1}{\omega_e} \int_0^{\omega_e} \left(X(e^{j\omega T}) - \frac{B_q(e^{j\omega T})}{A_p(e^{j\omega T})} \right) \frac{e^{jk\omega T}}{A_p^*(e^{j\omega T})} e\omega.
\end{aligned}
$$ (20.8)

We see that the equations $\frac{\partial \varepsilon}{\partial a^*[k]} = 0$ and $\frac{\partial \varepsilon}{\partial b^*[k]} = 0$ are nonlinear. Their resolution is difficult in general. For this reason, the least squares method is rarely used in this context and other approximations are preferred.

20.2   Padé Representation

The method consists to equal $x[n]$ and $\hat{x}[n]$ on a given time interval and taking $\hat{X}(z)$ in the form (20.4). Let us detail it:

$$\hat{X}(z) = \frac{B_q(z)}{A_p(z)} = \frac{\sum_{k=0}^{q} b_q[k]z^{-k}}{1 + \sum_{k=1}^{p} a_p[k]z^{-k}}. \tag{20.9}$$

The method operates in the time domain. Putting in the time domain we have:

$$\hat{x}[n] + \sum_{k=1}^{p} a_p[k]\,\hat{x}[n-k] = \sum_{k=0}^{q} b_q[k]\,\delta[n-k]. \tag{20.10}$$

Indeed by multiplying each term of this equation by z^{-n} and summing over n:

$$\sum_{n=-\infty}^{\infty} \hat{x}[n]\,z^{-n} + \sum_{k=1}^{p} a_p[k] \sum_{n=-\infty}^{\infty} \hat{x}[n-k]\,z^{-n} = \sum_{k=0}^{q} b_q[k] \sum_{n=-\infty}^{\infty} \delta[n-k]z^{-n},$$

which has the form:

$$A_p(z)\hat{X}(z) = B_q(z). \tag{20.11}$$

It is recognized in the temporal Eq. (20.10), the equation of an ARMA filter, $\hat{x}[n]$ appearing as the impulse response of this filter.

The first terms will be given by:

$$\hat{x}[0] = b_q[0],$$
$$\hat{x}[1] + a_p[1]\hat{x}[0] = b_q[1],$$
$$\hat{x}[2] + a_p[1]\hat{x}[1] + a_p[2]\hat{x}[0] = b_q[2],$$
$$\hat{x}[3] + a_p[1]\hat{x}[2] + a_p[2]\hat{x}[1] + a_p[3]\hat{x}[0] = b_q[3],$$

As mentioned above, the method consists in giving to the estimator $\hat{x}[n]$ the signal $x[n]$ values for some values of n. Since we limit the orders of functions $A_p(z)$ and $B_q(z)$ respectively to p and q, the equality of $x[n]$ and $\hat{x}[n]$ can only be imposed on a number of points limited to $p+q+1$. The moments are chosen in the interval $\{0, p+q\}$. In this interval we must have:

$$x[n] + \sum_{k=1}^{p} a_p[k]x[n-k] = \begin{cases} b_q[n]; & n = 0, 1, \ldots, q \\ 0; & n = q+1, \ldots, q+p \end{cases}. \tag{20.12}$$

This equation will allow to determine the parameters $a_p[k]$ and $b_q[k]$.
We can write this equation in matrix form:

$$
\begin{bmatrix}
x[0] & 0 & \cdots & 0 \\
x[1] & x[0] & \cdots & 0 \\
x[2] & x[1] & \cdots & 0 \\
\vdots & \vdots & & \vdots \\
x[q] & x[q-1] & \cdots & x[q-p] \\
\hline
x[q+1] & x[q] & \cdots & x[q-p+1] \\
\vdots & \vdots & & \vdots \\
x[q+p] & x[q+p-1] & \cdots & x[q]
\end{bmatrix}
\cdot
\begin{bmatrix}
1 \\
a_p[1] \\
a_p[2] \\
\vdots \\
a_p[p]
\end{bmatrix}
=
\begin{bmatrix}
b_q[0] \\
b_q[1] \\
b_q[2] \\
\vdots \\
b_q[q] \\
\hline
0 \\
\vdots \\
0
\end{bmatrix}.
$$

$$(20.13)$$

The resolution of this linear system is carried out in two stages. First we solve
the system consisting of the second part of the matrices in which the coefficients
$b_q[k]$ are not involved:

$$
\begin{bmatrix}
x[q+1] & x[q] & \cdots & x[q-p+1] \\
x[q+2] & x[q+1] & \cdots & x[q-p+2] \\
\vdots & \vdots & & \vdots \\
x[q+p] & x[q+p-1] & \cdots & x[q]
\end{bmatrix}
\cdot
\begin{bmatrix}
1 \\
a_p[1] \\
a_p[2] \\
\vdots \\
a_p[p]
\end{bmatrix}
=
\begin{bmatrix}
0 \\
0 \\
\vdots \\
0
\end{bmatrix}.
\qquad (20.14)
$$

This system can be rewritten as:

$$
\begin{bmatrix}
x[q+1] \\
x[q+2] \\
\vdots \\
x[q+p]
\end{bmatrix}
+
\begin{bmatrix}
x[q] & x[q-1] & \cdots & x[q-p+1] \\
x[q+1] & x[q] & \cdots & x[q-p+2] \\
\vdots & \vdots & & \vdots \\
x[q+p-1] & x[q+p-2] & \cdots & x[q]
\end{bmatrix}
\cdot
\begin{bmatrix}
a_p[1] \\
a_p[2] \\
\vdots \\
a_p[p]
\end{bmatrix}
=
\begin{bmatrix}
0 \\
0 \\
\vdots \\
0
\end{bmatrix},
$$

$$(20.15)$$

or:

$$
\begin{bmatrix}
x[q] & x[q-1] & \cdots & x[q-p+1] \\
x[q+1] & x[q] & \cdots & x[q-p+2] \\
\vdots & \vdots & & \vdots \\
x[q+p-1] & x[q+p-2] & \cdots & x[q]
\end{bmatrix}
\cdot
\begin{bmatrix}
a_p[1] \\
a_p[2] \\
\vdots \\
a_p[p]
\end{bmatrix}
= -
\begin{bmatrix}
x[q+1] \\
x[q+2] \\
\vdots \\
x[q+p]
\end{bmatrix}.
$$

$$(20.16)$$

We can write this system in a more condensed way:

$$X_q a_p = -x_{q+1}, \tag{20.17}$$

with

$$a_p = \left[a_p[1], a_p[2], \ldots, a_p[p]\right]^T, \tag{20.18}$$

$$x_{q+1} = \left[x[q+1], \ x[q+2], \ldots, x[q+p]\right]^T,$$

and

$$X_q = \begin{bmatrix} x[q] & x[q-1] & \cdots & x[q-p+1] \\ x[q+1] & x[q] & \cdots & x[q-p+2] \\ \vdots & \vdots & & \vdots \\ x[q+p-1] & x[q+p-2] & \cdots & x[q] \end{bmatrix}. \tag{20.19}$$

X_q is a Toeplitz matrix (matrix of which all elements along a parallel to the main diagonal are equal) unsymmetrical.

Depending on the properties of the matrix X_q three cases may be met:

1. The matrix X_q is non-singular. It thus has an inverse matrix and Eq. (20.16) can be solved by multiplying on the left by the inverse: $a_p = -X_q^{-1} x_{q+1}$.
2. The matrix X_q is singular. If there is a vector a_p that solves the system (20.16), this solution is not unique. As X_q is singular, the homogeneous system $X_q z = 0$ has nonzero solutions. In this case the vector $\tilde{a}_p = a_p + z$ is also a solution of (20.16).

   The solution which gives the vector having a reduced number of nonzero terms $a_p[k]$ is often chosen.
3. The matrix X_q is singular and no solution a_p of the system (20.16) exists. This system has a second member as $a[0] = 1$ is assumed. This assumption is erroneous. One must seek the solution a_p of the system $X_q a_p = 0$.

   Having determined the series of coefficients a_p, in a second step, the coefficients of the vector b_q are determined. To this we transfer these coefficients a_p in the upper part of the matrix Eq. (20.13) which is written:

$$\begin{bmatrix} x[0] & 0 & \cdots & 0 \\ x[1] & x[0] & \cdots & 0 \\ x[2] & x[1] & \cdots & 0 \\ \vdots & \vdots & & \vdots \\ x[q] & x[q-1] & \cdots & x[q-p] \end{bmatrix} \cdot \begin{bmatrix} 1 \\ a_p[1] \\ a_p[2] \\ \vdots \\ a_p[p] \end{bmatrix} = \begin{bmatrix} b_q[0] \\ b_q[1] \\ b_q[2] \\ \vdots \\ b_q[q] \end{bmatrix}. \tag{20.20}$$

This equation can also be written as a recurrence equation whose resolution is immediate:

$$x[n] + \sum_{k=1}^{p} a_p[k]\, x[n-k] = b_q[n] \quad \text{for} \quad n = 0,\, 1\ldots, q. \tag{20.21}$$

20.2.1 Padé Approximation

So far, we found the equations for finding the coefficients multiplying the data values $x[n-k]$ to satisfy the recurrence Eq. (20.21) on the $p+q+1$ first points. It is not an approximation but the resolution of a system of equation.

On the other hand, this system of equations is found to be the same as the one found in mathematics when it is desired to represent a function $f(x)$ of a continuous variable x by a rational fraction of polynomials by imposing that its first $p+q+1$ derivatives have the same values in $x=0$ than those of the function $f(x)$. This method can be qualified as an improved Taylor expansion to order $p+q+1$ of the function $f(x)$ in the vicinity of $x=0$. In this context the method is called the Padé approximation. It is known in numerical analysis that the Taylor expansion is good in the vicinity of the origin but strongly deviates from the function $f(x)$ when one moves away from the origin. The same phenomenon will be observed in the framework of digital signals when attempting to use the Padé development for values of n outside of the initial interval.

This fact leads us to rule out the Padé representation when searching an estimator of the signal $x[n]$.

20.2.2 All-pole Modeling

In the case where one seeks an all-pole modeling of the signal in the frame of the Padé representation, $\hat{X}(z)$ is sought in the form:

$$\hat{X}(z) = \frac{b[0]}{1 + \sum_{k=1}^{p} a_p[k] z^{-k}}. \tag{20.22}$$

The system giving a_p is written in this case:

$$
\begin{bmatrix}
x[0] & 0 & \cdots & 0 \\
x[1] & x[0] & \cdots & 0 \\
\vdots & \vdots & & \vdots \\
x[p-1] & x[p-2] & \cdots & x[0]
\end{bmatrix}
\cdot
\begin{bmatrix}
a_p[1] \\
a_p[2] \\
\vdots \\
a_p[p]
\end{bmatrix}
= -
\begin{bmatrix}
x[1] \\
x[2] \\
\vdots \\
x[p]
\end{bmatrix}. \tag{20.23}
$$

The latter matrix is triangular, the solution is obtained by simply solving the iteration:

$$a_p[k] = -\frac{1}{x[0]}\left[x[k] + \sum_{l=1}^{k-1} a_p[l]x[k-l]\right],$$ (20.24)

and $b[0] = x[0]$.

20.2.3 Examples

Example 1

We look to model a signal whose first values are:

$$x = [2, \quad -0.1, \quad 0.81, \quad 0.729, \quad 0.6561, \quad ...., \quad ....].$$

We place ourselves in a case where we look for a model with $p = 2$, $q = 1$. Equation (20.13) has the form:

$$\begin{bmatrix} x[0] & 0 & 0 \\ x[1] & x[0] & 0 \\ --- & --- & --- \\ x[2] & x[1] & x[0] \\ x[3] & x[2] & x[1] \end{bmatrix} \cdot \begin{bmatrix} 1 \\ a_p[1] \\ a_p[2] \end{bmatrix} = \begin{bmatrix} b_q[0] \\ b_q[1] \\ --- \\ 0 \\ 0 \end{bmatrix}.$$ (20.25)

First of all one must solve the system:

$$\begin{bmatrix} x[2] & x[1] & x[0] \\ x[3] & x[2] & x[1] \end{bmatrix} \cdot \begin{bmatrix} 1 \\ a_p[1] \\ a_p[2] \end{bmatrix} = \begin{bmatrix} 0 \\ 0 \end{bmatrix}.$$ (20.26)

Here:

$$\begin{bmatrix} 0.81 & -0.1 & 2 \\ 0.729 & 0.81 & -0.1 \end{bmatrix} \cdot \begin{bmatrix} 1 \\ a_p[1] \\ a_p[2] \end{bmatrix} = \begin{bmatrix} 0 \\ 0 \end{bmatrix},$$

or, using the form (20.23) to solve:

$$\begin{bmatrix} -0.1 & 2 \\ 0.81 & -0.1 \end{bmatrix} \cdot \begin{bmatrix} a_p[1] \\ a_p[2] \end{bmatrix} = -\begin{bmatrix} 0.81 \\ 0.729 \end{bmatrix}.$$

Using Matlab we get:

$$a_p[1] = -0.9559 \quad \text{and} \quad a_p[2] = -0.4528.$$

We then deduce the coefficients $b_q[0]$ and $b_q[1]$:

$$\begin{bmatrix} 2 & 0 & 0 \\ -0.1 & 2 & 0 \end{bmatrix} \cdot \begin{bmatrix} 1 \\ -0.9559 \\ -0.4528 \end{bmatrix} = \begin{bmatrix} b_q[0] \\ b_q[1] \end{bmatrix}.$$

We find using Matlab: $b_q[0] = 2.000$ and $b_q[1] = -2.0118$.

The elements of $x[n]$ are recalculated to analyze the result and verify the absence of miscalculation:

$$x[0] = 2,$$
$$x[1] = b_q[1] - a_p[1]x[0] = -2.0118 - (-0.9559) * 2 = -0.1,$$
$$x[2] = -a_p[1]x[1] - a_p[2]x[0] = 0.9559 * (-0.1) + 0.4528 * 2 = 0.81,$$
$$x[3] = -a_p[1]x[2] - a_p[2]x[1] = 0.9559 * 0.81 + 0.4528 * (-0.1) = 0.729,$$
$$\hat{x}[4] = -a_p[1]x[3] - a_p[2]x[2] = 0.9559 * 0.729 + 0.4528 * 0.81 = 1.036,$$

Note that the values for $n = 0, 1, 2, 3$, are found exactly.

This is the principle of Padé approximation. However, the estimated value of $x[4]$ is 1.036 while the exact value was 0.6561.

It appears that the Padé approximation does not ensure that the error is controlled outside of the interval used for the estimate. If orders p and q used are not good, the errors can become important outside the interval $\{0, p+q\}$. It is this feature that limits the interest of the Padé approximation.

We retry now the approximation in the previous example with another order for the model. Taking $p = q = 1$ Eq. (20.13) now takes the form:

$$\begin{bmatrix} x[0] & 0 \\ x[1] & x[0] \\ \hline x[2] & x[1] \end{bmatrix} \cdot \begin{bmatrix} 1 \\ a_p[1] \end{bmatrix} = \begin{bmatrix} b_q[0] \\ b_q[1] \\ \hline 0 \end{bmatrix}.$$

First of all, we solve the system:

$$[x[2] \quad x[1]] \cdot \begin{bmatrix} 1 \\ a_p[1] \end{bmatrix} = 0,$$

thus $a_p[1] = -\frac{x[2]}{x[1]} = -\frac{0.81}{-0.1} = 8.1$.

Example 2

$$x = [2, \quad 0.8, \quad 0.72, \quad 0.648, \quad 0.5832, \quad \ldots., \quad \ldots.].$$

We look for a model with $p = 2$ and $q = 1$. Equation (20.13) has the form (20.25). First of all, we solve the system: (20.26).

Here:

$$\begin{bmatrix} 0.72 & 0.8 & 2 \\ 0.648 & 0.72 & 0.8 \end{bmatrix} \cdot \begin{bmatrix} 1 \\ a_p[1] \\ a_p[2] \end{bmatrix} = \begin{bmatrix} 0 \\ 0 \end{bmatrix},$$

or, using the form (20.23), to solve the system:

$$\begin{bmatrix} 0.8 & 2 \\ 0.72 & 0.8 \end{bmatrix} \cdot \begin{bmatrix} a_p[1] \\ a_p[2] \end{bmatrix} = - \begin{bmatrix} 0.72 \\ 0.648 \end{bmatrix}.$$

Using Matlab, we find: $a_p[1] = -0.9$ and $a_p[2] = 0$. We then derive the coefficients $b_q[0]$ and $b_q[1]$ of the system:

$$\begin{bmatrix} 2 & 0 & 0 \\ 0.8 & 2 & 0 \end{bmatrix} \begin{bmatrix} 1 \\ -0.9 \\ 0 \end{bmatrix} = \begin{bmatrix} b_q[0] \\ b_q[1] \end{bmatrix}.$$

We find: $b_q[0] = 2.00$ and $b_q[1] = -1$.

To finish we recalculate the elements of $x[n]$:

$$x[0] = 2,$$
$$x[1] = b_q[1] - a_p[1]x[0] = -1 + 0.9 * 2 = 0.8,$$
$$x[2] = -a_p[1]x[1] - a_p[2]x[0] = 0.9 * 0.8 + 0 = 0.72,$$
$$x[3] = -a_p[1]x[2] - a_p[2]x[1] = 0.9 * 0.72 = 0.648,$$
$$\hat{x}[4] = -a_p[1]x[3] = 0.9 * 0.648 = 0.5832,$$

Again, we still find the exact values of $x[n]$ for $n = 0, 1, 2, 3 = p + q$ as it is expected.

But now the estimated value $\hat{x}[4]$ is 0.5832, the same as that of the signal (the index $n = 4$ is yet outside the range $\{0, 3\}$ that was used for the estimate). In this case, we found exactly the impulse response of the filter that was in the underlying signal $x[n]$.

Its transfer function is:

$$H(z) = H_1(z)H_2(z) = \left(2 - z^{-1}\right) \left(\frac{1}{1 - 0.9z^{-1}}\right).$$

The impulse response $h[n]$ of this filter is the convolution of the two impulse responses $h_1[n] = 2\delta[n] - \delta[n-1]$ and $h_2[n] = (0.9)^n U[n-1]$.

20.3 Prony's Approximation Method. Shanks Method

The limits of the Padé representation as an approximation method have been demonstrated using examples. The problem encountered is due to the fact that by completely canceling the error in a limited time interval, we allow this error to be important outside this range.

20.3.1 Prony's Method

The Prony's method distributes the error over the entire time axis by releasing the constraint in the interval $\{0, p+q\}$. We are still looking for modeling $x[n]$ using an ARMA filter ().

Having discussed above the aspect of the estimator as may be interpreted as an impulse response, we note the estimator

$$\hat{x}[n] = h[n]. \tag{20.27}$$

We write:

$$e[n] = x[n] - h[n]. \tag{20.28}$$

Using z transform, we write:

$$E(z) = X(z) - H(z) = X(z) - \frac{B_q(z)}{A_p(z)}. \tag{20.29}$$

Multiplying by $A_p(z)$, we have:

$$E(z)A_p(z) = X(z)A_p(z) - B_q(z).$$

Noting:

$$E'(z) = E(z)A_p(z),$$

We have

$$E'(z) = X(z)A_p(z) - B_q(z), \tag{20.30}$$

or, in the time domain:

$$e'[n] = x[n] + x[n] \otimes a_p[n] - b_q[n] \text{ with } p > 0. \tag{20.31}$$

We write in more detail this equation, using the fact that $b_q[n]$ is zero for $n > q$:

$$e'[n] = \begin{cases} x[n] + \sum_{k=1}^{p} a_p[k]\,x[n-k] - b_q[n]; & n = 0,\,1,\,\ldots,\,q \\[2mm] x[n] + \sum_{k=1}^{p} a_p[k]\,x[n-k]; & n > q \end{cases}. \tag{20.32}$$

The Prony's method is to determine the coefficients a_p such that the square error is minimum for $n > q$:

$$\varepsilon' = \sum_{n=q+1}^{\infty} |e'[n]|^2 = \sum_{n=q+1}^{\infty} \left| x[n] + \sum_{k=1}^{p} a_p[k]x[n-k] \right|^2. \tag{20.33}$$

This error is minimum when the partial derivatives of ε' with respect to parameters $a_p^*[k]$ are zero: $\frac{\partial \varepsilon'}{\partial a_p^*[k]} = 0$ for $k = 1,\,2,\ldots,\,p$.

$$\frac{\partial \varepsilon'}{\partial a_p^*[k]} = \sum_{n=q+1}^{\infty} x^*[n-k]\left[x[n] + \sum_{l=1}^{p} a_p[l]x[n-l] \right] = \sum_{n=q+1}^{\infty} x^*[n-k]e'[n] = 0. \tag{20.34}$$

$$\frac{\partial \varepsilon'}{\partial a_p^*[k]} = \sum_{n=q+1}^{\infty} x^*[n-k]x[n] + \left[\sum_{n=q+1}^{\infty} x^*[n-k]\sum_{l=1}^{p} a_p[l]x[n-l] \right] = 0$$

$$\text{or } \sum_{n=q+1}^{\infty} x^*[n-k]x[n] + \left[\sum_{l=1}^{p} a_p[l] \sum_{n=q+1}^{\infty} x^*[n-k]x[n-l] \right] \tag{20.35}$$

$$= 0;\ k = 1,\,2,\,\ldots,\,p;\ l = 1,\,2,\,\ldots,\,p.$$

We write:

$$r_{xx}[k,l] = \sum_{n=q+1}^{\infty} x^*[n-k]\,x[n-l]. \tag{20.36}$$

The function $r_{xx}[k,l]$ has the form of a **deterministic correlation function** of the signal $x[n]$. Beware though, $r_{xx}[k,l]$ is not the correlation function of $x[n]$ because the summation upon n does not go from minus infinity to plus infinity (or in the case of a causal signal from zero to infinity, which is the case for $x[n]$). Equations (20.35) are written also:

$$\sum_{l=1}^{p} a_p[l]\,r_{xx}[k,l] = -r_{xx}[k,0];\quad k = 1,\,2,\,\ldots,\,p. \tag{20.37}$$

This system of equations is called the ensemble of **Prony's normal equations**. It can be written in matrix form:

$$
\begin{bmatrix}
r_{xx}[1,1] & r_{xx}[1,2] & r_{xx}[1,3] & \cdots & r_{xx}[1,p] \\
r_{xx}[2,1] & r_{xx}[2,2] & r_{xx}[2,3] & \cdots & r_{xx}[2,p] \\
r_{xx}[3,1] & r_{xx}[3,2] & r_{xx}[3,3] & \cdots & r_{xx}[3,p] \\
\vdots & \vdots & \vdots & & \vdots \\
r_{xx}[p,1] & r_{xx}[p,2] & r_{xx}[p,3] & \cdots & r_{xx}[p,p]
\end{bmatrix}
\cdot
\begin{bmatrix}
a_p[1] \\
a_p[2] \\
a_p[3] \\
\vdots \\
a_p[p]
\end{bmatrix}
= -
\begin{bmatrix}
r_{xx}[1,0] \\
r_{xx}[2,0] \\
r_{xx}[3,0] \\
\vdots \\
r_{xx}[p,0]
\end{bmatrix}.
$$

(20.38)

These equations can be written in condensed form: $R_{xx}a_p = -r_{xx}$.

The matrix R_{xx} is square, $p \times p$, and has Hermitian symmetry (two elements symmetrical with respect to the main diagonal are complex conjugates). r_{xx} is a column vector.

We note:

$$
X_q =
\begin{bmatrix}
x[q] & x[q-1] & \cdots & x[q-p+1] \\
x[q+1] & x[q] & \cdots & x[q-p+2] \\
x[q+2] & x[q+1] & & x[q-p+3] \\
\vdots & \vdots & \cdots & \vdots
\end{bmatrix}.
$$

(20.39)

We can write:

$$
R_{xx} = X_q^H X_q.
$$

(20.40)

The error in the least square sense made in this approximation can be assessed:

$$
\varepsilon'_{LS} = \sum_{n=q+1}^{\infty} |e'[n]|^2 = \sum_{n=q+1}^{\infty} e'[n]e'^*[n] = \sum_{n=q+1}^{\infty} e'[n]\left[x[n] + \sum_{k=1}^{p} a_p[k]x[n-k] \right]^*
$$

$$
= \sum_{n=q+1}^{\infty} e'[n]x^*[n] + e'[n]\left[\sum_{k=1}^{p} a_p[k]x[n-k] \right]^*.
$$

The second term is zero, since the error is orthogonal to the various components $x^*[n-k]$ as shown in Eq. (20.34).

We have

$$
\varepsilon'_{LS} = \sum_{n=q+1}^{\infty} e'[n]\,x^*[n] = \sum_{n=q+1}^{\infty} \left[x[n] + \sum_{k=1}^{p} a_p[k]x[n-k] \right] x^*[n],
$$

(20.41)

and finally:

$$\varepsilon'_{LS} = r_{xx}[0,0] + \sum_{k=1}^{p} a_p[k] r_{xx}[0,k]. \tag{20.42}$$

The second step in the Prony's approximation method is to determine the coefficients $b_q[k]$ as is done in the Padé's method. For this, one uses the coefficients $a_p[k]$ determined by the resolution of (20.38) in Eq. (20.20) of the Padé's representation. At this stage, the only remaining unknowns are the coefficients $b_q[k]$:

$$x[n] + \sum_{k=1}^{p} a_p[k] x[n-k] = b_q[n]; \quad n = 0, 1, \dots, q. \tag{20.43}$$

As seen above, the coefficients $b_q[k]$ are obtained by iteration.

Another possible writing for the research of coefficients is to write the set of equations in the form of the **augmented Prony's normal equation**s.

The system (20.37) can be written in the form:

$$\begin{bmatrix} r_{xx}[1,0] & r_{xx}[1,1] & r_{xx}[1,2] & r_{xx}[1,3] & \cdots & r_{xx}[1,p] \\ r_{xx}[2,0] & r_{xx}[2,1] & r_{xx}[2,2] & r_{xx}[2,3] & \cdots & r_{xx}[2,p] \\ r_{xx}[3,0] & r_{xx}[3,1] & r_{xx}[3,2] & r_{xx}[3,3] & \cdots & r_{xx}[3,p] \\ \vdots & \vdots & \vdots & \vdots & & \vdots \\ r_{xx}[p,0] & r_{xx}[p,1] & r_{xx}[p,2] & r_{xx}[p,3] & \cdots & r_{xx}[p,p] \end{bmatrix} \cdot \begin{bmatrix} 1 \\ -\!-\!- \\ a_p[1] \\ a_p[2] \\ \vdots \\ a_p[p] \end{bmatrix} = \begin{bmatrix} 0 \\ 0 \\ 0 \\ \vdots \\ 0 \end{bmatrix}, \tag{20.44}$$

or, by introducing the error given by (20.42) in the system of equations:

$$\begin{bmatrix} r_{xx}[0,0] & r_{xx}[0,1] & r_{xx}[0,2] & \cdots & r_{xx}[0,p] \\ -\!-\!- & -\!-\!- & -\!-\!- & -\!-\!- & -\!-\!- \\ r_{xx}[1,0] & r_{xx}[1,1] & r_{xx}[1,2] & \cdots & r_{xx}[1,p] \\ r_{xx}[2,0] & r_{xx}[2,1] & r_{xx}[2,2] & \cdots & r_{xx}[2,p] \\ \vdots & \vdots & \vdots & & \vdots \\ r_{xx}[p,0] & r_{xx}[p,1] & r_{xx}[p,2] & \cdots & r_{xx}[p,p] \end{bmatrix} \cdot \begin{bmatrix} 1 \\ -\!-\!- \\ a_p[1] \\ a_p[2] \\ \vdots \\ a_p[p] \end{bmatrix} = \begin{bmatrix} \varepsilon' \\ -\!-\!- \\ 0 \\ 0 \\ \vdots \\ 0 \end{bmatrix}. \tag{20.45}$$

This system of equations can be written in condensed form:

$$R_{xx} a_p^1 = \varepsilon' u_1. \tag{20.46}$$

The matrix R_{xx} is square, $(p+1)$ x $(p+1)$. a_p^1 is the vector a_p increased with 1 and u_1 is the vector $u_1 = [1, \ 0, \ 0, \ \ldots, \ 0]^T$. The resolution of the system (20.45) leads to the determination of the coefficients $a_p[k]$.

Note: The Prony's method thus formulated requires exact knowledge of the deterministic correlation matrix R_{xx}. To calculate it, we will need to have knowledge of signal values $x[n]$ on the entire time axis, $\{-\infty, \ +\infty\}$. The adjustments to be made to the practical method when one only knows the signal values $x[n]$ on a finite time interval will be discussed later.

20.3.2   Shanks Method

It differs from the Prony's method at the evaluation of coefficients $b_q[k]$ stage. Instead of forcing strict equality between the values of the signal $x[n]$ and of its estimator $\hat{x}[n]$ in the interval $\{0, q\}$, we look for the coefficients $b_q[k]$ by minimizing the squared error over the entire time axis (or on a portion of it). Thus, the signal values $x[n]$ for $n > q$ are included in the processing.

The problem is now reformulated. The error is given by:

$$e[n] = x[n] - \hat{x}[n], \tag{20.47}$$

with:

$$\hat{X}(z) = H(z) = \frac{B_q(z)}{A_p(z)} = B_q(z)\left\{\frac{1}{A_p(z)}\right\}. \tag{20.48}$$

The shape of the latter term emphasizes the fact that the filter transfer function $H(z)$ results from the cascading of two filters with transfer functions $\frac{1}{A_p(z)}$ and $B_q(z)$.

The time equation of the impulse response $g[n]$ of the first filter $1/A_p(z)$ has the form:

$$g[n] = \delta[n] - \sum_{k=1}^{p} a_p[k]g[n - k]. \tag{20.49}$$

It is an all-pole filter attacked by a unit pulse $\delta[n]$. During the cascade, $g[n]$ becomes the input signal of the second filter $B_q(z)$ whose time equation is:

$$\hat{x}[n] = \sum_{k=0}^{q} b_q[k]g[n - k]. \tag{20.50}$$

The estimation error is:

$$e[n] = x[n] - \hat{x}[n] = x[n] - \sum_{k=0}^{q} b_q[k]g[n-k]. \qquad (20.51)$$

We minimize the squared error over the entire time axis:

$$\varepsilon = \sum_{n=0}^{\infty} |e[n]|^2 = \sum_{n=0}^{\infty} |x[n] - \hat{x}[n]|^2. \qquad (20.52)$$

This error is minimum when the partial derivatives of ε with respect to parameters $b_q^*[k]$ are zero:

$$\frac{\partial \varepsilon}{\partial b_q^*[k]} = 0 \quad \text{for} \quad k = 0, 1, \ldots, q.$$

or

$$\frac{\partial \varepsilon}{\partial b_q^*[k]} = \frac{\partial}{\partial b_q^*[k]} \left\{ \sum_{n=0}^{\infty} e^*[n]e[n] \right\}$$

$$= -\sum_{n=0}^{\infty} g^*[n-k] \left\{ x[n] - \sum_{l=0}^{q} b_q[l]g[n-l] \right\} = 0,$$

or even:

$$\sum_{l=0}^{q} b_q[l] \sum_{n=0}^{\infty} g^*[n-k]g[n-l] = \sum_{n=0}^{\infty} g^*[n-k]x[n]; \quad k = 0, 1, \ldots, q. \qquad (20.53)$$

As was done in the Prony's method, the deterministic correlation function is defined by:

$$r_{gg}[k,l] = \sum_{n=0}^{\infty} g^*[n-k]\, g[n-l]. \qquad (20.54)$$

In addition, we note

$$r_{xg}[k] = \sum_{n=0}^{\infty} g^*[n-k]\, x[n]. \qquad (20.55)$$

Equation (20.53) can be rewritten as:

$$\sum_{l=0}^{q} b_q[l]\, r_{gg}[k,l] = r_{xg}[k]; \quad k = 0, 1, \ldots, q. \qquad (20.56)$$

The latter relationship is now written in matrix form:

$$\begin{bmatrix} r_{gg}[0,0] & r_{gg}[0,1] & r_{gg}[0,2] & \cdots & r_{gg}[0,q] \\ r_{gg}[1,0] & r_{gg}[1,1] & r_{gg}[1,2] & \cdots & r_{gg}[1,q] \\ r_{gg}[2,0] & r_{gg}[2,1] & r_{gg}[2,2] & \cdots & r_{gg}[2,q] \\ \vdots & \vdots & \vdots & & \vdots \\ r_{gg}[q,0] & r_{gg}[q,1] & r_{gg}[q,2] & \cdots & r_{gg}[q,q] \end{bmatrix} \begin{bmatrix} b_q[0] \\ b_q[1] \\ b_q[2] \\ \vdots \\ b_q[q] \end{bmatrix} = \begin{bmatrix} r_{xg}[0] \\ r_{xg}[1] \\ r_{xg}[2] \\ \vdots \\ r_{xg}[q] \end{bmatrix}. \quad (20.57)$$

This last equation can be written in simplified form using an induction on the coefficients $r_{gg}[k,l]$.

Indeed:

$$r_{gg}[k+1,l+1] = \sum_{n=0}^{\infty} g^*[n-[k+1]]\, g[n-[l+1]]$$

$$= \sum_{n=-1}^{\infty} g^*[n-k]\, g[n-l] = \sum_{n=0}^{\infty} g^*[n-k]\, g[n-l] + g^*[-1-k] g[-1-l].$$

As $k \geq 0$ and $l \geq 0$, the filter $1/A_p(z)$ is causal, the second term in the last sum is zero and we can write:

$$r_{gg}[k+1,l+1] = r_{gg}[k,l]. \quad (20.58)$$

This same reason leads to write for convenience:

$$r_{gg}[k,l] = r_{gg}[k-l].$$

We can rewrite Eq. (20.56) in the form:

$$\sum_{l=0}^{q} b_q[l]\, r_{gg}[k-l] = r_{xg}[k]; \quad k = 0, 1, \ldots, q. \quad (20.59)$$

This equation is written in matrix form:

$$\begin{bmatrix} r_{gg}[0] & r_{gg}^*[1] & r_{gg}^*[2] & \cdots & r_{gg}^*[q] \\ r_{gg}[1] & r_{gg}[0] & r_{gg}^*[1] & \cdots & r_{gg}^*[q-1] \\ r_{gg}[2] & r_{gg}[1] & r_{gg}[0] & \cdots & r_{gg}^*[q-2] \\ \vdots & \vdots & \vdots & & \vdots \\ r_{gg}[q] & r_{gg}[q-1] & r_{gg}[q-2] & \cdots & r_{gg}[0] \end{bmatrix} \begin{bmatrix} b_q[0] \\ b_q[1] \\ b_q[2] \\ \vdots \\ b_q[q] \end{bmatrix} = \begin{bmatrix} r_{xg}[0] \\ r_{xg}[1] \\ r_{xg}[2] \\ \vdots \\ r_{xg}[q] \end{bmatrix},$$

$$(20.60)$$

that one can still write in condensed form:

$$\boldsymbol{R}_{gg}\boldsymbol{b}_q = \boldsymbol{r}_{xg}. \tag{20.61}$$

The matrix $\boldsymbol{R}_{gg}$ is square, $(q+1) \times (q+1)$, which has the Hermitian symmetry. $\boldsymbol{r}_{xg}$ is a column vector. The squared error is then:

$$\varepsilon = \sum_{n=0}^{\infty} |e[n]|^2 = \sum_{n=0}^{\infty} |x[n] - \hat{x}[n]|^2 = \sum_{n=0}^{\infty} \left[x[n] - \sum_{k=0}^{q} b_q[k]g[n-k] \right]^* e[n],$$

$$\varepsilon = \sum_{n=0}^{\infty} x^*[n]e[n] - \sum_{k=0}^{q} b_q^*[k] \sum_{n=0}^{\infty} g^*[n-k]\, e[n].$$

The squared error is minimal when the error is orthogonal to functions $g^*[n-k]$. The last term is zero and we have for the least square error term:

$$\varepsilon_{LS} = \sum_{n=0}^{\infty} x^*[n]e[n] = \sum_{n=0}^{\infty} x^*[n]x[n] - \sum_{k=0}^{q} b_q[k] \sum_{n=0}^{\infty} g[n-k]\, x^*[n],$$

or:

$$\varepsilon_{LS} = r_{xx}[0] - \sum_{k=0}^{q} b_q[k] r_{xg}[k]. \tag{20.62}$$

20.4 All-pole Modeling in the Context of the Prony's Method

In some situations, the modeling of a signal by an all-pole model is sufficient. This is the case when the physical signal can be considered as the result of filtering of a simple signal by a bank of resonators. The modeling of the voice by this method is very efficient. Modeling by all poles within the Prony's method is a special case of this method wherein

$$b_q[n] = b_0\delta[n]. \tag{20.63}$$

The formula giving the error is now:

$$e'[n] = \begin{cases} x[n] + \sum\limits_{k=1}^{p} a_p[k]\, x[n-k] - b_0\delta[n] & \text{for} \quad n = 0 \\[2mm] x[n] + \sum\limits_{k=1}^{p} a_p[k]\, x[n-k] & \text{for} \quad n > 0 \end{cases}, \tag{20.64}$$

which incidentally is just $e'[n] = x[n] + \sum_{k=1}^{p} a_p[k]\, x[n-k] - b_0 \delta[n] \,\forall\, n.$

The formula (20.36) becomes:

$$r_{xx}[k,l] = \sum_{n=1}^{\infty} x^*[n-k]\, x[n-l] \quad \text{with} \quad k = 1, 2, \ldots, p;\ l = 1, 2, \ldots, p.$$

(20.65)

Noting that for $n = 0$ the values of $x^*[n-k]$ and $x[n-l]$ are zero because the signal is causal and $k = 1, 2, \ldots, p$; $l = 1, 2, \ldots, p$, we can start the summation index at $n = 0$. We can rewrite the last sum:

$$r_{xx}[k,l] = \sum_{n=0}^{\infty} x^*[n-k]\, x[n-l].$$

(20.66)

One recognizes the deterministic autocorrelation function of the signal $x[n]$. We still have to solve the system

$$\begin{bmatrix} r_{xx}[1,1] & r_{xx}[1,2] & r_{xx}[1,3] & \cdots & r_{xx}[1,p] \\ r_{xx}[2,1] & r_{xx}[2,2] & r_{xx}[2,3] & \cdots & r_{xx}[2,p] \\ r_{xx}[3,1] & r_{xx}[3,2] & r_{xx}[3,3] & \cdots & r_{xx}[3,p] \\ \vdots & \vdots & \vdots & & \vdots \\ r_{xx}[p,1] & r_{xx}[p,2] & r_{xx}[p,3] & \cdots & r_{xx}[p,p] \end{bmatrix} \cdot \begin{bmatrix} a_p[1] \\ a_p[2] \\ a_p[3] \\ \vdots \\ a_p[p] \end{bmatrix} = - \begin{bmatrix} r_{xx}[1,0] \\ r_{xx}[2,0] \\ r_{xx}[3,0] \\ \vdots \\ r_{xx}[p,0] \end{bmatrix}.$$

(20.67)

But this time, we can use the fact that $r_{xx}[1,2] = r_{xx}[2,3] = r_{xx}[3,4]$, that implies that one can go from a double index notation to one with a single index:

$$r_{xx}[k,l] = r_{xx}[l - k].$$

(20.68)

The autocorrelation matrix used in that system (20.67) is Hermitian and Toeplitz. We can rewrite the system in the form:

$$\begin{bmatrix} r_{xx}[0] & r_{xx}[1] & r_{xx}[2] & \cdots & r_{xx}[p-1] \\ r_{xx}[-1] & r_{xx}[0] & r_{xx}[1] & \cdots & r_{xx}[p-2] \\ r_{xx}[-2] & r_{xx}[-1] & r_{xx}[0] & \cdots & r_{xx}[p-3] \\ \vdots & \vdots & \vdots & & \vdots \\ r_{xx}[1-p] & r_{xx}[2-p] & r_{xx}[3-p] & \cdots & r_{xx}[0] \end{bmatrix} \cdot \begin{bmatrix} a_p[1] \\ a_p[2] \\ a_p[3] \\ \vdots \\ a_p[p] \end{bmatrix} = - \begin{bmatrix} r_{xx}[-1] \\ r_{xx}[-2] \\ r_{xx}[-3] \\ \vdots \\ r_{xx}[-p] \end{bmatrix}.$$

(20.69)

In the case of a real signal $x[n]$, the autocorrelation matrix is real, symmetric and the linear system of equations becomes:

$$
\begin{bmatrix}
r_{xx}[0] & r_{xx}[1] & r_{xx}[2] & \cdots & r_{xx}[p-1] \\
r_{xx}[1] & r_{xx}[0] & r_{xx}[1] & \cdots & r_{xx}[p-2] \\
r_{xx}[2] & r_{xx}[1] & r_{xx}[0] & \cdots & r_{xx}[p-3] \\
\vdots & \vdots & \vdots & & \vdots \\
r_{xx}[p-1] & r_{xx}[p-2] & r_{xx}[p-3] & \cdots & r_{xx}[0]
\end{bmatrix}
\cdot
\begin{bmatrix}
a_p[1] \\
a_p[2] \\
a_p[3] \\
\vdots \\
a_p[p]
\end{bmatrix}
= -
\begin{bmatrix}
r_{xx}[1] \\
r_{xx}[2] \\
r_{xx}[3] \\
\vdots \\
r_{xx}[p]
\end{bmatrix}.
$$

$$(20.70)$$

In the case where the autocorrelation matrix is invertible, we get directly the coefficients of the AR filter model. The estimator of the signal will then be

$$
\text{For } n = 0, \ b_0 = x[0]; \text{ for } n \neq 0, \ \hat{x}[n] = - \sum_{k=1}^{p} a_p[k]\,\hat{x}[n-k]. \tag{20.71}
$$

These formulas give the estimator of the signal $x[n]$ by the all-pole model.

20.5   All-pole Modeling in the Case of a Finite Number of Data

The all-pole method must be set in the case where the signal $x[n]$ is unknown outside of the interval $\{0, N\}$. The model is called Linear Predictive Coding (LPC).

Two methods known as the autocorrelation method and autocovariance method names are used.

20.5.1   Autocorrelation Method

As seen above, in the all-pole case the Prony's method defines the error to minimize as:

$$
\varepsilon = \sum_{n=0}^{\infty} |e[n]|^2 \text{ with } e[n] = x[n] + \sum_{k=1}^{p} a_p[k]x[n-k]. \tag{20.72}
$$

The autocorrelation method makes the additional assumption that the signal is zero outside the interval $\{0, N\}$.

The resolution begins by defining a new signal $\tilde{x}[n]$ by the product of $x[n]$ with a rectangular window,

$$\tilde{x}[n] = x[n]w_R[n], \tag{20.73}$$

with

$$w_R[n] = \begin{cases} 1; & n = 0, 1, \ldots, N \\ 0; & \text{elsewhere} \end{cases}. \tag{20.74}$$

Correlations are now:

$$\begin{aligned} r_{xx}[k] &= \sum_{n=0}^{\infty} \tilde{x}^*[n-k]\tilde{x}[n] \\ &= \sum_{n=k}^{N} x^*[n-k]x[n]; \quad k = 0, 1, 2, .., p. \end{aligned} \tag{20.75}$$

The normal equations become:

$$\sum_{l=1}^{p} a_p[l]r_{xx}[k-l] = -r_{xx}[k]; \quad k = 1, 2, .., p. \tag{20.76}$$

The minimum square error is:

$$\varepsilon_{LS} = r_{xx}[0] + \sum_{k=1}^{p} a_p[k]r_{xx}^*[k]. \tag{20.77}$$

We note that there is contradiction between the application of an all-pole model with a finite number of poles, whose impulse response is of infinite duration (thus has nonzero values outside the range $\{0, N\}$), to model a signal that is zero outside the data range. This contradiction leads to a low quality of modeling.

20.5.2 Covariance Method

Rather than assigning zero values to the signal outside the range $\{0, N\}$, it is preferable in this method only evaluate the error on the data window. In addition, for greater generality the signal is not assumed causal. To avoid the transient effect created by the shutting in $n = 0$, we begin to take into account the error starting at index $n = p$, for which the correlation matrix does not include null values in the beginning.

The following squared error is minimized:

$$\varepsilon = \sum_{n=p}^{N} |e[n]|^2. \tag{20.78}$$

The system of equations to be solved is the same as in the Prony's method:

$$\begin{bmatrix} r_{xx}[1,1] & r_{xx}[1,2] & r_{xx}[1,3] & \cdots & r_{xx}[1,p] \\ r_{xx}[2,1] & r_{xx}[2,2] & r_{xx}[2,3] & \cdots & r_{xx}[2,p] \\ r_{xx}[3,1] & r_{xx}[3,2] & r_{xx}[3,3] & \cdots & r_{xx}[3,p] \\ \vdots & \vdots & \vdots & & \vdots \\ r_{xx}[p,1] & r_{xx}[p,2] & r_{xx}[p,3] & \cdots & r_{xx}[p,p] \end{bmatrix} \cdot \begin{bmatrix} a_p[1] \\ a_p[2] \\ a_p[3] \\ \vdots \\ a_p[p] \end{bmatrix} = - \begin{bmatrix} r_{xx}[1,0] \\ r_{xx}[2,0] \\ r_{xx}[3,0] \\ \vdots \\ r_{xx}[p,0] \end{bmatrix}, \tag{20.79}$$

but the values of the correlation coefficients are

$$r_{xx}[k,l] = \sum_{n=p}^{N} x^*[n-k]\, x[n-l]. \tag{20.80}$$

The correlation matrix is no longer symmetric nor Toeplitz.
The minimum square error is:

$$\varepsilon_{LS} = r_{xx}[0,0] + \sum_{k=1}^{p} a_p[k] r_{xx}[0,k]. \tag{20.81}$$

This method is considered to give better results than the correlation method in the case of an all-pole filter because it does not require signal zero values as does the autocorrelation method.

The all-pole modeling gives satisfactory results in the analysis of the voice. Figure 20.1 shows the spectrum of the French vowel a and the spectrum modeled by LPC with 32 coefficients $a_{32}[k]$. The agreement is satisfactory. The coefficients can be used back to synthesize the vowel a.

20.6 Adaptive Filter

This paragraph is a little apart in this chapter. It outlines a method for modeling a nonstationary signal by a finite impulse response filter. The filter coefficients are changed over time so as to adapt to slow changes in the parameters characterizing the signal. This technique, due to Widrow (1975), provides a very powerful noise removal algorithm when the useful signal is vitiated by an additive noise. It is, inter alia, the origin of the noise canceling techniques. An FIR finite impulse response

Fig. 20.1 Spectrum of
French vowel *a* and the
spectrum modeled by LPC

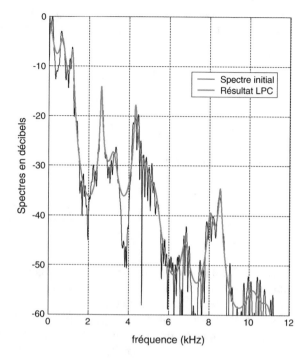

filter is used with a record entry for signal near the noise source. The impulse
response of the filter adjusts itself automatically. The output of this filter is used to
reduce unwanted noise from reception.

This method has its origin in the search of a way to remove the echo that appears
on a duplex phone line. To allow the passage of information in both directions on
the same line, a transformer is present at each end of the line. The impedance
mismatch at both ends of the line then causes reflections at these ends. This is
manifested for an interlocutor for an echo of his own voice. This echo is particularly
annoying if it occurs with a delay larger than a few tenths of seconds after emission.
Signal transmission taking place at speeds of the order of the speed of light, the
echo is annoying in the case of long distance propagation (for example
intercontinental).

For this presentation, although the processing is performed digitally, we adopt an
analog signal notation which allows a more intuitive description. Let us consider
the case of the problem of false echo cancelation. The useful signal is noted $x(t)$. At
this signal is superimposed additively a spurious signal $z(t)$ that is a delayed version
of $x(t)$. Typically, this signal $z(t)$ is not a simple delayed reproduction of the signal
$x(t)$ but it has undergone unknown transformations that we model by the action of
an unknown filter with impulse response $h(t)$.

We write then: $z(t) = x(t) \otimes h(t)$. $z(t)$ is the annoying echo that we try to delete
(Fig. 20.2).

Fig. 20.2 Superposition of a signal $x(t)$ and one echo

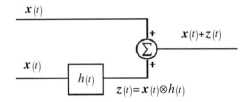

Fig. 20.3 Error signal, difference of one echo and its estimation

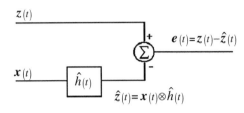

To eliminate $z(t)$, we seek to create a filter whose impulse response $\hat{h}(t)$ tends to estimate the exact impulse response $h(t)$. We subtract its response $\hat{z}(t)$ from $z(t)$, the ideal is to achieve a zero subtraction result.

The error signal is noted $e(t)$: $e(t) = z(t) - \hat{z}(t)$. (see Fig. 20.3).

We chose to model the action of the filters with digital filters with finite impulse response (FIR) with N elements. The convolutions then take a discrete form:

$$z(t) = z[n] = \sum_{i=0}^{N-1} h[i]x[n-i]. \tag{20.82}$$

Vector Analogy

$z(t)$ can be interpreted as the result of the dot product of two vectors: the vectors x and h defined by:

$$x = \begin{pmatrix} x[n] \\ x[n-1] \\ x[n-2] \\ \ldots \\ x[n-N+1] \end{pmatrix} \quad \text{and} \quad h = \begin{pmatrix} h[0] \\ h[1] \\ h[2] \\ \ldots \\ h[N-1] \end{pmatrix}. \tag{20.83}$$

The convolution appears as the inner product of these vectors. Note the time reversal in x.

We use Sondhi notation (Sondhi and Berkley 1980) to signify the transposition of a vector.

We can write the spurious signal z as the dot product $z = h'x$. Similarly $\hat{z} = \hat{h}'x$. With these notations, the error $e = z - \hat{z} = \left(h - \hat{h}\right)'x$ is a scalar.

The principle of adaptive estimator filter is to vary $\hat{h}$ over time to reduce the error. We will show below that the equation for determining the filter $\hat{h}$ is:

$$\frac{d\hat{h}}{dt} = Ke(t)x(t), \tag{20.84}$$

or

$$\frac{d\hat{h}}{dt} = KF(e)x(t). \tag{20.85}$$

K is a positive constant to adjust in practice and $F(e)$ is a non-increasing function of the error e.

It is emphasized here that the estimator is constructed from $x(t)$ which is not marred by the noise. It will be necessary in practice to have a sensor providing the pure signal $x(t)$.

We now show using the vector analogy why Eq. (20.84) is favorable (Fig. 20.4). Misalignment vector is noted r:

$$r = h - \hat{h}. \tag{20.86}$$

We see that $e(t) = r'x$.

The filter with impulse response h is constant during the filter adaptation time, we have $\frac{dh}{dt} = 0$. So $\frac{dr}{dt} = -\frac{d\hat{h}}{dt}$.

Assuming the equation of adaptive filter (20.84) we have

$$\frac{dr}{dt} = -K(r'x)x. \tag{20.87}$$

Fig. 20.4 Misalignment vector r and its projection on signal

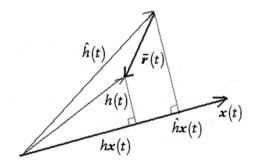

In this case the error norm tends to zero as time increases. To show this, we multiply the two members of the previous equation by r'. We have

$$r'\frac{dr}{dt} = -Ke^2, \quad \text{or} \quad r'\frac{dr}{dt} = \frac{1}{2}\frac{d}{dt}\|r\|^2 = -Ke^2. \tag{20.88}$$

The length of the misalignment vector r thus decreases. This is true for any $K > 0$. So:

$$\frac{d}{dt}(r'r) = -2Ke^2. \tag{20.89}$$

We integrate this relationship between 0 and any subsequent time τ:

$$rr'|_{t=0} - r'r\big|_{t=\tau} = 2K\int_0^\tau e^2 d\tau. \tag{20.90}$$

As $r'r$ is not growing, the left term is bounded. It will necessarily be the same for the right-hand side. So when $\tau \to \infty$ the standard error must approach zero ($|e| \to 0$) to ensure the convergence of the integral.

The validity of Eq. (20.84) has been shown to determine the filter $\hat{h}$.

Use of the form (20.85) **of the filter**: The general expression (20.85) of the adaptive filter with a non-increasing function $F(e)$ of the error e.

$$\frac{d\hat{h}}{dt} = KF(e)x.$$

Suppose that there exists a function $C(e)$ such that $F(e) = \frac{dC}{de}$.
As $\frac{de}{d\hat{h}} = -x$, we have

$$\nabla_{\hat{h}}C = \frac{dC}{de}\frac{de}{d\hat{h}} = -\frac{dC}{de}x = -F(e)x = \frac{d\hat{h}}{dt}. \tag{20.91}$$

Then $\frac{d\hat{h}}{dt}$ is chosen as a vector pointing in the direction of the gradient $C(e)$, where the decrease is the fastest. This is what gives the gradient method name to this technique.

Practical determination of $\hat{h}[n]$ in digital form:

The derivative $\frac{d\hat{h}}{dt}$ is replaced by the difference $\hat{h}[n+1,i] - \hat{h}[n,i]$ and then we have the equation of the filter:

$$\hat{h}[n+1,i] = \hat{h}[n,i] + Ke[n]x[n-1] \pm \text{ leakage term}. \tag{20.92}$$

We add a leakage term so that in case of misalignment, the vector $\hat{h}$ does not remain aligned in a wrong direction.

How do we choose K? We should not take it too big because the misalignment vector would land in a plane perpendicular to x and can no longer move although h and $\hat{h}$ had quite different directions. If the vector x moves slightly, the error will be important.

Summary

We have treated in this chapter the modeling of a digital causal signal in the time domain, i.e. we searched a finite number of coefficients as small as possible, which bring the possibility to reconstruct the signal using these coefficients. The signal is modeled here as the impulse response of a LTI ARMA system using a least squares approach. After having shown in the frequency domain that the general equations of the model are nonlinear making the resolution very difficult, we used less ambitious approaches. We developed the Padé representation of a signal, which is exact on a number of points equal to the number of coefficients chosen for the model, but whose estimate of the signal outside this range is very poor. The search of coefficients reduces to solving a linear system of equations whose solutions were sought by matrix methods.

In Prony's method, the least square method is performed on subsets of the time axis. The Shanks method is a refinement of that method. All-pole modeling (AR) with the Prony's method gives very good results in speech processing (it is known under the name LPC Linear Predictive Coding). We have exposed correlation and covariance methods which are used for time-limited signals. Adaptive filtering has been developed for nonstationary signals analysis. The filter coefficients are reassessed as the signal is changing. This type of filtering is efficient to block out known spurious signals, such as echoes on a transmission line.

Exercises
Calculating an inverse filter:

Let the digital filter defined by the recurrence relation: $g[n] = f[n] - 0.9f[n-1]$.

1. What is the impulse response $h_1[n]$ of this filter? Represent this function.
2. Give the expression of its transfer function $H_1(z)$. Deduce the frequency response. Knowing that the sampling rate is $f_e = 20\text{kHz}$ and is using a geometric argument, trace the evolution with frequency of the frequency response magnitude $|H_1(e^{j\omega T})|$. What is the character of the filter?
3. Let the causal filter whose transfer function is $H_2(z)$ be the inverse of $H_1(z)$. $(H_2(z)H_1(z) = 1)$. Trace the evolution with the frequency of the frequency response magnitude $|H_2(e^{j\omega T})|$. What is the character of the filter?
4. Give the expression of the impulse response $h_2[n]$ of the inverse filter $H_2(z)$. Calculate the convolution product $h_1[n] \otimes h_2[n]$. Explain simply the result.
5. We want to simulate the inverse filter with a finite impulse response filter, causal, whose impulse response $h[n]$ is limited to $N+1$ terms.

The input signal of this filter is $h_1[n]$. The output signal of the FIR is denoted $\hat{\delta}[n]$. It is an estimator of the pulse unit $\delta[n]$: $\hat{\delta}[n] = \sum_{k=0}^{N} b[k] h_1[n-k]$.
The error at instant n is denoted $e[n] = \delta[n] - \hat{\delta}[n]$.
The filter coefficients are obtained by a least squares method in which one seeks to minimize the squared error

$$\varepsilon = \sum_{n=0}^{\infty} |e[n]|^2 = \sum_{n=0}^{\infty} \left(\delta[n] - \hat{\delta}[n] \right)^2.$$

The solution of this problem is analogous to the method of Shanks.
Demonstrate formula to reach the filter coefficients $b[k]$.
6. Calculate the matrices of the linear system used in solving the case $N+1 = 6$.
7. The system solution provides a coefficients vector:
   B = [0.9304, 0.7601, 0. 5982, 0.4429, 0.2926, 0.1455]'

   Calculate the estimator $\hat{\delta}[n]$ and the quadratic error in the case of Question 6.
8. It is found that the error proves less, for the same number $N+1$ of terms, when creating an inverse filter whose output is a delayed pulse $\delta[n-n_0]$ instead of $\delta[n]$. What is the new expression of the linear system used in solving the problem? (N.A. $n_0 = 5$)

Solution:

1. $h_1[n] = \delta[n] - 0.9\delta[n-1]$.
2. $H_1(z) = 1 - 0.9z^{-1}$. $H_1(e^{j\omega T}) = 1 - 0.9e^{-j\omega T}$. It is a high-pass filter.
3. $H_2(z) = \frac{1}{H_1(z)} = \frac{1}{1-0.9z^{-1}}$. $H_2(e^{j\omega T}) = \frac{1}{1-0.9e^{-j\omega T}}$. It is a low-pass filter.
4. We assume a causal filter. $h_2[n] = 0.9^n U[n]$.
5. This error is minimum when the partial derivatives of ε_{MC} with respect to the parameters $b^*[k]$ are zero:

$$\frac{\partial \varepsilon}{\partial b^*[k]} = 0 \quad \text{for } k = 0, 1, \ldots, N.$$

Namely:

$$\frac{\partial \varepsilon}{\partial b^*[k]} = \frac{\partial}{\partial b^*[k]} \left\{ \sum_{n=0}^{\infty} e^*[n] e[n] \right\} = -\sum_{n=0}^{\infty} h_1[n-k] \left\{ \delta[n] - \sum_{l=0}^{N} b[l] h_1[n-l] \right\}$$
$$= 0,$$

or also:

$$\sum_{l=0}^{N} b[l] \sum_{n=0}^{\infty} h_1[n-k] h_1[n-l] = \sum_{n=0}^{\infty} h_1[n-k] \delta[n]; \quad k = 0, 1, \ldots, N.$$

As was done in the Prony's method, we define the deterministic correlation function: $r_{h_1h_1}[k, l] = \sum\limits_{n=0}^{\infty} h_1[n - k] h_1[n - l]$. Furthermore, we define $r_{\delta h_1}[k] = \sum\limits_{n=0}^{\infty} h_1[n - k] \delta[n]$.

The equation can be rewritten as:

$$\sum_{l=0}^{N} b[l]\, r_{h_1h_1}[k, l] = r_{\delta h_1}[k]; \ k = 0, 1, \ldots, N. \tag{20.93}$$

This relationship can be written in matrix form:

$$\begin{bmatrix} r_{h_1h_1}[0,0] & r_{h_1h_1}[0,1] & r_{h_1h_1}[0,2] & \cdots & r_{h_1h_1}[0,N] \\ r_{h_1h_1}[1,0] & r_{h_1h_1}[1,1] & r_{h_1h_1}[1,2] & \cdots & r_{h_1h_1}[1,N] \\ r_{h_1h_1}[2,0] & r_{h_1h_1}[2,1] & r_{h_1h_1}[2,2] & \cdots & r_{h_1h_1}[2,N] \\ \vdots & \vdots & \vdots & & \vdots \\ r_{h_1h_1}[N,0] & r_{h_1h_1}[N,1] & r_{h_1h_1}[N,2] & \cdots & r_{h_1h_1}[N,N] \end{bmatrix} \cdot \begin{bmatrix} b[0] \\ b[1] \\ b[2] \\ \vdots \\ b[N] \end{bmatrix} = \begin{bmatrix} r_{\delta h_1}[0] \\ r_{\delta h_1}[1] \\ r_{\delta h_1}[2] \\ \vdots \\ r_{\delta h_1}[N] \end{bmatrix}.$$

This last equation can be written in simplified form using an induction on the coefficients $r_{h_1h_1}[k, l]$. Indeed:

$$r_{h_1h_1}[k + 1, l + 1] = \sum_{n=0}^{\infty} h_1[n - [k + 1]] h_1[n - [l + 1]]$$

$$= \sum_{n=-1}^{\infty} h_1[n - k] h_1[n - l] = \sum_{n=0}^{\infty} h_1[n - k] h_1[n - l] + h_1[-1 - k] h_1[-1 - l].$$

As $k \geq 0$ and $l \geq 0$, and the filter causal, the second term is zero and we can write:

$$r_{h_1h_1}[k + 1, l + 1] = r_{h_1h_1}[k, l].$$

For this same reason we write for convenience:

$$r_{h_1h_1}[k, l] = r_{h_1h_1}[k - l].$$

We can rewrite Eq. (20.93) under the form:

$$\sum_{l=0}^{N} b[l]\, r_{h_1h_1}[k - l] = r_{\delta h_1}[k]; \quad k = 0, 1, \ldots, N. \tag{20.94}$$

The equation is then written:

$$
\begin{bmatrix}
r_{h_1 h_1}[0] & r_{h_1 h_1}[1] & r_{h_1 h_1}[2] & \cdots & r_{h_1 h_1}[N] \\
r_{h_1 h_1}[1] & r_{h_1 h_1}[0] & r_{h_1 h_1}[1] & \cdots & r_{h_1 h_1}[N-1] \\
r_{h_1 h_1}[2] & r_{h_1 h_1}[1] & r_{h_1 h_1}[0] & \cdots & r_{h_1 h_1}[N-2] \\
\vdots & \vdots & \vdots & & \vdots \\
r_{h_1 h_1}[N] & r_{h_1 h_1}[N-1] & r_{h_1 h_1}[N-2] & \cdots & r_{h_1 h_1}[0]
\end{bmatrix}
\cdot
\begin{bmatrix}
b[0] \\ b[1] \\ b[2] \\ \vdots \\ b[N]
\end{bmatrix}
=
\begin{bmatrix}
r_{\delta h_1}[0] \\ r_{\delta h_1}[1] \\ r_{\delta h_1}[2] \\ \vdots \\ r_{\delta h_1}[N]
\end{bmatrix}.
$$

$r_{h_1 h_1}[n] = h_1[n] \otimes h_1[-n]$.

$h_1[n] = \delta[n] - 0.9\delta[n-1]$.

$r_{h_1 h_1}[0] = 1 + 0.9 \times 0.9 = 1.81$, $r_{h_1 h_1}[1] = -0.9$, $r_{h_1 h_1}[n] = 0$ elsewhere,

$r_{\delta h_1}[0] = 1$, $r_{\delta h_1}[k] = 0$ if $k \neq 0$.

6. Calculation of the matrices used in the linear system resolution when $N + 1 = 6$. The system is written:

$$
\begin{bmatrix}
1.81 & -0.9 & 0 & \cdots & 0 \\
-0.9 & 1.81 & -0.9 & \cdots & 0 \\
0 & -0.9 & 1.81 & \cdots & 0 \\
\vdots & \vdots & \vdots & & \vdots \\
0 & 0 & 0 & \cdots & 1.81
\end{bmatrix}
\cdot
\begin{bmatrix}
b[0] \\ b[1] \\ b[2] \\ \vdots \\ b[5]
\end{bmatrix}
=
\begin{bmatrix}
1 \\ 0 \\ 0 \\ \vdots \\ 0
\end{bmatrix}.
$$

With Matlab we find:

$$
B = (0.9304 \quad 0.7601 \quad 0.5982 \quad 0.4429 \quad 0.2926 \quad 0.1455)^T
$$

7. Vector B components are the values of the estimator of the impulse response of the inverse system:

$$
\hat{h}_2[n] = \{0.9304, \ 0.7601, \ 0.5982, \ 0.4429, \ 0.2926, \ 0.1455, \ 0, \ 0, \ 0, \ 0\}, \ \text{for } n
$$
$$
= 0, 1, \ldots
$$

True inverse filter impulse response is $h_2[n] = 0.9^n U[n]$, being for first values of n: 1, 0.9, 0.81, 0.73, 0.66, 0.59.

The convolution of the vector B with the vector $h_1 = 1, -0.9, 0, 0$ gives

$$
(0.9304, \ -0.0773, \ -0.0859, \ -0.0954, \ -0.1060, \ -0.1178, \ -0.1309, \ 0, \ 0, \ 0, \ 0 \ldots)^T
$$

The result is satisfactory.

References

Widrow: Adaptive noise canceling. Principles and applications. In: Proceedings of the IEEE (1975)

Sondhi, M.M., Berkley, D.A.: Silencing echoes on the telephone network. In: Proceedings of the IEEE, vol. 68, No 8 (1980)

Chapter 21
Random Signals: Statistics Basis

This chapter recalls the basis of the statistics for a random variable with values in a continuum. We define the probability density function and the cumulative distribution function of a r.v.: expectancy, variance, the moments of a distribution, and the characteristic function. Particular attention is paid to the Gaussian distribution (normal distribution). The probability density function of a function of a random variable is determined.

In the second part of this chapter, we present the statistics of two random variables (also called second-order statistics). We define their joint probability density and the marginal probability densities. We give an overview of the Bayesian statistical aspect. We define the correlation coefficient, the orthogonality and independence in probability. These concepts are used for the study of two jointly Gaussian variables. It is then shown that the probability density function of the sum of two independent r.v. is the convolution of their probability densities. This result is extended qualitatively to the sum of a large number of independent random variables that appear to follow approximately a Gaussian distribution. This result is known as the central limit theorem. We finally consider the statistical distribution of complex variables and the correlation of two complex r.v.

A table of Gauss's law is given at the end of the chapter.

A random signal is a function of time for which the value at a given time is not certain. It depends on the value of a random event ξ. In general, ξ is not identified.

We denote $x(t, \xi)$ the signal or $x(t)$ too. The use of bold letter emphasizes the randomness of the variable. Thus, at a given time, $x(t)$ may generally take infinite possible values dependent on ξ. To follow the successive values of $x(t)$ over time means following a realization of $x(t)$.

Figure 21.1 shows three realizations of a random signal $x(t)$.

At a given time t, $x(t)$ is a random variable. To study this random variable (r.v.), we can use the classical statistical concepts: probability density function $f_{x(t)}(x)$, expectancy, variance, etc.

© Springer International Publishing Switzerland 2016
F. Cohen Tenoudji, *Analog and Digital Signal Analysis*,
Modern Acoustics and Signal Processing, DOI 10.1007/978-3-319-42382-1_21

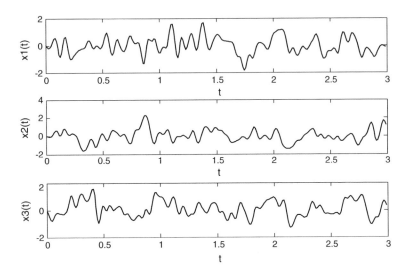

Fig. 21.1 Example of 3 realizations of $x(t)$

We will mainly studyin the following the case where $x(t)$ values belong to a continuum in $\mathbb{R}$ or $\mathbb{C}$. We only briefly talk about the case of a discrete distribution of values of $x(t)$, when we discuss the quantization problem of error made when converting an analog signal to digital using an analog/digital converter.

At every moment t, the probability density function of the variable $x(t)$ provides the statistical properties of this variable. However, this function, defined for each value of t, is insufficient when one seeks to describe the time evolution of $x(t)$. Probability densities should be used for this purpose involving several values of time:

$f_{x(t_1)x(t_2)x(t_3)...}(x(t_1), x(t_2), x(t_3), ...)$ and whatever these times. Most often in practice, these functions are not known. We then simply use parameters such as the average or the moments, and the expectancies of the products of signal values at different times.

21.1 First-Order Statistics

21.1.1 Case of a Real Random Variable

Probability density function of a real continuous random variable:

We assume that the values of $x(t)$ are real. The function $f_{x(t)}(x)$ defined in the following is called the probability density function (pdf) of this continuous random variable. To simplify the notation, we simply write x the variable $x(t)$ at time t. The probability density function will be noted $f_x(x)$ below.

By definition, $f_x(x)dx$ is the probability that $x(t)$ has values in the infinitely small interval $\{x, x + dx\}$ (Fig. 21.2):

$$P\{x < x(t) \leq x + dx\} = f_x(x)dx. \tag{21.1}$$

As a probability is positive or zero, the probability density function $f_x(x)$ is necessarily positive or zero.

On a finite length interval, the probability for x to realize in the interval $\{x_1, x_2\}$ is: (Fig. 21.3)

$$P\{x_1 < x \leq x_2\} = \int_{x_1}^{x_2} f_x(x)dx. \tag{21.2}$$

Normalization condition: Since the probability of getting x over the entire range of real numbers is equal to 1, we can write:

$$\int_{-\infty}^{+\infty} f_x(x)\,dx = 1. \tag{21.3}$$

Cumulative distribution function:

The cumulative distribution function $F_x(x)$ is the probability that the random variable x takes any value less than a value x. By definition (Fig. 21.4)

$$F_x(x) = P\{x \leq x\} = \int_{-\infty}^{x} f_x(x)\,dx. \tag{21.4}$$

Fig. 21.2 An example of a probability density function $f_x(x)$

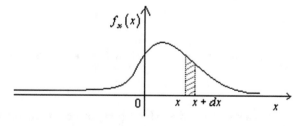

Fig. 21.3 The *grayed* area is the probability that x is found in the interval $\{x_1, x_2\}$

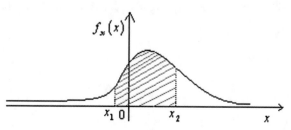

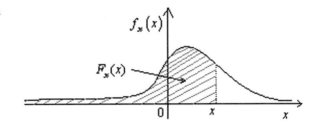

Fig. 21.4 The *grayed* area is
the cumulative distribution
function $F_x(x)$

Obviously we have:

$$F_x(\infty) = 1. \tag{21.5}$$

Expected value of the random variable x:
It is defined by:

$$E\{x\} = \int_{-\infty}^{+\infty} x f_x(x) dx = \eta_x. \tag{21.6}$$

The expected value η_x is also called expectation, average, mean or, better, ensemble average (to better distinguish it from a time average which will be studied in the following).

In the general case, the expectation of a variable in a random process is time dependent.

Variance: It is the mean (the ensemble average) of the squared deviation of the random variable from its mean value.

It is defined by

$$v_x = E\left\{(x - \eta_x)^2\right\} = \int_{-\infty}^{+\infty} (x - \eta_x)^2 f_x(x) \, dx. \tag{21.7}$$

The standard deviation σ_x is defined as the square root of the variance:

$$\sigma_x = \sqrt{v_x}. \tag{21.8}$$

Property: An important result in practice is that the variance of a r.v. is equal to the square expectation minus the squared mean of the r.v.:

Indeed

$$
v_x = \int_{-\infty}^{+\infty} (x - \eta_x)^2 f_x(x) \, dx = \int_{-\infty}^{+\infty} (x^2 - 2x\eta_x + \eta_x^2) f_x(x) \, dx,
$$

$$
v_x = \int_{-\infty}^{+\infty} x^2 f_x(x) \, dx - 2\eta_x \int_{-\infty}^{+\infty} x f_x(x) \, dx + \eta_x^2 \int_{-\infty}^{+\infty} f_x(x) \, dx.
$$

(21.9)

Therefore

$$
v_x = E\{x^2\} - (E\{x\})^2.
$$

(21.10)

Skewness and kurtosis:
First of all we define the moment of order n of a distribution

$$
\mu_n = E\{(x - \eta_x)^n\}.
$$

(21.11)

We call skewness (asymmetry) the third-order moment normalized by the division by the standard deviation raised to the third power:

$$
m_3 = \frac{E\left\{(x - \eta_x)^3\right\}}{\sigma_x^3}.
$$

(21.12)

The kurtosis is the fourth normalized moment

$$
m_4 = \frac{E\left\{(x(t) - \eta_x)^4\right\}}{\sigma_x^4}.
$$

(21.13)

Chebyshev inequality:
Let x be a real r.v. Its variance is given by

$$
\sigma^2 = \int_{-\infty}^{\infty} (x - \eta)^2 f(x) \, dx = \int_{-\infty}^{\infty} x_c^2 f(x_c) \, dx_c.
$$

We used the centered random variable $x_c = x - \eta_x$.
Let a be any positive constant. We have

$$
\sigma^2 = \int_{-\infty}^{\infty} x_c^2 f(x_c) \, dx_c = \int_{-\infty}^{-a} x_c^2 f(x_c) \, dx_c + \int_{-a}^{a} x_c^2 f(x_c) \, dx_c + \int_{a}^{\infty} x_c^2 f(x_c) \, dx_c.
$$

In the first and the third term $x_c^2 > a^2$.

We then have the inequality: $\sigma^2 \geq a^2 \int_{-\infty}^{-a} f(x_c)\,dx_c + \int_{-a}^{a} x_c^2 f(x_c)\,dx_c + a^2 \int_a^{\infty} f(x_c)\,dx_c$.

Alternatively, since the second term is positive:

$$\sigma^2 \geq a^2 \int\limits_{-\infty}^{-a} f(x_c)\,dx_c + a^2 \int\limits_a^{\infty} f(_c x)\,d_c x.$$

So finally we have the Chebyshev inequality

$$\Pr[|x - \eta| \geq a] \leq \frac{\sigma^2}{a^2}. \qquad (21.14)$$

Exercise

Let x be a real random variable uniformly distributed in the interval $\left\{-\frac{a}{2}, \frac{a}{2}\right\}$.

By definition, the probability density function $f_x(x)$ is constant between and $\left\{-\frac{a}{2}, \frac{a}{2}\right\}$ and zero elsewhere: (See Fig. 21.5).

$$f_x(x) = \begin{vmatrix} C & \text{if } |x| \leq \frac{a}{2} \\ 0 & \text{elsewhere} \end{vmatrix}. \qquad (21.15)$$

The normalization condition $\int_{-\infty}^{+\infty} f_x(x)\,dx = 1$ leads to find

$$C = \frac{1}{a}. \qquad (21.16)$$

The expectation of x is

$$E\{x\} = \eta_x = \int\limits_{-\infty}^{+\infty} x f_x(x)\,dx = \frac{1}{a} \int\limits_{-\frac{a}{2}}^{\frac{a}{2}} x\,dx = 0. \qquad (21.17)$$

Fig. 21.5 Uniform probability density function in the interval $\left\{-\frac{a}{2}, \frac{a}{2}\right\}$

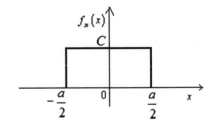

The variable x is centered; its variance is

$$v_x = \int_{-\infty}^{+\infty} (x - \eta_x)^2 f_x(x)\, dx = \int_{-\infty}^{+\infty} x^2 f_x(x)\, dx = \frac{1}{a} \int_{-\frac{a}{2}}^{\frac{a}{2}} x^2\, dx = \frac{1}{a} \frac{x^3}{3}\Big|_{-\frac{a}{2}}^{\frac{a}{2}} = \frac{a^2}{12}.$$

$$(21.18)$$

21.1.2   Gaussian Distribution (Normal Law)

We say that $x = x(t)$ is a Gaussian variable when its probability density function is given by (Fig. 21.6)

$$f_x(x) = \frac{1}{\sqrt{2\pi\sigma^2}} e^{-\frac{(x-\eta)^2}{2\sigma^2}}. \tag{21.19}$$

We will show by calculating the Gaussian integrals detailed in the following that the distribution parameters appearing in the probability density function are the mean and the variance: $E\{x\} = \eta$ and $v(x) = \sigma^2$.

It is also said that $x(t)$ follows a normal distribution $N(\eta, \sigma)$. This distribution has great theoretical and practical interests. It depends only on two parameters η and σ.

As shown in (21.19), the probability density function is maximum when the x value equals the expectation η. It is symmetrical with respect to η. One can see that if the variable x differs from the average value by some σ, the exponential becomes small. Thus, σ characterizes the distribution spreading. A more quantitative property of σ will be given subsequently.

The method of calculating a probability in the case of a Gaussian distribution is now studied on a numerical application.

We assume that, due to the thermal noise, the voltage across a resistor at time t is a Gaussian random variable with mean value $\eta = 1.1$ mV and standard deviation 2.3 mV.

Fig. 21.6 Gaussian probability density function

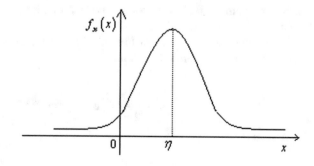

Fig. 21.7 The *shaded* area is the probability for x to realize in the interval $\{x_1, x_2\}$

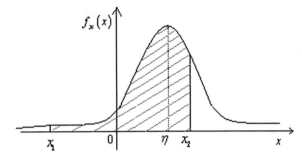

We want to evaluate the probability of measuring a voltage between $-2\,\text{mV}$ and $2\,\text{mV}$. As seen above, the probability that the Gaussian r.v. x is realized in the range $\{x_1, x_2\}$ is given in (Fig. 21.7)

$$P = \{x_1 < x \le x_2\} = \frac{1}{\sqrt{2\pi\sigma^2}} \int_{x_1}^{x_2} e^{-\frac{(x-\eta)^2}{2\sigma^2}} \, dx. \tag{21.20}$$

The integral of the Gaussian exponential is not a simple function; the result cannot be obtained directly.

The following change of variable is generally carried out:

$$z = \frac{x - \eta}{\sigma}. \tag{21.21}$$

Then: $dx = \sigma \, dz$,

$$P\{x_1 < x \le x_2\} = \frac{1}{\sqrt{2\pi}} \int_{z_1}^{z_2} e^{-\frac{z^2}{2}} dz = P\{z_1 < z \le z_2\} = \int_{z_1}^{z_2} f_z(z) \, dz, \tag{21.22}$$

with

$$f_z(z) = \frac{1}{\sqrt{2\pi}} e^{-\frac{z^2}{2}}. \tag{21.23}$$

z is a random variable following a normal distribution $N(0, 1)$, centered with a standard deviation equal to 1.

The cumulative distribution function $F(z)$ of the centered r.v. z is connected to the commonly used error function.

$$F(z_1) = P\{z < z_1\} = \frac{1}{\sqrt{2\pi}} \int_{-\infty}^{z_1} e^{-\frac{z^2}{2}} dz = 0.5 + \frac{1}{\sqrt{2\pi}} \int_{0}^{z_1} e^{-\frac{z^2}{2}} dz. \tag{21.24}$$

The error function is defined by

$$\text{erf}(t_1) = \frac{2}{\sqrt{\pi}} \int_0^{t_1} e^{-t^2} dt. \tag{21.25}$$

The integral on z in (21.24) is connected to the error function $\text{erf}(z)$ by the change of variable $t = \frac{z}{\sqrt{2}}$. So

$$\frac{1}{\sqrt{2\pi}} \int_0^{z_1} e^{-\frac{z^2}{2}} dz = \frac{1}{\sqrt{\pi}} \int_0^{t_1} e^{-t^2} dt, \quad \text{with } t_1 = \frac{z_1}{\sqrt{2}}. \tag{21.26}$$

$$F(z_1) = 0.5 \left(1 + \frac{2}{\sqrt{\pi}} \int_0^{t_1} e^{-t_1^2} dt \right) = 0.5 \left(1 + \text{erf}\left(\frac{z_1}{\sqrt{2}}\right) \right). \tag{21.27}$$

This relationship is used to calculate, using a spreadsheet, the integral values of the reduced centered variable z on the range going from $-\infty$ to any value z_1. These values are given in the table at the end of this chapter.

The evaluation of the integral (21.20) passes by the calculation of the boundaries z_1 and z_2. Here $z_1 = \frac{-2-1.1}{2.3} = -1.35$; $z_2 = \frac{2-1.1}{2.3} = 0.39$. (See Fig. 21.8).

We then assess the probability sought (shaded area) from the values read from the table: $P = 0.651 - (1 - 0.911) = 0.562$. There are 56.2 chances out of 100 that the measured voltage is within the range $\{-2\text{mV}, 2\text{mV}\}$.

Prediction interval with a 5 % risk of a Gaussian variable:

We now calculate the **prediction interval** (symmetric) with a 5 % risk of the random variable x distributed according to the law $N(\eta, \sigma)$, that is to say, the range in which the r.v. has 95 % chances to be observed. In other words, we search the interval such that

$$P\{x_1 \le x < x_2\} = P\{\eta - a \le x < \eta + a\} = 0.95. \tag{21.28}$$

Fig. 21.8 *Shaded* area under the pdf of centered reduced r.v.z

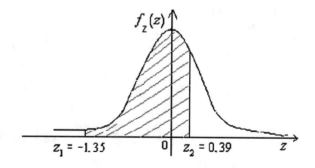

We are interested in the reduced variable

$$z = \frac{x - \eta}{\sigma}. \tag{21.29}$$

We look for z_1 such that

$$P\{-z_1 \leq z < z_1\} = 0.95. \tag{21.30}$$

Due to the symmetry of the distribution, z_1 is such that

$$P\{z < z_1\} = 0.975. \tag{21.31}$$

Reading the table, we find $z_1 = 1.96$.
So

$$P\{\eta - 1.96\,\sigma \leq x < \eta + 1.96\,\sigma\} = 0.95. \tag{21.32}$$

We note that (See Fig. 21.9).

$$PI_{5\%}(x) = \{\eta - 1.96\,\sigma, \eta + 1.96\,\sigma\}. \tag{21.33}$$

We conclude that the variable x has 95 % chances of being realized in an interval with approximate width 4σ centered on the mean η .

21.1.2.1   Moments of the Gaussian Distribution

We recall now the general methodology for calculating Gaussian integrals. First we calculate the integral:

$$I(\alpha) = \int\limits_{-\infty}^{+\infty} e^{-\alpha x^2} dx \quad \text{with } \alpha > 0 \text{ real.} \tag{21.34}$$

Fig. 21.9 The value of the *shaded* area is 0.95. Its boundaries give the 95 % prediction interval

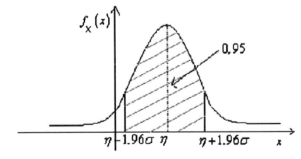

For this calculation, we evaluate the square of this integral:

$$I^2(\alpha) = \int_{-\infty}^{+\infty} \int_{-\infty}^{+\infty} e^{-\alpha(x^2+y^2)} dx\,dy = \int_0^\infty \int_0^{2\pi} e^{-\alpha r^2} r\,dr\,d\theta.$$

(21.35)

$$I^2(\alpha) = 2\pi \int_0^\infty e^{-\alpha r^2} r\,dr.$$

We note $r^2 = u$; $du = 2r\,dr$

$$I^2(\alpha) = \pi \int_0^\infty e^{-\alpha u} du = -\frac{\pi}{\alpha} [e^{-\alpha u}]_0^\infty = \frac{\pi}{\alpha}.$$

(21.36)

So

$$I(\alpha) = \sqrt{\frac{\pi}{\alpha}}.$$

(21.37)

The moments of the Gaussian distribution are calculated using the derivatives of $I(\alpha)$:

$$\frac{dI(\alpha)}{d\alpha} = -\int_{-\infty}^{+\infty} x^2 e^{-\alpha x^2} dx = \sqrt{\pi} \left(-\frac{1}{2}\right) \alpha^{-\frac{3}{2}}.$$

(21.38)

This shows that

$$\int_{-\infty}^{+\infty} x^2 e^{-\alpha x^2} dx = \frac{1}{2} \sqrt{\frac{\pi}{\alpha^3}}.$$

(21.39)

Similarly, by differentiating twice with respect to α we get the fourth moment

$$\frac{d^2 I}{d\alpha^2} = \int_{-\infty}^{+\infty} x^4 e^{-\alpha x^2} dx = \sqrt{\pi} \frac{3}{4} \alpha^{-\frac{5}{2}}.$$

(21.40)

It is thus seen that by successive differentiations, one can reach all integrals containing even powers of x, i.e., calculate the different even moments of the Gaussian distribution. The odd moments are zero as the functions to integrate are odd in the range $-\infty, +\infty$.

Application: asymmetry and flatness of the Gaussian distribution:

As the moments of odd order are zero, the third-order moment, then the asymmetry, is zero.

On the other hand,

$$\mu_4 = \frac{1}{\sqrt{2\pi\sigma^2}} \int\limits_{-\infty}^{\infty} (x-\eta)^4 e^{-\frac{(x-\eta)^2}{2\sigma^2}} dx = \frac{1}{\sqrt{2\pi\sigma^2}} \int\limits_{-\infty}^{\infty} x_c^4 e^{-\frac{x_c^2}{2\sigma^2}} dx_c. \qquad (21.41)$$

Using the result (21.40) we can write

$$\mu_4 = \frac{1}{\sqrt{2\pi\sigma^2}} \int\limits_{-\infty}^{\infty} x_c^4 e^{-\frac{x_c^2}{2\sigma^2}} dx_c = \frac{1}{\sqrt{2\pi\sigma^2}} \sqrt{\pi}\frac{3}{4}\left(2\sigma^2\right)^{5/2} = 3\sigma^4. \qquad (21.42)$$

The flatness of a Gaussian distribution is then

$$m_4 = \frac{\mu_4}{\sigma^4} = 3. \qquad (21.43)$$

We have calculated in Chap. 7 the Fourier transform of a Gaussian. The result was

$$\int\limits_{-\infty}^{+\infty} e^{-\alpha x^2} e^{-ikx} dx = e^{-\frac{k^2}{4\alpha}} \sqrt{\frac{\pi}{\alpha}}. \qquad (21.44)$$

We deduce the expression of the Fourier transform of a Gaussian probability density function of a centered variable:

$$\boxed{\text{Let } f(x) = \frac{1}{\sqrt{2\pi\sigma^2}} e^{-\frac{x^2}{2\sigma^2}}, \text{ its FT is : } F(k) = e^{-\frac{\sigma^2 k^2}{2}}.} \qquad (21.45)$$

It is thus seen that the Fourier transform of a Gaussian is a Gaussian.

For a Gaussian variable with a nonzero expected value η, the pdf appears as the previous density translated by η. Using the shift theorem connecting the FT of a translated function to that of the un-translated function by multiplication by a phase factor, we have

$$\boxed{\text{Let } f(x) = \frac{1}{\sqrt{2\pi\sigma^2}} e^{-\frac{(x-\eta)^2}{2\sigma^2}}, \text{ its FT is : } F(k) = e^{-ik\eta} e^{-\frac{\sigma^2 k^2}{2}}.} \qquad (21.46)$$

21.1.2.2 Characteristic Function

We call characteristic function $\Phi(k)$ of a r.v. with probability density function $f(x)$ the function given by the integral

$$\Phi(k) = \int_{-\infty}^{+\infty} f(x)e^{ikx}dx. \tag{21.47}$$

It appears that the characteristic function is connected to the Fourier transform $F(k)$ of the probability density function of the r.v. by the relationship:

$$\Phi(k) = F(-k).$$

So for a Gaussian variable

$$\boxed{\text{Let } f(x) = \frac{1}{\sqrt{2\pi\sigma^2}}e^{-\frac{(x-\eta)^2}{2\sigma^2}}, \text{ the characteristic function is : } \Phi(k) = e^{ik\eta}e^{-\frac{\sigma^2 k^2}{2}}.} \tag{21.48}$$

21.1.3 Probability Density Function of a Function of a Random Variable

Let y be a function of the random variable x. How, knowing the probability law of x can we deduce that of y?

In a domain where the function $y(x)$ is monotonous, every event giving to x the value x, will give to y the value y. The probability to find x between x and $x + dx$ is equal to the probability to find y between y and $y + dy$. We can then write:

$$|f_x(x)dx| = |f_y(y)dy|. \tag{21.49}$$

Note the absolute values that serve to ensure the necessarily positive character of the probabilities. Densities being positive, we may write

$$f_y(y) = f_x(x)\frac{1}{\left|\frac{dy}{dx}\right|}. \tag{21.50}$$

In the case where a same value of y can be obtained for several values of x, we split the interval of variation of x in intervals where the function $y(x)$ is monotonous.

The probability $|f_y(y)dy|$ is the sum of the probabilities of realization of x: $|f_x(x)dx|_i$ on the different intervals with a one to one correspondence between x and y:

$$|f_y(y)dy| = |f_x(x)dx|_1 + |f_x(x)dx|_2 + \ldots \tag{21.51}$$

Figure 21.10 shows the example where $y = x^2$. Two intervals of x give the realization of y between y and $y + dy$.

Application: χ^2 (Chi-square) variable with one degree of freedom.

This variable plays an important role in statistics since it is largely used to evaluate the noise in a signal, the squared error of a model or the deviation of a distribution of variables with a theoretical distribution (for that last application, see Chap. 22). Let us study here its properties.

Let x be a Gaussian centered reduced r.v. $N(0, 1)$. One seeks the distribution of the variable $y = x^2$ (called a χ^2 variable with one degree of freedom):

We have:

$$f_x(x) = \frac{1}{\sqrt{2\pi}} e^{-\frac{x^2}{2}}, \ y = x^2 \text{ and } \frac{dy}{dx} = 2x. \tag{21.52}$$

$$f_y(y) = 2\frac{1}{\sqrt{2\pi}} e^{-\frac{y}{2}} \frac{1}{2\sqrt{y}} U(y) = \frac{1}{\sqrt{2\pi y}} e^{-\frac{y}{2}} U(y). \tag{21.53}$$

We have multiplied by 2 because there are 2 values of x for a given value of y.

$$E\{y\} = \int_{-\infty}^{\infty} y f_y(y)dy = \frac{1}{\sqrt{2\pi}} \int_0^{\infty} \sqrt{y} e^{-\frac{y}{2}}dy. \tag{21.54}$$

We note $y = x^2$; $dy = 2xdx$; $\sqrt{y} = x$.

$$E\{y\} = \frac{1}{\sqrt{2\pi}} \int_0^{\infty} x e^{-\frac{x^2}{2}} 2xdx = \frac{1}{\sqrt{2\pi}} \int_{-\infty}^{+\infty} x^2 e^{-\frac{x^2}{2}}dx = 1. \tag{21.55}$$

The calculus of the variance of x is recognized: $v(y) = E\{y^2\} - (E\{y\})^2$.

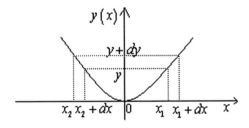

Fig. 21.10 In this example, $|f_y(y)dy| = |f_x(x)dx|_1 + |f_x(x)dx|_2$

In a first step, we calculate $E\{y^2\} = \frac{1}{\sqrt{2\pi}} \int_0^\infty y^2 e^{-\frac{y}{2}} \frac{1}{\sqrt{y}} dy$.

We note $y = x^2$. $E\{y^2\} = \frac{1}{\sqrt{2\pi}} \int_0^\infty x^4 e^{-\frac{x^2}{2}} \frac{1}{x} 2x dx = \frac{1}{\sqrt{2\pi}} \int_{-\infty}^{+\infty} x^4 e^{-\frac{x^2}{2}} dx$.

We must integrate the product of the fourth power of x with the Gaussian function. Using the results recalled above on Gaussian integrals, we have

$$E\{y^2\} = \frac{1}{\sqrt{2\pi}} \sqrt{\pi} \frac{3}{4} \sqrt{32} = 3.$$

Therefore $v(y) = 3 - 1 = 2$.
So

$$E\{\chi^2\} = 1; \quad v(\chi^2) = 2. \tag{21.56}$$

Characteristic function of the χ^2 variable with one degree of freedom:

$$\Phi(k) = \int_{-\infty}^{+\infty} \frac{1}{\sqrt{2\pi y}} e^{-\frac{y}{2}} U(y) e^{iky} dy = \frac{1}{\sqrt{2\pi}} \int_{-\infty}^{+\infty} y^{-\frac{1}{2}} e^{-\frac{y}{2}} U(y) e^{iky} dy.$$

Using the result in the table of Fourier transforms given in Chap. 7, we get

$$\Phi(k) = \frac{1}{\sqrt{2\pi}} \left(-jk + \frac{1}{2} \right)^{-\frac{1}{2}} \sqrt{\pi} = \frac{1}{(-2jk+1)^{1/2}}. \tag{21.57}$$

21.2   Second-Order Statistics

21.2.1   Case of Two Real Random Variables

We are interested in the result of the values $x(t_1)$ and $x(t_2)$ obtained for two times t_1 and t_2. They are both random variables. It is clear that if the signal $x(t)$ is obtained at the output of a physical system and if the times t_1 and t_2 are quite close, the measured values are generally not independent. For example, if the value at time t_1 is strongly negative, the value obtained at a close time t_2 will most likely be also negative, etc.

We are led to study statistics of two r.v. which will be also called second-order statistics. For convenience, we write $x(t_1) = x$ and $x(t_2) = y$.

21.2.1.1   Joint Probability Density Function

The probability that x and y are realized in the neighborhoods dx and dy of x and y is (See Fig. 21.11):

$$P\{x < x \leq x + dx, y < y \leq y + dy,\} = f_{xy}(x, y)dxdy. \qquad (21.58)$$

This probability is proportional to the infinitesimal area $dx\,dy$ of a rectangle whose vertices is the point (x, y) in the plane xOy. It is weighted by the function $f_{xy}(x, y)$ that acts as a mass density in the plane xOy.

The function $f(x(t_1), x(t_2)) = f_{xy}(x, y)$ is the **joint probability density function**.

21.2.1.2   Joint Cumulative Distribution Function

It is defined as the probability that $x \leq x$ and $y \leq y$:

$$P(x \leq x, y \leq y) = F_{xy}(x, y) = \int_{-\infty}^{x} \int_{-\infty}^{y} f_{xy}(x, y)dxdy. \qquad (21.59)$$

The domain of integration is hatched in gray (Fig. 21.12).

$F_{xy}(x, y)$ is the joint cumulative distribution function.

Fig. 21.11 Infinitesimal area
$dx\,dy$

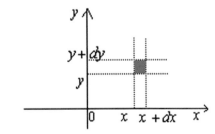

Fig. 21.12 Domain of
integration is hatched in *gray*

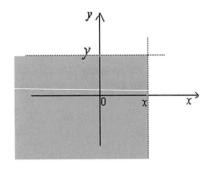

We have evidently

$$f_{xy}(x, y) = \frac{\partial^2 F_{xy}}{\partial x \partial y}.$$ (21.60)

21.2.1.3 Marginal Cumulative Distribution Function

The marginal cumulative distribution function is defined as the probability that $x \leq x$ whatever the value of y

$$P(x \leq x) = P\{x \leq x, \forall y\} = \int_{-\infty}^{x} \int_{-\infty}^{+\infty} f_{xy}(x, y)\mathrm{d}x\mathrm{d}y = F_{xy}(x, \infty).$$ (21.61)

The domain of integration is hatched in gray (Fig. 21.13).
The marginal probability density function $f_x(x)$ is given by $f_x(x) = \frac{\partial F_{xy}(x,\infty)}{\partial x}$.
It comes

$$f_x(x) = \int_{-\infty}^{+\infty} f_{xy}(x, y)\mathrm{d}y.$$ (21.62)

Correlation of the random variables x and y:
By definition, the correlation of variables x and y (assumed real) is the expectation of their product

$$E\{xy\} = \int_{-\infty}^{+\infty} \int_{-\infty}^{+\infty} xy f_{xy}(x, y)\mathrm{d}x\mathrm{d}y.$$ (21.63)

Fig. 21.13 The domain of integration is the half-plane $x \leq x$

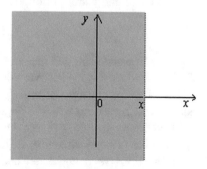

Covariance of r.v. x and y:

By definition, the covariance C_{xy} is the correlation of the centered variables. It is defined as

$$C_{xy} = E\{(x - \eta_x)(y - \eta_y)\} = E\{x_c y_c\}. \tag{21.64}$$

Property: One can easily show that

$$C_{xy} = E\{xy\} - E\{x\}E\{y\} = E\{xy\} - \eta_x \eta_y. \tag{21.65}$$

Correlation coefficient of variables x and y:

The correlation coefficient r_{xy} of variables x and y is defined by

$$r_{xy} = \frac{C_{xy}}{\sigma_x \sigma_y}. \tag{21.66}$$

Important property:

$$|r_{xy}| \leq 1. \tag{21.67}$$

To show this property, we construct a quadratic form where λ is any real number and we take the expectation of this form. This expectation of a square quantity is necessarily positive or zero:

$$E\left\{ \left[\lambda(x - \eta_x) + (y - \eta_y)\right]^2 \right\} = \lambda^2 \sigma_x^2 + \sigma_y^2 + 2\lambda C_{xy} \geq 0. \tag{21.68}$$

We recognize a polynomial of degree 2 in λ. In order that its value be always positive or zero regardless λ, it is necessary that its discriminant is negative or zero, so that there is no root accompanied by a change of sign of the polynomial.

It is thus necessary that $C_{xy}^2 - \sigma_x^2 \sigma_y^2 \leq 0$. Then $\frac{|C_{xy}|}{\sigma_x \sigma_y} \leq 1$, or: $|r_{xy}| \leq 1$.

Orthogonality and non-correlation:

Two r.v. x and y are said orthogonal if their correlation is zero:

$$E\{xy\} = 0. \tag{21.69}$$

Two r.v. x and y are said uncorrelated if their covariance is zero, that is to say if $C_{xy} = 0$ (and then $r_{xy} = 0$). We will have in this case: $E\{xy\} = \eta_x \eta_y$.

Note that the concepts of orthogonality and no correlation imposed by use are not in good agreement with the original semantic definitions.

Discussion: If two r.v. are uncorrelated, centered variables $(x - \eta_x)$ and $(y - \eta_y)$ are orthogonal: $x - \eta_x \perp y - \eta_y$.

If two r.v. are uncorrelated and their expectations are zero, then they are orthogonal: $x \perp y$

Characteristic function of two r.v. x and y:

The characteristic function $\Phi(k_1, k_2)$ of joint random variables x and y is defined from the double integral of the joint probability density function $f_{xy}(x, y)$:

$$\Phi(k_1, k_2) = \int\limits_{-\infty}^{+\infty} \int\limits_{-\infty}^{+\infty} f_{xy}(x, y)e^{ik_1 x}e^{ik_2 y}\,dx\,dy. \tag{21.70}$$

The characteristic function is related to the two-dimensional FT $F(k_1, k_2)$ of the joint probability density function $f_{xy}(x, y)$:

$$\Phi(k_1, k_2) = F(-k_1, -k_2). \tag{21.71}$$

21.2.1.4 Conditional Probability Density Function

The conditional probability density function of a random variable y is defined as the probability density function of that r.v. knowing that the r.v. x has been realized and took a value x. This density is noted $f_{y|x}(y)$. One says: $f_{y|x}(y)$ is the probability density of y if x.

There is an important relationship between the joint probability density and marginal and conditional probability density functions expressed in the following theorem:

Theorem *The joint probability density function is equal to the product of the marginal probability density function of x by the conditional probability density function of y.*

$$f_{xy}(x, y) = f_x(x)f_{y|x}(y). \tag{21.72}$$

We have, of course, also

$$f_{xy}(x, y) = f_y(y)f_{x|y}(x). \tag{21.73}$$

Definition The random variables x and y are called **independent** if their joint probability density function is equal to the product of their marginal densities:

$$f_{xy}(x, y) = f_x(x)f_y(y). \tag{21.74}$$

We can see that, in the case of independence, the conditional probability densities are equal to the marginal probability density functions.

$$f_{y|x}(y) = f_y(y) \text{ and } f_{x|y}(x) = f_x(x).$$ (21.75)

Property *If the random variables* x *and* y *are* **independent**, *according to* (22.17) *their characteristic function is given by*

$$\Phi(k_1, k_2) = \int_{-\infty}^{+\infty} \int_{-\infty}^{+\infty} f_{xy}(x, y) e^{ik_1 x} e^{ik_2 y} dx\, dy = \int_{-\infty}^{+\infty} \int_{-\infty}^{+\infty} f_x(x) f_y(y) e^{ik_1 x} e^{ik_2 y} dx\, dy,$$

$$\Phi(k_1, k_2) = \int_{-\infty}^{+\infty} f_x(x) e^{ik_1 x} dx \int_{-\infty}^{+\infty} f_y(y) e^{ik_2 y} dy = \Phi_x(k_1) \Phi_y(k_2).$$

(21.76)

In this case, the characteristic function of the couple (x, y) of joint variables is the product of the marginal characteristic functions of x and of y.

21.2.1.5 Bayesian Statistical Aspect

In recent years, the use of Bayesian statistics has been rapidly developing in signal processing. It is used in many areas such as signal quality enhancement or source of signal localization. It may be seen as the evaluation of the probability of the cause of a phenomenon. It is an aspect of statistical inference where one looks to statistical properties by analyzing the data. The treatment of this topic is beyond the scope of this book and the reader is advised to get insights of this field in the literature. To expose the concept of Bayesian statistics, rather than focus on the probability densities, we use discrete events probabilities.

Consider two events A and B with probabilities $P(A)$ and $P(B)$. The likelihood of the joint realization of both events A and B, $P(A \text{ and } B)$ is noted $P(A \cap B)$.

The conditional probability of B, the event A having been realized, is defined by the ratio of the joint probability $P(A \cap B)$ by $P(A)$. So:

$$P(B|A) = \frac{P(A \cap B)}{P(A)}.$$ (21.77)

Obviously we can write $P(A \cap B) = P(B)P(A|B)$, and $P(A) = \frac{P(A \cap B)}{P(B|A)}$, and also $P(A|B) = \frac{P(A \cap B)}{P(B)}$; $P(B) = \frac{P(A \cap B)}{P(A|B)}$.

Bayes rule:
Since $P(A \cap B) = P(A)P(B|A) = P(B)P(A|B)$, we have

$$\boxed{P(A|B) = \frac{P(B|A)P(A)}{P(B)}.}$$ (21.78)

For conditional probability densities defined in Sect. 21.4.4, the Bayes rule would be written as

$$f_{y|x}(y) = \frac{f_{x|y}(x)f_y(y)}{f_x(x)}.$$ (21.79)

The rule given in (21.78) is also called the rule of the probability of the causes. To explain this term, we change the notation. We call D the event that has been observed as a data of the particular problem (D is used to say that it is a data). We note H the event that is assumed to have been realized (hypothesis H). The Eq. (21.78) takes the form:

$$P(H|D) = \frac{P(D|H)P(H)}{P(D)}.$$ (21.80)

Let us call H' the event corresponding to the non realization of H.
We can write

$$P(D) = P(D|H)P(H) + P(D|H')P(H'),$$ (21.81)

or:

$$P(D) = P(D|H)P(H) + P(D|H')(1 - P(H)).$$ (21.82)

Equation (21.80) becomes:

$$P(H|D) = \frac{P(D|H)P(H)}{P(D|H)P(H) + P(D|H')(1 - P(H))}.$$ (21.83)

This last expression expresses the probability that the measurement D has been caused by the realization of the event H.

Two exercises from problems encountered in medical statistics are given at the end of this chapter.

21.2.2   Two Joint Gaussian r.r.

21.2.2.1   Probability Densities

By definition, x and y are jointly Gaussian if their joint probability density function is given by

$$f_{xy}(x,y) = \frac{1}{2\pi\sigma_x\sigma_y\sqrt{1-r^2}} e^{-\frac{1}{2(1-r^2)}\left(\frac{(x-\eta_x)^2}{\sigma_x^2} - \frac{2r(x-\eta_x)(y-\eta_y)}{\sigma_x\sigma_y} + \frac{(y-\eta_y)^2}{\sigma_y^2}\right)}. \qquad (21.84)$$

Further, it will be shown that r is the correlation coefficient of x and y.

In the following, the study is restricted to the case where the means of x and y are null and where their variances are equal. This particular case makes it simple calculation, but is still worth in practice. It helps to understand the nature of the problem.

In that case

$$f_{xy}(x,y) = \frac{1}{2\pi\sigma^2\sqrt{1-r^2}} e^{-\frac{1}{2\sigma^2(1-r^2)}(x^2 - 2rxy + y^2)}. \qquad (21.85)$$

We calculate the marginal probability density of x: $f_x(x) = \int_{-\infty}^{+\infty} f_{xy}(x,y)dy$.

For this, we write initially $f_{xy}(x,y)$ in another form. We can pose $\alpha = \frac{1}{2\sigma^2(1-r^2)}$ and the term is factored. There is then a term $e^{-\alpha(y^2 - 2rxy)}$ in which the first two terms of a squared difference is recognized.

We write then

$$y^2 - 2rxy = (y - rx)^2 - r^2x^2. \qquad (21.86)$$

So

$$f_{xy}(x,y) = \frac{1}{2\pi\sigma^2\sqrt{1-r^2}} e^{-\frac{x^2(1-r^2)}{2\sigma^2(1-r^2)}} e^{-\frac{(y-rx)^2}{2\sigma^2(1-r^2)}}, \qquad (21.87)$$

or:

$$f_{xy}(x,y) = \frac{1}{\sqrt{2\pi\sigma^2}} e^{-\frac{x^2}{2\sigma^2}} \frac{1}{\sqrt{2\pi\sigma^2(1-r^2)}} e^{-\frac{(y-rx)^2}{2\sigma^2(1-r^2)}}. \qquad (21.88)$$

We recognize the product of the marginal probability density of x by the conditional density of y. Indeed

$$f_x(x) = \int_{-\infty}^{+\infty} f_{xy}(x,y)\,dy = \frac{1}{\sqrt{2\pi\sigma^2}}e^{-\frac{x^2}{2\sigma^2}}\int_{-\infty}^{+\infty}\frac{1}{\sqrt{2\pi\sigma^2(1-r^2)}}e^{\frac{(y-rx)^2}{2\sigma^2(1-r^2)}}dy.\quad(21.89)$$

We recognize in this expression the integral of a Gaussian probability density function from minus infinity to plus infinity. This integral is 1.
So we have

$$f_{xy}(x,y) = f_x(x)f_{y|x}(y).\qquad(21.90)$$

with:

$$f_x(x) = \frac{1}{\sqrt{2\pi\sigma^2}}e^{-\frac{x^2}{2\sigma^2}} \text{ and } f_{y|x}(y) = \frac{1}{\sqrt{2\pi\sigma^2(1-r^2)}}e^{-\frac{(y-rx)^2}{2\sigma^2(1-r^2)}}.\qquad(21.91)$$

Interpreting the shape of this density:
The value x being observed, the r.v. y is conditioned by this observation. It is r.v. with e.v. $\eta_{y|x} = rx$ and variance $\sigma_{y|x}^2 = \sigma^2(1-r^2) < \sigma^2$.
The observed value of x has "pulled" the statistics of y. For example, if r is positive, it is more likely to observe a negative value for y if a negative value was observed for x.
We also note that the conditioned variance of y is equal or less (as $r^2 \leq 1$) to that of y unconditioned. The statistical range of probable values of y is reduced. We see in Fig. 21.14 that the width of conditional probability density $f_{y|x}(y)$ is smaller than that of the marginal density $f_y(y)$:

Fig. 21.14 Marginal pdf $f_y(y)$ and conditioned pdf $f_{y|x}(y)$

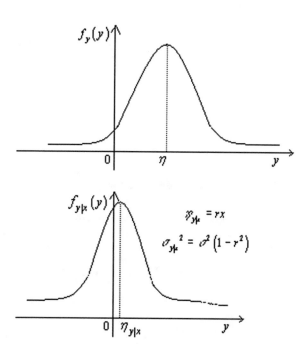

Limit cases:

$r = 0$ (no correlation); the joint probability density is

$$f_{xy}(x, y) = \frac{1}{\sqrt{2\pi\sigma^2}} e^{-\frac{x^2}{2\sigma^2}} \frac{1}{\sqrt{2\pi\sigma^2}} e^{-\frac{y^2}{2\sigma^2}}. \tag{21.92}$$

The two r.v. x and y are thus also independent. For Gaussian variables, the non-correlation is equivalent to independence.

$r = 1$ (full correlation); the variance of $y|x$ is zero. If a given value of x is observed, the value of $y|x$ is definitely the value of x.

We verify now that the coefficient r is the correlation coefficient of x and y defined by

$$r_{xy} = \frac{C_{xy}(x, y)}{\sigma_x \sigma_y} = \frac{E\{(x - \eta_x)(y - \eta_y)\}}{\sigma_y \sigma_y} = \frac{E\{xy\}}{\sigma_x \sigma_y} \tag{21.93}$$

in the case where $\eta_x = \eta_y = 0$.

Indeed

$$E\{xy\} = \int\limits_{-\infty}^{+\infty} \frac{1}{\sqrt{2\pi\sigma^2}} x e^{-\frac{x^2}{2\sigma^2}} dx \int\limits_{-\infty}^{+\infty} \frac{1}{\sqrt{2\pi\sigma^2(1 - r^2)}} y e^{-\frac{(y-rx)^2}{2\sigma^2(1-r^2)}} dy. \tag{21.94}$$

The second integral is the average of y conditioned by x, that is $\eta_{(y|x)} = rx$.
So:

$$E\{xy\} = \frac{r}{\sqrt{2\pi\sigma^2}} \int\limits_{-\infty}^{+\infty} x^2 e^{-\frac{x^2}{2\sigma^2}} dx = r\sigma^2. \tag{21.95}$$

We thus get

$$r = \frac{E\{xy\}}{\sigma^2}. \tag{21.96}$$

21.2.2.2   Characteristic Function

We place first in the case previously treated where the expectations of the variables is zero and variances are both equal to σ^2. Their correlation coefficient is r.

$$f_{xy}(x, y) = \frac{1}{2\pi\sigma^2\sqrt{1 - r^2}} e^{-\frac{1}{2\sigma^2(1-r^2)}(x^2 - 2rxy + y^2)}, \tag{21.97}$$

$$f_{xy}(x,y) = \frac{1}{\sqrt{2\pi\sigma^2}} e^{-\frac{x^2}{2\sigma^2}} \frac{1}{\sqrt{2\pi\sigma^2(1-r^2)}} e^{-\frac{1}{2\sigma^2(1-r^2)}(y-rx^2)}. \tag{21.98}$$

$$\Phi(k_1,k_2) = \int_{-\infty}^{+\infty} \frac{1}{\sqrt{2\pi\sigma^2}} e^{-\frac{x^2}{2\sigma^2}} e^{ik_1 x} dx \int_{-\infty}^{+\infty} \frac{1}{\sqrt{2\pi\sigma^2(1-r^2)}} e^{-\frac{1}{2\sigma^2(1-r^2)}(y-rx^2)} e^{ik_2 y} dy.$$

$$\tag{21.99}$$

The characteristic function of the Gaussian conditional variable $y|x$ is recognized in the second integral. It is $e^{ik_2 rx} e^{-\frac{\sigma^2(1-r^2)k_2^2}{2}}$.

We now have to evaluate

$$\Phi(k_1,k_2) = \frac{1}{\sqrt{2\pi\sigma^2}} e^{-\frac{\sigma^2(1-r^2)k_2^2}{2}} \int_{-\infty}^{+\infty} e^{-\frac{x^2}{2\sigma^2}} e^{i(k_1+k_2 r)x} dx. \tag{21.100}$$

We finally have

$$\Phi(k_1,k_2) = e^{-\frac{\sigma^2(1-r^2)k_2^2}{2}} e^{-\frac{\sigma^2(k_1+rk_2)^2}{2}}. \tag{21.101}$$

21.2.3   *Properties of the Sum of Two r.v*

21.2.3.1   Probability Density Function

Consider

$$z = x+y, \tag{21.102}$$

the probability density function $f_z(z)$ can be deduced from the joint probability density $f_{xy}(x,y)$. For that we calculate the cumulative distribution function of z:

$$F_z(z) = \int_{x+y\leq z} f_{xy}(x,y)dx\,dy = \int_{-\infty}^{+\infty} dx \int_{-\infty}^{z-x} f_{xy}(x,y)\,dy. \tag{21.103}$$

The integration domain is the half-plane below the line of equation $y = z - x$ which is parallel to the second bisector (see Fig. 21.15).

Fig. 21.15 Half plane of
integration below the line
$y = z - x$

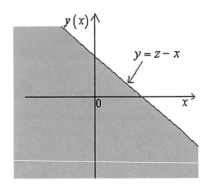

In the particular case where the r.v. x and y are **independent**:

$$f_{xy}(x, y) = f_x(x)f_y(y).$$ (21.104)

So

$$F_z(z) = \int_{-\infty}^{+\infty} f_x(x)\, dx \int_{-\infty}^{z-x} f_y(y)\, dy = \int_{-\infty}^{+\infty} f_x(x)F_y(z - x)dx.$$ (21.105)

$$f_z(z) = \frac{dF_z(z)}{dz} = \int_{-\infty}^{+\infty} f_x(x)f_y(z - x)\, dx = f_x(z) \otimes f_y(z).$$ (21.106)

Thus

$$f_z(z) = f_x(z) \otimes f_y(z).$$ (21.107)

It is thus seen that the probability density function of the sum of two independent r.v. is equal to the convolution product of the probability density functions of these r.v..

Expectation of the sum of two r.v.:

If $z = x + y$, $E\{z\} = E\{x\} + E\{y\} = \eta_x + \eta_y$. The expectation of the sum is the sum of expectations.

If x_1 and x_2 have a same average η, the expectation of their arithmetic mean m is:

$$E\{m\} = \frac{1}{2}(E\{x_1\} + E\{x_2\}) = \eta.$$ (21.108)

Variance of the sum of two r.v.

If

$$z = x + y, \text{var}(z) = E\left\{(x + y - (E\{x\} + E\{y\}))^2\right\} = E\left\{(x_c + y_c)^2\right\},$$
$$= E\{x_c^2\} + E\{y_c^2\} + 2E\{x_c y_c\}$$

If the two variables are independent $E\{x_c y_c\} = E\{x_c\}E\{y_c\} = 0$.
In that case

$$\text{var}(z) = E\{x_c^2\} + E\{y_c^2\} = \text{var}(x) + \text{var}(y). \tag{21.109}$$

In this case, the variance of the sum of the two r.v. is the sum of their variances.
Thus, let $m = \frac{1}{N}\sum_{i=1}^{N} x_i$, the arithmetic mean of N independent random variables with identical expected values and variances (η and σ^2). So,

$$E\{m\} = \eta \text{ and var}(m) = \frac{1}{N}\sigma^2. \tag{21.110}$$

We see that the arithmetic mean m of r.v. will approach more the expectation value if the number N of variables is large, since its variance decreases as N increases.

21.2.3.2 Central Limit Theorem

This theorem is presented here with an example. Suppose that x and y are two independent uniformly distributed r.v. within the interval $\left\{-\frac{a}{2}, \frac{a}{2}\right\}$. The probability density function of their sum z being the auto convolution of rectangular functions densities is a triangular function. The length of the base of the triangle, which is the sum of the supports of the functions that are convoluted, has the value $2a$.

If now x_1, x_2, x_3 and x_4, are four independent variables uniformly distributed in an interval $\left\{-\frac{a}{2}, \frac{a}{2}\right\}$, by grouping terms two by two, it is seen that the density of the sum z, convolution of two triangular shapes densities has a bell shape where the connection to zero values at the ends of the interval is parabolic.

In pursuing these summations of r.v., and if we make the sum of a large number of independent r.v. distributed identically on a bounded support, it is found that the shape of the density of the sum of these variables approaches a Gaussian shape. We understand that it would be the same if the initial density had a form other than rectangular, triangular, for example.

This property is very useful in statistics, and is known as the **central limit theorem**.

Exercise
Calculation of the probability density function of a function of two r.v.
Let

$$z = \sqrt{x^2 + y^2}, \tag{21.111}$$

to calculate the probability density function of the r.v. z, it is again more convenient to pass through the calculation of the cumulative distribution function of z,

$$F_z(z) = \int\int_D f_{xy}(x, y) dx dy. \tag{21.112}$$

The domain of integration D is the set of points in the plane xOy that realizes the condition $z < z$. It is the disk centered in O with radius z (Fig. 21.16).

We assume that x and y are Gaussian variables, independent with zero expectations and variances σ^2, their joint probability density is given by

$$f_{xy}(x, y) = f_x(x) f_y(y) = \frac{1}{\sqrt{2\pi\sigma^2}} e^{-\frac{x^2}{2\sigma^2}} \frac{1}{\sqrt{2\pi\sigma^2}} e^{-\frac{y^2}{2\sigma^2}}. \tag{21.113}$$

$$F_z(z) = \frac{1}{2\pi\sigma^2} \int\int_D e^{-\frac{(x^2+y^2)}{2\sigma^2}} dx \, dy \quad (D : z \le z). \tag{21.114}$$

The integration is performed in polar coordinates. The infinitesimal surface element becomes $z \, dz \, d\theta$.

$$F_z(z) = \frac{1}{2\pi\sigma^2} \int_0^z e^{-\frac{x^2+y^2}{2\sigma^2}} z \, dz \int_0^{2\pi} d\theta = \frac{1}{\sigma^2} \int_0^z e^{-\frac{z^2}{2\sigma^2}} z \, dz. \tag{21.115}$$

Fig. 21.16 Disc of
integration $z < z$

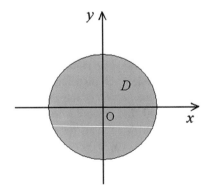

So

$$f_z(z) = \frac{1}{\sigma^2} z e^{-\frac{z^2}{2\sigma^2}} U(z).$$

(21.116)

z is distributed according to a <u>Rayleigh law</u>.

Let us calculate the mean and variance of z:

$$E\{z\} = \frac{1}{\sigma^2} \int_0^\infty e^{-\frac{z^2}{2\sigma^2}} z^2 dz = \frac{1}{2\sigma^2} \int_{-\infty}^{+\infty} z^2 e^{-\frac{z^2}{2\sigma^2}} dz$$

(21.117)

$$= \frac{1}{2\sigma^2} \sqrt{2\pi\sigma^2}\sigma^2 = \sigma\sqrt{\frac{\pi}{2}}.$$

(21.118)

To calculate the variance of z,, we use the relationship $v(z) = E\{z^2\} - (E\{z\})^2$.

The first term is: $E\{z^2\} = \frac{1}{2\sigma^2} \int_0^\infty e^{-\frac{z^2}{2\sigma^2}} z^3 dz = \frac{1}{\sigma^2} \int_0^\infty e^{-\frac{u}{2\sigma^2}} u \frac{du}{2}$

We denoted $z^2 = u$, and so $2zdz = du$.

Integrating by parts

$$E\{z^2\} = \frac{1}{2\sigma^2} \left\{ \left[-2\sigma^2 u e^{-\frac{u}{2\sigma^2}} \right]_0^\infty + 2\sigma^2 \int_0^\infty e^{-\frac{u}{2\sigma^2}} du \right\}.$$

(21.119)

The first term of this sum is zero. It comes: $E\{z^2\} = -2\sigma^2 \left[e^{-\frac{u}{2\sigma^2}} \right]_0^\infty = 2\sigma^2$.

Finally: $v(z) = 2\sigma^2 - \sigma^2 \frac{\pi}{2} = \sigma^2 \left(2 - \frac{\pi}{2} \right)$.

Then

$$\rho = \frac{E\{z\}}{\sqrt{v(z)}} = \frac{\sqrt{\frac{\pi}{2}}}{\sqrt{\left(2 - \frac{\pi}{2}\right)}} = \sqrt{\frac{\pi}{4 - \pi}} = 1.91306.$$

(21.120)

21.2.4 Complex Random Variables

21.2.4.1 Probability Density Function of a Complex r.v.

Let x_R and x_I be two random variables. The quantity $x = x_R + jx_I$ is a complex random variable, with $j = \sqrt{-1}$.

The statistical properties of x are governed by the joint probability density function:

$$P\{x_R < x_R \le x_R + dx_R, x_I < x_I \le x_I + dx_I\} = f_{x_R x_I}(x_R, x_I) dx_R dx_I. \tag{21.121}$$

Expectancy: Expectancy of $x = x_R + jx_I$ is the sum of expectations of x_R and jx_I.

$$E\{x\} = E\{x_R\} + jE\{x_I\}. \tag{21.122}$$

Variance: Before defining the variance, first let us calculate the following expression:

$$E\left\{(x - E\{x\})^2\right\} = E\left\{(x_R + jx_I - E\{x_R\} - jE\{x_I\})^2\right\} \tag{21.123}$$

$$= E\left\{(x_{Rc} + jx_{Ic})^2\right\} = E\{x_{Rc}^2 - x_{Ic}^2 + 2jx_{Rc}x_{Ic}\}. \tag{21.124}$$

A variance is expected to be a real nonnegative number. The real part of the previously calculated quantity is

$$E\{x_{Rc}^2 - x_{Ic}^2\} = E\{x_{Rc}^2\} - E\{x_{Ic}^2\}. \tag{21.125}$$

This expression could be negative if $E\{x_{Rc}^2\} < E\{x_{Ic}^2\}$. On the other hand, it has an eventually nonzero imaginary part: $2E\{x_{Rc}x_{Ic}\}$.

So we must define the variance in another way by using the complex conjugates:

$$v(x) = E\{(x - E\{x\})(x - E\{x\})^*\}. \tag{21.126}$$

Indeed

$$v(x) = E\{(x - E\{x\})(x - E\{x\})^*\} = E\{(x_{Rc} + jx_{Ic})(x_{Rc} + jx_{Ic})^*\},$$
$$v(x) = E\{(x_{Rc} + jx_{Ic})(x_{Rc} - jx_{Ic})\} = E\{x_{Rc}^2 + x_{Ic}^2\}. \tag{21.127}$$

This expectation is real, positive, or zero, as expected of a variance.

21.2.4.2 Correlation of Two Complex R.V

Consider two complex random variables:

$$x_1 = x_{1R} + jx_{1I} \text{ and } x_2 = x_{2R} + jx_{2I}. \tag{21.128}$$

Their correlation is defined as

$$E\{x_1 x_2^*\} = E\{(x_{1R} + jx_{1I})(x_{2R} - jx_{2I})^*\}. \tag{21.129}$$

We have

$$E\{x_1 x_2^*\} = E\{x_{1R}x_{2R} + x_{1I}x_{2I}\} + jE\{x_{1I}x_{2R} - x_{1R}x_{2I}\}. \tag{21.130}$$

This correlation is a complex number, sum of four correlations. Its real and imaginary parts are each the sum of two correlations. $E\{x_1 x_2^*\}$ carries only incomplete information about correlations.

There is a case where the correlation gives all the information, it is the case where some of the correlations contained in (22.133) are equal. Specifically if

$$E\{x_{1R}x_{2R}\} = E\{x_{1I}x_{2I}\} \text{ and } E\{x_{1I}x_{2R}\} = -E\{x_{1R}x_{2I}\}. \tag{21.131}$$

In this case we have

$$E\{x_1 x_2^*\} = 2E\{x_{1R}x_{2R}\} + 2jE\{x_{1I}x_{2R}\}. \tag{21.132}$$

It is noted that then

$$E\{x_1 x_2\} = E\{x_{1R}x_{2R}\} - E\{x_{1I}x_{2I}\} + jE\{x_{1I}x_{2R}\} + jE\{x_{1R}x_{2I}\} = 0. \tag{21.133}$$

We always assume in the following that the complex random variables under study have this property. This is related to the physical nature of signals which are the general subject of our study.

Example 1
Let the random signal

$$x(t) = A_0 e^{j\omega t} e^{j\varphi}. \tag{21.134}$$

The amplitude A_0 is a certain, real positive number. The phase φ is a random number uniformly distributed between $-\pi$ and $+\pi$.

The values of the signal at two instants t_1 and t_2 are two random variables.

$$x_1 = x(t_1) = A_0 e^{j\omega t_1} e^{j\varphi}; \; x_2 = x(t_2) = A_0 e^{j\omega t_2} e^{j\varphi}.$$
$$E\{x_1\} = E\{A_0 e^{j\omega t_1} e^{j\varphi}\} = A_0 e^{j\omega t_1} E\{e^{j\varphi}\} = 0. \tag{21.135}$$

The expectancy is null, as

$$E\{e^{j\varphi}\} = \frac{1}{2\pi} \int_{-\pi}^{\pi} e^{j\varphi} d\varphi = 0. \tag{21.136}$$

The signal has the property (21.134) emphasized in the previous paragraph:

$$E\{x_1 x_2\} = E\{A_0 e^{j\omega t_1} e^{j\varphi} A_0 e^{j\omega t_2} e^{j\varphi}\} = A_0^2 e^{j\omega(t_1 + t_2)} E\{e^{j2\varphi}\} = 0. \tag{21.137}$$

We calculate the correlation:

$$E\{x_1 x_2^*\} = E\{A_0 e^{j\omega t_1} e^{j\varphi} A_0 e^{-j\omega t_2} e^{-j\varphi}\} = A_0^2 e^{j\omega(t_1 - t_2)} E\{1\} = A_0^2 e^{j\omega(t_1 - t_2)}. \tag{21.138}$$

Example 2

We discuss these properties on an example from acoustics. Let a plane acoustic wave propagating in a fluid in the direction x.

The sound pressure can be written as $p = p_0 e^{i\omega(t - \frac{x}{c})}$.

It is assumed that due to homogeneities in the medium, the speed c varies randomly around an average value c_0. One writes the index in the form $n = \frac{c_0}{c}$. The index is a random variable that can be written in the form $n = 1 + \varepsilon$. The r.v. ε has zero expectation. It is assumed in the sequel that n is Gaussian and its standard deviation is noted σ. Similarly n is Gaussian with mean 1 and variance σ^2.

Sound pressure is thus a function of the random variable n. We now determine the mean and variance of the acoustic pressure: $p = p_0 e^{i\omega t} e^{-i\omega \frac{nx}{c_0}}$.

$$E\{p\} = p_0 e^{i\omega t} E\left\{e^{-i\omega \frac{nx}{c_0}}\right\} = p_0 e^{i\omega t} E\{e^{-i\alpha n}\} = p_0 e^{i\omega t} \int_{-\infty}^{+\infty} e^{-i\alpha n} f(n) dn.$$

We denoted $\alpha = \frac{\omega x}{c_0}$. One sees the Fourier transform of the probability density function (Gaussian) of n. Using the formula (21.46) we obtain

$$E\{p\} = p_0 e^{i\omega t} e^{-i\frac{\omega x}{c_0}} e^{-\frac{\sigma^2 \alpha^2}{2}} = p_0 e^{i\omega t} e^{-i\frac{\omega x}{c_0}} e^{-\frac{\sigma^2 \omega^2 x^2}{2c_0^2}}, \tag{21.139}$$

or, using the average wave number $k_0 = \frac{\omega}{c_0}$, $E\{p\} = p_0 e^{i\omega t} e^{-ik_0 x} e^{-\frac{\sigma^2 k_0^2 x^2}{2}}$.

Limit cases:

At low frequencies $E\{\boldsymbol{p}\} = p_0 e^{i\omega t} e^{-ik_0 x}$. At high frequencies $E\{\boldsymbol{p}\} = 0$.
The variance of $\boldsymbol{p}$ is defined by

$$\mathrm{var}(\boldsymbol{p}) = E\{(\boldsymbol{p} - E\{\boldsymbol{p}\})(\boldsymbol{p}^* - E\{\boldsymbol{p}^*\})\}.$$

It is

$$\mathrm{var}(\boldsymbol{p}) = p_0^2 + p_0^2 e^{-\sigma^2 k_0^2 x^2} - 2E\{\boldsymbol{p}\}E\{\boldsymbol{p}^*\} = p_0^2(1 - e^{-\sigma^2 k_0^2 x^2}).$$

At low frequencies the variance is low. It goes to 0 with frequency going to 0.
The variance becomes equal to p_0^2 at high frequencies. The amplitude of the
variation of $\boldsymbol{p}$ is then equal to the pressure modulus p_0 .

Summary

We presented the main statistical properties of a random variable with values in a
continuum. We have defined the probability density function and the cumulative
distribution function of a r.v.: expectancy, variance, the moments of a distribution,
and the characteristic function. The Gaussian distribution (normal distribution) has
been studied thoroughly. The probability density function of a function of a random
variable has been studied with chi-square law as an example. In the second part of
this chapter, we have presented the second-order statistics.

The joint probability density, the marginal probability densities were defined.
Bayesian aspect of conditional statistics has been introduced. The correlation
coefficient, orthogonality and independence in probability concepts were introduced. Two jointly Gaussian variables are exposed as an illustration. It has been
shown that the probability density function of the sum of two independent r.v. is the
convolution of their probability densities. This result is extended to the sum of a
large number of independent random variables that appear to follow approximately
a Gaussian distribution (central limit theorem). Basic properties of complex r.v.
have been exposed.

A table of Gauss's law is given at the end of the chapter.

Exercises

I. If the size of an individual in a population can be considered as a Gaussian r.v.
with expectancy 1.75 m and standard deviation 0.1 m, what is the probability
that the size of an individual taken at random deviates from the mean by more
than 10 cm?; that it exceeds 1.95 m?

II. Let x be a random variable uniformly distributed between $-\pi$ and π. Let $y = \sin x$. Show that the pdf of y is $f_y(y) = \frac{1}{\pi}\frac{1}{\sqrt{1-y^2}}$.

III. Two r.v. x and y are jointly Gaussian. Their expectancies are null and their variances are equal. We assume that $\sigma_x = 5.64 \times 10^{-4}$ and that their correlation coefficient is $r = 0.8415$. Knowing that $x = 10^{-3}$ V was measured, what is the probability of measuring a negative value for y?

Solution: $y|x$ is a Gaussian random variable with $\sigma_{y|x} = \sqrt{\sigma_x^2(1 - r^2)} = 3.05 \times 10^{-4}$.

We can apply the formula of the conditional probability:

$$f_{y|x}(y) = \frac{1}{\sqrt{2\pi\sigma^2(1 - r^2)}}e^{-\frac{(y-rx)^2}{2\sigma^2(1-r^2)}} \text{ with } \eta_{y|x} = rx = 0.8415 \times 10^{-3}.$$

We make the change of variable $z = \frac{y - \eta_{y|x}}{\sigma_{y|x}}$. Writing $y_1 = 0$, the corresponding boundary of the reduced centered variable is $z_1 = \frac{-\eta_{y|x}}{\sigma_{y|x}} = -\frac{0.8415 \times 10^{-3}}{3.05 \times 10^{-4}} = 2.762$. Referring to the table of Gauss's law, we see that there is $1 - 0.9971 = 0.0029$, less than three chances in a thousand for obtaining a negative value for y.

IV. Let x and y be two independent random variables distributed according to a Gaussian distribution. Show that their sum is also distributed according to a Gaussian law.

Hint: Use characteristic functions and the theorem of the FT of a convolution product.
Explain qualitatively why values of y near its average are more probable.

V. A box labeled 1 contains 75 red and 25 black balls. A second box 2 contains 40 red and 40 black balls. Knowing that a black ball was drawn from one of the two boxes selected at random, what is the probability that it was drawn from the box 1?

We note $P(N)$ the probability that a ball is black and $P(R)$ that it is red. Since the box is chosen at random: $P(1) = P(2) = 0.5$.
The probability that a ball is red or black is necessarily equal to 1:

$$P(N \cup R) = P(N) + P(R) - P(N \cap R) = 1.$$

A ball cannot be both red and black, so $P(N \cap R) = 0$; then $P(N) + P(R) = 1$.
A black ball necessarily belonging to the box 1 or 2, since a black ball cannot be from both urns 1 and 2 we have:

$$P(N) = P(N \cap 1) + P(N \cap 2) - P(N \cap 1 \cap 2) = P(N \cap 1) + P(N \cap 2).$$
$$P(N) = P(1)P(N|1) + P(2)P(N|2).$$

Since $P(1) = P(2) = 0.5$, $P(N) = 0.5 \cdot 0.75 + 0.5 \cdot 0.5 = 0.625$.
Similarly $P(R) = 0.5 \cdot 0.25 + 0.5 \cdot 0.5 = 0.375$.
Using Bayes' theorem we can write

$$P(1|N) = \frac{P(N|1)P(1)}{P(N)} = \frac{0.75 \cdot 0.5}{0.625} = 0.6.$$

The probability that the black ball was drawn from the box 1 is 0.6.
Similarly $P(2|N) = \frac{P(N|2)P(2)}{P(N)} = \frac{0.5 \cdot 0.5}{0.625} = 0.4$.
It is normal to obtain that $P(1|N) + P(2|N) = 1$, the black ball could not have been drawn otherwise that from box 1 or 2.

VI. A person belongs to a population whose risk of developing cancer is 5 %. This person makes a test whose probability of detection of an existing cancer is 80 %. This test, however, gives a wrongly positive result in 20 % of cases. The test happens to be positive for that person. What is the probability that the person has cancer?

We use the formula (21.83).
D is the event that the test is positive. H is the event: the person has cancer.
$P(H)$ is the probability that the person has cancer: $P(H) = 0.05$.
$P(D|H)$ is the probability that the test is positive, the person having cancer $P(D|H) = 0.8$.
$P(D|H')$ is the probability that the test is positive, the person having not cancer $P(D|H') = 0.2$.
We seek the probability that the person has cancer when the test has been positive
Formula (21.83) is written: $P(H|D) = \frac{0.8 \cdot 0.05}{0.8 \cdot 0.05 + 0.2 \cdot 0.95} = 0.1739$.
There are less than one in five chances that the person has cancer. The need for treatment will be assessed knowingly by the therapist.

The following table gives the values of the function: $F(z_1) = P\{z < z_1\} = \frac{1}{\sqrt{2\pi}} \int_{-\infty}^{z_1} e^{-\frac{z^2}{2}} dz$.

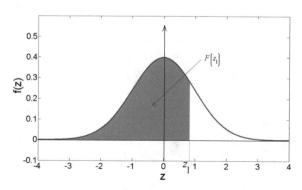

Table of the normal centered reduced law

z1	0	0.01	0.02	0.03	0.04	0.05	0.06	0.07	0.08	0.09
0	0.50000	0.50399	0.50798	0.51197	0.51595	0.51994	0.52392	0.52790	0.53188	0.53586
0.1	0.53983	0.54380	0.54776	0.55172	0.55567	0.55962	0.56356	0.56749	0.57142	0.57535
0.2	0.57926	0.58317	0.58706	0.59095	0.59483	0.59871	0.60257	0.60642	0.61026	0.61409
0.3	0.61791	0.62172	0.62552	0.62930	0.63307	0.63683	0.64058	0.64431	0.64803	0.65173
0.4	0.65542	0.65910	0.66276	0.66640	0.67003	0.67364	0.67724	0.68082	0.68439	0.68793
0.5	0.69146	0.69497	0.69847	0.70194	0.70540	0.70884	0.71226	0.71566	0.71904	0.72240
0.6	0.72575	0.72907	0.73237	0.73565	0.73891	0.74215	0.74537	0.74857	0.75175	0.75490
0.7	0.75804	0.76115	0.76424	0.76730	0.77035	0.77337	0.77637	0.77935	0.78230	0.78524
0.8	0.78814	0.79103	0.79389	0.79673	0.79955	0.80234	0.80511	0.80785	0.81057	0.81327
0.9	0.81594	0.81859	0.82121	0.82381	0.82639	0.82894	0.83147	0.83398	0.83646	0.83891
1	0.84134	0.84375	0.84614	0.84849	0.85083	0.85314	0.85543	0.85769	0.85993	0.86214
1.1	0.86433	0.86650	0.86864	0.87076	0.87286	0.87493	0.87698	0.87900	0.88100	0.88298
1.2	0.88493	0.88686	0.88877	0.89065	0.89251	0.89435	0.89617	0.89796	0.89973	0.90147
1.3	0.90320	0.90490	0.90658	0.90824	0.90988	0.91149	0.91309	0.91466	0.91621	0.91774
1.4	0.91924	0.92073	0.92220	0.92364	0.92507	0.92647	0.92785	0.92922	0.93056	0.93189
1.5	0.93319	0.93448	0.93574	0.93699	0.93822	0.93943	0.94062	0.94179	0.94295	0.94408
1.6	0.94520	0.94630	0.94738	0.94845	0.94950	0.95053	0.95154	0.95254	0.95352	0.95449
1.7	0.95543	0.95637	0.95728	0.95818	0.95907	0.95994	0.96080	0.96164	0.96246	0.96327
1.8	0.96407	0.96485	0.96562	0.96638	0.96712	0.96784	0.96856	0.96926	0.96995	0.97062
1.9	0.97128	0.97193	0.97257	0.97320	0.97381	0.97441	0.97500	0.97558	0.97615	0.97670
2	0.97725	0.97778	0.97831	0.97882	0.97932	0.97982	0.98030	0.98077	0.98124	0.98169
2.1	0.98214	0.98257	0.98300	0.98341	0.98382	0.98422	0.98461	0.98500	0.98537	0.98574

(continued)

(continued)

z1	0	0.01	0.02	0.03	0.04	0.05	0.06	0.07	0.08	0.09
2.2	0.98610	0.98645	0.98679	0.98713	0.98745	0.98778	0.98809	0.98840	0.98870	0.98899
2.3	0.98928	0.98956	0.98983	0.99010	0.99036	0.99061	0.99086	0.99111	0.99134	0.99158
2.4	0.99180	0.99202	0.99224	0.99245	0.99266	0.99286	0.99305	0.99324	0.99343	0.99361
2.5	0.99379	0.99396	0.99413	0.99430	0.99446	0.99461	0.99477	0.99492	0.99506	0.99520
2.6	0.99534	0.99547	0.99560	0.99573	0.99585	0.99598	0.99609	0.99621	0.99632	0.99643
2.7	0.99653	0.99664	0.99674	0.99683	0.99693	0.99702	0.99711	0.99720	0.99728	0.99736
2.8	0.99744	0.99752	0.99760	0.99767	0.99774	0.99781	0.99788	0.99795	0.99801	0.99807
2.9	0.99813	0.99819	0.99825	0.99831	0.99998	0.99841	0.99846	0.99851	0.99856	0.99861

Values of $F(z_1)$ for the great values of z_1:

z1	3	3.1	3.2	3.3	3.4	3.5	3.6	3.7	3.8	3.9
	0.998650	0.999032	0.999313	0.999517	0.999663	0.999767	0.999841	0.999892	0.999928	0.999952

The table read is as follows: We assume that $z_1 = 1.35$. To obtain the value of $F(1.35)$, we move down in the column until $z_1 = 1.3$ then move on the horizontal until the column 00.05. We read $F(1.35) = 0.91149$.

Chapter 22
Multiple Random Variables—Linear Regression Maximum Likelihood Estimation

This chapter is dealing with the statistics of multiple random variables. First, we recall the theoretical results of the statistics of a Chi-square variable with any number of degrees of freedom. The Chi-square distribution law is used for comparing two statistical distributions. One example of application is used to test the applicability of the central limit theorem to the sum of up to 26 r.v. We study the linear regression of the data of multiple observations on r.v. After recalling the simple method of regression, we encounter more elaborate methods based on approximation methods used in linear algebra. We give the principle of Tikhonov regularization of the problem which is useful when the involved matrix is ill conditioned. The useful, empirical L-curve method for obtaining the regularizing parameter is presented. A simple example is given to present the different aspects of the problem. In the following, we discuss the maximum likelihood aspect of statistical parameter estimation and introduce the Cramér-Rao bound and its properties.

22.1 χ_v^2 (Chi-Square) Variable with v Degrees of Freedom

By definition, it is the sum of v independent Gaussian reduced centered variables z_i, squared: $\chi_v^2 = \sum_{i=1}^{v} z_i^2$. It is a random variable. χ_v^2 is the sum of independent random variables; the probability density of its distribution is the convolution of the probability density functions of the elements z_i^2 of the sum.

Characteristic function: Accordingly, its characteristic function is the product of the characteristic functions of all variables. According to formula (21.57) the characteristic function of each variable is: $\Phi_{z_i^2}(k) = \frac{1}{(-2jk+1)^{1/2}}$. We have then

$$\Phi_{\chi_v^2}(k) = \frac{1}{(-2jk+1)^{v/2}}. \tag{22.1}$$

© Springer International Publishing Switzerland 2016
F. Cohen Tenoudji, *Analog and Digital Signal Analysis,*
Modern Acoustics and Signal Processing, DOI 10.1007/978-3-319-42382-1_22

Probability density function: The probability density function of χ_v^2 is the inverse Fourier transform of $\Phi_{\chi_v^2}(-k) = \frac{1}{(2jk+1)^{v/2}}$. According to the formula given in the table of Fourier transforms in Chap. 7, with the notation $\chi_v^2 = x$, we have

$$f_x(x) = \frac{1}{2\Gamma\left(\frac{v}{2}\right)} e^{-\frac{x}{2}} \left(\frac{x}{2}\right)^{\frac{v}{2}-1} U(x).$$ (22.2)

$\Gamma(x)$ is the gamma function; it is an extension of the factorial function. $U(x)$ is the Heaviside function reflecting the fact that the probability density function is zero for $x < 0$.

Depending on the parity of v, we have

$$f_x(x) = \frac{e^{-\frac{x}{2}} x^{\frac{v}{2}-1}}{2^{\frac{v}{2}}\left(\frac{v}{2}-1\right)\left(\frac{v}{2}-2\right)\ldots\left(\frac{1}{2}\right)\sqrt{\pi}} U(x) \text{ if } v \text{ is odd,}$$ (22.3)

and

$$f_x(x) = \frac{e^{-\frac{x}{2}} x^{\frac{v}{2}-1}}{2^{\frac{v}{2}}\left(\frac{v}{2}-1\right)\left(\frac{v}{2}-2\right)\ldots(2)(1)} U(x) \text{ for } v \text{ even.}$$ (22.4)

Properties: Using the independence assumption, we have from (21.56):

$$\text{Mean} : E\{\chi_v^2\} = E\left\{\sum_{i=1}^{v} z_i^2\right\} = \sum_{i=1}^{v} E\{z_i^2\} = v.1 = v.$$ (22.5)

$$\text{Variance} : \text{var}\left(\chi_v^2\right) = \text{var}\left(\sum_{i=1}^{v} z_i^2\right) = \sum_{i=1}^{v} \text{var}\left(z_i^2\right) = v.2 = 2v.$$ (22.6)

χ^2 **test**: This test is used to compare two distributions of data, particularly, to compare a distribution of observed values to a theoretical distribution.

If the r.v. is continuous, one separates its range of variation in r classes (usually $r \sim 15$–25). When performing N measurements of the variable, the frequency of occurrence $f_k = \frac{N_k}{N}$ is determined, ratio of the number N_k observed in each class k to the total number of measurements. We denote N_k' the expected number in the class according to the proposed theoretical distribution. Most often N_k is different from N_k'. This difference may result from a random deviation from the theoretical value or because the assumed theoretical statistical distribution is not the one that governs the issue.

It is shown that if the theoretical law is satisfied and the deviations are due to chance, the following variable

$$X^2 = \sum_{k=1}^{r} \frac{(N_k - N_k')^2}{N_k'}. \tag{22.7}$$

follows a χ^2 law. The number of degrees of freedom is r at maximum. It diminishes by the number of parameters of the observed distribution used to build the theoretical distribution.

It is $r - 1$ if we impose a common mean, and is $r - 2$ if we impose the same mean and the same variance.

The principle of the test is the following:

We assume that the observed data are distributed along the expected theoretical distribution (this assumption is called H_0 hypothesis, or also null hypothesis). The observed difference is in that event due to chance. The probability of observing a very high value of X^2 is low. We adopt the following decision rule: if the value X^2 is less than a threshold value noted χ_0^2, we accept the null hypothesis. Above this value the hypothesis is rejected. The threshold value depends also on the degree of certainty that we seek to achieve.

The following table gives the values χ_0^2 that have a 5 % probability to be exceeded with a number of degrees of freedom going from 1 to 30 for a χ^2 r.v.

v	1	2	3	4	5	6	7	8	9	10
χ_0^2	3.84	5.99	7.81	9.49	11.07	12.59	14.07	15.51	16.92	18.31
v	11	12	13	14	15	16	17	18	19	20
χ_0^2	19.67	21.03	22.36	23.68	25	26.3	27.59	28.87	30.14	31.41
v	21	22	23	24	25	26	27	28	29	30
χ_0^2	32.67	33.92	35.17	36.41	37.65	38.88	40.11	41.34	42.56	43.77

The operating mode of the test is studied on two examples:

Example 1

We launched 1000 times a coin and observed 550 tails and 450 heads. Can we accept that the coin is fair (i.e., that the probability is ½ for each side)?

Solution: We make the hypothesis H_0 that the coin is fair. With the above notation, the numbers observed in both classes are: $N_1 = 550$ and $N_2 = 450$ while the expected numbers are $N_1' = 500$ and $N_2' = 500$. The X^2 difference between the two distributions is:

$$X^2 = \frac{(550 - 500)^2}{500} + \frac{(450 - 500)^2}{500} = 2\frac{2500}{500} = 10.$$

As we use a distribution with an assumed mean of 0.5, the number of degrees of freedom is $r - 1 = 1$. The value χ_0^2 at the 95 % confidence level is 3.84. The 95 % confidence level is widely used to define the threshold of acceptability.

The deviation observed X^2 is equal to 10, exceeding (quite clearly) the threshold value; we reject the null hypothesis, that is to say, we reject the hypothesis that the observed difference is due to odds. In conclusion, we consider that the coin is unfair.

Example 2

We test the central limit theorem by posing the following problem: if one makes the sum of M independent variables uniformly distributed in an interval around 0 in the same way, from what value of M can we accept that the sum of these variables is Gaussian?

The results of a test performed numerically with Matlab are exposed in the following. The rand() function is used. It provides a draw for a uniformly distributed variable between 0 and 1. First, 0.5 is removed from each draw to obtain a centered variable distributed between -0.5 and 0.5. According to the result (21.18), the variance of this variable is $\frac{1}{12}$. The variance of the sum of these M identical independent variables is $\frac{M}{12}$.

To calculate the expected frequencies of the sample having a Gaussian distribution, we use the erf() Matlab function. It provides the integral from 0 to x of the probability density function of a centered reduced Gaussian variable (zero mean, standard deviation 1). In the second step, we use the diff() function to get an approximation of its derivative.

95 % of the values of the reduced centered Gaussian lie between -1.96 and 1.96. We choose to define 24 classes between -3.45 and $+3.55$ and use the hist() function to generate the histogram of the experimental distribution. It will be necessary to divide the sum of the M uniformly distributed r.v. by its standard deviation $\sqrt{\frac{M}{12}}$ to obtain a variable with standard deviation equal to 1 to perform the comparison with the Gaussian population.

We do the sum of 10^5 draws of the sum of M uniformly distributed r.v. for different values of M. The following table summarizes the values of X^2 that we call $X^2_{\text{PseudoGaussian}}$ (2nd row).

For comparison we note on the third row the values of X^2 obtained for 10^5 draws of a Gaussian r.v.

For 24 classes, or $v = 23$ degrees of freedom, the value χ^2_0 at the 95 % confidence level is 35.17.

M	2	4	6	8	10	12	14	16	18	20	22	24	26
X^2_{PseudoG}	2900	408	196	94	52	50	66	33	33	29	42	9	18
X^2_{Gauss}	64	42	40.5	44.5	37	32	28	52	54	36	21	53	42

X^2_{PseudoG} decreases when M increases. For $M < 16$, $X^2_{\text{PseudoG}} > \chi^2_0$. In consequence we reject in these cases the H_0 hypothesis, we will not accept that this variable has a Gaussian character. From around $M = 20$, we can accept the H_0 hypothesis and admit the Gaussian character stated by the central limit theorem.

In the third line we see that X_{Gauss}^2 exceeds χ_0^2 fairly regularly. This abnormality is due to imperfections in the numerical test conducted, in particular the class treatment of values at the ends of the range of variation.

22.2　Least Squares Linear Regression

22.2.1　Simple Method

We assume that an unknown linear relationship relates two variables x and y. We perform m measurements $\{x_i, y_i\}$ of these variables. Several possible sources of errors are the cause that a perfect linear relationship is not observed. Each measurement provides:

$$y_i = \alpha x_i + \beta + \varepsilon_i. \tag{22.8}$$

We assume that the error ε_i is random and distributed evenly above or under the line.

We look here for the "best" line relating the data, in other words, we look for the line minimizing the sum of squared errors

$$\varepsilon^2 = \sum_{i=1}^m \varepsilon_i^2 = \sum_{i=1}^m (y_i - \alpha x_i - \beta)^2. \tag{22.9}$$

We look for the "best" α and β. They are obtained by minimizing ε^2 versus α and β.

$$\frac{\partial \varepsilon^2}{\partial \beta} = -\sum_{i=1}^m 2(y_i - \alpha x_i - \beta) = 0, \text{ or } \beta = \frac{1}{m}\sum_{i=1}^m (y_i - \alpha x_i). \tag{22.10}$$

In the following, the arithmetic means of the data are noted as:

$$\bar{y} = \frac{1}{m}\sum_{i=1}^m y_i \text{ and } \bar{x} = \frac{1}{m}\sum_{i=1}^m x_i. \tag{22.11}$$

It comes

$$\beta = \bar{y} - \alpha\bar{x}. \tag{22.12}$$

$$\frac{\partial \varepsilon^2}{\partial \alpha} = -\sum_{i=1}^m 2(y_i - \alpha x_i - \beta)x_i = 0. \tag{22.13}$$

Also

$$\sum_{i=1}^{m} \left(y_i - \alpha x_i - \frac{1}{m} \sum_{j=1}^{m} (y_j - \alpha x_j) \right) x_i = 0,$$

or:

$$\sum_{i=1}^{m} \left(y_i x_i - \alpha x_i x_i - \frac{1}{m} \sum_{j=1}^{m} (y_j x_i - \alpha x_j x_i) \right) = 0.$$

Then

$$\alpha = \frac{\frac{1}{m} \sum_{i,j=1}^{m} y_j x_i - \sum_{i=1}^{m} y_i x_i}{\frac{1}{m} \sum_{i,j=1}^{m} x_j x_i - \sum_{i=1}^{m} x_i x_i} = \frac{\sum_{i=1}^{m} y_i x_i - m \bar{x} \bar{y}}{\sum_{i=1}^{m} x_i^2 - m \bar{x}^2} = \frac{\frac{1}{m} \sum_{i=1}^{m} y_i x_i - \bar{x} \bar{y}}{\frac{1}{m} \sum_{i=1}^{m} x_i^2 - \bar{x}^2}. \quad (22.14)$$

Last formula is sometimes written $\alpha = \frac{\text{cov}(x,y)}{\text{var}(x)}$. This notation is abusive as $\text{cov}(x, y)$ and $\text{var}(x)$ are estimates which result here from arithmetic sums and are not statistical expectations.

Example
For four successive times, $x_i = 1, 2, 3, 4$, four values of a quantity y have been measured: $y_i = 2, 3, 7, 8$. The regression line parameters given by (22.12) and (22.14) are $\alpha = 2.2$ and $\beta = -0.5$. We see in Fig. 22.1 the four measured pairs $\{x_i, y_i\}$ and the regression line with parameters $\alpha = 2.2$ and $\beta = -0.5$.

Fig. 22.1 Four measurements of the pair $\{x_i, y_i\}$ (*stars*) and the regression line

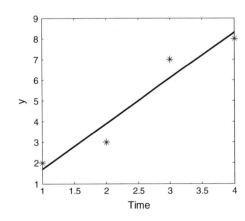

22.2.2   *Elaborate Method*

We write the problem met in the previous example in a linear application matrix form $Ax = b$.

Here the unknown vector is

$$x = \begin{pmatrix} \alpha \\ \beta \end{pmatrix}, \text{ with } A = \begin{pmatrix} 1 & 1 \\ 2 & 1 \\ 3 & 1 \\ 4 & 1 \end{pmatrix} \text{ and } b = \begin{pmatrix} 2 \\ 3 \\ 7 \\ 8 \end{pmatrix}. \tag{22.15}$$

We verify now that the determination of the regression line is equivalent to the research of the solutions of the system is $Ax = b$, equivalently

$$\alpha \begin{pmatrix} 1 \\ 2 \\ 3 \\ 4 \end{pmatrix} + \beta \begin{pmatrix} 1 \\ 1 \\ 1 \\ 1 \end{pmatrix} = \begin{pmatrix} 2 \\ 3 \\ 7 \\ 8 \end{pmatrix}. \tag{22.16}$$

The discussion here follows the developments given in Appendix 2 on linear algebra.

We are in the case A_{mn} where $m = 4$ and $n = 2$. The system is overdetermined. The two column vectors of matrix A are linearly independent. They are vectors in $\mathbb{R}^4$ but cannot form a basis of $\mathbb{R}^4$. The rank of A is 2, number equal to the count of its independent columns. The matrix is full rank. The vector $b \in \mathbb{R}^4$ cannot in this case be written as a linear combination of the columns vectors of A. The system has no solution. The linear regression is the result of the search of an approximate solution of the system in the least mean square sense. This approximate solution has been derived in Appendix 2, it is given by formula (A2.34).

It is:

$$x_0 = (A^H A)^{-1} A^H b. \tag{22.17}$$

With the use of Matlab, it comes $x_0 = \begin{pmatrix} 2.2 \\ -0.5 \end{pmatrix}$. We find again the values of the slope 2.2 and the y-intercept -0.5 of the regression line which were found by the simple method.

It is clear that for more than two coefficients for vector x, the matrix method is much easier when one uses a numerical computing software as Matlab.

22.3   Linear Regression with Noise on Data—Tikhonov Regularization

The solution given by the formula (22.16) is conceivable in the case encountered above since the inversion of matrix $A^H A$ is possible because it is full rank ($r = 2$). The presence of noise on the matrix elements will influence the stability of the result and thus the closeness of the result of the inversion with the ideal solution. The noise sensitivity is reflected in the condition number of matrix $A^H A$, that is to say the ratio of its maximum to its minimum eigenvalues. In case the matrix is not full rank, at least one of the eigenvalues of $A^H A$ is zero, which corresponds to the limit case of an infinite condition number and the matrix $A^H A$ is not invertible. Let us discuss here from the spectral decomposition theorem consequences of robustness to noise through the condition number the influence of the eigenvalues of matrix $A^H A$. Following the formula (A2.52) in Appendix 2, the spectral theorem states that $A^H A = \sum_{i=1}^{n} \lambda_i u_i u_i^H$ the orthonormal vectors u_i are the eigenvectors of matrix $A^H A$ and λ_i are the associated eigenvalues.

Assuming that the matrix $A^H A$ is invertible, which is the case when all the eigenvalues are different from 0, and following formula (A2.19) in Appendix 2 we have $(A^H A)^{-1} = \sum_{i=1}^{n} \frac{1}{\lambda_i} u_i^H u_i$. It is found again that the matrix is not invertible if one of its eigenvalues is zero. It is seen that the magnitude of the vector solution x_0 given by Eq. (22.17) is highly dependent upon the lowest eigenvalues.

The noise present on matrix A will influence the eigenvalues of $A^H A$; accordingly, the noise on the smaller eigenvalues may cause a significant error on the amplitude of the solution x_0. It is understood that adding a small quantity to all eigenvalues will change only slightly the values of significant eigenvalues but is a great change to small eigenvalues; in consequence that addition stabilizes the result by reducing the contribution to the solution of the small eigenvalues as this solution depends upon the inverse matrix. In other words, it reduces the amplitude of the solution which may blow up due to errors caused by noise on the smaller eigenvalues.

In its simplest form, Tikhonov regularization takes the following form; rather than trying to minimize the quantity $\|Ax - b\|^2$ (called norm L_2), we seek to minimize the square error

$$\varepsilon^2 = \|Ax - b\|^2 + \mu \|x\|^2. \tag{22.18}$$

Since

$$\varepsilon^2 = \|Ax - b\|^2 + \mu \|x\|^2 = (Ax - b)^H (Ax - b) + \mu x^H x,$$

$\frac{\partial \varepsilon^2}{\partial x^H} = A^H (Ax - b) + \mu x = 0$. The solution is such that $A^H A x_\mu - A^H b + \mu x_\mu = 0$, so $(A^H A + \mu I) x_\mu = A^H b$, where I is the $(n \times n)$ identity matrix.

Finally, the best approximate least square solution is

$$x_\mu = (A^H A + \mu I)^{-1} A^H b. \tag{22.19}$$

When the matrix A is real, that solution is

$$x_\mu = (A^T A + \mu I)^{-1} A^T b. \tag{22.20}$$

It is seen that the chosen form of the error given by (22.17) leads to heighten the low eigenvalues by addition of the term μI before calculating the inverse matrix, as was discussed above. This operation is called prewhitening because it tends to equalize the spectrum of values, to whiten it.

In case the matrix A is rectangular, the discussion must be conducted from the spectral theorem which involves the SVD. We can also say that the orthogonal vectors of the identity matrix I form a conceivable basis for the development of a vector $\in \mathbb{R}^n$. Some of these vectors are supplement to the absent vectors in the development of $A^H A$ according to the spectral theorem.

The question that arises at this level is: what is the value that we must give the regularization parameter μ for the best estimate of x_μ in the presence of noise? A large value of μ has the effect of lowering the x_μ norm but tends to obscure the role of the actual eigenvalues and thus lose its proximity with the right solution, so to lose in resolution. Conversely, too low a value of μ leads to a noisy solution. A compromise must be found. Different methods of finding the μ optimal value have been proposed. We will discuss here only the method of using the elbow of the L-shaped curve, when the logarithms of the norm of the solution versus the residual norm are represented.

To test the μ parameter values, the curve looks as follows (Fig. 22.2). As recorded in the literature on the subject, the experience has shown that the optimal value of μ occurs at the elbow which is identified in the figure by an arrow.

Fig. 22.2 *L-curve* used for the determination of the optimum value of the regularizing parameter (From P.C. Hansen, The L-curve and its use in the numerical treatment of inverse problems)

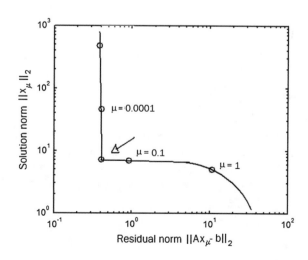

If μ is too small, there is an important noise on the solution causing a sharp increase of the norm solution when diminishing μ. If μ is too large, the distance between the estimated vector Ax_μ and the goal b becomes too large.

Note: We have discussed the simplest form of Tikhonov regularization. More elaborate methods have been proposed. One example is the case where instead of adding the matrix μI before inversion as given by formula (22.20) where this diagonal matrix has the same value along the diagonal, a richer matrix is added which can, in principle lead to an estimation closer to the best value.

We illustrate the method of regularization described above (Formula 22.22) by the following simple numerical simulation derived from the previous example. We use a matrix with a large A condition number to accentuate the problem related to the high discrepancy of the eigenvalues of the matrix. If both columns were equal, the condition number would be infinite; if we slightly modify the elements of the second column we avoid the infinite and get an important condition number:

$$A = \begin{pmatrix} 1 & 1.1 \\ 2 & 1.9 \\ 3 & 2.95 \\ 4 & 4.15 \end{pmatrix}.$$ The singular value decomposition of A gives the two sin-

gular values 7.7926 and 0.1422. The condition number 7.7926/0.1422 is fairly high, leading to an ill-conditioned matrix and making the inversion sensitive to noise.

For a vector $x_{00} = \begin{pmatrix} -0.5 \\ 2 \end{pmatrix}$, the vector $Ax_{00} = b_0$ is $b_0 = \begin{pmatrix} 1.92 \\ 3.18 \\ 4.99 \\ 7.13 \end{pmatrix}$.

To simulate the presence of zero mean noise on the b_0 measurement, we set

$$b = \begin{pmatrix} 2 \\ 3 \\ 5 \\ 7 \end{pmatrix}.$$

The least square optimal unregularized estimation of x is $x_0 = (A^T A)^{-1} A^T b = \begin{pmatrix} -0.8961 \\ 2.566 \end{pmatrix}$.

The noise added to b_0 has caused the discrepancy between x_0 and x_{00}.

$$Ax_0 = \begin{pmatrix} 1.9267 \\ 3.0835 \\ 4.8819 \\ 7.0652 \end{pmatrix}.$$

Following Tikhonov regularization procedure, we look for a solution x_μ closer to x_{00}. We use the relationship (22.22) in function of the regularizing factor μ.

The results are given in Fig. 22.3. We see in Fig. 22.3a that the squared norm $\|Ax_\mu - b_0\|^2$ has a minimum for $\mu = 5.5 \times 10^{-3}$. For this value of μ, we have $x_\mu = \begin{pmatrix} -0.5248 \\ 2.1992 \end{pmatrix}$.

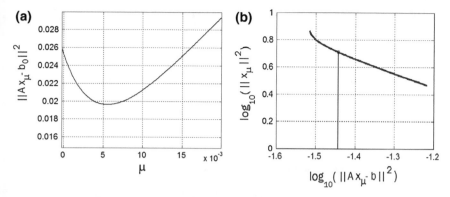

Fig. 22.3 a Squared distance between the estimated b and the vector b_0 without noise; **b** Squared solution norm versus squared residual norm in log-log plot

Using the regularizing factor μ, we reach a vector x_μ fairly close to x_{00}.

In Fig. 22.3b, the vector norm squared $\|x_\mu\|^2$ is plotted in function of the residual norm squared $\|Ax_\mu - b\|^2$ in logarithmic scale. The classic L-shaped curve is not found but nevertheless as marked on the figure the value of $\mu = 5.5 \times 10^{-3}$ corresponds to the transition of the curve between the linear regime on the right and the curved part on its left.

22.4   Parametric Estimation

22.4.1   Issues of the Estimation

A physical measured signal depends on one or more parameters that one seeks to estimate by experience. This is the case of the depth of a geological formation, the distance and/or speed of a radar target, the frequency of a sound signal, etc.

We note θ the unknown parameter sought after and $\hat{\theta}$ an estimator for this parameter that is built from the physical knowledge of the problem and the measured values of the signal. In the general case, the data are affected by random uncertainties due to noise on the signal. As a result, the estimator is a random variable. It will be noted $\hat{\boldsymbol{\theta}}$ in bold in the following text. We call bias of the estimator the difference of the expectation of the estimator and of the true value of the parameter.

It is noted

$$b = E\{\hat{\boldsymbol{\theta}}\} - \theta. \tag{22.21}$$

Since we do not know θ, this bias is unknown.

The mean square error of the estimate is defined by

$$\varepsilon^2 = E\left\{(\hat{\boldsymbol{\theta}} - \theta)^2\right\}. \tag{22.22}$$

The squared error is the sum of the signal variance and of the squared bias. To show this, we write:

$$\varepsilon^2 = E\left\{(\hat{\boldsymbol{\theta}} - \theta)^2\right\} = E\left\{(\hat{\boldsymbol{\theta}} - E\{\hat{\boldsymbol{\theta}}\} + E\{\hat{\boldsymbol{\theta}}\} - \theta)^2\right\}.$$

The reader should infer from the previous that

$$\varepsilon^2 = \text{var}(\hat{\boldsymbol{\theta}}) + b^2. \tag{22.23}$$

Then, the mean square error of an estimator is superior or equal to its variance. The mean square error is minimum if the estimator is unbiased.

Chebyshev inequality expressed by Eq. (22.14) is written for the estimator $\hat{\boldsymbol{\theta}}$:

$$\Pr\left[|\hat{\boldsymbol{\theta}} - \theta| \geq \varepsilon\right] \leq \frac{\text{var}(\hat{\boldsymbol{\theta}})}{\varepsilon^2}. \tag{22.24}$$

We deduce from (22.23) that between two estimators having the same variance, the unbiased estimator has the smallest squared error. It is useful to note here that one can define different unbiased estimators for the parameter θ and that their variances will generally be different. A natural method of research of an optimal $\hat{\theta}$ is minimizing the mean square error. Unfortunately, if the estimator is biased (so it is not known a priori), this minimization usually does not lead to the expression of a valid optimal estimator whatever the value of the parameter θ. The method of maximum likelihood estimation outlined in the following section provides a means of determining the minimum variance unbiased estimator.

Example: Sample of N drawings of a r.v. x

We measure N values x_i. Formulas (21.112) give the mean and variance of the arithmetic mean of the measured values x_i. They prove that it is an unbiased estimate of the ensemble average η.

Now we discuss the variance estimation of the statistical distribution.

If the expectation η is known, the following quantity is an unbiased estimator of the variance

$$\hat{\sigma}^2 = \frac{1}{N}\sum_{i=1}^{N}(x_i - \eta)^2. \tag{22.25}$$

Indeed

$$E\{\hat{\sigma}^2\} = E\left\{\frac{1}{N}\sum_{i=1}^{N}(x_i - \eta)^2\right\} = \frac{1}{N}\sum_{i=1}^{N}E\{(x_i - \eta)^2\} = \frac{1}{N}N\sigma^2 = \sigma^2. \quad (22.26)$$

In the case of a Gaussian variable, the centered differences $\frac{x_{ci}}{\sigma}$ are reduced centered Gaussian variables and we can apply the results of the χ^2 distribution with N degrees of freedom. We deduce the variance of the estimator (22.25):

$$\text{var}(N\hat{\sigma}^2) = \sum_{i=1}^{N}\text{var}(x_i^2) = N\text{var}(x_1^2) = N\text{var}\left(\sigma^2\frac{x_{c1}^2}{\sigma^2}\right) = N\sigma^4\text{var}\left(\frac{x_{c1}^2}{\sigma^2}\right) = 2N\sigma^4.$$

Therefore, $\text{var}(N\hat{\sigma}^2) = N^2\text{var}(\hat{\sigma}^2) = 2N\sigma^4$ from which

$$\text{var}(\hat{\sigma}^2) = \frac{2\sigma^4}{N}. \quad (22.27)$$

In the case where the expectation is unknown, it can be estimated by $m = \frac{1}{N}\sum_{i=1}^{N}x_i$.

The following variance estimator is unbiased

$$\hat{\sigma}^2 = \frac{1}{N-1}\sum_{i=1}^{N}(x_i - m)^2. \quad (22.28)$$

Indeed

$$E\{\hat{\sigma}^2\} = E\left\{\frac{1}{N-1}\sum_{i=1}^{N}(x_i - m)^2\right\} = \frac{1}{N-1}\sum_{i=1}^{N}E\left\{\left(x_i - \frac{1}{N}\sum_{j=1}^{N}x_j\right)^2\right\},$$

$$E\{\hat{\sigma}^2\} = \frac{1}{N-1}\sum_{i=1}^{N}E\left\{\left(\frac{x_i}{N} + \frac{x_i}{N} + \ldots - \frac{1}{N}\sum_{j=1}^{N}x_j\right)^2\right\}.$$

$$E\{\hat{\sigma}^2\} = \frac{1}{N-1}\frac{1}{N^2}\sum_{i=1}^{N}E\left\{\left(x_i + x_i + \ldots - \sum_{j=1}^{N}x_j\right)^2\right\},$$

$$E\{\hat{\sigma}^2\} = \frac{1}{N-1}\frac{1}{N^2}\sum_{i=1}^{N}E\left\{((x_i - x_1) + (x_i - x_2) + \ldots + (x_i - x_N))^2\right\}.$$

It is easily shown that we can introduce the centered variables $x_{ic} = x_i - \eta$ and get

$$E\{\hat{\sigma}^2\} = \frac{1}{N-1}\frac{1}{N^2}\sum_{i=1}^{N} E\Big\{\big((x_{ic} - x_{1c}) + (x_{ic} - x_{2c}) + \ldots + (x_{ic} - x_{Nc})\big)^2\Big\}.$$

The term $(x_{ic} - x_{ic})$ being zero, there are $N-1$ nonzero terms in the brace. In the squaring operation, there are $N-1$ square terms and $\frac{(N-1)(N-2)}{2}$ cross terms.

Expectation of one square term $E\{(x_{ic} - x_{jc})^2\} = E\{x_{ic}^2\} + E\{x_{jc}^2\} - 2E\{x_{ic}x_{jc}\}$. If measurements are independent, $E\{x_{ic}x_{jc}\} = E\{x_{ic}\}E\{x_{jc}\} = 0$.

In that case $E\{(x_{ic} - x_{jc})^2\} = 2\sigma^2$.

Next, we evaluate the expectation of a cross-term when the measures are independent

$$E\big\{(x_{ic} - x_{jc})(x_{ic} - x_{kc})\big\} = E\{x_{ic}^2\} = \sigma^2.$$

It comes $E\{\hat{\sigma}^2\} = \frac{1}{(N-1)}\frac{1}{N^2}N\Big((N-1)2\sigma^2 + 2\frac{(N-1)(N-2)}{2}\sigma^2\Big)$,

and finally $E\{\hat{\sigma}^2\} = \frac{1}{N}(2\sigma^2 + (N-2)\sigma^2) = \sigma^2$.

This proves that the estimator (22.28) is unbiased.

Analogously to the evaluation of the variance of the estimator in the previous case where the expectation was known, it is shown that the variance of that estimator is

$$\mathrm{var}(\hat{\sigma}^2) = \frac{2\sigma^4}{N-1}. \tag{22.29}$$

22.4.2   Maximum Likelihood Parametric Estimation

Likelihood function

We consider successively one, then several, random variables.

When the probability density function is considered as a function of the parameter θ, it is called the likelihood function. It is noted $f_{x;\theta}(x; \theta)$.

For the purposes of estimating an unknown parameter maximum likelihood, the parameter is estimated by the value that maximizes the probability density. Note On.

This property can be interpreted qualitatively using the following example: An extraterrestrial arrives on earth in the desert and has no idea of the size of a human: 10 m? 20 cm? The first human he met measures 1.75 m. It is natural for the alien to assume that the average size of humans is that one, because if that is the case the probability of observing a size in a given neighborhood of the average is maximum (the assumption that the distribution is Gaussian is the underlying reasoning).

For estimating an unknown parameter in the sense of maximum likelihood, the parameter θ is estimated by the value that maximizes the probability density function. We note it $\hat{\theta}_{mv}$.

Gaussian case: The likelihood function for the mean of the Gaussian process is

$$f_{x;\eta}(x;\eta) = \frac{1}{\sqrt{2\pi\sigma_0^2}} e^{-\frac{(x-\eta)^2}{2\sigma_0^2}}. \tag{22.30}$$

It is assumed that a value x_i of x has been observed and that it is known that the process is Gaussian without however knowing η. We look for an estimation of η.

The value of the estimator $\hat{\eta}$ which makes the likelihood function maximum is obtained by canceling the derivative of $f_{x;\eta}$ or that of $\ln f_{x;\eta}$ with respect to η (since the function ln is monotonic).

The function $\ln f_{x;\eta}(x;\eta)$ is called log-likelihood function.

The latter function is derived and one seeks the value of $\hat{\eta}$ that makes it maximum

$$\frac{\partial \ln f_{x;\eta}(x;\eta)}{\partial \eta} = -\frac{\partial \frac{(x_i-\eta)^2}{2\sigma_0^2}}{\partial \eta} = \frac{(x_i-\eta)}{\sigma_0^2} = 0, \quad \text{or} \quad \hat{\eta}_{mv} = x_i. \tag{22.31}$$

Thus, the estimator of the maximum likelihood expectation is the observed value x_i of the variable. This estimator is not biased as $E\{x_i\} = \eta$.

If is carried out N independent observations x_i of the r.v. x, the likelihood function is defined as the product of probability density functions. In the case of Gaussian variables, likelihood function of expectation is

$$f_{x;\eta}(x;\eta) = \prod_{i=1}^{N} \frac{1}{\sqrt{2\pi\sigma_0^2}} e^{-\frac{(x_i-\eta)^2}{2\sigma_0^2}}. \tag{22.32}$$

The logarithm of this function is

$$\ln f_{x;\eta}(x;\eta) = -N \ln\sqrt{2\pi\sigma_0^2} - \sum_{i=1}^{N} \frac{(x_i-\eta)^2}{2\sigma_0^2}. \tag{22.33}$$

This function can be written in expanded form

$$\ln f_{x;\eta}(x;\eta) = -\frac{N}{2\sigma_0^2}\eta^2 + \left(\sum_{i=1}^{N}\frac{x_i}{\sigma_0^2}\right)\eta - \left(N\ln\sqrt{2\pi\sigma_0^2} + \sum_{i=1}^{N}\frac{x_i^2}{2\sigma_0^2}\right). \tag{22.34}$$

We seek the maximum of the log-likelihood function

$$\frac{\partial \ln f_{x;\eta}(x;\eta)}{\partial \eta} = -\frac{N}{\sigma_0^2}\eta + \frac{1}{\sigma_0^2}\sum_{i=1}^{N} x_i = 0. \tag{22.35}$$

The estimator of the expectation with maximum likelihood will be in this case

$$\hat{\eta}_{mv} = \frac{1}{N}\sum_{i=1}^{N} x_i. \tag{22.36}$$

It is the arithmetic average of the observed values of the r.v.
It is noted that this estimator is unbiased since $E\{\hat{\eta}_{mv}\} = \eta$.
The variance of the maximum likelihood estimator is given by

$$\text{var}(\hat{\eta}_{mv}) = E\left\{(\hat{\eta}_{mv} - \eta)^2\right\} = E\left\{\left(\sum_{i=1}^{N}\left(\frac{x_i - \eta}{N}\right)\right)^2\right\} = \frac{1}{N^2}\sum_{i=1}^{N} E\left\{(x_i - \eta)^2\right\}$$

$$= \frac{\sigma_0^2}{N}. \tag{22.37}$$

We used the fact that as two successive observations are uncorrelated, the expectation of the cross terms is zero $E\{x_{ic}x_{jc}\} = 0$ if $i \neq j$.

Search formulas for maximum likelihood

$$\left.\frac{\partial f_{x;\theta}(x;\theta)}{\partial \theta}\right|_{\theta=\hat{\theta}_{mv}} = 0 \quad \text{or} \quad \left.\frac{\partial \ln f_{x;\theta}(x;\theta)}{\partial \theta}\right|_{\theta=\hat{\theta}_{mv}} = 0. \tag{22.38}$$

For complex random variables we would have

$$\left.\nabla_{\theta^*} f_{x;\theta}(x;\theta)\right|_{\theta=\hat{\theta}_{mv}} = 0 \quad \text{or} \quad \left.\nabla_{\theta^*} \ln f_{x;\theta}(x;\theta)\right|_{\theta=\hat{\theta}_{mv}} = 0. \tag{22.39}$$

22.4.3 Cramér-Rao Bound

Let $\hat{\theta}$ be an unbiased estimator of θ. We show in the following that its variance should verify the inequality

$$\text{var}(\hat{\theta}) \geq \frac{1}{E\left\{\left(\frac{\partial \ln f_{x;\theta}(x;\theta)}{\partial \theta}\right)^2\right\}}. \tag{22.40}$$

An estimator is said to be efficient if its variance is minimal. This minimum variance is called the Cramér-Rao bound. We have thus for an efficient estimator

$$\text{var}(\hat{\boldsymbol{\theta}}) = \frac{1}{E\left\{\left(\frac{\partial \ln f_{x;\theta}(x;\theta)}{\partial \theta}\right)^2\right\}}. \tag{22.41}$$

As will be shown below, we have in this case

$$\hat{\theta}(x) - \theta = K(\theta) \cdot \frac{\partial \ln f_{x;\theta}(x;\theta)}{\partial \theta}. \tag{22.42}$$

If an unbiased minimum variance estimator exists, it must satisfy the Eq. (22.42).

By definition, the estimator in the sense of maximum likelihood of θ must satisfy

$$\frac{\partial \ln f_{x;\theta}(x;\theta)}{\partial \theta}\bigg|_{\theta=\hat{\theta}_{mv}} = 0.$$

Substituting θ by $\hat{\theta}_{mv}$ in (22.42) we get

$$\hat{\theta}(x) - \hat{\theta}_{mv} = K(\hat{\theta}_{mv}) \cdot \frac{\partial \ln f_{x;\theta}(x;\theta)}{\partial \theta}\bigg|_{\theta=\hat{\theta}_{mv}} = 0. \tag{22.43}$$

Therefore,

$$\hat{\theta}(x) = \hat{\theta}_{mv}. \tag{22.44}$$

In conclusion, the unbiased maximum likelihood estimator is the minimum variance estimator.

Demonstration of the Cramér-Rao formula

It is assumed that the estimator is unbiased $E\{\hat{\boldsymbol{\theta}}(x)\} = \theta$. We can write in this case:

$$E\left\{\left(\hat{\boldsymbol{\theta}}(x) - \theta\right)\right\} = \int_{-\infty}^{\infty} f_{x;\theta}(x;\theta)\left(\hat{\theta}(x) - \theta\right) dx = 0 \tag{22.45}$$

We derive this expression under the integral sign with respect to θ:

$$\int_{-\infty}^{\infty} \left(\frac{\partial f_{x;\theta}}{\partial \theta}\left(\hat{\theta}(x) - \theta\right) - f_{x;\theta}(x;\theta)\right) dx = 0, \tag{22.46}$$

or even:

$$\int\limits_{-\infty}^{\infty} \frac{\partial f_{x;\theta}}{\partial \theta} \left(\hat{\theta}(x) - \theta \right) dx = 1. \tag{22.47}$$

The above formula can be rewritten

$$\int\limits_{-\infty}^{\infty} \frac{\partial \ln f_{x;\theta}}{\partial \theta} f_{x;\theta} \left(\hat{\theta}(x) - \theta \right) dx = 1, \tag{22.48}$$

or

$$1 = \left(\int\limits_{-\infty}^{\infty} \left(\frac{\partial \ln f_{x;\theta}}{\partial \theta} f_{x;\theta}^{1/2} \right) \left(\left(\hat{\theta}(x) - \theta \right) f_{x;\theta}^{1/2} \right) dx \right)^2. \tag{22.49}$$

Using the Schwarz inequality we can write,

$$1 \le \int\limits_{-\infty}^{\infty} \left(\left(\frac{\partial \ln f_{x;\theta}}{\partial \theta} \right)^2 f_{x;\theta} \right) dx \int\limits_{-\infty}^{\infty} \left(\hat{\theta}(x) - \theta \right)^2 f_{x;\theta} dx. \tag{22.50}$$

It is recognized in the second term the variance of θ and we can write

$$\text{var}(\hat{\boldsymbol{\theta}}) \ge \frac{1}{\int_{-\infty}^{\infty} \left(\left(\frac{\partial \ln f_{x;\theta}}{\partial \theta} \right)^2 f_{x;\theta} \right) dx}. \tag{22.51}$$

This last expression is the inequality (22.40).

The Schwarz inequality becomes equality when the two terms in the integral (22.49) are proportional

$$\left(\hat{\theta}(x) - \theta \right) f_{x;\theta}^{1/2} = K(\theta) \cdot \left(\frac{\partial \ln f_{x;\theta}(x; \theta)}{\partial \theta} f_{x;\theta}^{1/2} \right), \tag{22.52}$$

Or even $\left(\hat{\theta}(x) - \theta \right) = K(\theta) \cdot \frac{\partial \ln f_{x;\theta}(x;\theta)}{\partial \theta}$ which is the Eq. (22.42).

Other form of Cramér-Rao inequality

$$\text{var}(\hat{\boldsymbol{\theta}}) \ge \frac{1}{-E\left\{ \frac{\partial^2 \ln f_{x;\theta}(x;\theta)}{\partial \theta^2} \right\}}. \tag{22.53}$$

Indeed, we start from the integral $\int_{-\infty}^{\infty} f_{x;\theta}(x;\theta)\mathrm{d}x = 1$ and take its derivative with respect to θ:

$$\int\limits_{-\infty}^{\infty} \frac{\partial f_{x;\theta}}{\partial \theta}\mathrm{d}x = \int\limits_{-\infty}^{\infty} \frac{\partial \ln f_{x;\theta}}{\partial \theta} f_{x;\theta}\mathrm{d}x = 0. \tag{22.54}$$

Deriving a second time with respect to θ:

$$\int\limits_{-\infty}^{\infty} \left(\frac{\partial^2 \ln f_{x;\theta}}{\partial \theta^2} f_{x;\theta} + \left(\frac{\partial \ln f_{x;\theta}}{\partial \theta} \right)^2 f_{x;\theta} \right) \mathrm{d}x = 0. \tag{22.55}$$

Or, having recognized expectations

$$E\left\{ \left(\frac{\partial \ln f_{x;\theta}}{\partial \theta} \right)^2 \right\} = -E\left\{ \left(\frac{\partial^2 \ln f_{x;\theta}}{\partial \theta^2} \right) \right\}. \tag{22.56}$$

Example
In the example of Gaussian variables encountered earlier, we had

$$\frac{\partial \ln f_{x;\eta}(x;\eta)}{\partial \eta} = \sum_{i=1}^{N} \frac{(x_i - \eta)}{\sigma_0^2}. \tag{22.57}$$

We take the expectation of the square of this function. Since observations are not correlated, the cross terms are zero and we have

$$E\left\{ \left(\frac{\partial \ln f_{x;\eta}(x;\eta)}{\partial \eta} \right)^2 \right\} = \sum_{i=1}^{N} \frac{E\left\{ (x_i - \eta)^2 \right\}}{\sigma_0^4} = \frac{N\sigma_0^2}{\sigma_0^4} = \frac{N}{\sigma_0^2}. \tag{22.58}$$

From this result, we deduce from the inequality of Cramér-Rao (22.40) that any unbiased estimator of the average should satisfy the relationship

$$\mathrm{var}(\hat{\eta}) \geq \frac{\sigma_0^2}{N}. \tag{22.59}$$

The variance of the estimator of the maximum likelihood expectation calculated in (22.37) satisfies precisely to equality, as it should.

Summary
We have first recalled in this chapter the important Chi-square law for testing the statistical distribution of a collection of data. We have tested the tendency of the sum of many r.v. to follow a Gaussian distribution. We have studied the classical

methods of linear regression of a collection of measurements and presented the Tikhonov regularizing method. The method of using the L-curve elbow to determine the value of the regularizing parameter value has been qualitatively discussed. The chapter ends with the estimation of statistical parameters in the sense of maximum likelihood.

Exercises

I. A random variable x is distributed according to a normal distribution. Its mean is known but its variance is not.

1. We measure x and find the value x. Give an estimator of the variance of x at maximum likelihood.
2. We make N measurements and found the values x_i. Give an estimator of the variance of x at maximum likelihood.
3. In the context of Question 2, give the Cramér-Rao lower bound of variance.

II. An unknown parameter x is measured with two equipments having different precisions. The errors are random, Gaussian. The result of the first measurement is x_1; the measurement standard error is σ_1. The second equipment delivers the value x_2 with a standard error σ_2. (1) Given a maximum likelihood estimation (MLE) of x. What is the standard error of the estimation?
(2) Can we say that the MLE of x is the minimum variance estimator of x.
Solution:

(1) The pdf of random measurements x_1 and x_2 are:

$$f_1(x_1) = \frac{1}{\sqrt{2\pi\sigma_1^2}} e^{-\frac{(x_1-x)^2}{2\sigma_1^2}} \text{ and } f_2(x_2) = \frac{1}{\sqrt{2\pi\sigma_2^2}} e^{-\frac{(x_2-x)^2}{2\sigma_2^2}}.$$

The measurement being independent, the joint pdf is $f(x_1,x_2) = \frac{1}{\sqrt{2\pi\sigma_1^2\sigma_2^2}} e^{-\frac{(x_1-x)^2}{2\sigma_1^2}} e^{-\frac{(x_2-x)^2}{2\sigma_2^2}}$. This function may be interpreted as the likelihood function of x.

$$f_x(x) = f(x_1,x_2) = \frac{1}{\sqrt{2\pi\sigma_1^2\sigma_2^2}} e^{-\frac{(x_1-x)^2}{2\sigma_1^2}} e^{-\frac{(x_2-x)^2}{2\sigma_2^2}}.$$

To get the maximum likelihood estimation of x we apply condition (22.38): $\frac{\partial \log f_x(x)}{\partial x} = 0$. Thus $\frac{(x_1-x)}{\sigma_1^2} + \frac{(x_2-x)}{\sigma_2^2} = 0$ leading to $\bar{x} = x_1 \frac{\sigma_2^2}{\sigma_1^2 + \sigma_2^2} + x_2 \frac{\sigma_1^2}{\sigma_1^2 + \sigma_2^2}$.
We may interpret this result in saying that if, for example, the standard error σ_2 is smaller than σ_1, the weight of x_1 is smaller than that of x_2, the MLE of x will be closer to the measurement given by the best instrument (x_2 in this example).

x being the true, unknown value of the parameter, the estimation $\bar{x}$ will be a Gaussian variable with average x and variance $\sigma_{\bar{x}}^2$. We have

$$f_{\bar{x}}(\bar{x}) = \frac{1}{\sqrt{2\pi\sigma_{\bar{x}}^2}} e^{-\frac{(\bar{x}-x)^2}{2\sigma_{\bar{x}}^2}} = \frac{1}{\sqrt{2\pi\sigma_1^2\sigma_2^2}} e^{-\frac{(x_1-x)^2}{2\sigma_1^2}} e^{-\frac{(x_2-x)^2}{2\sigma_2^2}}. \text{ This last relation must be}$$

true for any value of x, for example 0 and any couple of measurements (x_1, x_2), for example $x_1 = 1$ and $x_2 = 1$. The variance of the MLE of x is thus given by: $\frac{1}{\sigma_{\bar{x}}^2} = \frac{1}{\sigma_1^2} + \frac{1}{\sigma_2^2}$.

If $\sigma_2 \ll \sigma_1$, $\frac{1}{\sigma_{\bar{x}}^2} \approx \frac{1}{\sigma_2^2}$, $\sigma_{\bar{x}}^2 \approx \sigma_2^2$, $\bar{x}$ will be very close to x_2, within a small margin error σ_2.

(2) The estimator $\bar{x} = x_1 \frac{\sigma_2^2}{\sigma_1^2 + \sigma_2^2} + x_2 \frac{\sigma_1^2}{\sigma_1^2 + \sigma_2^2}$ is unbiased since

$$E\{\bar{x}\} = E\{x_1\} \frac{\sigma_2^2}{\sigma_1^2 + \sigma_2^2} + E\{x_2\} \frac{\sigma_1^2}{\sigma_1^2 + \sigma_2^2} = x \frac{\sigma_1^2 + \sigma_2^2}{\sigma_1^2 + \sigma_2^2} = x.$$

Property (22.44) states that $\bar{x}$ is the minimum variance estimator of x.

Chapter 23
Correlation and Covariance Matrices of a Complex Random Vector

We introduce in this chapter the correlation and covariance matrices of a complex random vector. The Hermitian nature of these matrices allows their diagonalization in the basis of their orthogonal eigenvectors. These concepts are discussed on jointly Gaussian variables. We study the principal component analysis of a vector of observations and the optimum Karhunen-Loève development.

23.1 Definition of Correlation and Covariance Matrices

Let the complex random vector

$$x = \begin{bmatrix} x_1 \\ x_2 \\ \vdots \\ x_N \end{bmatrix}. \tag{23.1}$$

Its correlation matrix is defined by

$$R_{xx} = E\{xx^H\} = \begin{bmatrix} E\{|x_1|^2\} & E\{x_1x_2^*\} & \cdots & E\{x_1x_N^*\} \\ E\{x_2x_1^*\} & E\{x_2x_2^*\} & \cdots & E\{x_2x_N^*\} \\ \vdots & \vdots & \vdots & \vdots \\ E\{x_Nx_1^*\} & E\{x_Nx_2^*\} & \cdots & E\{|x_N|^2\} \end{bmatrix}. \tag{23.2}$$

This matrix is square with dimensions $N \times N$.

© Springer International Publishing Switzerland 2016
F. Cohen Tenoudji, *Analog and Digital Signal Analysis*,
Modern Acoustics and Signal Processing, DOI 10.1007/978-3-319-42382-1_23

The covariance matrix is defined by

$$C_{xx} = E\{(x - \eta_x)(x - \eta_x)^H\}. \tag{23.3}$$

We have the relationship

$$R_{xx} = C_{xx} + \eta_x \eta_x^H. \tag{23.4}$$

Correlation and covariance matrices have the Hermitian symmetry, i.e., they are equal to their transposed conjugate:

$$R_{xx} = R_{xx}^H \quad \text{and} \quad C_{xx} = C_{xx}^H. \tag{23.5}$$

23.1.1   Properties of Correlation Matrix

1.
$$\left| E\{x_i x_j^*\} \right| \leq \sqrt{E\{|x_i|^2\}} \sqrt{E\{|x_j|^2\}}. \tag{23.6}$$

Indeed, following a method that has been used previously herein we calculate

$$\begin{aligned}
E\{|(x_i + \lambda x_j)|^2\} &= E\{(x_i + \lambda x_j)(x_i + \lambda x_j)^*\} \\
&= E\{(|x_i|^2 + |\lambda|^2 |x_j|^2 + x_i \lambda^* x_j^* + x_i^* \lambda x_j)\}.
\end{aligned} \tag{23.7}$$

This expectation must be positive or zero, since it is the expectation of a square modulus which is a positive or zero number and whatever the value of the parameter λ is.

We note $E\{x_i x_j^*\} = A e^{j\varphi}$. λ is chosen such that

$$\lambda = |\lambda| e^{j\varphi}. \tag{23.8}$$

Then

$$\begin{aligned}
E\{(|x_i|^2 + |\lambda|^2 |x_j|^2 + x_i \lambda^* x_j^* + x_i^* \lambda x_j)\} &= E\{(|x_i|^2)\} + E\{(|\lambda|^2 |x_j|^2)\} \\
&\quad + E\{(x_i \lambda^* x_j^* + x_i^* \lambda x_j)\}, \\
E\{(|x_i|^2)\} + E\{(|\lambda|^2 |x_j|^2)\} + E\{(x_i \lambda^* x_j^* + x_i^* \lambda x_j)\} &= E\{(|x_i|^2)\} + |\lambda|^2 E\{(|x_j|^2)\} + 2A|\lambda|.
\end{aligned}$$

This polynomial with the modulus of λ as a variable must be positive or zero regardless $|\lambda|$, for that its discriminant must be negative or zero, i.e.,

$$A^2 - E\left\{\left(|x_i|^2\right)\right\}E\left\{\left(|x_j|^2\right)\right\} \leq 0, \tag{23.9}$$

we then find the desired formula: $\left|E\left\{x_i x_j^*\right\}\right| \leq \sqrt{E\left\{|x_i|^2\right\}}\sqrt{E\left\{|x_j|^2\right\}}$.

2. The correlation matrix is positive-semidefinite. That is to say that, for any vector a with dimension N we have

$$a^H R_{xx} a \geq 0. \tag{23.10}$$

It may be noted that the above expression implies that this quantity is a scalar since it is compared to 0.

Indeed

$$a^H R_{xx} a = a^H E\{xx^H\}a = E\{a^H xx^H a\} = E\left\{|x^H a|^2\right\}, \tag{23.11}$$

which is scalar, real, always positive or zero. The quantity $x^H a$ is scalar and $a^H x$ is its complex conjugate.

23.2   Linear Transformation of Random Vectors

Let x be a complex random vector. Let y resulting from a linear application $y = Ax$.

The expectancy calculation operation is itself a linear operation, the expectation vector of y is written $\eta_y = A\eta_x$.

The correlation matrix of the r.v. y is given by

$$R_{yy} = E\{yy^H\} = E\{(Ax)(Ax)^H\} = AE\{xx^H\}A^H. \tag{23.12}$$

So

$$R_{yy} = AR_{xx}A^H. \tag{23.13}$$

Diagonalization of the Correlation Matrix

We denote e an eigenvector of the correlation matrix

$$R_{xx}e = \lambda e. \tag{23.14}$$

Since the $(N \times N)$ matrix R_{xx} is Hermitian, it is possible to find N orthonormal eigenvectors.

$$e_l^H e_k = \delta[l - k] = \begin{cases} 1 & \text{if} & l = k \\ 0 & \text{if} & l \neq k \end{cases}. \tag{23.15}$$

Each eigenvector satisfies a relation of the type

$$R_{xx} e_k = \lambda_k e_k. \tag{23.16}$$

So we have

$$e_l^H R_{xx} e_k = \begin{cases} \lambda_k & \text{if} & l = k \\ 0 & \text{if} & l \neq k \end{cases}. \tag{23.17}$$

The matrix E whose columns are the eigenvectors of R_{xx} is defined as follows:

$$E = \begin{bmatrix} | & | & & | \\ e_1 & e_2 & \cdots & e_N \\ | & | & & | \end{bmatrix}. \tag{23.18}$$

$EE^H = I$. The matrix E is unitary.

We can write

$$E^H R_{xx} E = \begin{bmatrix} -- & e_1^H & -- \\ -- & e_2^H & -- \\ & \vdots & \\ -- & e_N^H & -- \end{bmatrix} R_{xx} \begin{bmatrix} | & | & & | \\ e_1 & e_2 & \cdots & e_N \\ | & | & & | \end{bmatrix} = \begin{bmatrix} \lambda_1 & 0 & & 0 \\ 0 & \lambda_2 & & 0 \\ \cdots & \cdots & \cdots & \cdots \\ 0 & & 0 & \lambda_N \end{bmatrix}$$
$$= \Lambda. \tag{23.19}$$

Let us define the vector x' by

$$x' = E^H x = \begin{bmatrix} -- & e_1^H & -- \\ -- & e_2^H & -- \\ & \vdots & \\ -- & e_N^H & -- \end{bmatrix} x. \tag{23.20}$$

The random variables components of x' are two by two orthogonal. Indeed

$$E\{x' x'^H\} = R_{x'x'} = E\{E^H x x^H E\} = E^H E\{x x^H\} E = E^H R_{xx} E = \Lambda \tag{23.21}$$

The correlation matrix $R_{x'x'}$ is diagonal. It is composed by the eigenvalues of R_{xx}.

We can also write

$$EE^H R_{xx} EE^H = R_{xx} = E\Lambda E^H. \tag{23.22}$$

This relationship can be used to invert the matrix R_{xx}. Since the matrix is unitary

$$E^{-1} = E^H. \tag{23.23}$$

We have

$$R_{xx}^{-1} = \left(E\Lambda E^H\right)^{-1} = E\Lambda^{-1} E^H. \tag{23.24}$$

Eigenvalues of the Correlation Matrix

As seen above, the eigenvalues of the correlation matrix are the expectations of square moduli of the random components of the vector x'.

$$E\{x_i' x_i'^*\} = E\left\{|x_i'|^2\right\} = \lambda_i \tag{23.25}$$

As being the expectations of a square modulus, the eigenvalues of the correlation matrix will therefore be positive or zero.

If the correlation matrix is not singular, its discriminant is nonzero. As the determinant remains unchanged in the unitary base change generated by the matrix E, its value is equal to the product of the eigenvalues, which will therefore be all positive (nonzero) in this case.

The trace of the correlation matrix also remains unchanged in the base change. It is therefore equal to the sum of the eigenvalues of the matrix.

23.3   Multivariate Gaussian Probability Density Functions

x is a real Gaussian random vector with dimension N when its probability density function has the form:

$$f_x(x) = \frac{1}{(2\pi)^{\frac{N}{2}} |C_{xx}|^{\frac{1}{2}}} e^{-\frac{1}{2}(x-\eta_x)^T C_{xx}^{-1}(x-\eta_x)}. \tag{23.26}$$

C_{xx} is the covariance matrix of the vector x. This matrix is symmetric
C_{xx}^{-1} is the inverse matrix of the covariance matrix of x. It is also symmetrical
$(x - \eta)^T$ is the transposed vector of centered vector x_c

If x is a complex Gaussian random vector of dimension N, its probability density function has the form

$$f_x(x) = \frac{1}{\pi^N |C_{xx}|} e^{-(x-\eta_x)^H C_{xx}^{-1}(x-\eta_x)}. \tag{23.27}$$

$(x - \eta)^H$ is the conjugate transpose (said Hermitian conjugate) of the centered vector x_c

The covariance matrix C_{xx} of the vector x has Hermitian symmetry; it is equal to its conjugate transpose. The inverse matrix C_{xx}^{-1} of the covariance matrix is also Hermitian.

Conditional Probability Density Functions

If x and y are real Gaussian random vectors with N and M dimensions, respectively, the probability density of y conditioned by x has the form

$$f_{y|x}(y|x) = \frac{1}{(2\pi)^{\frac{M}{2}} |C_{y|x}|^{\frac{1}{2}}} e^{-\frac{1}{2}(y-\eta_{y|x})^T C_{y|x}^{-1}(y-\eta_{y|x})}. \tag{23.28}$$

If x and y are complex Gaussian random vectors with N and M dimensions, respectively, the probability density of y conditioned by x has the form:

$$f_{y|x}(y|x) = \frac{1}{\pi^M |C_{y|x}|} e^{-(y-\eta_{y|x})^H C_{y|x}^{-1}(y-\eta_{y|x})}. \tag{23.29}$$

Example of Two Real Jointly Gaussian Random Vectors

Let x be a random vector whose two components are jointly Gaussian random variables. By definition, the vector x probability density function is given by

$$f_x(x) = \frac{1}{(2\pi) |C_{xx}|^{\frac{1}{2}}} e^{-\frac{1}{2}(x-\eta_x)^T C_{xx}^{-1}(x-\eta_x)}. \tag{23.30}$$

The vector of expectations is

$$\eta = \begin{pmatrix} \eta_1 \\ \eta_2 \end{pmatrix}. \tag{23.31}$$

We note x_c the centered vector. We have

$$f_{x_c}(x_c) = \frac{1}{(2\pi) |C_{xx}|^{\frac{1}{2}}} e^{-\frac{1}{2}(x_c)^T C_{xx}^{-1}(x_c)}. \tag{23.32}$$

To illustrate these properties, we take the following example of a covariance matrix:

$$C_{xx} = \begin{pmatrix} 8 & 2 \\ 2 & 2 \end{pmatrix}.$$

Its inverse is $C_{xx}^{-1} = \begin{pmatrix} 0.1667 & -0.1667 \\ -0.1667 & 0.6667 \end{pmatrix}.$

The eigenvalues of the covariance matrix are 8.606 and 1.394, and the eigenvalues of the inverse matrix are 0.1162 and 0.7171. These values are the inverse of the eigenvalues of the covariance matrix.

The matrix C_{xx}^{-1} is diagonalizable $\mathrm{Diag}_{C_{xx}^{-1}} = \begin{pmatrix} 0.1162 & 0 \\ 0 & 0.7171 \end{pmatrix}.$

The eigenvectors with norm 1 of the matrix C_{xx}^{-1} corresponding to the eigenvalue 0.1162 and 0.7171 are, respectively, $\begin{pmatrix} 0.9571 \\ 0.2898 \end{pmatrix}$ and $\begin{pmatrix} -0.2898 \\ 0.9571 \end{pmatrix}.$

It is verified that these two vectors are orthogonal. These eigenvectors are also eigenvectors of the matrix C_{xx}.

Passage from Vector x to Vector x'

As seen above, we define vector x' by $x' = E^H x$,

Where $E = \begin{bmatrix} | & | \\ e_1 & e_2 \\ | & | \end{bmatrix}$ is the matrix composed of the eigenvectors of C_{xx},

and $E^H = \begin{bmatrix} - & e_1^H & - \\ - & e_2^H & - \\ - & - & - \end{bmatrix}$, the Hermitian conjugate matrix of these vectors.

Thus continuing the previous example

$$E = \begin{bmatrix} 0.9571 & -0.2898 \\ 0.2898 & 0.9571 \end{bmatrix}; \quad E^H = \begin{bmatrix} 0.9571 & 0.2898 \\ -0.2898 & 0.9571 \end{bmatrix}.$$

The relationship between x and x' is given by $\begin{pmatrix} x_1' \\ x_2' \end{pmatrix} = \begin{pmatrix} 0.9571 & 0.2898 \\ -0.2898 & 0.9571 \end{pmatrix} \begin{pmatrix} x_1 \\ x_2 \end{pmatrix}.$

The quadratic form appearing in the exponential who wrote

$$Q = (x_1 \quad x_2) \begin{pmatrix} 0.1667 & -0.1667 \\ -0.1667 & 0.6667 \end{pmatrix} \begin{pmatrix} x_1 \\ x_2 \end{pmatrix}.$$

becomes $Q = (x_1' \quad x_2') \begin{pmatrix} 0.1162 & 0 \\ 0 & 0.7171 \end{pmatrix} \begin{pmatrix} x_1' \\ x_2' \end{pmatrix}.$

So,

$$Q = 0.1162x_1'^2 + 0.7171x_2'^2 = \frac{x_1'^2}{8.606} + \frac{x_2'^2}{1.394}. \tag{23.33}$$

Because $C_{x_1'x_2'} = 0$ we conclude that the components x_1' and x_2' are not correlated. This implies independence for two Gaussian r.v.. It is thus possible, by a change of basis, to diagonalize the inverse of the covariance matrix and create a random vector whose components are Gaussian and independent in probability. In (23.33) we recognize the equation of an ellipse with the semi-axes equal to 2.9336 and 1.181.

We have $x = \left(E^H\right)^{-1}x'$ with $\left(E^H\right)^{-1} = \begin{pmatrix} 0.9571 & 0.2898 \\ -0.2898 & 0.9571 \end{pmatrix}^{-1} = \begin{pmatrix} 0.9571 & -0.2898 \\ 0.2898 & 0.9571 \end{pmatrix}.$

One can express the direction cosines of the unitary vectors of the new base in the former

$$\begin{pmatrix} \cos\alpha \\ \sin\alpha \end{pmatrix} = \begin{pmatrix} 0.9571 & -0.2898 \\ 0.2898 & 0.9571 \end{pmatrix}\begin{pmatrix} 1 \\ 0 \end{pmatrix} = \begin{pmatrix} 0.9571 \\ 0.2898 \end{pmatrix}.$$

$$\begin{pmatrix} \cos\beta \\ \sin\beta \end{pmatrix} = \begin{pmatrix} 0.9571 & -0.2898 \\ 0.2898 & 0.9571 \end{pmatrix}\begin{pmatrix} 0 \\ 1 \end{pmatrix} = \begin{pmatrix} -0.2898 \\ 0.9571 \end{pmatrix}.$$

23.4 Estimation of the Correlation Matrix from Observations

In some cases, we do not know the correlation matrix but we have only K vectors $x^{(K)}$ that are independent realizations of the random vector x.

These data are used to estimate the correlation function by noting that the expectation of any function $\psi(x)$ of x can be estimated by averaging the values of this function on the K realizations

$$E\{\psi(x)\} \cong \frac{1}{K}\sum_{k=1}^{K}\psi\left(x^{(K)}\right). \tag{23.34}$$

Example

$$x^{(1)} = \begin{bmatrix} 1 \\ 0 \end{bmatrix}; \quad x^{(2)} = \begin{bmatrix} -2 \\ 1 \end{bmatrix}; \quad x^{(3)} = \begin{bmatrix} 2 \\ -2 \end{bmatrix}; \quad x^{(4)} = \begin{bmatrix} 0 \\ 2 \end{bmatrix}. \tag{23.35}$$

Estimation of the expectation vector

$$\hat{\eta}_x \cong \frac{1}{K}\sum_{k=1}^{K}x^{(K)} = \frac{1}{4}\left\{\begin{bmatrix} 1 \\ 0 \end{bmatrix} + \begin{bmatrix} -2 \\ 1 \end{bmatrix} + \begin{bmatrix} 2 \\ -2 \end{bmatrix} + \begin{bmatrix} 0 \\ 2 \end{bmatrix}\right\} = \begin{bmatrix} \frac{1}{4} \\ \frac{1}{4} \end{bmatrix}. \tag{23.36}$$

Estimation of the correlation matrix

$$
\begin{aligned}
\hat{R}_{xx} &= \frac{1}{K}\sum_{k=1}^{K} x^{(K)} x^{(K)H} \\
&= \frac{1}{4}\left\{ \begin{bmatrix} 1 \\ 0 \end{bmatrix} [1\ \ 0] + \begin{bmatrix} -2 \\ 1 \end{bmatrix}[-2\ \ 1] + \begin{bmatrix} 2 \\ -2 \end{bmatrix}[2\ \ -2] + \begin{bmatrix} 0 \\ 2 \end{bmatrix}[0\ \ 2] \right\} \\
&= \begin{bmatrix} \frac{9}{4} & -\frac{3}{2} \\ -\frac{3}{2} & \frac{9}{4} \end{bmatrix}.
\end{aligned}
$$

$$(23.37)$$

We can formulate the calculation in a more condensed manner. We define the data matrix X composed of sampled vectors

$$
X = \begin{bmatrix} - & x^{(1)H} & - \\ - & x^{(2)H} & - \\ & \vdots & \\ & \vdots & \\ - & x^{(K)H} & - \end{bmatrix}.
$$

$$(23.38)$$

The estimator of the correlation function can be written

$$
\hat{R}_{xx} = \frac{1}{K} X^{H} X.
$$

$$(23.39)$$

In the above example the last two expressions take the form

$$
X = \begin{bmatrix} 1 & 0 \\ -2 & 1 \\ 2 & -2 \\ 0 & 2 \end{bmatrix};\quad
\hat{R}_{xx} = \frac{1}{4} \begin{bmatrix} 1 & -2 & 2 & 0 \\ 0 & 1 & -2 & 2 \end{bmatrix} \begin{bmatrix} 1 & 0 \\ -2 & 1 \\ 2 & -2 \\ 0 & 2 \end{bmatrix} = \begin{bmatrix} \frac{9}{4} & -\frac{3}{2} \\ -\frac{3}{2} & \frac{9}{4} \end{bmatrix}.
$$

$$(23.40)$$

It is noted that the data matrix can be partitioned in columns vectors

$$
X = \begin{bmatrix} \vdots & \vdots & & \vdots \\ \vdots & \vdots & & \vdots \\ x_1 & x_2 & \cdots & x_N \\ \vdots & \vdots & & \vdots \\ \vdots & \vdots & & \vdots \end{bmatrix}.
$$

$$(23.41)$$

An element of the matrix of the estimator of the correlation function can be written

$$\hat{r}_{kl} = \frac{1}{K} x_k^H x_l.$$

23.5 Karhunen-Loève Development

23.5.1 Example of Using the Correlation and Covariance Matrices

This method is also called principal component development. We choose to begin the presentation of this method with an example:

We assume that we measure the radiation emitted by vegetation by remote sensing using four optical sensors operating in different light spectral bands. The amplitudes appear as jointly Gaussian random variables. The flow rate of the communication channel between the satellite and the earth is limited, it is desired to transmit only to earth the data that contain the most information and are statistically independent.

The results of a preliminary calibration of such sensors have shown that the r.v. are real and that the expectation vector is

$$\eta_x = \begin{bmatrix} 0.17 \\ 0.7 \\ -0.13 \\ 0.21 \end{bmatrix}, \tag{23.42}$$

and that the correlation matrix between the four channels is

$$R_{xx} = \begin{bmatrix} 5.3289 & 5.019 & -4.8221 & 0.3357 \\ 5.019 & 9.59 & -1.091 & -3.953 \\ -4.8221 & -1.091 & 9.2169 & -3.8273 \\ 0.3357 & -3.953 & -3.8273 & 4.3441 \end{bmatrix}. \tag{23.43}$$

Note that the correlation matrix is symmetrical as expected. We seek to diagonalize the covariance matrix which is

$$C_{xx} = R_{xx} - \eta_x \eta_x^H = \begin{bmatrix} 5.3 & 4.9 & -4.8 & 0.3 \\ 4.9 & 9.1 & -1.0 & -4.1 \\ -4.8 & -1.0 & 9.2 & -3.8 \\ 0.3 & -4.1 & -3.8 & 4.3 \end{bmatrix}. \tag{23.44}$$

The diagonalization of the covariance matrix gives the matrix of the eigenvalues and the corresponding eigenvectors, respectively,

$$
\Lambda = \begin{bmatrix}
15.0 & 0 & 0 & 0 \\
0 & 0.7379 & 0 & 0 \\
0 & 0 & 0.0268 & 0 \\
0 & 0 & 0 & 12.1353
\end{bmatrix};
$$

$$
E = \begin{bmatrix}
-0.5774 & -0.6285 & -0.5211 & -0.01 \\
-0.5774 & 0.0467 & 0.5723 & 0.5805 \\
0.5774 & -0.5818 & -0.0512 & 0.5705 \\
0 & -0.5140 & 0.6312 & -0.5809
\end{bmatrix}.
$$

(23.45)

Note that two eigenvalues are clearly distinguishable $\lambda_1 = 15.0$ and $\lambda_4 = 12.1353$. The other two eigenvalues are much smaller.

The eigenvectors of the covariance matrix define four orthogonal directions in space. The traces of matrices C_{xx} and Λ are equal. But the values are concentrated (the power carried by the random signal components is concentrated) on the values 1 and 4.

The relationship $x' = E^H x = \begin{bmatrix} -- & e_1^H & -- \\ -- & e_2^H & -- \\ & \vdots & \\ -- & e_N^H & -- \end{bmatrix} x$ may be interpreted as a

rotation in the four-dimensional space effected on the data vector x.

The components of x' represent the components of the random measurements vector along the four orthogonal directions. The first and fourth components are those that have the greatest variance, so that will hold the most power, so the more information.

Practically, having obtained a measured value x_1, x_2, x_3 and x_4, from the four sensors we apply to that vector x the application E^H,

$$
\begin{bmatrix} x_1' \\ x_2' \\ x_2' \\ x_{4'}' \end{bmatrix} = E^H \begin{bmatrix} x_1 \\ x_2 \\ x_2 \\ x_4 \end{bmatrix} = \begin{bmatrix}
-0.5774 & -0.5774 & 0.5774 & 0 \\
-0.6285 & 0.0467 & -0.5818 & -0.5140 \\
-0.5211 & 0.5723 & -0.0512 & 0.6312 \\
-0.01 & 0.5805 & 0.5705 & -0.5809
\end{bmatrix} \begin{bmatrix} x_1 \\ x_2 \\ x_2 \\ x_4 \end{bmatrix}.
$$

(23.46)

We will transmit on the communications channel only those components x_1' and x_4' which are the independent random variables (equivalence between independence and orthogonality for Gaussian variables) that have the largest variances. The compression ratio of the data is two.

23.5.2 Theoretical Aspects

The above example lies within the scope of the Karhunen-Loève expansion that we detail now.

Let us consider a segment of a random sequence

$$\{x[n]; \quad n = 0, 1, \ldots, N - 1\}. \tag{23.47}$$

This segment can be developed on any base formed of a sequence $\varphi_i[n]$ of deterministic orthonormal functions

$$\{x[n] = \kappa_1 \varphi_1[n] + \kappa_2 \varphi_2[n] + \cdots + \kappa_N \varphi_N[n]\}, \tag{23.48}$$

With the functions $\varphi_i[n]$ satisfying the relationship

$$\sum_{n=0}^{N-1} \varphi_i^*[n]\varphi_j[n] = \begin{cases} 1 & i = j \\ 0 & i \neq j \end{cases}. \tag{23.49}$$

The coefficients κ_i can then be calculated by

$$\kappa_i = \sum_{n=0}^{N-1} \varphi_i^*[n]x[n]. \tag{23.50}$$

We now want a particular set of functions $\varphi_i[n]$ to perform the statistical orthogonality

$$E\{\kappa_i \kappa_j^*\} = \begin{cases} \zeta_i^2 & \text{if} \quad i = j \\ 0 & \text{if} \quad i \neq j \end{cases}. \tag{23.51}$$

The vector of coefficients is defined as

$$\kappa = \begin{bmatrix} \kappa_1 \\ \kappa_2 \\ \vdots \\ \kappa_N \end{bmatrix}. \tag{23.52}$$

And the matrix

$$\Phi = \begin{bmatrix} | & | & & | \\ \varphi_1 & \varphi_2 & \cdots & \varphi_N \\ | & | & & | \end{bmatrix}, \quad \text{with} \quad \varphi_i = \begin{bmatrix} \varphi_i[0] \\ \varphi_i[1] \\ \vdots \\ \varphi_i[N-1] \end{bmatrix}, i = 1, 2, \ldots, N \tag{23.53}$$

Note that the column vectors of Φ are orthonormal and satisfy

$$\varphi_i^H \varphi_j = \begin{vmatrix} 1 & \text{if} & i = j \\ 0 & \text{if} & i \neq j \end{vmatrix}. \tag{23.54}$$

The matrix is Φ unitary. Equation (23.50) can be rewritten in matrix form:

$$x = \begin{bmatrix} | & | & & | \\ \varphi_1 & \varphi_2 & \cdots & \varphi_N \\ | & | & & | \end{bmatrix} \begin{bmatrix} \kappa_1 \\ \kappa_2 \\ \\ \kappa_N \end{bmatrix} = \Phi \kappa, \tag{23.55}$$

and

$$\kappa = \begin{bmatrix} -- & \varphi_1^H & -- \\ -- & \varphi_2^H & -- \\ & \vdots & \\ -- & \varphi_N^H & -- \end{bmatrix} x = \Phi^H x. \tag{23.56}$$

One can give the following interpretation: If we consider the sequence $x[n]$ as a vector in a N-dimensional space, the coefficients κ_i appear to be the components of the same vector in a coordinate system obtained by rotation.

The eigenvectors of the correlation matrix having the orthogonality property may be selected as vectors φ_i. Then:

$$R_{xx} \varphi_i = \lambda_i \varphi_i. \tag{23.57}$$

If the vector $x[n]$ can be viewed as consisting of N values of a wide sense stationary signal, the correlation matrix is Hermitian and Toeplitz and we can write

$$\sum_{k=0}^{N-1} R_{xx}[l - k] \varphi_i[k] = \lambda_i \varphi_i[l], \quad i = 0, 1, \ldots, N - 1. \tag{23.58}$$

We will now show that the vectors φ_i may be used to achieve the Eq. (23.51) of the stochastic orthogonality of coefficients κ_i.

We note

$$\kappa_i = \sum_{n=0}^{N-1} \varphi_i^*[n] x[n] \quad \text{and} \quad \kappa_j^* = \sum_{m=0}^{N-1} \varphi_j[m] x^*[m]. \tag{23.59}$$

$$E\left\{ \kappa_i \kappa_j^* \right\} = \sum_{n=0}^{N-1} \sum_{m=0}^{N-1} \varphi_i^*[n] E\{x[n] x^*[m]\} \varphi_j[m] = \sum_{n=0}^{N-1} \sum_{m=0}^{N-1} \varphi_i^*[n] R_{xx}[n - m] \varphi_j[m]. \tag{23.60}$$

or, based on the above relationship

$$E\left\{\kappa_i\kappa_j^*\right\} = \lambda_j \sum_{n=0}^{N-1} \varphi_i^*[n]\varphi_j[n] = \begin{vmatrix} \lambda_j & \text{if} & i=j \\ 0 & \text{if} & i \neq j \end{vmatrix}. \tag{23.61}$$

In this context, the Eq. (23.50) is called the Karhunen-Loève development.

It is also shown that the development (23.50) in eigenfunctions of the correlation function is the only one that satisfies the Eq. (23.51).

23.5.3 Optimality of Karhunen-Loève Development

We want to approach the vector $x[n]$ by a linear combination of functions $\varphi_i[n]$ but with a number $M < N$ of these functions. $x[n]$ estimator is noted $\hat{x}[n]$

$$\hat{x}[n] = \sum_{i=1}^{M} \kappa_i \varphi_i[n]; \quad M < N. \tag{23.62}$$

The error sequence is defined by

$$\varepsilon[n] = x[n] - \hat{x}[n]. \tag{23.63}$$

The problem is to find the coefficients κ_i and the base functions $\varphi_i[n]$ that minimize the error. We chose to minimize the squared error

$$\varepsilon = E\left\{\sum_{n=0}^{N-1} |\varepsilon|^2[n]\right\}. \tag{23.64}$$

We can write

$$x = \sum_{i=1}^{N} \kappa_i \varphi_i = \underbrace{\sum_{i=1}^{M} \kappa_i \varphi_i}_{\hat{x}} + \underbrace{\sum_{i=M+1}^{N} \kappa_i \varphi_i}_{\varepsilon}. \tag{23.65}$$

$$\varepsilon = E\{\varepsilon^H \varepsilon\} = E\left\{\left(\sum_{i=M+1}^{N} \kappa_i^* \varphi_i^H\right)\left(\sum_{i=M+1}^{N} \kappa_i \varphi_i\right)\right\} = \sum_{i=M+1}^{N} E\left\{|\kappa_i|^2\right\}. \tag{23.66}$$

The error may be written in the form

$$\varepsilon = \sum_{i=M+1}^{N} \varphi_i^H R_{xx} \varphi_i. \tag{23.67}$$

We must now minimize (23.67) within the constraints

$$\varphi_i^H \varphi_i = 1 \quad \text{for} \quad i = M+1, M+2, \ldots, N. \tag{23.68}$$

This problem is that of a constrained minimization. The Lagrange multipliers method is used. We define the function

$$L = \sum_{i=M+1}^{N} \varphi_i^H R_{xx} \varphi_i + \sum_{i=M+1}^{N} \lambda_i \left(1 - \varphi_i^H \varphi_i \right). \tag{23.69}$$

Taking the gradient with respect to φ_i^H and get

$$\nabla_{\varphi_i^H} L = R_{xx} \varphi_i - \lambda_i \varphi_i = 0, \tag{23.70}$$

or:

$$R_{xx} \varphi_i = \lambda_i \varphi_i \quad \text{for} \quad i = M+1, M+2, \ldots, N. \tag{23.71}$$

This implies that the vectors φ_i should be eigenvectors of the correlation matrix. The error is

$$\varepsilon = \sum_{i=M+1}^{N} \varphi_i^H R_{xx} \varphi_i = \sum_{i=M+1}^{N} \varphi_i^H \lambda_i \varphi_i = \sum_{i=M+1}^{N} \lambda_i. \tag{23.72}$$

So, when choosing the M eigenvectors among those with the largest eigenvalues, we minimize the error committed in the approximation. This is the principle of the optimal Karhunen-Loève development.

Summary
We have introduced in this chapter the correlation and covariance matrices of a complex random vector. We have demonstrated that due to their Hermitian nature, matrices may be diagonalized in the basis of their orthogonal eigenvectors. We have used Gaussian variables to illustrate these concepts. We studied the principal component analysis of a vector of observations and established the optimum Karhunen-Loève development.

Chapter 24
Correlation Functions, Spectral Power Densities of Random Signals

To keep continuity with the previous part of the book on digital signals, we choose to show first in this chapter the properties of digital random signals, knowing that the treatments today are mainly performed on digital signals. We encounter in this chapter two functions which are fundamental for signal analysis: the correlation function and the power spectral density (PSD). These functions are defined for wide sense stationary (WSS) random signals, signals whose first two moments of the signal values at two different instants are constant over time. The PSD is defined as the Fourier transform of the correlation function. We study the filtering of WSS signals by LTI systems and give the theorems linking correlations and DSP of input and output signals. White noise filtering by a first order autoregressive system is treated as an example. The coherence function defined afterward is a powerful tool to identify and quantify in a noisy signal the sources constituting the noise. At the end of this chapter, we give a brief definition of correlation functions and power spectral densities of analog signals. This gives more intuitive demonstrations of certain applications such as the influence of a filter for increasing the signal-to-noise ratio or matched filtering of a noisy signal with random noise. Many exercises with worked solutions at the end of this chapter will help the reader become familiar with the results, important in signal analysis.

24.1 Correlation Function of a Random Signal

Let $x[n]$ be a random signal, which is assumed here complex for more generality.

For any two times n_1 and n_2, $x[n_1]$ and $x[n_2]$ are r.v. each with their first order statistics. In particular, their expectations are

$$E\{x[n_1]\} = \eta_{x[n_1]} \quad \text{and} \quad E\{x[n_2]\} = \eta_{x[n_2]}. \tag{24.1}$$

A priori, these expectations may be different.

© Springer International Publishing Switzerland 2016
F. Cohen Tenoudji, *Analog and Digital Signal Analysis*,
Modern Acoustics and Signal Processing, DOI 10.1007/978-3-319-42382-1_24

Their variances are

$$\text{var}(x[n_1]) = \text{cov}(x[n_1], x[n_1]) = E\left\{\left|x[n_1] - \eta_{x[n_1]}\right|^2\right\}, \tag{24.2}$$

$$\text{var}(x[n_2]) = \text{cov}(x[n_2], x[n_2]) = E\left\{\left|x[n_2] - \eta_{x[n_2]}\right|^2\right\}. \tag{24.3}$$

which also differ a priori.

From Eq. (24.2) it follows that

$$\text{var}(x[n_1]) = E\{x[n_1]x^*[n_1]\} - \eta_{x[n_1]}\eta^*{}_{x[n_1]}. \tag{24.4}$$

Note the complex conjugation which allows treatment of the general case of complex signals.

Let us consider the two r.v. $x[n_1]$ and $x[n_2]$; We call correlation function the expectancy of the product

$$R_{x[n_1]x[n_2]}[n_1, n_2] = E\{x[n_1]x^*[n_2]\}. \tag{24.5}$$

In the general case $R_{xx}[n_1, n_2]$ is a function of the two times n_1 and n_2.

24.1.1   Correlation Function of a Wide Sense Stationary (WSS) Signal

A very important special case occurs when the function $R_{xx}[n_1, n_2]$ depends only on the proximity of times n_1 and n_2 regardless of the absolute position n_1. This is the case of wide sense stationary signals (WSS).

We say that a signal is **wide sense stationary** (WSS) iff:

(a) The expectation $E\{x[n]\} = \eta_{x[n]}$ is independent of time. Therefore

$$\eta_{x[n]} = \eta_x. \tag{24.6}$$

(b) The correlation function $R_{x[n_1]x[n_2]}[n_1, n_2]$ depends only on the time difference $m = n_1 - n_2$.

Note:

$$R_{xx}[n_1, n_2] = R_{xx}[n_2 + m, n_2] = E\{x[n_2 + m]x^*[n_2]\} = E\{x[n + m]x^*[n]\} = R_{xx}[m]. \tag{24.7}$$

These signals are often encountered in practice.

The correlation function $R_{xx}[m]$ is also called autocorrelation function of the signal $x[n]$.

Power of a WSS Signal

By definition, the power of a wide sense stationary signal is given by

$$P[n] = E\{x[n]x^*[n]\}. \tag{24.8}$$

This is a priori a function of time in the general case.

The time interval between the two values of x is zero, we have

$$P[n] = E\{x[n]x^*[n]\} = R_{xx}[n-n] = R_{xx}[0]. \tag{24.9}$$

It is noted that the power is constant over time $P[n] = P = Cte$.

24.1.2 Properties of the Correlation Function

(a) The maximum of the autocorrelation function of a WSS signal is located at the origin.

We have

$$|R_{xx}[m]| \leq R_{xx}[0] \tag{24.10}$$

To demonstrate this property, in the following, the expectation of the square is calculated with a complex parameter λ:

$$E\left\{|x[n+m] - \lambda x[n]|^2\right\} = E\left\{|x[n+m]|^2\right\} - E\{\lambda x[n]x^*[n+m] - \lambda^* x^*[n]x[n+m]\}$$
$$+ E\left\{|\lambda x[n]|^2\right\}.$$
$$E\left\{|x[n+m] - \lambda x[n]|^2\right\} = \left(1 + |\lambda|^2\right)R_{xx}[0] - \lambda R_{xx}^*[m] - \lambda^* R_{xx}[m].$$

$$\tag{24.11}$$

This expectation of a square modulus is necessarily positive or zero, regardless of the parameter λ value.

We note $R_{xx}[m] = |R_{xx}[m]|e^{j\varphi[m]}$. We choose λ such that $\lambda = |\lambda|e^{j\varphi[m]}$. Expression (24.11) becomes

$$\left(1 + |\lambda|^2\right)R_{xx}[0] - 2|\lambda||R_{xx}[m]| \geq 0. \tag{24.12}$$

For the polynomial to be always positive or zero, it is necessary that the discriminant of the second degree polynomial in $|\lambda|$ is negative or zero. So we will have

$|R_{xx}[m]|^2 - R_{xx}^2[0] \leq 0$, so, as $R_{xx}[0]$ is positive $|R_{xx}[m]| \leq R_{xx}[0]$.

(b)   The autocorrelation function is a symmetric conjugate function

$$R_{xx}[m] = E\{x[n+m]x^*[n]\} = R_{xx}^*[-m]. \tag{24.13}$$

If the signal $x[n]$ is real, $R_{xx}[m] = R_{xx}[-m]$, the correlation function is even.

24.1.3   Centered White Noise

By definition, a centered **white noise** is a real random signal of zero mean, wide sense stationary, whose correlation function is given by the Kronecker function $\delta[m]$:

$$R_{xx}[m] = N\delta[m]. \tag{24.14}$$

24.2   Filtering a Random Signal by a LTI Filter

24.2.1   Expected Values

The random signal to the filter input is noted $x[n]$. The output signal $y[n]$ has the form

$$y[n] = \sum_{m=-\infty}^{\infty} x[n-m]h[m] = x[n] \otimes h[n]. \tag{24.15}$$

The expectation of the output signal $y[n]$ is

$$E\{y[n]\} = \eta_y[n] = \sum_{m=-\infty}^{\infty} E\{x[n-m]\}h[m] = \sum_{m=-\infty}^{\infty} \eta_x[n-m]h[m]. \tag{24.16}$$

If $\eta_x[n-m]$ is constant (especially, if $x[n]$ is stationary), we have

$$\eta_y = \eta_x \sum_{m=-\infty}^{\infty} h[m] = \eta_x H(1). \tag{24.17}$$

It follows that if expectancy of $x[n]$ is zero, the expectancy of $y[n]$ will be zero too.

24.2.2 Correlation Functions of Input and Output Signals

Assume now that signal $x[n]$ is WSS. We calculate the cross-correlation functions of the input and output signals:

$$R_{xy}[n+m,n] = E\{x[n+m]y^*[n]\},\tag{24.18}$$

$$R_{xy}[n+m,n] = \sum_{m'=-\infty}^{\infty} E\{x[n+m]x^*[n-m']\}h^*[m'],\tag{24.19}$$

$$R_{xy}[n+m,n] = \sum_{m'=-\infty}^{\infty} R_{xx}[m+m']\,h^*[m'].\tag{24.20}$$

We note $m'' = -m'$; $R_{xy}[n+m,n] = R_{xy}[m] = \sum_{m''=-\infty}^{\infty} R_{xx}[m-m'']h^*[-m'']$.

$$R_{xy}[m] = R_{xx}[m] \otimes h^*[-m].\tag{24.21}$$

$R_{xy}[m]$ is the convolution of $R_{xx}[m]$ and $h^*[-m]$.
Similarly, we have

$$R_{yx}[n+m,n] = E\{y[n+m]x^*[n]\}.\tag{24.22}$$

$$R_{yx}[n+m,n] = \sum_{m'=-\infty}^{\infty} E\{x[n+m-m']x^*[n]\}\,h[m'],\tag{24.23}$$

$$R_{yx}[n+m,n] = \sum_{m'=-\infty}^{\infty} R_{xx}[m-m']h[m'] = R_{xx}[m] \otimes h[m].\tag{24.24}$$

Calculation of the Autocorrelation Function of the Output Signal

$$R_{yy}[n+m,n] = E\{y[n+m]y^*[n]\}.\tag{24.25}$$

$$R_{yy}[n+m,n] = E\{y[n+m]y^*[n]\} = \sum_{m'=-\infty}^{\infty} E\{x[n+m-m']y^*[n]\}\,h[m'],$$

$$R_{yy}[n+m,n] = \sum_{m'=-\infty}^{\infty} R_{xy}[m-m']h[m'] = R_{xy}[m] \otimes h[m].\tag{24.26}$$

So $R_{yy}[n+m,n]$ no longer depends on n and we will write

$$R_{yy}[m] = R_{xy}[m] \otimes h[m]. \tag{24.27}$$

It follows that the filter output signal is also wide sense stationary.
Finally a double convolution leads from $R_{xx}[m]$ to $R_{yy}[m]$

$$R_{yy}[m] = R_{xx}[m] \otimes h^*[-m] \otimes h[m]. \tag{24.28}$$

24.3   Power Spectral Density of a WSS Signal

The signal $x[n]$ is assumed wide sense stationary. The power spectral density of $x[n]$ is, by definition, the Fourier transform of the autocorrelation function

$$S_{xx}\left(e^{j\omega T}\right) = \sum_{n=-\infty}^{\infty} R_{xx}[n]e^{-jn\omega T}. \tag{24.29}$$

The spectral density is real. Indeed

$$S_{xx}^*\left(e^{j\omega T}\right) = \left(\sum_{n=-\infty}^{\infty} R_{xx}[n]e^{-jn\omega T}\right)^* = \sum_{n=-\infty}^{\infty} R_{xx}^*[n]e^{jn\omega T} = \sum_{n=-\infty}^{\infty} R_{xx}[-n]e^{jn\omega T}$$

$$= \sum_{n'=-\infty}^{\infty} R_{xx}[n']e^{-jn'\omega T} = S_{xx}\left(e^{j\omega T}\right).$$

So,

$$S_{xx}^*\left(e^{j\omega T}\right) = S_{xx}\left(e^{j\omega T}\right). \tag{24.30}$$

Conversely, one can evaluate the correlation function by taking the inverse Fourier transform of the power spectral density:

$$R_{xx}[n] = \frac{1}{\omega_e} \int_{-\frac{\omega_e}{2}}^{+\frac{\omega_e}{2}} S_{xx}\left(e^{j\omega T}\right) e^{jn\omega T} d\omega. \tag{24.31}$$

The cross-spectrum $S_{xy}\left(e^{j\omega T}\right)$ is defined by:

$$S_{xy}\left(e^{j\omega T}\right) = \sum_{n=-\infty}^{\infty} R_{xy}[n]e^{-jn\omega T}. \tag{24.32}$$

By taking the Fourier transform of the convolution product we have

$$S_{xy}\left(e^{j\omega T}\right) = \sum_{n=-\infty}^{\infty} (R_{xx}[n] \otimes h^*[-n])e^{-jn\omega T},$$

$$S_{xy}\left(e^{j\omega T}\right) = S_{xx}\left(e^{j\omega T}\right) \sum_{n=-\infty}^{\infty} h^*[-n]e^{-jn\omega T} = S_{xx}\left(e^{j\omega T}\right)H^*\left(e^{j\omega T}\right). \tag{24.33}$$

Similarly, we calculate

$$S_{yy}\left(e^{j\omega T}\right) = S_{xx}\left(e^{j\omega T}\right)H^*\left(e^{j\omega T}\right)H\left(e^{j\omega T}\right) = S_{xx}\left(e^{j\omega T}\right)\left|H\left(e^{j\omega T}\right)\right|^2. \tag{24.34}$$

This important expression gives the relationship between the spectral densities of the input and output signals of a filter with frequency response $H(e^{j\omega T})$.

More generally we define the z function

$$S_{xx}(z) = \sum_{n=-\infty}^{\infty} R_{xx}[n]z^{-n}. \tag{24.35}$$

Property

$$S_{xx}^*(1/z^*) = \left(\sum_{n=-\infty}^{\infty} R_{xx}[n]z^{*n}\right)^* = \sum_{n=-\infty}^{\infty} R_{xx}^*[n]z^n = \sum_{n=-\infty}^{\infty} R_{xx}[-n]z^n$$

$$= \sum_{n'=-\infty}^{\infty} R_{xx}[n']z^{-n'} = S_{xx}(z),$$

thus

$$S_{xx}^*(1/z^*) = S_{xx}(z). \tag{24.36}$$

We have also

$$S_{xy}(z) = \sum_{n=-\infty}^{\infty} (R_{xx}[n] \otimes h^*[-n])z^{-n} = S_{xx}(z)\sum_{n=-\infty}^{\infty} h^*[-n]z^{-n} = S_{xx}(z)H^*\left(\frac{1}{z^*}\right),$$

$$\tag{24.37}$$

and also

$$S_{yy}(z) = S_{xx}(z)H^*\left(\frac{1}{z^*}\right)H(z). \tag{24.38}$$

This last expression is the generalization of relationship (24.34) to the entire z-plane. It is widely used in filter modeling as in Wiener filter modeling.

Power Spectral Density of Centered White Noise

Let $R_{xx}[m] = N\delta[m]$. Its spectral density is

$$S_{xx}\left(e^{j\omega T}\right) = \sum_{n=-\infty}^{\infty} R_{xx}[n]e^{-jn\omega T} = N \sum_{n=-\infty}^{\infty} \delta[n]e^{-jn\omega T} = N. \tag{24.39}$$

This density is uniform throughout the frequency axis; It is this property that gives it its name, by analogy with the spectrum of white light.

We Infer an Important Property

A (power spectral density) is real, positive or zero

$$S_{xx}\left(e^{j\omega T}\right) \geq 0. \tag{24.40}$$

To show this, we calculate

$$E\left\{|y[n]|^2\right\} = R_{yy}[0] = \frac{1}{\omega_e} \int_{-\frac{\omega_e}{2}}^{+\frac{\omega_e}{2}} S_{xx}\left(e^{j\omega T}\right)\left|H\left(e^{j\omega T}\right)\right|^2 d\omega. \tag{24.41}$$

We assume $y[n]$ to be the output of a filter attacked by $x[n]$ and with frequency response $H\left(e^{j\omega T}\right)$ such that

$$H\left(e^{j\omega T}\right) = \begin{vmatrix} 1 & a < \omega < b \\ 0 & \text{elsewhere in the period} \end{vmatrix}. \tag{24.42}$$

We will have

$$R_{yy}[0] = \frac{1}{\omega_e} \int_{a}^{b} S_{xx}\left(e^{j\omega T}\right) d\omega. \tag{24.43}$$

Now $R_{yy}[0]$ is always positive or zero since it is the expectation of a square modulus. The two boundaries a and b of the filter frequency band were selected in any manner, it follows that the right side must be ≥ 0 $\forall (a$ and $b)$, which requires that $S_{xx}\left(e^{j\omega T}\right) \geq 0$.

In particular we have

$$R_{xx}[0] = \frac{1}{\omega_e} \int_{-\frac{\omega_e}{2}}^{+\frac{\omega_e}{2}} S_{xx}\left(e^{j\omega T}\right) d\omega. \tag{24.44}$$

Recall that $R_{xx}[0]$ is the signal power. This explains why $S_{xx}(e^{j\omega T})$ that is real, positive or zero, and whose integral over frequency gives the power of the WSS signal is called power spectral density (PSD).

24.4 Filtering a Centered White Noise with a First Order Filter

Let the real centered white noise $x[n]$, wide sense stationary of zero expectation and autocorrelation function $R_{xx}[n] = N\delta[n]$ with $\delta[n]$ the function of Kronecker.
   This signal is used as input to a causal first order filter of temporal equation

$$y[n] = Ky[n-1] + x[n].\tag{24.45}$$

The impulse response of this filter is

$$h[n] = K^n U[n].\tag{24.46}$$

and the system response is

$$y[n] = h[n] \otimes x[n] = \sum_{m=-\infty}^{\infty} h[m]x[n-m].\tag{24.47}$$

The expectation of $y[n]$ is

$$\boldsymbol{\eta}_y[n] = E\{y[n]\} = E\left\{\sum_{m=-\infty}^{\infty} h[m]x[n-m]\right\} = \sum_{m=-\infty}^{\infty} h[m]E\{x[n-m]\} = 0.$$
$$\tag{24.48}$$

The cross-correlation function of $x[n]$ and $y[n]$ is calculated with

$$R_{xy}[m] = R_{xx}[m] \otimes h^*[-m] = Nh^*[-m].\tag{24.49}$$

The autocorrelation of $y[n]$ is written

$$R_{yy}[m] = R_{xx}[m] \otimes h^*[-m] \otimes h[m] = N h^*[-m] \otimes h[m] = N \sum_{m'=-\infty}^{\infty} h^*[-m']h[m-m']$$

$$= N \sum_{m''=-\infty}^{\infty} h^*[m'']h[m+m''] = N \sum_{m'=-\infty}^{\infty} h^*[m']h[m+m']$$

$$= N \sum_{m'=-\infty}^{\infty} K^{*m'} U[m']K^{(m+m')} U[m+m'].$$

$$\tag{24.50}$$

First, we consider the case $m > 0$

$$R_{yy}[m] = NK^m \sum_{m'=0}^{\infty} |K|^{2m'} U[m+m']. \qquad (24.51)$$

Within the sum, the step function is different from zero if $m + m' \geq 0$, namely for $m' \geq -m$. In the range of variation of m, the step function has always value 1.

$$R_{yy}[m] = NK^m \sum_{m'=0}^{\infty} |K|^{2m'} = N\frac{K^m}{1-|K|^2}. \qquad (24.52)$$

(If $|K| < 1$, as it is assumed a priori to ensure stability of the system).
We now deal with the case $m < 0$

$$R_{yy}[m] = N \sum_{m'=-\infty}^{\infty} K^{*m'} U[m']K^{(m+m')} U[m+m']. \qquad (24.53)$$

The function $U[m+m']$ is nonzero if $m + m' \geq 0$, that is to say for $m' \geq -m$. The starting index of the sum is now $-m$ which is positive.

$$R_{yy}[m] = NK^m \sum_{m'=-\infty}^{\infty} |K^{2m'}| = NK^m \frac{|K|^{-2m}}{1-|K|^2} = NK^{*(-m)}\frac{1}{1-|K|^2} \qquad (24.54)$$

We verify that we have

$$R_{yy}[m] = R_{yy}^*[-m]. \qquad (24.55)$$

Power Spectral Density of the Output Filter Noise

$$
\begin{aligned}
S_{yy}\left(e^{j\omega T}\right) &= \sum_{m=-\infty}^{m=\infty} R_{yy}[m]e^{-jm\omega T} \\
&= N\frac{1}{1-|K|^2}\left(\sum_{m=0}^{m=\infty} K^m e^{-jm\omega T} - 1 + \sum_{m=-\infty}^{m=0} K^{*(-m)} e^{-jm\omega T}\right). \quad (24.56)
\end{aligned}
$$

The first sum is

$$\sum_{m=0}^{m=\infty} K^m e^{-jm\omega T} = \frac{1}{1-Ke^{-j\omega T}}. \qquad (24.57)$$

The second sum is

$$\sum_{m=-\infty}^{m=0} K^{*(-m)}e^{-jm\omega T} = \sum_{m'=0}^{m'=\infty} K^{*m'}e^{jm'\omega T} = \frac{1}{1 - K^*e^{j\omega T}}. \tag{24.58}$$

$$S_{yy}\left(e^{j\omega T}\right) = N\frac{1}{1 - |K|^2}\left(\frac{1}{1 - Ke^{-j\omega T}} + \frac{1}{1 - K^*e^{j\omega T}} - 1\right),$$

$$S_{yy}\left(e^{j\omega T}\right) = N\frac{1}{1 - |K|^2}\left(\frac{\left(1 - K^*e^{j\omega T}\right) + \left(1 - Ke^{-j\omega T}\right) - \left(1 - Ke^{-j\omega T}\right)\left(1 - K^*e^{j\omega T}\right)}{\left(1 - Ke^{-j\omega T}\right)\left(1 - K^*e^{j\omega T}\right)}\right),$$

$$S_{yy}\left(e^{j\omega T}\right) = N\frac{1}{1 - |K|^2}\left(\frac{2 - K^*e^{j\omega T} - Ke^{-j\omega T} - 1 + K^*e^{j\omega T} + Ke^{-j\omega T} - |K|^2}{\left(1 - Ke^{-j\omega T}\right)\left(1 - K^*e^{j\omega T}\right)}\right).$$

Finally,

$$S_{yy}\left(e^{j\omega T}\right) = \frac{N}{\left(1 - Ke^{-j\omega T}\right)\left(1 - K^*e^{j\omega T}\right)}. \tag{24.59}$$

It is noted that since the filter frequency response is

$$H\left(e^{j\omega T}\right) = \frac{1}{\left(1 - Ke^{-j\omega T}\right)}, \tag{24.60}$$

We have

$$S_{yy}(e^{j\omega T}) = S_{xx}(e^{j\omega T})H(e^{j\omega T})H^*(e^{j\omega T}). \tag{24.61}$$

These results are consistent with the general relationship (24.34).

24.5 Coherence Function

A method widely used for identifying noise sources is the method of the coherence function. This method lies in the comparison of a measured noise with that created by a possible source of the noise. At each frequency one can measure the fraction of the noise assignable to that source.

We define the **coherence function** between two WSS random signals $s[n]$ and $x[n]$ by the expression

$$\gamma_{sx}\left(e^{j\omega T}\right) = \frac{\left|S_{sx}\left(e^{j\omega T}\right)\right|}{\sqrt{S_{xx}(e^{j\omega T})}\sqrt{S_{ss}(e^{j\omega T})}} \tag{24.62}$$

Where $S_{sx}\left(e^{j\omega T}\right)$ is the cross-spectral density (cross-spectrum, FT of $R_{sx}[m]$) of $s[n]$ and $x[n]$. $S_{xx}\left(e^{j\omega T}\right)$ and $S_{ss}\left(e^{j\omega T}\right)$ are the respective spectral densities of $s[n]$ and $x[n]$.

Properties:

1. Let us assume that $s[n]$ is the output of a LTI system of which $x[n]$ is the input signal and $h[n]$ the impulse response. We calculate functions appearing in the coherence function $\gamma_{sx}\left(e^{j\omega T}\right)$:

$$S_{sx}\left(e^{j\omega T}\right) = \sum_{n=-\infty}^{\infty} R_{sx}[n]e^{-jn\omega T} = \sum_{n=-\infty}^{\infty} R_{sx}[n]e^{-jn\omega T} = \sum_{n=-\infty}^{\infty} R_{xx}[n] \otimes h[n]e^{-jn\omega T}.$$

(24.63)

$$S_{sx}\left(e^{j\omega T}\right) = S_{xx}\left(e^{j\omega T}\right)H\left(e^{j\omega T}\right).$$

(24.64)

Referring to this result in (24.62), we have

$$\gamma_{sx}\left(e^{j\omega T}\right) = \frac{\left|S_{xx}\left(e^{j\omega T}\right)\right|\left|H\left(e^{j\omega T}\right)\right|}{\sqrt{S_{xx}\left(e^{j\omega T}\right)}\sqrt{S_{ss}\left(e^{j\omega T}\right)}}.$$

(24.65)

A. A spectral density being always ≥ 0, this equation becomes

$$\gamma_{sx}\left(e^{j\omega T}\right) = \frac{\left|S_{xx}\left(e^{j\omega T}\right)\right|\left|H\left(e^{j\omega T}\right)\right|}{\sqrt{S_{xx}\left(e^{j\omega T}\right)}\sqrt{S_{ss}\left(e^{j\omega T}\right)}} = \frac{S_{xx}\left(e^{j\omega T}\right)\left|H\left(e^{j\omega T}\right)\right|}{\sqrt{S_{xx}\left(e^{j\omega T}\right)}\sqrt{S_{xx}\left(e^{j\omega T}\right)|H\left(e^{j\omega T}\right)|^2}} = 1.$$

(24.66)

2. Assume now that the signal $s[n]$ consists of the sum of the output signal $y[n]$ of a LTI system (whose input is $x[n]$) and a second random signal $z[n]$ of unknown origin and which is considered as a noise, $y[n]$ is considered the signal. It is assumed that the signals $x[n]$ are $z[n]$ are not correlated. Signals are also assumed real with zero mean: $s[n] = y[n] + z[n]$.

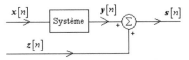

Let us calculate in this case the coherence function $\gamma_{sx}\left(e^{j\omega T}\right)$. The calculus begins by the determination of the correlation functions. Using the non correlation of $x[n]$ and $z[n]$ we can write

$$R_{sx}[m] = R_{yx}[m] + R_{zx}[m] = R_{xx}[m] \otimes h[m] + R_{zx}[m] = R_{xx}[m] \otimes h[m]. \quad (24.67)$$

$$S_{sx}(e^{j\omega T}) = S_{sx}(e^{j\omega T})H(e^{j\omega T}). \quad (24.68)$$

$R_{ss}[m]$ is then calculated. Because $z[n]$ is not correlated to $x[n]$ it is not to $y[n]$. We have therefore

$$R_{ss}[m] = R_{yy}[m] + R_{yz}[m] + R_{zy}[m] + R_{zz}[m] = R_{yy}[m] + R_{zz}[m]. \quad (24.69)$$

Taking the Fourier transform of $R_{ss}[m]$ we then have

$$S_{ss}(e^{j\omega T}) = S_{yy}(e^{j\omega T}) + S_{zz}(e^{j\omega T}) = S_{xx}(e^{j\omega T})|H(e^{j\omega T})|^2 + S_{zz}(e^{j\omega T}). \quad (24.70)$$

$$\gamma_{sx}(e^{j\omega T}) = \frac{|S_{sx}(e^{j\omega T})H(e^{j\omega T})|}{\sqrt{S_{xx}(e^{j\omega T})}\sqrt{S_{xx}(e^{j\omega T})|H(e^{j\omega T})|^2 + S_{zz}(e^{j\omega T})}}, \quad (24.71)$$

$$\gamma_{sx}(e^{j\omega T}) = \frac{|H(e^{j\omega T})|}{\sqrt{|H(e^{j\omega T})|^2 + \frac{S_{zz}(e^{j\omega T})}{S_{xx}(e^{j\omega T})}}} = \frac{1}{\sqrt{1 + \frac{S_{zz}(e^{j\omega T})}{|H(e^{j\omega T})|^2 S_{xx}(e^{j\omega T})}}}, \quad (24.72)$$

or

$$\gamma_{sx}(e^{j\omega T}) = \frac{1}{\sqrt{1 + \frac{S_{zz}(e^{j\omega T})}{S_{yy}(e^{j\omega T})}}}. \quad (24.73)$$

Since the spectral densities are real, positive or zero, we see that the coherence function $\gamma_{sx}(e^{j\omega T})$ is between 0 and 1.

$$0 \le \gamma_{sx}(e^{j\omega T}) \le 1. \quad (24.74)$$

The coherence function will be close to 1 for the frequencies where the spectral density of the signal $y[n]$ is large compared with that of the noise $z[n]$.

The ratio of power spectral densities of the noise and of the signal at frequencies ω is noted:

$$r(e^{j\omega T}) = \frac{S_{zz}(e^{j\omega T})}{S_{yy}(e^{j\omega T})}. \quad (24.75)$$

Equation (24.73) is then written

$$\gamma_{sx}\left(e^{j\omega T}\right) = \frac{1}{\sqrt{1 + r(e^{j\omega T})}}, \tag{24.76}$$

or even

$$r\left(e^{j\omega T}\right) = \frac{1}{\gamma_{sx}^2\left(e^{j\omega T}\right)} - 1. \tag{24.77}$$

24.6 Autocorrelation Matrix of a Random Signal

Let $x = [x[0], x[1], \ldots, x[p]]^T$ a vector consisting of $p + 1$ values of the random signal $x[n]$. We further assume that the signal $x[n]$ is wide sense stationary.

The exterior product of x by its Hermitian conjugate x^H is a matrix $(p + 1) \times (p + 1)$:

$$xx^H = \begin{bmatrix} x[0]x^*[0] & x[0]x^*[1] & \cdots & x[0]x^*[p] \\ x[1]x^*[0] & x[1]x^*[1] & \cdots & x[1]x^*[p] \\ \vdots & \vdots & \vdots & \vdots \\ x[p]x^*[0] & x[p]x^*[1] & \cdots & x[p]x^*[p] \end{bmatrix}. \tag{24.78}$$

If $x[n]$ is WSS, the autocorrelation matrix is obtained by taking the expectations of each term.

Having noticed that since the signal is SSL, $E\{x[n + m]x^*[n]\} = R_{xx}[m]$. since $R_{xx}[m] = R_{xx}^*[-m]$,

$$R_{xx} = E\{xx^H\} = \begin{bmatrix} R_{xx}[0] & R_{xx}^*[1] & \cdots & R_{xx}^*[p] \\ R_{xx}[1] & R_{xx}[0] & \cdots & R_{xx}^*[p-1] \\ \vdots & \vdots & \vdots & \vdots \\ R_{xx}[p] & R_{xx}[p-1] & \cdots & R_{xx}[0] \end{bmatrix}. \tag{24.79}$$

R_{xx} is a Hermitian matrix. It is also a Toeplitz matrix, that is to say that the elements along the main diagonal are equal. This matrix is also definite, not negative. Its eigenvalues are real and not negative. The eigenvectors associated with different eigenvalues are orthogonal.

Property: The eigenvalues of an autocorrelation matrix $n \times n$ of a stationary signal at large are bounded below and above by the minimum and maximum values of the power spectral density):

$$\min S_{xx}\left(e^{j\omega T}\right) \leq \lambda_i \leq \max S_{xx}\left(e^{j\omega T}\right). \tag{24.80}$$

24.7 Beamforming

Consider a signal $x[n]$ resulting from a recording by a sensor of a random acoustic signal wave propagating in a fluid. Assume now that one has an antenna consisting of an array of identical sensors uniformly distributed on a line and distant from each other by a. In the case of propagation in the form of a plane wave, the signal received by the sensor $p+1$ is the signal received by the sensor p multiplied by a delay $s = \frac{a\sin\theta}{c}$. c is the speed of the wave in the medium, θ is the angle of the wave vector with the perpendicular axis to the network:

$$x_{p+1}[n] = x_p[n-s] = x_p[n] \otimes \delta[n-s]. \tag{24.81}$$

The next development aims to find the direction of the source of the acoustic wave from the data recorded by the sensors. Considering the matrix $R_{xx}[m]$ of cross-correlation functions between the signals of the various sensors:

$$R_{xx}[m] = \begin{bmatrix} R_{x_1x_1}[m] & R_{x_1x_2}[m] & \cdots & R_{x_1x_M}[m] \\ R_{x_2x_1}[m] & R_{x_2x_2}[m] & \cdots & R_{x_2x_M}[m] \\ \vdots & \vdots & \vdots & \vdots \\ R_{x_Mx_1}[m] & R_{x_Mx_2}[m] & \cdots & R_{x_Mx_M}[m] \end{bmatrix}. \tag{24.82}$$

It follows from the above assumptions that the cross-correlation function of two successive sensor signals is related to the autocorrelation function of the signal from a sensor by the relationship

$$R_{x_{p+1}x_p}[m] = E\{x_{p+1}[n+m]x_p[n]\} = E\{x_p[n+m-s]x_p[n]\} = R_{x_px_p}[m-s].$$

Or even

$$R_{x_{p+1}x_p}[m] = R_{x_px_p}[m] \otimes \delta[m-s]. \tag{24.83}$$

The matrix $R_{xx}[m]$ is then

$$R_{xx}[m] = R_{x_nx_n}[m]$$

$$\otimes \begin{bmatrix} \delta[m] & \delta[m-s] & \cdots & \delta[m-(M-1)s] \\ \delta[m+s] & \delta[m] & \cdots & \delta[m-(M-2)s] \\ \vdots & \vdots & \vdots & \vdots \\ \delta[m+(M-1)s] & \delta[m+(M-2)s] & \cdots & \delta[m] \end{bmatrix}. \tag{24.84}$$

We now consider the matrix of cross-spectra, consisting of FT of the different terms of the matrix of correlation functions. By applying the shift theorem to the different elements of the matrix and by denoting $S_0(e^{j\omega T})$ the power spectral density of the signal of one sensor, then

$$S_{xx}(e^{j\omega T}) = S_0(e^{j\omega T}) \begin{bmatrix} 1 & e^{-js\omega T} & \cdots & e^{-j(M-1)s\omega T} \\ e^{js\omega T} & 1 & \cdots & e^{-j(M-2)s\omega T} \\ \vdots & \vdots & \vdots & \vdots \\ e^{j(M-1)s\omega T} & e^{j(M-2)s\omega T} & \cdots & 1 \end{bmatrix}. \quad (24.85)$$

The matrix of phase factors is Hermitian and Toeplitz. Generally, its eigenvalues are real or null and its eigenvectors are orthogonal. Here, the column vectors of this matrix are not independent. They are deduced from each other by multiplication by a complex term of the form $e^{-jrs\omega T}$. It follows that the determinant of the matrix is zero. The matrix is singular and rank 1. The eigenvalues are all zero except one whose value is M. The corresponding eigenvector is the column vector $\begin{pmatrix} 1 & e^{-js\omega T} & \cdots & e^{-j(M-1)s\omega T} \end{pmatrix}^H$. The determination of this vector can be used to trace back the delay s and thus the direction of propagation of the plane wave and therefore the direction of the source.

By studying the case of superposition of two signals from two propagation directions (in practice two remote sources), the matrix of cross-spectra has two nonzero eigenvalues. This method gives access to the knowledge of the number of sources. However the two associated eigenvectors are orthogonal and are not independently associated with the direction of each source. The inverse Fourier transform of the first row of the matrix, however, allows to determine the values of the delays s_1 and s_2 and so find the directions of the sources.

Note: The results would be the same if the signal $x[n]$ was certain. In that case we would be dealing with deterministic correlation functions and their Fourier transforms.

24.8   Analog Random Signals

It is appropriate at this stage to give the main results of the analysis of random signals in continuous time. The formulas are similar to those digital signals but some developments given hereinafter are easier and more intuitive to establish on analog signals. Consider $x(t)$ an analog random signal. The correlation function for this signal is defined by:

$$R_{x(t_1)x(t_2)}(t_1, t_2) = E\{x(t_1)x^*(t_2)\}. \quad (24.86)$$

The signal will be wide sense stationary iff:

(a) His expectation is independent of time $E\{x(t)\} = \eta_{x(t)} = \eta_x$.

(b) Its correlation function $R_{x(t_1)x(t_2)}(t_1, t_2)$ depends only on the time difference $\tau = t_1 - t_2$. We shall note $R_{xx}(\tau)$ the signal $x(t)$ correlation function.

The power of the WSS signal $x(t)$ is given by: $P(t) = E\{x(t)x^*(t)\}$.

We have in this case $P(t) = P = R_{xx}(0)$.

An analog centered white noise is a real random signal, zero expectancy, WSS, whose correlation function is given by the Dirac distribution $\delta(\tau)$: $R_{xx}(\tau) = N\delta(\tau)$.

The correlation function of the output signal $y(t)$ of LTI filter with impulse response $h(t)$ is

$$R_{yy}(\tau) = R_{xx}(\tau) \otimes h^*(\tau) \otimes h(\tau). \tag{24.87}$$

The power spectral density) of the WSS signal $x(t)$ is the Fourier transform of its autocorrelation function

$$S_{xx}(\omega) = \int_{-\infty}^{\infty} R_{xx}(\tau) e^{-j\omega\tau} d\tau. \tag{24.88}$$

A power spectral density is real, positive, or zero:

$$S_{xx}(\omega) \geq 0 \tag{24.89}$$

Considering a real signal $x(t)$, sum of a deterministic signal $s(t)$ and a random signal $b(t)$: $x(t) = s(t) + b(t)$. The signal-to-noise ratio for this signal is defined as the power of the certain signal to that of the random signal. We write

$$\rho(t) = \frac{P_{s(t)}}{P_{b(t)}} = \frac{s^2(t)}{E\{b^2(t)\}}. \tag{24.90}$$

We see that the signal-to-noise ratio can be increased by decreasing the power of the noise by filtering.

Passage of White Noise in an Ideal Low Pass Filter

Consider a white noise $b(t)$. Its correlation function is $R_{bb}(\tau) = N\delta(\tau)$. Its power spectral density is $S_{bb}(\omega) = \int_{-\infty}^{+\infty} R_{bb}(\tau) e^{-j\omega\tau} d\tau = \int_{-\infty}^{+\infty} N\delta(\tau) e^{-j\omega\tau} d\tau = N$. It is constant. The power of white noise is infinite as $P_b = R_{bb}(0) = \infty$. Note that since an infinite power is physically impossible, white noise should not have infinite power. It is a mathematical fiction. In practice, a white noise is a very broadband noise.

Now we input this white noise in a low pass filter ideal (unity gain within the bandwidth) with cutoff frequency ω_c.

In the time domain we have $b_1(t) = b(t) \otimes h(t)$.

The spectral density of the output signal is

$$S_{b_1 b_1}(\omega) = |H(\omega)|^2 S_{bb}(\omega). \tag{24.91}$$

The power of the output signal is

$$R_{b_1 b_1}(0) = \frac{1}{2\pi} \int_{-\infty}^{\infty} S_{b_1 b_1}(\omega)d\omega = \frac{1}{2\pi} \int_{-\infty}^{\infty} |H(\omega)|^2 S_{bb}(\omega)d\omega$$

$$= \frac{1}{2\pi} \int_{-\omega_c}^{\omega_c} N\,d\omega = N\frac{\omega_c}{\pi}. \tag{24.92}$$

It is seen that the output power from the filter is proportional to the bandwidth of this filter.

We have a similar result when a band-pass filter is used. It is assumed that the band-pass filter, assumed ideal, has unity gain within a bandwidth $2\omega_c$ centered in ω_0.

The power of the output signal is

$$R_{b_1 b_1}(0) = \frac{1}{2\pi} \int_{-\omega_0 - \omega_c}^{-\omega_0 + \omega_c} N\,d\omega + \frac{1}{2\pi} \int_{\omega_0 + \omega_c}^{\omega_0 - \omega_c} N\,d\omega = N\frac{2\omega_c}{\pi}. \tag{24.93}$$

The practical value of this filtering appears in the following situation. It is assumed that in the noisy signal $x(t) = s(t) + b(t)$, the noise is white.

At the outlet of the filter we have $x_1(t) = (s(t) + b(t)) \otimes h(t) = s_1(t) + b_1(t)$.

$$\xrightarrow{s(t) + b(t)} \boxed{H} \xrightarrow{s_1(t) + b_1(t)}$$

Assume that the bandwidth of the ideal filter is adjusted to that of the signal $s(t)$. The signal $s(t)$ remains unchanged in the filter $x_1(t) = s(t) + b_1(t)$, but the noise power is decreased. The signal-to-noise ratio at the output of the filter is in this case

$$\rho(t) = \frac{s^2(t)}{E\{b_1^2(t)\}} = \frac{s^2(t)}{N\frac{2\omega_c}{\pi}}. \tag{24.94}$$

It thus appears that the signal-to-noise ratio that can be achieved with noisy signals is all the more larger as the bandwidth of the signal is weak.

In space telecommunications, signals from very distant sensors of the earth have extremely low amplitude. The signals are very noisy. To benefit from the result set and can effectively filter the noise at reception, we choose to communicate information using very narrow-band signals. In reception the signals are amplified by masers which are amplifiers with very high gain and very narrow band. This improves considerably the signal-to-noise ratio at reception. Research on masers accompanied those on lasers that can be considered very narrow-band amplifiers operating in bands of optical frequencies.

24.9   Matched Filter

In the former case, it was sought to improve the signal-to-noise ratio by filtering, preserving as well as possible the temporal shape of the certain signal. Another situation is where one seeks to detect the presence of a signal within a noisy signal. This is the case where a radar signal processing first looks for the presence of a target by detecting the echo. The shape of the expected echo is known: it is identical to that of the transmitted signal which we denote here $s(t)$. We show in the following that there is a filter which optimizes the detection of the signal $s(t)$ in the noise. Let us note the impulse response $h(t)$ of this desired filter. It is assumed that the noise $b(t)$ is WSS.

The output of the filter is $x_1(t) = (s(t) + b(t)) \otimes h(t) = s_1(t) + b_1(t)$.

$$\xrightarrow{\quad s(t)+b(t) \quad} \boxed{H} \xrightarrow{\quad s_1(t)+b_1(t) \quad}$$

The impulse response of the system is determined so that at a certain time t_1 the signal-to-noise ratio as defined below is maximized. At the output of the filter, this ratio is

$$\rho(t_1) = \frac{s_1^2(t_1)}{E\{b_1^2(t_1)\}}. \tag{24.95}$$

We have $S_1(\omega) = S(\omega)H(\omega)$. So, at time t_1: $s_1(t_1) = \frac{1}{2\pi} \int_{-\infty}^{+\infty} S(\omega)H(\omega)e^{j\omega t_1} d\omega$. For a real signal

$$s_1^2(t_1) = |s_1(t_1)|^2 = \frac{1}{4\pi^2} \left| \int_{-\infty}^{+\infty} S(\omega)H(\omega)e^{j\omega t_1} d\omega \right|^2. \tag{24.96}$$

The denominator is estimated. Since $S_{b_1 b_1}(\omega) = S_{bb}(\omega)|H(\omega)|^2$, taking its inverse FT we obtain $R_{b_1 b_1}(\tau) = \frac{1}{2\pi} \int_{-\infty}^{+\infty} S_{bb}(\omega)|H(\omega)|^2 e^{j\omega\tau} d\omega$.

$$E\{b_1^2(t_1)\} = R_{b_1 b_1}(0) = \frac{1}{2\pi} \int_{-\infty}^{+\infty} S_{bb}(\omega)|H(\omega)|^2 d\omega. \tag{24.97}$$

It is assumed that the input is white noise with $S_{bb}(\omega) = N$.

$$\rho(t_1) = \frac{\frac{1}{4\pi^2} \left| \int_{-\infty}^{+\infty} S(\omega)H(\omega)e^{j\omega t_1} d\omega \right|^2}{\frac{N}{2\pi} \int_{-\infty}^{+\infty} |H(\omega)|^2 d\omega}. \tag{24.98}$$

Recall Schwarz's inequality

$$\left| \int_{-\infty}^{+\infty} F_1(\omega)F_2(\omega)d\omega \right|^2 \leq \int_{-\infty}^{+\infty} |F_1(\omega)|^2 d\omega \int_{-\infty}^{+\infty} |F_2(\omega)|^2 d\omega$$

There is equality iff $F_1(\omega) = kF^*(\omega)$, where k is any real constant.
This result is used in the problem considered

$$\left| \int_{-\infty}^{+\infty} S(\omega)H(\omega)e^{j\omega t_1} d\omega \right|^2 \leq \int_{-\infty}^{+\infty} |S(\omega)|^2 d\omega \int_{-\infty}^{+\infty} |H(\omega)|^2 d\omega,$$

thus

$$\rho(t_1) \leq \frac{1}{2\pi N} \int_{-\infty}^{+\infty} |S(\omega)|^2 d\omega. \tag{24.99}$$

The maximum value of the signal-to-noise ratio will be achieved if and only if

$$H(\omega) = k S^*(\omega)e^{-j\omega t_1}. \tag{24.100}$$

Resulting in the time domain

$$h(t) = k\frac{1}{2\pi} \int_{-\infty}^{+\infty} S^*(\omega)e^{-j\omega t_1} e^{j\omega t} d\omega.$$

As $S(\omega) = \int_{-\infty}^{+\infty} s(t)\, e^{-j\omega t} dt$, and since the signal is supposedly real,

$$S^*(\omega) = \int_{-\infty}^{+\infty} s(t)\, e^{j\omega t} dt. \tag{24.101}$$

It comes by making the change of variable $t' = -t$ $S^*(\omega) = \int_{-\infty}^{+\infty} s(-t')$
$e^{-j\omega t'} dt'$.

So $S^*(\omega)$ is the Fourier transform of $s(-t)$.
Thus $s(-t) = \frac{1}{2\pi} \int_{-\infty}^{+\infty} S^*(\omega)\, e^{j\omega t} d\omega$ so $s(t_1 - t) = \frac{1}{2\pi} \int_{-\infty}^{+\infty} S^*(\omega)\, e^{-j\omega t_1} e^{j\omega t} d\omega$,
It finally comes

$$h(t) = ks(t_1 - t). \tag{24.102}$$

Therefore the optimal filter impulse response is a replica of the signal $s(t)$
obtained by performing

1. a time reversal of the transmitted signal
2. a temporal translation of the reversed signal by a delay equal to t_1 that is the time of the round trip of the backscattered signal from the target.

The convolution operation with a returned signal being a correlation, this type of radar is called radar by correlation.

This type of radar is only a stage in the evolution of remote sensing techniques. It provides information on the position of an object but not its speed. Measuring the velocity of the target is based on the measurement of the Doppler effect which affected the echo. This effect is a shift in frequency of the signal due to the relative speed of the target and the transmitter–receiver antenna. The change in frequency is low; it is necessary that the signal duration is relatively long to allow determination. The signal $s(t)$ of radar that simultaneously measures the position and frequency must meet two apparently contradictory requirements: be short enough to permit accurate localization and discrimination between targets and yet be sufficiently long to permit measurement of speed. Modern techniques optimize the shape of the signal used $s(t)$. The reader is encouraged to refer to books on the subject to deepen this theme.

Summary

We encountered in this chapter two functions which are fundamental for signal analysis: the correlation function and the power spectral density of wide sense stationary (WSS) random signals. We studied the filtering of WSS signals by a LTI system and gave the theorems linking correlations and DSP of input and output signals. The coherence function defined afterward is a powerful tool to identify in a noisy signal the sources constituting the noise. At the end of this chapter, we gave the equivalent formula for correlation functions and power spectral densities of analog signals. We have seen how one can increase the signal-to-noise ratio with a filter. We have exposed the principle of the matched filter, widely used in radar and sonar. It increases the probability of detection of an echo in a noisy environment. Many corrected exercises at the end of this chapter will help the reader become familiar with these important results.

Exercises

I. Matched digital filter. It is assumed that the digital signals are obtained by analog–digital conversion at a sampling frequency $f_s = 20$ kHz.

A. Let the digital filter defined by the following time equation:

$$g[n] = -f[n] - f[n-1] + f[n-2] + f[n-3].$$

1. Give the impulse response $h[n]$ of the filter and represent this function. Is the filter causal?
2. Calculate the system transfer function $H(z)$. Having noticed that $z = 1$ is a root of $H(z) = 0$, determine the notable points of $H(z)$ and represent them in the z plane.

3. Deduct the shape of the frequency response modulus of $H(e^{j\omega T})$ from the position of these notable points. What are the frequencies of the signals blocked by the filter? Calculate the expressions of the frequency response $H(e^{j\omega T})$ and of its module.
4. We use at the filter input a signal $f_0[n]$ which is time reversal of $h[n]$ $f_0[n] = h[-n]$. Calculate the filter output signal $g_0[n]$. What is the Fourier transform of this signal?

B. We show now that the previous filter is well suited to the detection of the signal $f_0[n]$ when it is vitiated by an additional noise $b[n]$. It is assumed in the following that the digital noise $b[n]$ is white, wide sense stationary, Gaussian, with zero mean and autocorrelation function $R_{bb}[m] = \delta[m]$ ($\delta[m]$ is the Kronecker function).
1. What is the standard deviation of the noise signal $b[n]$? Give the prediction interval (symmetric) at risk 5 % of noise voltage $b[n]$. What is the power spectral density of $b[n]$?
2. It is first assumed that the noise $b[n]$ is the question I filter input. We note $R_{yy}[m]$ the output signal. What is the expectation of the signal $y[n]$? What is its variance? Give the prediction interval (symmetric) at 5 % risk of noise $y[n]$ at the filter output. Give the expression of the autocorrelation function $R_{yy}[m]$. What is the power spectral density of $y[n]$?
3. We note $f[n]$ the signal $f_0[n]$ (met in question I.4) added with noise $b[n]$:

   $f[n] = f_0[n] + b[n]$. What are the expectation and variance of $f[n]$?
   This signal is presented at the input of the filter met in A. The output signal is noted $g[n]$. What are the expectation and variance of $g[n]$?

4. We call power (instantaneous) of a deterministic signal, its square. The power of a random signal is the expectation of its square. Thus $P_{f_0}[n] = f_0^2[n]$ and $P_b[n] = E\{b^2[n]\}$.

   The signal-to-noise ratio of the signal $f[n]$ is $\rho[n] = \dfrac{P_{f_0}[n]}{E\{b^2[n]\}}$. Calculate the value of this ratio for different values of n.

5. Calculate the signal-to-noise ratio in the signal $g[n]$. Show that the fact of having used to filter the noisy signal a filter whose impulse response is the time reverse of the signal $f_0[n]$ that is to be detected has increased the probability of detection of this signal (increased ratio signal to noise at the crossing of the matched filter).

   Solution: A 1. $h[n] = -\delta[n] - \delta[n-1] + \delta[n-2] + \delta[n-3]$. The filter is causal because the impulse response is zero for $n < 0$.

2. $H(z) = -1 - z^{-1} + z^{-2} + z^{-3} = z^{-3}(-z^3 - z^2 + z + 1)$.
   $H(1) = (-1 - 1 + 1 + 1) = 0$.

By the division of polynomials we get

$H(z) = -z^{-3}(z-1)(z^2 + 2z + 1) = -z^{-3}(z-1)(z+1)^2$, which has a triple pole in $z = 1$, a single zero in $z = 1$ and a double zero in $z = -1$. $H(e^{j\omega T}) = -1 - e^{-j\omega T} + e^{-j2\omega T} + e^{-j3\omega T}$.

3. The filter is band-pass. The gain is zero for $\omega = 0$ and $\omega = \frac{\pi}{T}$(for $f = \frac{f_c}{2}$).
4. $g_0[n] = f_0[n] \otimes h[n] = h[-n] \otimes h[n]$. That is to say, the autocorrelation function of $h[n]$.

$g_0[n] = f_0[n] \otimes h[n] = h[-n] \otimes h[n]$.
$g_0[n] = -\delta[n+3] - 2\delta[n+2] + \delta[n+1] + 4\delta[n] + \delta[n-1] - 2\delta[n-2] - \delta[n-3]$.

As expected, this function is even and has its maximum in $n = 0$.

$$G(e^{j\omega T}) = |H(e^{j\omega T})|^2.$$

B1. $\sigma_b^2 = R_{bb}[0] = \delta[0] = 1$. The standard deviation $\sigma_b = 1$. $PI_{5\%} = \{-1.96, 1.96\}$.

2. The expectation of $b[n]$ being zero, that of $R_{yy}[m]$ will be also.

$$R_{yy}[m] = R_{bb}[m] \otimes h[m] \otimes h[-m] = \delta[m] \otimes h[m] \otimes h[-m] = h[m] \otimes h[-m];$$
$$\sigma_y^2 = R_{yy}[0] = 4.$$

3. $E\{f[n]\} = E\{f_0[n]\} + E\{b[n]\} = f_0[n] = h[-n]$, for the expectation of a certain value is itself and the expectation of $b[n]$ is zero by hypothesis.

$$g[n] = f[n] \otimes h[n] = (f_0[n] + b[n]) \otimes h[n]; E\{g[n]\} = f_0[n] \otimes h[n] = g_0[n].$$

The variance of $g[n]$ is that of $R_{yy}[m]$: $\sigma_g^2 = 4$.

4. For the signal $f[n]$ the signal-to-noise ratio is $\rho_f[n] = \frac{f_0^2[n]}{1} = f_0^2[n]$ equal to zero or 1.

5. For the signal $g[n]$ the signal-to-noise ratio is $\rho_g[n] = \frac{g_0^2[n]}{4}$. In particular $\rho_g[0] = 4$. At time $n = 0$ the signal-to-noise ratio of the filtered signal $g[n]$ is at least fourtimes greater than that of $f[n]$ increasing the probability of signal $f_0[n]$ detection in noise.

II. MA digital filter

1. Let the digital filter defined by the temporal equation $y[n] = \frac{1}{N}\sum_{l=-\left(\frac{N-1}{2}\right)}^{\frac{N-1}{2}} x[n+l]$ with N odd.

a. What is the impulse response of the filter? Represent this when $N = 9$.
b. Calculate the filter transfer function $H(z)$. Locate its zeros in the complex plane.

From the position of these zeros predict appearance of the filter frequency gain modulus.

c. Give the expression of the frequency response $H\left(e^{j\omega T}\right)$ (T is the sampling step).

Why could we predict that this function was real? Which are the frequencies of signals blocked by the filter in the case where the sampling frequency is $f_e = 1\,\text{MHz}$?
Accurately represent the frequency response.

2. It is now assumed that the signal $x[n]$ is random, WSS and for each value of n, $x[n]$ has an even distribution between and -1 and $+1$.

a. Specify the mean and the standard deviation of $x[n]$.
b. Assuming that two successive values of the signal are independent, give the distribution of $g[n] = x[n] + x[n+1]$. Calculate the variance of $g[n]$.
c. Extend these results to the filter defined in question 1.

Solution: 1. a. $h[n] = \frac{1}{N}\sum_{l=-\left(\frac{N-1}{2}\right)}^{\frac{N-1}{2}} \delta[n+l]$. If $N = 9$, $h[n] = \frac{1}{9}\sum_{l=-4}^{4} \delta[n+l]$.

b. $H(z) = \frac{1}{9}\sum_{l=-4}^{4} z^{-l} = \frac{1}{9}\frac{z^4 - z^{-5}}{1 - z^{-1}} = \frac{1}{9}z^{-5}\frac{z^9 - 1}{1 - z^{-1}} = \frac{1}{9}z^{-4}\frac{z^9 - 1}{z - 1}$.

The numerator has 9 zeros regularly placed on the unit circle $z_k = e^{jk\frac{2\pi}{9}}$ ($k = 0, 1, .., 8$). The zero in $z = 1$ is balanced by the zero at denominator. The filter will let pass the DC component and block frequencies $f_k = k\frac{f_e}{9}$ with ($k = 0, 1, .., 8$).

c $H\left(e^{j\omega T}\right) = \frac{1}{9}(1 + 2\cos\omega T + 2\cos 2\omega T + 2\cos 4\omega T)$. This function is real since the impulse response is even. The blocked frequencies are $f_k = k\frac{10^6}{9}$ ($k = 1, .., 8$).

2. a. $f_x = \frac{1}{2}$ for $-1 \le x < 1$ and zero elsewhere. $E\{x[n]\} = 0$. $\sigma_x^2 = \frac{1}{3}$.

b. The two values $x[n]$ and $x[n+1]$ being independent, the PDF of $g[n]$ is $f_g = f_x(x) \otimes f_x(x)$, convolution of two rectangular windows. It is a triangular window with base $\{-2, 2\}$. The variance of $g[n]$ is the sum of the two variances, $\sigma_g^2 = \frac{2}{3}$.

c. For the sum upon nine elements, the probability density function is the convolution of several triangles. This is a bell curve that starts to look like a Gaussian (see the central limit theorem). The variance is $9\frac{1}{3}$.

III. Sinusoidal signal with a random phase

Let $x[n] = A\cos(\omega_0 nT + \Theta)$ the random signal where A and ω_0 are constants and Θ is a random variable uniformly distributed between $-\pi$ and π.

1. Give the PDF f_Θ of Θ.
2. What are the expectations of Θ and $x[n]$?
3. Calculate the autocorrelation function of the signal $x[n]$.
4. Same questions if now $x[n] = Ae^{j(\omega_0 nT + \Theta)}$.

Solution: 1. $f_\Theta = \frac{1}{2\pi}$ between 0 and 2π.

2. $E\{\Theta\} = 0$. $E\{x[n]\} = E\{A(\cos(\omega_0 nT)\cos\Theta - \sin(\omega_0 nT)\sin\Theta)\}$.

$$E\{x[n]\} = A\cos(\omega_0 nT)E\{\cos\Theta\} - A\sin(\omega_0 nT)E\{\sin\Theta\} = 0.$$

3. $E\{x^*[n]x[n+m]\} = E\{A\cos(\omega_0 nT + \Theta)A\cos(\omega_0(n+m)T + \Theta)\}$

$$= \frac{A^2}{2}E\{\cos(\omega_0 mT) + \cos(\omega_0(2n+m)T + 2\Theta)\} = \frac{A^2}{2}\cos(\omega_0 mT).$$

4. $E\{x[n]\} = 0$; $E\{x^*[n]x[n+m]\} = E\{Ae^{-j(\omega_0 nT + \Theta)}Ae^{j(\omega_0(n+m)T + \Theta)}\} = A^2 E\{e^{j(\omega_0 mT + \Theta)}\} = 0$.

IV. Statistics of first and second order Gaussian signals.

Let $b(t)$ be an analog Gaussian noise, with zero expectancy and whose correlation function is $R_{bb}(\tau) = B\delta(\tau)$. (where $\delta(\tau)$ is the Dirac distribution and $B = 10^{-12}$).

This random signal is the input of an ideal, gain 1, low pass filter for $\{|\omega| < \omega_0 = 10^6\}$ and zero elsewhere. The filter output is noted $x(t)$.

1. Calculate the correlation function of signal $x(t)$. What is the joint probability function of the couple of variables $\{x(t_1), x(t_2)\}$, where t_1 are t_2 two times separated by $\tau = 10^{-6}$?
2. Knowing that at time t_1, was measured $x(t_1) = 10^{-3}$ V was measured, what is the probability of measuring a negative value 1 µs later?

3. The signal $x(t)$ is presented to the input of a quadratic electronic circuit (the output of the circuit is the square of the input signal). The output is noted $y(t)$.

a. What is the PDF of $y(t)$.
b. Calculate the correlation function and the psd of $y(t)$.

   Solution $S_{bb}(\omega) = B$. So $S_{xx}(\omega) = B$ for $\{|\omega| < \omega_0 = 10^6\}$ and zero elsewhere.

1.
$$R_{xx}(\tau) = \frac{1}{2\pi} \int\limits_{-\omega_0}^{+\omega_0} B e^{j\omega\tau} d\omega = \frac{B \sin \omega_0\tau}{\pi} \frac{}{\tau} = \frac{B\omega_0}{\pi} \mathrm{sinc}(\omega_0\tau).$$

2. Since the filter input signal $b(t)$ is Gaussian, $x(t)$ is also Gaussian. It is also WSS as $b(t)$ is. His expectancy is zero like that of $b(t)$. Its variance is

$$\sigma_x^2 = R_{xx}(0) = \frac{B\omega_0}{\pi} .\mathrm{N.A.}\sigma_x^2 = \frac{1}{\pi} 10^{-6}.\sigma_x = 5.64\,10^{-4}.$$

The PDF of $x(t)$ is $f_x(x) = \frac{1}{\sqrt{2\pi\sigma_x^2}} e^{-\frac{x^2}{2\sigma_x^2}}$.

The correlation coefficient of $x(t_1)$ and $x(t_2)$ is given by $r(\tau) = \frac{R_{xx}(\tau)}{R_{xx}(0)} = \mathrm{sinc}(\omega_0\tau)$.

N. A. For $\tau = 10^{-6}$, $r = \mathrm{sinc}(1) = 0.8415$.

Denoting $x = x(t_1)$ and $y = x(t_2)$, the joint PDF is

$$f_{xy}(x,y) = \frac{1}{2\pi\sigma_x^2\sqrt{1-r^2}} e^{-\frac{1}{2(1-r^2)\sigma_x^2}(x^2 - 2rxy + y^2)} \text{ with } r = 0.8415 \text{ and } \sigma_x^2 = \frac{1}{\pi} 10^{-6}.$$

3. The result is given by the exercise I of Chap. 22.

   Reference should be made to Chap. 21. We have

a. $f_y(y) = \frac{1}{\sqrt{2\pi\sigma_x^2}} \frac{1}{\sqrt{y}} e^{-\frac{y}{2\sigma_x^2}} U(y)$.

b. Correlation of $y(t)$ is $R_{yy}(\tau) = E\{y^*(t)y(t+\tau)\} = E\{x^2(t)x^2(t+\tau)\}$

   To return to the notations of Chap. 22, we set $x(t) = x$ and $x(t+\tau) = y$.
   We look for $E\{x^2y^2\} = \int_{-\infty}^{\infty}\int_{-\infty}^{\infty} x^2y^2 f_{xy}(x,y)dxdy$. Using the conditional PDF

of y, $E\{x^2y^2\} = \int_{-\infty}^{\infty}\int_{-\infty}^{\infty} x^2y^2 \frac{1}{\sqrt{2\pi\sigma_x^2}} e^{-\frac{x^2}{2\sigma_x^2}} \frac{1}{\sqrt{2\pi\sigma_x^2(1-r^2)}} e^{-\frac{(y-rx)^2}{2\sigma_x^2(1-r^2)}}dxdy$.

   In the integral over y, the conditioned expectation of y^2 conditioned is recognized. The expectation of a square being equal to the variance plus the square of expectancy, this integral is $\sigma_x^2(1-r^2) + r^2x^2$.

Thus: $E\{x^2y^2\} = \frac{1}{\sqrt{2\pi\sigma_x^2}}\int_{-\infty}^{\infty} x^2\left(\sigma_x^2(1-r^2)+r^2x^2\right)e^{-\frac{x^2}{2\sigma_x^2}}dx$. Integrals giving the moments of order two and four of the Gaussian function appear. The result will be of the form $R_{yy}(\tau) = a + br^2 = a + b\left(\frac{\sin\omega_0\tau}{\omega_0\tau}\right)^2$

so : $S_{yy}(\omega) = \int\limits_{-\infty}^{+\infty} a + b\left(\frac{\sin\omega_0\tau}{\omega_0\tau}\right)^2 e^{-j\omega\tau}d\tau = 2\pi a\delta(\tau) + bF\left(\frac{\sin\omega_0\tau}{\omega_0\tau}\right)^2.$

The FT of sinc squared is the convolution (divided by 2π) of the FT of sinc by itself. As it is a rectangular function, the result will be a triangular window on the interval $\{-2\omega_0, 2\omega_0\}$.

V. Consider a WSS random Gaussian signal with zero mean $x(t)$ and autocorrelation $R_{xx}(\tau)$. Let $y(t) = x^2(t)$. Give the correlation function and spectral density of $y(t)$; its mean and variance.

Solution:

$$R_{yy}(\tau) = E\{y(t)y(t+\tau)\} = E\{x^2(t)x^2(t+\tau)\}.$$

We use the following property valid for the mean of the product of four real Gaussian variables b_1, b_2, b_3, b_4: $E\{b_1b_2b_3b_4\} = E\{b_1b_2\}E\{b_3b_4\} + E\{b_1b_4\}E\{b_2b_3\} + E\{b_1b_3\}E\{b_2b_4\}.$

$R_{yy}(\tau) = E\{x(t)x(t)\}E\{x(t+\tau)x(t+\tau)\} + 2E\{x(t)x(t+\tau)\}E\{x(t)x(t+\tau)\},$

$R_{yy}(\tau) = R_{xx}^2(0) + 2R_{xx}^2(\tau)$. $S_{yy}(\omega) = \sigma^4\delta(\omega) + \frac{1}{\pi}S_{xx}(\omega) \otimes S_{xx}(\omega)$, with σ^2 the variance of $x(t)$: $R_{xx}^2(0) = \sigma^4.$

$$E\{y(t)\} = E\{x^2(t)\} = R_{xx}(0) = \sigma^2. \quad E\{y^2(t)\} = E\{x^4(t)\} = 3\sigma^4.$$

Variance of $y(t) = E\{y^2(t)\} - e^2\{y(t)\} = 3\sigma^4 - \sigma^4 = 2\sigma^4.$

VI. We have three simultaneous recording of random signals. It is assumed that the signals are wide sense stationary and ergodic. The record length has $N = 8192$ points. We calculate the sums $\sum_{n=1}^{N} s_i[n]$ for $i = 1, 2, 3$ and find approximately zero. The following sums are calculated: $R_{ij} = \frac{1}{N}\sum_{n=1}^{N} s_i[n]\,s_j[n]$ for $i, j = 1, 2, 3$ and find $R_{11} = 43.0384$; $R_{12} = 7.4729$; $R_{13} = -10.5892$; $R_{22} = 26.5357$; $R_{23} = -31.7579$; $R_{33} = 42.9762.$

1. Build the correlation matrix of signal values at time n and recall the general properties of this type of matrix.

Using Matlab we look for the eigenvalues and eigenvectors of this matrix and find:

$$\text{Eigenvectors: } V = \begin{pmatrix} -0.0149 & 0.9184 & 0.3953 \\ -0.7889 & -0.2537 & 0.5598 \\ -0.6144 & 0.3035 & -0.7283 \end{pmatrix},$$

$$\text{Eigenvalues: } \Lambda = \begin{pmatrix} 1.9429 & 0 & 0 \\ 0 & 37.4753 & 0 \\ 0 & 0 & 73.1322 \end{pmatrix}.$$

2. Show that the information carried by these three signals can be approached by the contents of two orthogonal signals which are calculated from the three signals.

Estimate the error in this approximation.

Chapter 25
Ergodicity; Temporal and Spectral Estimations

In many practical cases, we have only one realization of a random signal $x[n]$ and we cannot operate an ensemble average. This is the case, for example, when transmitting a signal on a noisy communication channel, or the case of a satellite image of a terrain area, etc. We are led to try to estimate the statistical properties from the behavior of the process $x[n]$ using this single realization. It is conceivable that over time, the values of $x[n]$ can browse all the possible values of the measured variable, and there may be equivalence between expectancy at a given time and the time average. Time averaging should be performed on a large enough interval to allow that almost all probable values of $x[n]$ could be attained. The signal should necessarily possess qualities of stationarity. We will talk of ergodicity if the equivalence between expectation and time average exists. But even in the case where a signal is ergodic, difficulty is encountered in practice, because a record of this signal has necessarily a limited length. Ergodicity can only be reached asymptotically when the sample length tends to infinity. A precise discussion is needed when one studies the role of the duration of the realization. Temporal integrals can only provide estimates of intrinsic statistical properties. It is for this reason that we speak of estimators.

We study the estimation of the mean of a random signal by the sum of consecutive samples and the variance of the estimator. We discuss some conditions for the ergodicity of the signal regarding the mean. We present two estimators of the correlation functions and discuss their variances. An estimator of the power spectral density is taken as the Fourier transform of one estimator of the correlation function. We show that the poor quality of the raw estimator of the psd results from the poor estimation of the correlation function for large time delays. We present the methods that are used to improve largely the spectral estimation.

The chapter ends with the presentation of methods for extracting one or several harmonic components in a noisy spectrum. The Capon maximum likelihood super resolution method and Pisarenko method are discussed.

© Springer International Publishing Switzerland 2016
F. Cohen Tenoudji, *Analog and Digital Signal Analysis*,
Modern Acoustics and Signal Processing, DOI 10.1007/978-3-319-42382-1_25

25.1   Estimation of the Average of a Random Signal

For example, we define an estimate of the expected value (average) of the signal as
the sum of $2N + 1$ consecutive values of the signal. The caret above the obtained
value highlights the fact that we are dealing with an estimator.

$$\hat{\eta}_{2N+1}[n] = \frac{1}{2N+1} \sum_{m=-N}^{N} x[n-m]. \qquad (25.1)$$

We must realize that this estimator is a random variable, and the same operation
on a different realization sample or another time interval would give a different
value. Like any random variable this estimator has an expected value and a variance
that we will seek to evaluate.

25.1.1   Expectation of the Average Estimator

When the process is stationary, the expectancy of the signal is constant. In this case
the statistical average and time average operations being linear and independent,
their order of application can be switched, and then:

$$E\{\hat{\eta}_{2N+1}[n]\} = \frac{1}{2N+1} \sum_{m=-N}^{N} E\{x[n-m]\} = \frac{1}{2N+1} \sum_{m=-N}^{N} \eta_x = \eta_x. \qquad (25.2)$$

The expectation of the estimator is equal to the quantity which it is desired to
estimate. We then say that the estimator is unbiased.

25.1.2   Variance of the Average Estimator

We study now how this sum over $2N + 1$ values approaches the mathematical
expectation. We realize that if the successive values are highly correlated for small
time differences and not for large time laps, we will have to perform the sum on
large enough time intervals in order that the function $x[n]$ might take all possible
values and that chance plays its role. The length of the summation interval enters
into account. We cannot hope to have $\hat{\eta}_{2N+1} = \eta_x$ on any summation interval of
finite length.

It is however possible only by a passage to the limit $N \to \infty$, the estimator of the
mean tends towards the expected value: $\lim_{N\to\infty} \hat{\eta}_{2N+1}[n] = \eta_x$. We can then
consider that the successive values of $x[n]$ run through all the possibilities offered
by chance. We then say that the signal is ergodic with respect to the average.

We will have a value with certainty only if the variance of the estimator is equal to 0 for $N \to \infty$. So we will have:

$$\lim_{N\to\infty} \text{var}\left(\hat{\boldsymbol{\eta}}_{2N+1}[n]\right) = 0. \tag{25.3}$$

The variance of this estimator is:

$$\text{var}\left(\hat{\boldsymbol{\eta}}_{2N+1}[n]\right) = E\left\{ \left(\hat{\boldsymbol{\eta}}_{2N+1}[n] - \eta_x\right)^2 \right\}. \tag{25.4}$$

We want to show that:

$$\text{var}\left(\hat{\boldsymbol{\eta}}_{2N+1}[n]\right) = \frac{1}{2N+1} \sum_{m=-2N}^{2N} C_{xx}[m]\left(1 - \frac{|m|}{2N+1}\right). \tag{25.5}$$

Indeed, setting

$$\boldsymbol{x}[n] - \eta_x = \boldsymbol{x}_c[n], \tag{25.6}$$

$$\text{var}\left(\hat{\boldsymbol{\eta}}_{2N+1}[n]\right) = E\left\{ \left(\hat{\boldsymbol{\eta}}_{2N+1}[n] - \eta_x\right)^2 \right\}. \tag{25.7}$$

$$\text{var}\left(\hat{\boldsymbol{\eta}}_{2N+1}[n]\right) = E\left\{ \left(\frac{1}{2N+1} \sum_{m=-N}^{N} (\boldsymbol{x}[n-m] - \eta_x) \right)^2 \right\}$$

$$= E\left\{ \left(\frac{1}{2N+1} \sum_{m=-N}^{N} \boldsymbol{x}_c[n-m] \right)^2 \right\}$$

$$= E\left\{ \left(\frac{1}{2N+1} \sum_{m=-N}^{N} \boldsymbol{x}_c[n-m] \right)\left(\frac{1}{2N+1} \sum_{m'=-N}^{N} \boldsymbol{x}_c[n-m'] \right) \right\}$$

$$= \frac{1}{(2N+1)^2} \left(\sum_{m=-N}^{N} \sum_{m'=-N}^{N} E\{\boldsymbol{x}_c[n-m']\boldsymbol{x}_c[n-m]\} \right).$$

So:

$$\text{var}\left(\hat{\boldsymbol{\eta}}_{2N+1}[n]\right) = \frac{1}{(2N+1)^2} \left(\sum_{m=-N}^{N} \sum_{m'=-N}^{N} C_{xx}[m-m'] \right). \tag{25.8}$$

Since the signal is assumed wss its covariance $C_{xx}[m-m']$ depends only on the difference $m - m'$.

We write out $m - m' = m''$. As can be seen in Fig. 25.1, in the plane indicated by the axes m' and m, the relationship $m - m' = Cte$ is the equation of a line parallel to the first bisector which can be written as $m = m'' + m'$.

Fig. 25.1 Summation line
$m = m'' + m'$ in the plane
m', m''

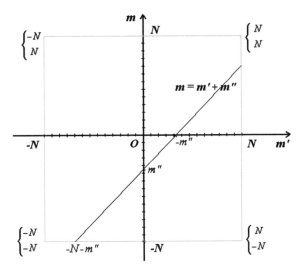

The number of points on this line segment is $2N + 1 - |m''|$.
We can therefore write:

$$\text{var}\left(\hat{\eta}_{2N+1}[n]\right) = \frac{1}{(2N+1)^2} \left(\sum_{m''=-2N}^{2N} C_{xx}[m''](2N + 1 - |m''|) \right), \qquad (25.9)$$

$$\text{var}\left(\hat{\eta}_{2N+1}[n]\right) = \frac{1}{2N+1} \left(\sum_{m''=-2N}^{2N} C_{xx}[m''] \left(1 - \frac{|m''|}{2N+1}\right) \right). \text{Q.E.D.} \qquad (25.10)$$

25.1.3 Ergodicity Conditions

One cannot infer from the expression (25.10) a necessary and sufficient condition
for ergodicity. However, we can find sufficient conditions by analyzing the behavior
of this sum to the limit where $N \rightarrow \infty$.

We first notice that the factor $\frac{1}{2N+1}$ tends to zero as $N \rightarrow \infty$. It is thus sufficient that
the sum converges in order that this factor provides a zero limit value for the variance.

Example of white noise:

The white noise covariance is the Kronecker function by definition. We can
write: $C_{xx}[m] = B\delta[m]$. The sum shown in Eq. (25.10) is equal to B.

We have then:

$$\lim_{N \rightarrow \infty} \text{var}\left(\hat{\eta}_{2N+1}[n]\right) = \lim_{N \rightarrow \infty} \frac{B}{2N+1} = 0. \qquad (25.11)$$

The variance of the estimator of the mean tends to zero as $N \rightarrow \infty$.

The white noise average can therefore be estimated with a time sum if the condition $\frac{B}{2N+1} \ll 1$ is verified.

Similarly if the covariance of the signal $x[n]$ is absolutely summable, the signal will be ergodic with respect to the average. This is the case for example if $C_{xx}[m]$ and its support are bounded.

25.2   Estimation of the Correlation Function

Assume now that one has at our disposal a limited number of signal values, and that one does not know the correlation function of the signal or its spectral power density. We place ourselves in a real case where we have only values of random signal $x[n]$ on an interval $\{0, N-1\}$.

The correlation function (unknown) is defined as $R_{xx}[m] = E\{x[n+m]x^*[n]\}$.

Consider the following estimator of the correlation function:

$$\hat{R}'_{xx}[m] = \frac{1}{N-m} \sum_{n=0}^{N-m-1} x[n+m]x^*[n] \quad \text{for } 0 \leq m < N, \tag{25.12}$$

and

$$\hat{R}'_{xx}[-m] = \hat{R}'^*_{xx}[m] \quad \text{for } -N < m < 0,$$

to satisfy the property of Hermitian conjugation of the correlation function.

It is noted that the summing interval decreases when $m > 0$ increases. For values of m approaching N, the summation interval becomes small. It is expected in this case that the variance of the estimator becomes large since the arithmetic average of the products $x[n+m]x^*[n]$ is done on a small number of these products.

The expectation of this estimator is for $0 \leq m < N$:

$$E\{\hat{R}'_{xx}[m]\} = \frac{1}{N-m} \sum_{n=0}^{N-m-1} E\{x[n+m]x^*[n]\} = \frac{N-m}{N-m} R_{xx}[m] = R_{xx}[m]. \tag{25.13}$$

This estimate of the correlation function is unbiased.

We can define a second estimator:

$$\hat{R}_{xx}[m] = \frac{1}{N} \sum_{n=0}^{N-m-1} x[n+m]x^*[n] \quad \text{for } 0 \leq m < N, \tag{25.14}$$

and

$$\hat{R}_{xx}[-m] = \hat{R}_{xx}^*[m] \quad \text{for} \quad -N < m < 0.$$

The expectation of this estimator is for $0 \leq m < N$:

$$E\{\hat{R}_{xx}[m]\} = \frac{1}{N} \sum_{n=0}^{N-m-1} E\{x[n+m]x^*[n]\} = \frac{N-m}{N} R_{xx}[m]. \qquad (25.15)$$

This estimator is biased. But it has the following advantages: for values of m approaching N, it gives a low value. In this case, this tends to provide low values for variance. A second advantage is that it lends itself much better to the calculations as is outlined below. It is widely used for that reason.

Variance of the estimator $\hat{R}'_{xx}[m]$:

Its calculation involves calculating the expectation of the square modulus of the estimator.

$$E\{|\hat{R}'_{xx}[m]|^2\} = \frac{1}{(N-m)^2} \sum_{n_1=0}^{N-m-1} \sum_{n_2=0}^{N-m-1} E\{x[n_1+m]x^*[n_1]x^*[n_2+m]x[n_2]\}. \qquad (25.16)$$

One sees in the summation a moment of order four, which makes it impractical calculation in the general case. In case the signal is Gaussian the calculation can be continued using the following property valid for complex Gaussian variables b_1, b_2, b_3, b_4: the fourth order moment can be expressed from moments of order 2:

$$E\{b_1 b_2^* b_3 b_4^*\} = E\{b_1 b_2^*\}E\{b_3 b_4^*\} + E\{b_1 b_4^*\}E\{b_2^* b_3\}. \qquad (25.17)$$

In this case one can write the Eq. (25.16) in the form:

$$E\{|\hat{R}'_{xx}[m]|^2\} = \frac{1}{(N-m)^2} \sum_{n_1=0}^{N-m-1} \sum_{n_2=0}^{N-m-1} (R_{xx}[m]R_{xx}[-m] + R_{xx}[n_1 - n_2]R_{xx}[n_2 - n_1]),$$

$$E\{|\hat{R}'_{xx}[m]|^2\} = \frac{1}{(N-m)^2} \sum_{n_1=0}^{N-m-1} \sum_{n_2=0}^{N-m-1} \left(|R_{xx}[m]|^2 + |R_{xx}[n_1 - n_2]|^2\right).$$

For the variance we must subtract from the last expression the square modulus of expectancy. It comes:

$$\text{var}(\hat{R}'_{xx}[m]) = \frac{1}{(N-m)^2} \sum_{n_1=0}^{N-m-1} \sum_{n_2=0}^{N-m-1} |R_{xx}[n_1 - n_2]|^2. \qquad (25.18)$$

By performing a similar calculation to that of the variance of the expectation of the average, one obtains:

$$\text{var}\left(\hat{R}'_{xx}[m]\right) = \frac{1}{N - |m|} \sum_{k=-(N-1-|m|)}^{(N-1-|m|)} \left(1 - \frac{|k|}{N - |m|}\right) |R_{xx}[k]|^2. \qquad (25.19)$$

As for the mean we can find ergodicity conditions for the estimator of the correlation function.

The variance of the unbiased estimator is deduced from the variance (25.19) by the relation:

$$\text{var}\left(\hat{R}_{xx}[m]\right) = \left(\frac{N - |m|}{N}\right)^2 \text{var}\left(\hat{R}'_{xx}[m]\right). \qquad (25.20)$$

For a non-Gaussian process $x[n]$, previous results are satisfactory in the approximation when $m \ll N$.

It is noted that one cannot calculate estimation beyond the summing interval. This estimate is equivalent to postulate that the values of the correlation function outside the range are zero. This is not the only way to proceed. Extrapolation methods of the correlation function beyond this range have been developed.

25.3 Spectral Estimation

The frame of spectral estimates is the evaluation of the power spectral density (PSD) of a random signal or the detection and evaluation of monochromatic components in a noisy signal. The signal is assumed to be wide sense stationary and ergodic to allow estimation of the time correlation function.

Different existing techniques can be distinguished in parametric and non-parametric methods. Among the non-parametric methods can be found:

- The methods of reduction of the spectral variance by averaging (Bartlett method), smoothing of spectra (Blackman and Tukey) or by a combination of both (Welch's method).
- The extraction methods of monochromatic components. They are based on the concept of orthogonal subspaces: signal subspace and noise subspace (Pisarenko, MUSIC, ESPRIT methods)

Among the parametric methods we find methods based on the use of the Yule Walker equations and other methods of constrained minimization:

- MA modeling
- AR modeling; Equivalent to Burg maximum entropy method
- ARMA modeling
- The method of "maximum likelihood" of Capon

25.3.1 Raw Estimator of the Power Spectral Density or Periodogram

Now it is assumed that the signal is ergodic for the correlation function.

The raw estimator of the power spectral density or periodogram is defined by the Fourier transform of $\hat{R}_{xx}[m]$:

$$\hat{S}_{xx}\left(e^{j\omega T}\right) = \sum_{m=-\infty}^{\infty} \hat{R}_{xx}[m]e^{-jm\omega T}. \tag{25.21}$$

The relationship (25.22) can be interpreted as containing a convolution product where appears the signal limited in time by a rectangular window $w_r[n]$:

$$x_r[n] = x[n]w_r[n], \tag{25.22}$$

with

$$w_r[n] = \begin{vmatrix} 1; & 1 \le n \le N \\ 0; & \text{elsewhere} \end{vmatrix}. \tag{25.23}$$

The convolution is written:

$$\hat{R}_{xx}[m] = \frac{1}{N}x_r[m] \otimes x_r^*[-m]. \tag{25.24}$$

The Fourier transform of this convolution is used to write the raw estimate of the PSD in the form:

$$\hat{S}_{xx}\left(e^{j\omega T}\right) = \frac{1}{N}\left| \sum_{m=-\infty}^{\infty} x_r[m]e^{-jm\omega T} \right|^2 = \frac{1}{N}\left|X_r\left(e^{j\omega T}\right)\right|^2. \tag{25.25}$$

The estimator is easily obtained by calculating the FFT of the data $x[n]$.

The variance of this estimator is large. As discussed below, the standard deviation of a value of the spectral estimator at a given frequency is of the order of magnitude of the correct value (this corresponds to a 100 % error margin). This problem is caused by the poor quality of the estimator of the correlation function for large offset values. Indeed, as we have noted earlier, for large offsets, the number of products that we sum is reduced.

25.3.2   *Statistical Properties of the Periodogram*

Expectancy of the Periodogram

The periodogram is defined by:

$$\hat{P}_{xx}\left(e^{j\omega T}\right) = \hat{S}_{xx}\left(e^{j\omega T}\right) = \frac{1}{N}\left|\sum_{m=-\infty}^{\infty} x_r[m]e^{-jm\omega T}\right|^2. \qquad (25.26)$$

The expectation of the periodogram is given by:

$$E\left\{\hat{P}_{xx}\left(e^{j\omega T}\right)\right\} = \sum_{l=-N+1}^{N-1} E\left\{\hat{R}_{xx}[l]\right\}e^{-jl\omega T} = \sum_{l=-N+1}^{N-1} \frac{N-|l|}{N} R_{xx}[l]e^{-jl\omega T}. \quad (25.27)$$

We use in this calculation the Bartlett window:

$$w_B\left(e^{j\omega T}\right) = \left|\begin{matrix} \frac{N-|l|}{N}; & |l| < N \\ 0; & \text{elsewhere} \end{matrix}\right.. \qquad (25.28)$$

It comes:

$$E\left\{\hat{P}_{xx}\left(e^{j\omega T}\right)\right\} = \sum_{l=-\infty}^{\infty} R_{xx}[l]w_B[l]e^{-jl\omega T}. \qquad (25.29)$$

In the frequency domain, we write:

$$E\left\{\hat{P}_{xx}\left(e^{j\omega T}\right)\right\} = W_B\left(e^{j\omega T}\right) \otimes S_{xx}\left(e^{j\omega T}\right), \qquad (25.30)$$

where $W_B\left(e^{j\omega T}\right)$ is the Fourier transform of the Bartlett window:

$$W_B\left(e^{j\omega T}\right) = \frac{1}{N}\frac{\sin^2(N\omega T/2)}{\sin^2(\omega T/2)}. \qquad (25.31)$$

The convolution is written as:

$$W_B\left(e^{j\omega T}\right) \otimes S_{xx}\left(e^{j\omega T}\right) = \frac{1}{\omega_e}\int_{-\frac{\pi}{T}}^{\frac{\pi}{T}} W_B\left(e^{j\omega' T}\right) S_{xx}\left(e^{j(\omega-\omega')T}\right)d\omega'. \qquad (25.32)$$

We see that the periodogram $\hat{P}_{xx}\left(e^{j\omega T}\right)$ is a biased estimator since the expectancy of this estimator is not equal to the amount $S_{xx}\left(e^{j\omega T}\right)$ which it is desired to estimate.

However this estimator is consistent (it tends toward the function to estimate in the limit $N \to \infty$), because when the width of the window tends to infinity, the function $W_B(e^{j\omega T})$ becomes very narrow, resulting in that the result of the convolution in (25.32) tends toward the function $S_{xx}(e^{j\omega T})$.

$$\lim_{N \to \infty} W_B(e^{j\omega T}) \otimes S_{xx}(e^{j\omega T}) = S_{xx}(e^{j\omega T}). \qquad (25.33)$$

Variance and Covariance of the Periodogram

1. The signal $x[n]$ is a Gaussian white noise

To address this problem it is assumed that the signal $x[n]$ is a Gaussian white noise complex $b[n]$ with variance σ_b^2.

Initially we calculate the correlation between two values of the periodogram obtained for two angular frequencies ω_1 and ω_2. It is given by:

$$E\{\hat{P}_{bb}(e^{j\omega_1 T})\hat{P}_{bb}(e^{j\omega_2 T})\} = \frac{1}{N^2}E\{|B(e^{j\omega_1 T})|^2|B(e^{j\omega_2 T})|^2\}, \qquad (25.34)$$

where $B(e^{j\omega T})$ is the Fourier transform of the data sample. We develop the previous expression:

$$\begin{aligned}
E\{\hat{P}_{bb}(e^{j\omega_1 T})\hat{P}_{bb}(e^{j\omega_2 T})\} &= \frac{1}{N^2}E\left\{\sum_{n_1=0}^{N-1} b[n_1]e^{-jn_1\omega_1 T}\sum_{k_1=0}^{N-1} b^*[k_1]e^{jk_1\omega_1 T}\right. \\
&\quad \left. \sum_{n_2=0}^{N-1} b[n_2]e^{-jn_2\omega_2 T}\sum_{k_2=0}^{N-1} b^*[k_2]e^{jk_2\omega_2 T}\right\} \\
&= \frac{1}{N^2}\sum_{n_1=0}^{N-1}\sum_{k_1=0}^{N-1}\sum_{n_2=0}^{N-1}\sum_{k_2=0}^{N-1} E\{b[n_1]b^*[k_1]b[n_2]b^*[k_2]\} \\
&\quad e^{-jn_1\omega_1 T}e^{jk_1\omega_1 T}e^{-jn_2\omega_2 T}e^{jk_2\omega_2 T}.
\end{aligned}$$

We now use the property of complex Gaussian variables given in formula (25.17):

$$E\{b_1 b_2^* b_3 b_4^*\} = E\{b_1 b_2^*\}E\{b_3 b_4^*\} + E\{b_1 b_4^*\}E\{b_2^* b_3\}. \qquad (25.35)$$

We have then:

$$E\{b[n_1]b^*[k_1]b[n_2]b^*[k_2]\} = \begin{cases} \sigma_b^4 & \text{if } n_1 = k_1 \text{ and } n_2 = k_2 \text{ or } n_1 = k_2 \text{ and } n_2 = k_1 \\ 0 & \text{elsewhere} \end{cases}.$$

$$(25.36)$$

By replacing in the correlation we get:

$$E\{\hat{\boldsymbol{P}}_{bb}(e^{j\omega_1 T})\boldsymbol{P}_{bb}(e^{j\omega_2 T})\} = \sigma_b^4\left[1 + \left(\frac{\sin(N(\omega_1 - \omega_2)T/2)}{N\sin((\omega_1 - \omega_2)T/2)}\right)^2\right]. \qquad (25.37)$$

The covariance is given by:

$$\begin{aligned}\mathrm{cov}\left[\hat{\boldsymbol{P}}_{bb}(e^{j\omega_1 T})\hat{\boldsymbol{P}}_{bb}(e^{j\omega_2 T})\right] &= E\{\hat{\boldsymbol{P}}_{bb}(e^{j\omega_1 T})\hat{\boldsymbol{P}}_{bb}(e^{j\omega_2 T})\} \\ &\quad - E\{\hat{\boldsymbol{P}}_{bb}(e^{j\omega_1 T})\}E\{\hat{\boldsymbol{P}}_{bb}(e^{j\omega_2 T})\}.\end{aligned} \qquad (25.38)$$

Since we have:

$$\begin{aligned}E\{\hat{\boldsymbol{P}}_{bb}(e^{j\omega_1 T})\} &= \sum_{l=-N+1}^{N-1}\frac{N - |l|}{N}R_{bb}[l]e^{-jl\omega_1 T} \\ &= \sum_{l=-N+1}^{N-1}\frac{N - |l|}{N}\sigma_b^2\delta[l]e^{-jl\omega_1 T} = \sigma_b^2,\end{aligned} \qquad (25.39)$$

It comes:

$$\mathrm{cov}\{\hat{\boldsymbol{P}}_{bb}(e^{j\omega_1 T})\hat{\boldsymbol{P}}_{bb}(e^{j\omega_2 T})\} = \sigma_b^4\left(\frac{\sin(N(\omega_1 - \omega_2)T/2)}{N\sin((\omega_1 - \omega_2)T/2)}\right)^2. \qquad (25.40)$$

The variance of the periodogram is obtained by letting $\omega_1 \to \omega_2$ in the covariance:

$$\mathrm{var}\left[\hat{\boldsymbol{P}}_{bb}(e^{j\omega T})\right] = \sigma_b^4. \qquad (25.41)$$

Thus, the standard deviation σ_b^2 of periodogram is equal to the PSD σ_b^2 which one seeks to estimate.

This estimator is not consistent to the extent that, by letting the sample length go to infinity, we do not find the PSD. An increase of the length of the record does not bring any improvement, as we accept the large values of the offset in the calculation of the correlation function.

2. The signal $x[n]$ is a regular random signal

The previous proof is valid for a Gaussian white noise signal. In the event that $x[n]$ is a regular Gaussian noise (see the definition in Chap. 26), noting $H_{ca}(e^{j\omega T})$ the frequency response of the causal filter used to model the signal $x[n]$, we can write:

$$\begin{aligned}\hat{\boldsymbol{P}}_{xx}(e^{j\omega T}) &= \frac{1}{N}|X(e^{j\omega T})|^2 = \frac{1}{N}|H_{ca}(e^{j\omega T})|^2|B(e^{j\omega T})|^2 \\ &= |H_{ca}(e^{j\omega T})|^2\hat{\boldsymbol{P}}_{bb}(e^{j\omega T}).\end{aligned} \qquad (25.42)$$

The covariance writes:

$$\text{cov}\left[\hat{\boldsymbol{P}}_{xx}\left(e^{j\omega_1 T}\right)\hat{\boldsymbol{P}}_{xx}\left(e^{j\omega_2 T}\right)\right] = \left|H_{ca}\left(e^{j\omega_1 T}\right)\right|^2\left|H_{ca}\left(e^{j\omega_2 T}\right)\right|^2\text{cov}\left[\hat{\boldsymbol{P}}_{bb}\left(e^{j\omega_1 T}\right)\hat{\boldsymbol{P}}_{bb}\left(e^{j\omega_2 T}\right)\right],$$

$$(25.43)$$

or:

$$\text{cov}\left[\hat{\boldsymbol{P}}_{xx}(e^{j\omega_1 T})\hat{\boldsymbol{P}}_{xx}(e^{j\omega_2 T})\right] = \left|H_{ca}(e^{j\omega_1 T})\right|^2\left|H_{ca}(e^{j\omega_2 T})\right|^2\sigma_b^4\left(\frac{\sin(N(\omega_1 - \omega_2)T/2)}{N\sin((\omega_1 - \omega_2)T/2)}\right)^2$$

$$= S_{xx}(e^{j\omega_1 T})S_{xx}(e^{j\omega_2 T})\left(\frac{\sin(N(\omega_1 - \omega_2)T/2)}{N\sin((\omega_1 - \omega_2)T/2)}\right)^2.$$

$$(25.44)$$

The variance is then:

$$\text{var}\left[\hat{\boldsymbol{P}}_{xx}(e^{j\omega T})\right] \approx S_{xx}^2(e^{j\omega T}).$$

$$(25.45)$$

It is observed that the variance of the periodogram is of the order of magnitude of the square of the spectral density. In other words, the standard deviation of periodogram is the magnitude of the spectral density which it is desired to estimate. The error in an estimate of the spectral density of the periodogram is of the order of 100 %.

25.4   Improvement of the Spectral Estimation

Several techniques for improving the estimation are now described.
Bartlett method:
If the length of the data sample is sufficient, the data are best used by partitioning the sample in L parts and averaging the estimators obtained on each slice. L is determined by tests.

The variance of the estimator is reduced through the averaging. The averaged estimator $\hat{S}_{xx}^M(e^{j\omega T})$ is

$$\hat{S}_{xx}^M(e^{j\omega T}) = \frac{1}{L}\sum_{i=1}^{L}\hat{S}_{xx}^i(e^{j\omega T}).$$

$$(25.46)$$

Blackman and Tukey method:
This method treats the source of the problem which is the poor quality of the estimator of the correlation function for large offset values. The method lies in the multiplication of the estimator of the correlation function given by (25.12) by a window $w[m]$ which retains only the most reliable values, the values obtained for

small offsets. The Hann window is suitable. The width of the window is determined by tests. We note $\hat{R}^a_{xx}[m]$ this estimator, called smoothed estimator.

We have:

$$\hat{R}^a_{xx}[m] = \hat{R}_{xx}[m]w[m]. \tag{25.47}$$

The smoothed corresponding estimator $\hat{S}^a_{xx}(e^{j\omega T})$ of the spectral density is then given by the FT of (25.47) leading to the circular convolution:

$$\hat{S}^a_{xx}(e^{j\omega T}) = \frac{1}{\omega_e} \int_{-\frac{\omega_e}{2}}^{\frac{\omega_e}{2}} \hat{S}_{xx}\left(e^{j\omega' T}\right) W\left(e^{j(\omega-\omega')T}\right) d\omega'. \tag{25.48}$$

The convolution carries out a smoothing in the frequency domain. The variance of the spectrum is significantly reduced.

Welch method:

This method combines the contributions of the two previous methods. The record is divided in L slices. The estimators smoothed by windowing the signal directly in the time domain are calculated for each slice, and the average is used:

$$\hat{S}^{Welch}_{xx}(e^{j\omega T}) = \frac{1}{L}\sum_{i=1}^{L} \hat{S}^{ai}_{xx}(e^{j\omega T}) \tag{25.49}$$

This estimator is best for this class of methods.

The following example illustrates in Fig. 25.2 the different stages of spectral analysis in the following situation: A signal consisting of 2048 samples of a digital white noise (Fig. 25.2b) is the input of an AR filter having the frequency response shown in Fig. 25.2a.

From the results in Chap. 24, the spectral density of the filter output signal is the squared modulus of the frequency response. The periodogram, raw estimator of this spectral density, given by the squared modulus of the Fourier transform of the signal resulting from the filtering of 2048 samples is shown in Fig. 25.2c. We see the randomness of this estimate for which the standard deviation of the estimation of a value is of the order of magnitude of that value. Figure 25.2d is the estimator of the correlation function calculated by inverse FT of the periodogram. In Fig. 25.2e, we see the result of the product of the correlation function by a time window. Figure 25.2f is the smoothed estimator of the spectral density obtained by the FT of this windowed correlation function. Note the greater regularity of this estimator.

The result would have been even better by averaging estimators obtained on slices of the signal. But beware; the slices lengths should not be too small to avoid the appearance of biased values in the spectral estimate. The smoothing of a spectrum allows more reliable detection of harmonic components of low amplitude (Application to detection of ships in underwater acoustics for example).

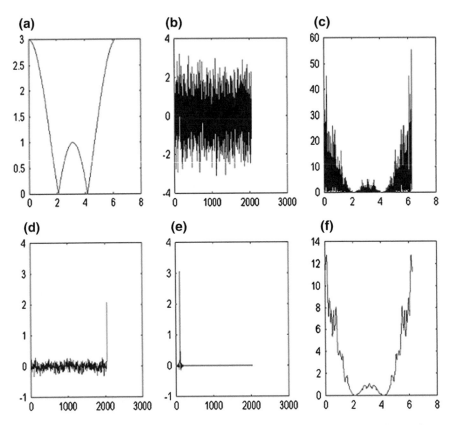

Fig. 25.2 **a** Frequency response of the initial filter; **b** Input white Gaussian noise; **c** Raw estimate of the output PSD; **d** Raw correlation function; **e** Translated, windowed correlation function; **f** Smooth estimate of the PSD

25.5   Search for Harmonic Components

In this section, we are looking for methods of extracting one or several monochromatic components embedded in noise.

25.5.1   Capon Method ("Maximum Likelihood")

In this method, we look for a FIR filter with N terms driven by a white noise whose frequency response is constrained to be 1 at the desired frequency of analysis and whose variance of the output signal is as low as possible. Thus the power of the output signal is mainly that of the signal at frequency ω_0.

We define the vectors:

$$
\boldsymbol{h}_0 = \begin{bmatrix} h_{\omega_0}[0] \\ h_{\omega_0}[1] \\ h_{\omega_0}[2] \\ \vdots \\ h_{\omega_0}[N-1] \end{bmatrix} ; \boldsymbol{s}_0 = \begin{bmatrix} 1 \\ e^{j\omega_0 T} \\ e^{j2\omega_0 T} \\ \vdots \\ e^{j(N-1)\omega_0 T} \end{bmatrix} ; \boldsymbol{x} = \begin{bmatrix} x[n-N+1] \\ \vdots \\ x[n-1] \\ x[n] \end{bmatrix}.
$$

The filter output is:

$$
y[n] = \sum_{k=0}^{N-1} h_{\omega_0}[k] x[n-k] = \boldsymbol{h}_0^T \tilde{\boldsymbol{x}},
$$

where $\tilde{\boldsymbol{x}}$ is the time-reversed of $\boldsymbol{x}$ and $\boldsymbol{h}_0^T$ is the transpose of $\boldsymbol{h}_0$.

We desire to minimize the output power:

$$
P = E\left\{|y[n]|^2\right\} = \boldsymbol{h}_0^T E\{\tilde{\boldsymbol{x}} \tilde{\boldsymbol{x}}^H\} \boldsymbol{h}_0^* = \boldsymbol{h}_0^T \tilde{R}_{xx} \boldsymbol{h}_0^* = \boldsymbol{h}_0^H R_{xx} \boldsymbol{h}_0, \tag{25.50}
$$

while satisfying the constraint:

$$
H_{\omega_0}\left(e^{j\omega_0 T}\right) = \sum_{n=0}^{N-1} h_{\omega_0}[n] e^{-jn\omega_0 T} = \boldsymbol{s}_0^H \boldsymbol{h}_0 = 1. \tag{25.51}
$$

We introduce the Lagrange function:

$$
L = \boldsymbol{h}_0^H R_{xx} \boldsymbol{h}_0 + \mu\left(1 - \boldsymbol{s}_0^H \boldsymbol{h}_0\right) + \mu^*\left(1 - \boldsymbol{h}_0^H \boldsymbol{s}_0\right). \tag{25.52}
$$

Taking the gradient with respect to $\boldsymbol{h}_0^H$ and zeroing it:

$$
\nabla_{\boldsymbol{h}_0^*} L = R_{xx} \boldsymbol{h}_0 - \mu^* \boldsymbol{s}_0 = 0, \tag{25.53}
$$

so:

$$
\boldsymbol{h}_0 = \mu^* R_{xx}^{-1} \boldsymbol{s}_0. \tag{25.54}
$$

Copying this result in (25.51), we have:

$$
\boldsymbol{s}_0^H \boldsymbol{h}_0 = \mu^* \boldsymbol{s}_0^H R_{xx}^{-1} \boldsymbol{s}_0 = 1, \tag{25.55}
$$

so:

$$\mu^* = \mu = \frac{1}{s_0^H R_{xx}^{-1} s_0}. \tag{25.56}$$

Thus Eq. (25.54) writes:

$$h_0 = \frac{R_{xx}^{-1} s_0}{s_0^H R_{xx}^{-1} s_0}. \tag{25.57}$$

Finally, the output power at frequency ω_0 is:

$$P = h_0^H R_{xx} h_0 = \frac{s_0^H R_{xx}^{-1} R_{xx} R_{xx}^{-1} s_0}{\left(s_0^H R_{xx}^{-1} s_0 \right)^2} = \frac{1}{s_0^H R_{xx}^{-1} s_0}. \tag{25.58}$$

By repeating the reasoning for different frequencies, we write the maximum likelihood spectrum estimator:

$$\hat{S}_{xx}^{ML} \left(e^{j\omega T} \right) = \frac{1}{s^H R_{xx}^{-1} s}. \tag{25.59}$$

25.5.2 Pisarenko Method

It is a method of extracting monochromatic components embedded in noise.

If we study a complex exponential with fixed frequency $As[n]$ but whose amplitude A is complex random because the phase ϕ is random, the phase is assumed uniformly distributed over the interval $[0, 2\pi]$:

$$As[n] = |A| e^{j\phi} e^{jn\omega_0 T}. \tag{25.60}$$

To this signal is added a white noise $w[n]$ with variance σ_0^2 and uncorrelated with signal $As[n]$. We note $x[n]$ the sum of these signals:

$$x[n] = As[n] + w[n]. \tag{25.61}$$

We assume that N consecutive values of the signal were measured. We write:

$$x = \begin{bmatrix} x[0] \\ x[1] \\ \vdots \\ x[N-1] \end{bmatrix}; \quad s = \begin{bmatrix} 1 \\ e^{j\omega_0 T} \\ e^{j2\omega_0 T} \\ \vdots \\ e^{j(N-1)\omega_0 T} \end{bmatrix}; \quad w = \begin{bmatrix} w[0] \\ w[1] \\ \vdots \\ w[N-1] \end{bmatrix}. \tag{25.62}$$

Since the signal and noise are uncorrelated, the correlation matrix has the form:

$$R_{xx} = E\{As(As)^H\} + E\{ww^H\} = P_0 ss^H + \sigma_0^2 I, \qquad (25.63)$$

with

$$P_0 = E\{AA^*\}$$

An eigenvector of the matrix R_{xx} is the vector s.
Indeed:

$$R_{xx}s = P_0 ss^H s + \sigma_0^2 s = (NP_0 + \sigma_0^2)s. \qquad (25.64)$$

We used in that $s^H s = N$.
The corresponding eigenvalue is

$$NP_0 + \sigma_0^2. \qquad (25.65)$$

Since the eigenvectors of R_{xx} are orthogonal, all its other eigenvectors are orthogonal to s. We note e_i one of these eigenvectors. We must have:

$$R_{xx}e_i = P_0 ss^H e_i + \sigma_0^2 e_i = \sigma_0^2 e_i. \qquad (25.66)$$

These eigenvectors will therefore have an eigenvalue smaller than that of s.
So the method is to search among the eigenvalues of R_{xx} the largest. The corresponding eigenvector allows the determination of the frequency ω_0 of the harmonic component.

In the case of several harmonic components, and only one component of noise, the eigenvector e_N is determined which corresponds to the lower eigenvalue. The eigenvectors corresponding to harmonic components will all be orthogonal to this vector and we have:

$$s_i^H e_N = 0. \qquad (25.67)$$

The following quantity which takes large values when the denominator approaches zero will detect the frequencies present in the signal:

$$P = \frac{1}{|s_i^H e_N|^2}. \qquad (25.68)$$

Summary
In practical cases, the statistics of a signal is often unknown and one has to use the data to estimate that statistics. It is possible when the signal has ergodic properties. We have studied in the chapter the estimators of the average and of the correlation functions. We have given their mean and variances and discussed the ergodicity

conditions. The estimator of the power spectral density is taken as the Fourier transform of the estimator of the correlation function. We have shown that the regularity of the estimated spectrum is improved by the regularization of the estimator of the correlation function for large offset times. The principles of the highly effective methods of Bartlett, Blackman and Tukey and Welch are given and illustrated by an example. The chapter ends with the presentation of methods for extracting one or several harmonic components in a noisy spectrum. Capon maximum likelihood super resolution method, Pisarenko method have been discussed.

Chapter 26
Parametric Modeling of Random Signals

This chapter deals with parametric modeling of random signals. Initially, we demonstrate the Paley–Wiener condition on the power spectral density of a signal. If this condition is verified, it is possible to factor the z power-density of a signal in the form of a product where appears the transfer function of a causal and stable system which has a causal and stable inverse. In that case, the noise is called regular. It appears that a regular random process can be seen as the output signal of a minimum phase filter driven by white noise. In the following, we study the filtering of white noise by an ARMA filter. We arrive at the Yule-Walker equations system connecting the values of the correlation function of the output signal to the filter coefficients. These equations make it possible to extrapolate the correlation function beyond the time interval used as the basis of the system of equations, or, in the case where the filter coefficients are unknown, to determine the coefficients of this filter. Calculating the coefficients of the MA part of the filter is delicate; one often seeks a more simple representation of a regular noise by an AR model. Then we arrive at a smoothed estimate of the power spectral density of the noise. The chapter concludes by modeling a regular noise by MA filtering of a white noise.

26.1 Paley–Wiener Condition

Random regular process
It is assumed here that the spectrum of a wide-sense stationary random signal under consideration does not include lines (no periodic signal time components).
    We show next that if the following condition is satisfied

$$\int_{-\frac{\pi}{T}}^{\frac{\pi}{T}} \left| \ln S_{xx}(e^{j\omega T}) \right| < \infty, \tag{26.1}$$

© Springer International Publishing Switzerland 2016
F. Cohen Tenoudji, *Analog and Digital Signal Analysis*,
Modern Acoustics and Signal Processing, DOI 10.1007/978-3-319-42382-1_26

the spectral density can then be factorized as

$$S_{xx}(z) = K_0 H_{ca}(z) H_{ca}^*(1/z^*),$$ (26.2)

where K_0 is a positive constant and $H_{ca}(z)$ is the transfer function of a stable causal system with a causal and stable inverse. The random process is then called **regular**.

For $z = e^{j\omega T}$, the power spectral density has the form

$$S_{xx}(e^{j\omega T}) = K_0 |H_{ca}(e^{j\omega T})|^2.$$ (26.3)

When $S_{xx}(z)$ is a rational fraction of polynomials, $H_{ca}(z)$ represents a minimum phase filter.

We first remark that the spectrum $S_{xx}(e^{j\omega T})$ can vanish at isolated points, but cannot be strictly limited to a frequency band. Indeed, when $S_{xx}(e^{j\omega T})$ vanishes, its logarithm becomes infinite. This infinite value does not prevent the convergence of the integral, when it occurs for isolated points; the singularity introduced by the logarithm is weak. However, it prevents the convergence when it manifests itself on a continuum of points.

Thus, it appears that a regular random process can be regarded as the output signal of a minimum phase filter driven by white noise. (K_0 represents the variance of white noise at the input of the filter).

$$\xrightarrow[\; [S_{bb(z)}=K_0] \;]{\text{white noise } b[n]} \boxed{H_{ca}(z)} \xrightarrow{x[n]}$$

Inversely: $\xrightarrow{x[n]} \boxed{H_{ca}^{-1}(z)} \xrightarrow{\text{white noise } b[n]}$ (innovations process).

Demonstration of Paley–Wiener condition

Assume the condition (26.1) satisfied. The spectral density therefore exists. The z-density is necessarily defined in a domain containing the unit circle. Recall that the spectral density is a nonnegative real function. The logarithm of this quantity is a function defined when $S_{xx}(e^{j\omega T})$ is positive. This function can be written as a convergent Fourier series.

$$\ln S_{xx}(e^{j\omega T}) = \sum_{k=-\infty}^{\infty} c_k e^{-jk\omega T} \text{ with } c_{-k} = c_k^*, \text{ since } \ln S_{xx}(e^{j\omega T}) \text{ is real.}$$

Now consider the development in power of z:

$$\ln S_{xx}(z) = \sum_{k=-\infty}^{\infty} c_k z^{-k}.$$ (26.4)

This series must converge in a circular ring centered in $z = 0$ and including the unit circle:

$$R < |z| < 1/R.$$

We can rewrite Eq. (26.4) in the form

$$\ln S_{xx}(z) = c_0 + \sum_{k=1}^{\infty} c_k z^{-k} + \sum_{k=-\infty}^{-1} c_k z^{-k} = c_0 + \sum_{k=1}^{\infty} c_k z^{-k} + \sum_{k'=1}^{\infty} c_{k'}^* z^{k'}. \quad (26.5)$$

Taking the exponential of the last expression

$$S_{xx}(z) = e^{c_0} e^{\sum_{k=1}^{\infty} c_k z^{-k}} e^{\sum_{k'=1}^{\infty} c_{k'}^* z^{k'}} = e^{c_0} \left(e^{\sum_{k=1}^{\infty} c_k z^{-k}} \right) \left(e^{\sum_{k=1}^{\infty} c_k (1/z^*)^{-k}} \right)^*. \quad (26.6)$$

The first term is a positive constant. The second term is the sum of a causal series that converges on the unit circle. We recognize in (26.6) the form of Eq. (26.2).

The Paley–Wiener condition involves the logarithm modulus. A second equation identical to (26.4) can be written for the function $\ln(1/S_{xx}(z))$ whose series must also be convergent, causal, and stable. Thus, the inverse of the causal and stable system $H(z)$ must also be causal and stable as specified in the statement accompanying the Eq. (26.2).

It can be shown, furthermore, that the Paley–Wiener condition is necessary, i.e., the Eq. (26.2) induces inequality (26.1).

Example

Let $S_{xx}(z) = \frac{15z^4 + 34z^2 + 15}{-15z^3 + 34z^2 - 15z}$. Its zeros and poles can be determined with Matlab.

Zeros : $(0 + 1.2910\mathrm{i}); (0 - 1.2910\mathrm{i}); (0 + 0.7746\mathrm{i}); (0 - 0.7746\mathrm{i})$.
Poles : $(0 + 0); (1.6667 + 0\mathrm{i}); (0.6000 + 0\mathrm{i})$.

These points have the necessary symmetry for a regular process, because, as it was seen above, if z_0 is a zero or a pole, $1/z_0^*$ should be a zero or a pole too.

For example, for $z_0 = 1.2910\mathrm{i}$, $1/z_0^* = (1/1.2910\mathrm{i})^* = 0.7746\mathrm{i}$ is also a zero.

To construct $H(z)$, we select from the zeros and the poles of $S_{xx}(z)$ those which ensure causality and stability of $H(z)$, which are thus within the unit circle.

$$H(z) = z^{-1} \frac{(z - 0.7746\mathrm{i})(z + 0.7746\mathrm{i})}{z - 0.6}.$$

It remains to be verified that the constant K_0 appearing in Eq. (26.3) is positive. A pole in $z = 0$ has been positioned to obtain a causal filter.

We obtain the impulse response of this filter with Matlab:

$h[0] = 0.9991-0.0000i; h[1] = 0.5988-0.0006i; h[2] = 0.9592+0.0019i;$
$h[3] = 0.5759+0.0017i; h[4] = 0.3459+0.0014i; h[5] = 0.2079+0.0010i;$
$h[6] = 0.1250+0.0007i; h[7] = 0.0753+0.0005i; h[8] = 0.0454+0.0004i;$
$h[9] = 0.0274+0.0002i;$

The appearance of an imaginary part (small) is due to imprecision in the calculation.

26.2 Parametric Modeling of Random Signals

26.2.1 Yule-Walker Equations

We study the case of filtering a wide-sense stationary white noise $w[n]$ with zero mean, by a causal ARMA filter. We are interested in the properties of different correlation functions. We denote $x[n]$ the output signal of the filter whose impulse response is noted $h[n]$.

As shown before, the random signal $x[n]$ is also wide-sense stationary. By assumption, the filter transfer function has the form $H(z) = \frac{B_q(z)}{A_p(z)}$.

In the time domain we have

$$x[n] = w[n] \otimes h[n]. \tag{26.7}$$

We can write formally

$$X(z) = H(z)W(z) = \frac{B_q(z)}{A_p(z)} W(z). \tag{26.8}$$

or

$$X(z)A_p(z) = B_q(z)W(z).$$

In the time domain we have

$$x[n] + \sum_{l=1}^{p} a_p[l]x[n-l] = \sum_{l=0}^{q} b_q[l]w[n-l]. \tag{26.9}$$

Let us multiply both members of the Eq. (26.9) by $x^*[n-k]$ and take the expectations of each term. We have

$$E\{x[n]x^*[n-k]\} + \sum_{l=1}^{p} a_p[l]E\{x[n-l]x^*[n-k]\} = \sum_{l=0}^{q} b_q[l]E\{w[n-l]x^*[n-k]\}. \tag{26.10}$$

Namely

$$R_{xx}[k] + \sum_{l=1}^{p} a_p[l]R_{xx}[k-l] = \sum_{l=0}^{q} b_q[l]R_{wx}[k-l], \tag{26.11}$$

since

$$R_{wx}[k-l] = E\{w[n-l]x^*[n-k]\} = E\left\{ w[n-l]\left[\sum_{m=-\infty}^{\infty} w[n-k-m]h[m]\right]^* \right\}$$

$$= E\left\{ w[n-l]\left[\sum_{m=-\infty}^{\infty} w[n-k-m]h[m]\right]^* = \sigma_w^2 \sum_{m=-\infty}^{\infty} \delta[k+m-l]h^*[m] \right\}$$

$$= \sigma_w^2 h^*[l-k],$$

we can rewrite (26.11) in the form

$$R_{xx}[k] + \sum_{l=1}^{p} a_p[l]R_{xx}[k-l] = \sigma_w^2 \sum_{l=0}^{q} b_q[l]h^*[l-k]. \tag{26.12}$$

Since the filter is supposed causal, each term of the second member is zero for $k > l$. For any value of k, the lower boundary of the sum on l will be k. We set the second term equal to $\sigma_w^2 c_q[k]$ with

$$c_q[k] = \sum_{l=k}^{q} b_q[l]h^*[l-k]. \tag{26.13}$$

Note that this term is zero for $k > q$.
In summary, we will therefore write

$$\boxed{R_{xx}[k] + \sum_{l=1}^{p} a_p[l] R_{xx}[k-l] = \left\{ \begin{array}{ll} \sigma_w^2 c_q[k]; & 0 \le k \le q \\ 0; & k > q \end{array} \right.}$$
(26.14)

This system is known as the **Yule–Walker system of equations**.
Let us write this system in matrix form:

$$
\begin{bmatrix}
R_{xx}[0] & R_{xx}[-1] & \cdots & R_{xx}[-p] \\
R_{xx}[1] & R_{xx}[0] & \cdots & R_{xx}[-p+1] \\
\vdots & \vdots & & \vdots \\
R_{xx}[q] & R_{xx}[q-1] & \cdots & R_{xx}[q-p] \\
\hdashline
R_{xx}[q+1] & R_{xx}[q] & \cdots & R_{xx}[q-p+1] \\
\vdots & \vdots & & \vdots \\
R_{xx}[q+p] & R_{xx}[q+p-1] & \cdots & R_{xx}[q]
\end{bmatrix}
\begin{bmatrix}
1 \\
a_p[1] \\
a_p[2] \\
\vdots \\
a_p[p]
\end{bmatrix}
$$

$$
= \sigma_w^2
\begin{bmatrix}
c_q[0] \\
c_q[1] \\
\vdots \\
c_q[q] \\
\hdashline
0 \\
\vdots \\
0
\end{bmatrix}.
$$
(26.15)

Several cases may be treated as follows:

- Extrapolation of the correlation function: If the values of coefficients $a_p[k]$ and $b_q[k]$ are known and if the correlation function $R_{xx}[k]$ is known up to the order p, its values can be deduced for higher values of time lags.
  For example if $p \ge q$, according to (26.14) we have

$$R_{xx}[k] = -\sum_{l=1}^{p} a_p[l] R_{xx}[k-l] \text{ for } k \ge p.$$

- If the coefficients $a_p[k]$ and $b_q[k]$ are unknown and if the correlation function $R_{xx}[k]$ is known, we can calculate the filter coefficients (case of modeling). However, the Yule-Walker equations are nonlinear with respect to the calculation of the coefficients $a_p[k]$ and $b_q[k]$. We may carry out linearization methods similar to those encountered for deterministic signals. These methods are studied in the following.

26.2.2 Search of the ARMA Model Coefficients for a Regular Process

The Yule–Walker Eq. (26.15) can be used to estimate the unknown parameters $a_p[k]$ and $b_q[k]$ of the model. It is assumed that the random process is regular. According to the Paley–Wiener condition, it can be modeled as the output of a minimum phase filter ARMA with white noise input. **We assume here that we know the correlation function of the process being modeled.** Without loss of generality, it is assumed that the input white noise model filter has a unit variance.

In a first step, using the lower submatrices in the system (26.15), we write, after a manipulation similar to that which was used in the Prony method

$$
\begin{bmatrix}
R_{xx}[q] & R_{xx}[q-1] & \cdots & R_{xx}[q-p+1] \\
R_{xx}[q+1] & R_{xx}[q] & \cdots & R_{xx}[q-p+2] \\
\vdots & \vdots & & \vdots \\
R_{xx}[q+p-1] & R_{xx}[q+p-2] & \cdots & R_{xx}[q]
\end{bmatrix}
\cdot
\begin{bmatrix}
a_p[1] \\
a_p[2] \\
\vdots \\
a_p[p]
\end{bmatrix}
$$

$$
= -
\begin{bmatrix}
R_{xx}[q+1] \\
R_{xx}[q+2] \\
\vdots \\
R_{xx}[q+p]
\end{bmatrix} .
\tag{26.16}
$$

This system called the **Yule-Walker modified equations** permits the evaluation of parameters $a_p[k]$.

The second step is the evaluation of the coefficients $b_q[k]$. The linear system must be solved from the upper part of the system (26.15):

$$
\begin{bmatrix}
R_{xx}[0] & R_{xx}^*[1] & \cdots & R_{xx}^*[p] \\
R_{xx}[1] & R_{xx}[0] & \cdots & R_{xx}^*[p-1] \\
\vdots & \vdots & & \vdots \\
R_{xx}[q] & R_{xx}[q-1] & \cdots & R_{xx}^*[p-q]
\end{bmatrix}
\cdot
\begin{bmatrix}
1 \\
a_p[1] \\
\vdots \\
a_p[p]
\end{bmatrix}
= \sigma_w^2
\begin{bmatrix}
c_q[0] \\
c_q[1] \\
\vdots \\
c_q[q]
\end{bmatrix} .
\tag{26.17}
$$

Recall here that $c_q[k] = \sum_{l=k}^{q} b_q[l]h^*[l-k]$, zero for $k > q$.

The function $c_q[k]$ appears as the convolution of $b_q[k]$ with $h^*[-k]$. Its z-transform is given by

$$
C_q(z) = B_q(z)H^*(1/z^*) = B_q(z)\frac{B_q^*(1/z^*)}{A_p^*(1/z^*)} .
\tag{26.18}
$$

At this stage, $c_q[k]$ is known for all values of $k \geq 0$ since it is calculated by the Eq. (26.17).

The causal part of the z-transform of $c_q[k]$ is thus known.

We note it $[C_q(z)]_+ = \sum_{k=0}^{\infty} c_q[k]z^{-k}$.

The anticausal part of $C_q(z)$ only contains positive powers of z.

An argument based on the factorization of power spectral densities is now used. Since the signal $x[n]$ is assumed regular, we can write

$$S_{xx}(z) = H(z)H^*(1/z^*) = \frac{B_q(z)\,B_q^*(1/z^*)}{A_p(z)\,A_p^*(1/z^*)}. \qquad (26.19)$$

It is assumed that the signal $x[n]$ is filtered by the filter whose transfer function $A_p(z)$ is known at this stage of calculation, and the output signal of this filter is noted $y[n]$. The spectral density of this signal:

$$S_{yy}(z) = S_{xx}(z)A_p(z)A_p^*(1/z^*) = B_q(z)B_q^*(1/z^*). \qquad (26.20)$$

According to (26.18) we have

$$S_{yy}(z) = C_q(z)A_p^*(1/z^*) = \left[C_q(z)\right]_+ A_p^*(1/z^*) + \left[C_q(z)\right]_- A_p^*(1/z^*). \qquad (26.21)$$

Since the function $A_p^*(1/z^*)$ contains only positive powers of z the causal portion of $S_{yy}(z)$ is written as

$$\left[S_{yy}(z)\right]_+ = \left[C_q(z)\right]_+ A_p^*(1/z^*). \qquad (26.22)$$

Knowing the coefficients $c_q[k]$ for $k > 0$ and the coefficients $a_p[k]$, we can calculate $\left[S_{yy}(z)\right]_+$ directly, and thus by identification of negative powers of z coefficients, the values of $R_{yy}[n]$ for $n \geq 0$.

We have $S_{yy}(z) = \sum_{n=-\infty}^{\infty} R_{yy}[n]z^{-n}$. Given that the autocorrelation function has the property of being symmetric conjugate: $R_{yy}[-n] = R_{yy}^*[n]$, and knowing the causal part of the power series in development of $S_{yy}(z)$, it is possible, by symmetry of powers of z, to restore the anticausal part of its development and therefore recover the function $S_{yy}(z)$.

We get finally $B_q(z)$ using the decomposition $S_{yy}(z) = B_q(z)B_q^*(1/z^*)$.

26.2.3 AR Modeling of a Regular Random Signal

This is a special case of the previous model.

Here

$$H(z) = \frac{b_q[0]}{1 + \sum_{k=1}^{p} a_p[k]z^{-k}}. \qquad (26.23)$$

The Yule–Walker equations then take the form

$$R_{xx}[k] + \sum_{l=1}^{p} a_p[l] R_{xx}[k-l] = |b_q[0]|^2 \delta[k]; \quad k \geq 0. \tag{26.24}$$

We write this system in matrix form

$$\begin{bmatrix} R_{xx}[0] & R_{xx}^*[1] & \cdots & R_{xx}^*[p-1] \\ R_{xx}[1] & R_{xx}[0] & \cdots & R_{xx}^*[p-2] \\ \vdots & \vdots & & \vdots \\ R_{xx}[p-1] & R_{xx}[p-2] & \cdots & R_{xx}[0] \end{bmatrix} \cdot \begin{bmatrix} a_p[1] \\ a_p[2] \\ \vdots \\ a_p[p] \end{bmatrix} = - \begin{bmatrix} R_{xx}[1] \\ R_{xx}[2] \\ \vdots \\ R_{xx}[p] \end{bmatrix}. \tag{26.25}$$

The resolution of this system provides the model parameters $a_p[k]$. To get $|b_q[0]|$ we use the Yule–Walker equation for $k = 0$.

$$|b_q[0]|^2 = R_{xx}[0] + \sum_{l=1}^{p} a_p[l] R_{xx}[-l] = r_{xx}[0] + \sum_{l=1}^{p} a_p[l] R_{xx}^*[l].$$

If the correlation function is known a priori as it has been assumed, the power spectral density is simply given by the FT of the correlation function.

If the correlation function is an estimate, the starting point is the Yule–Walker equations where appear estimators. Using an estimate of the correlation function, these equations take the form

$$\hat{R}_{xx}[k] + \sum_{l=1}^{p} a_p[l] \hat{R}_{xx}[k-l] = b_q[0] h^*[-k]; \quad \forall k. \tag{26.26}$$

We obtain the estimator of the spectral density by taking the FT of the estimator of the correlation function.

After determining the coefficients by solving the linear system of Yule-Walker, we can write the spectral density in a particular form where appear the coefficients of the AR filter. For this, taking the FT of the last equation that is obtained by multiplying by $e^{-jk\omega T}$ and summing on k

$$\hat{S}_{xx}(e^{j\omega T}) + \sum_{l=1}^{p} a_p[l] \hat{S}_{xx}(e^{j\omega T}) e^{-jl\omega T} = b_q[0] \sum_{k=-\infty}^{\infty} h^*[-k] e^{-jk\omega T}.$$

or

$$\hat{S}_{xx}(e^{j\omega T}) = \frac{b_q[0]\sum_{k=-\infty}^{\infty}h^*[-k]e^{-jk\omega T}}{1+\sum_{l=1}^{P}a_p[l]e^{-jl\omega T}} = \frac{b_q[0]H^*(e^{-j\omega T})}{1+\sum_{l=1}^{P}a_p[l]e^{-jl\omega T}}.$$

$$= \frac{b_q^2[0]}{1+\sum_{l=1}^{P}a_p[l]e^{-jl\omega T}}\frac{1}{1+\sum_{l=1}^{P}a_p^*[l]e^{jl\omega T}}. \qquad (26.27)$$

At this stage, it is interesting to use a vector notation showing scalar products:

$$\hat{S}_{xx}(e^{j\omega T}) = b_q^2[0]\frac{1}{s^H a_p^1}\frac{1}{a_p^{1H}s} = b_q^2[0]\frac{1}{\left|a_p^{1H}s\right|^2}, \qquad (26.28)$$

where the vectors are

$$s = \begin{bmatrix} 1 \\ e^{j\omega T} \\ e^{j2\omega T} \\ \vdots \\ e^{jp\omega T} \end{bmatrix} \text{ and } a_p^1 = \begin{bmatrix} 1 \\ a_p[1] \\ a_p[2] \\ \vdots \\ a_p[p] \end{bmatrix}. \qquad (26.29)$$

26.2.4 MA Modeling of a Regular Random Signal

It is also a special case of the model developed in 26.2. We recall that $w[n]$ is a white wss noise, with zero mean.

Now we have

$$x[n] = \sum_{k=0}^{q} b_q[k]w[n-k]. \qquad (26.30)$$

Noting that $h[k] = b_q[k]$, Eq. (26.14) becomes

$$R_{xx}[k] = \sum_{l=0}^{q} b_q[l]h^*[l-k] = b_q[k] \otimes b_q^*[-k]. \qquad (26.31)$$

This equation is nonlinear. To solve the problem, we pass by the power spectral densities by taking the z-transform of Eq. (26.31):

$$S_{xx}(z) = \sum_{k=-q}^{q} R_{xx}[k]z^{-k} = B_q(z)B_q^*(1/z^*). \qquad (26.32)$$

with

$$B_q(z) = \sum_{k=0}^{q} b_q[k] z^{-k}.$$

The zeros of $B_q(z)$ are noted β_k. We can write:

$$S_{xx}(z) = B_q(z) B_q^*(1/z^*) = |b_q[0]|^2 \prod_{k=1}^{q} (1 - \beta_k z^{-1}) \prod_{k=1}^{q} (1 - \beta_k^* z). \qquad (26.33)$$

Sorting zeros and respecting the symmetry, $B_q(z)$ can be traced back to a filter. Since the process to be modeled is regular, $B_q(z)$ takes the form of a minimum phase filter. Indeed, it was shown that a regular process could be written as

$$S_{xx}(z) = \sigma_0^2 Q(z) Q^*(1/z^*) = \sigma_0^2 \prod_{k=1}^{q} (1 - \alpha_k z^{-1}) \prod_{k=1}^{q} (1 - \alpha_k^* z), \qquad (26.34)$$

where $Q(z)$ is a minimum phase polynomial. Thus, the coefficients α_k are within the unit circle. The filter order q is determined by the study of the physical problem or by tests.

Note the analogy with Blackman and Tukey method of spectral estimation where we would use a narrow rectangular window with $2q + 1$ nonzero elements.

In the case where the correlation function $R_{xx}[k]$ and the power spectral density are not known a priori, the estimators are used for these functions on the assumption that the random process is regular.

Example 1

A simulation is done in which the filter has two zeros: $z_0 = 0.95 \, e^{j\frac{2\pi}{3}}$ and $z_0^* = 0.95 \, e^{-j\frac{2\pi}{3}}$. $N = 4096$ samples are drawn according to a normal $N(0, 1)$ law. The raw estimate of the power spectral density of the input signal is $\hat{S}_{xx}(e^{j\omega T}) = \frac{1}{N} |X(e^{j\omega T})|^2$. The raw estimate of the spectral power-density of the output signal is $\hat{S}_{yy}(e^{j\omega T}) = \hat{S}_{xx}(e^{j\omega T}) |H(e^{j\omega T})|^2$.

We calculate the raw estimate of the correlation function of the output signal by taking the inverse Fourier transform: $\hat{R}_{yy}[m] = F^{-1}(\hat{S}_{yy}(e^{j\omega T}))$.

Assuming (rightly) a filter having two zeros, the polynomial is created having 4 zeros and an axis of symmetry at value 1 (index 1 in Matlab corresponds to time zero) to respect the parity of the correlation function. The polynomial coefficients are $\hat{R}_{yy}[3], \hat{R}_{yy}[2], \hat{R}_{yy}[1], \hat{R}_{yy}[2], \hat{R}_{yy}[3]$. We select the zeros of this polynomial that are inside the unit circle to construct the estimated filter.

The results for a signal draw are shown in Fig. 26.1 On the top we see the position of the zeros of the polynomials. The zeros inside the unit circle are selected to build the causal estimated filter. In the bottom figure, the frequency response of the initial filter is shown in bold and the frequency response of the estimated filter in light.

Fig. 26.1 *Top* zeros of the polynomial; *bottom* filter frequency response (*bold*) and its estimate (*light*)

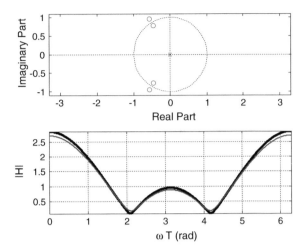

Fig. 26.2 *Top* zeros of the estimated filter; *bottom* filter frequency response and its estimates

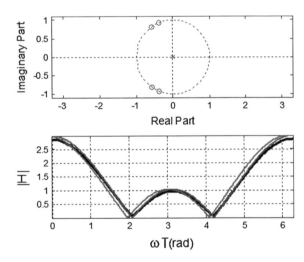

Results vary from a draw to another. The results of a second draw are presented in Fig. 26.2. It is seen on the top that the zero positions do not correspond to the proviso that if z_0 is a zero of the polynomial, $1/z_0^*$ must also be a root.

On the bottom, it is seen that the frequency response minima do not match, this is due to noise on the estimate of the zeros arguments. The outcome is improved by taking for the argument the average of the arguments of nearby zeros. There is an improvement on the position of the filter zeros in frequency seen on the graph where the initial response and the estimated frequency are almost superimposed.

Fig. 26.3 *Top* zeros of the estimated filter; *bottom* filter frequency response (*bold*) and its estimate (*light*)

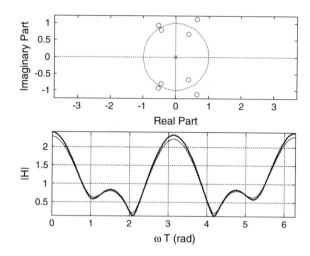

Example 2

In this example, two pairs of zeros of the filter lie inside the disk with radius $R = 1$.
$z_0, z_0^* = 0.95(\cos(2\pi/3) \pm j \sin(2\pi/3)); z_1, z_1^* = 0.8(\cos(\pi/3) \pm j \sin(\pi/3))$.

The polynomial coefficients estimates from data are $\hat{R}_{yy}[5], \hat{R}_{yy}[4], \hat{R}_{yy}[3], \hat{R}_{yy}[2], \hat{R}_{yy}[1], \hat{R}_{yy}[2], \hat{R}_{yy}[3], \hat{R}_{yy}[4], \hat{R}_{yy}[5]$.

In Fig. 26.3, we see on the top the zeros of the polynomial; on the bottom in bold the true frequency response and in light, its estimate.

Summary

We have demonstrated the Paley–Wiener condition on the power spectral density of a signal. When it is verified, it allows the factorization of the z power-density of a random signal in a product of transfer functions of causal and stable systems with inverse causal and stable. A regular random process can be seen as the output signal of a minimum phase filter driven by white noise. We studied the filtering of white noise by an ARMA filter and arrived to Yule–Walker equations These equations allow extrapolation of the correlation function or, in the case where the filter coefficients are unknown, the determination of the coefficients of this filter. We have studied the simple representations of a regular noise by an AR model or a MA model. We arrived at smoothed estimates of the power spectral density of the noise.

Exercises

Linear Prediction: We wish to model a wide-sense stationary random digital signal $y[n]$ as the response of an autoregressive second-order digital filter attacked by white noise with unit variance. On a sample of 8192 data points, with duration

163.84×10^{-6} s, we obtain the following estimates of the first elements of the correlation function in the vicinity of n = 0:

$$\hat{R}_{yy}[0] = 47.4429; \hat{R}_{yy}[1] = 33.6281; \hat{R}_{yy}[2] = 2.3623; \hat{R}_{yy}[3] = -27.3656;$$
$$\hat{R}_{yy}[4] = -39.1847......$$

1. Determine the filter model coefficients (Demonstrate the formulas used).
2. What are the resonance frequency and the bandwidth at −3 dB of the filter?

Chapter 27
Optimal Filtering; Wiener and Kalman Filters

This final chapter is devoted to the modeling and filtering of noisy signals by seeking an optimum estimator in the least squares sense. The basics of this analysis were laid by N. Wiener. These techniques are now an essential part of signal processing. First, we define stochastic orthogonality of two r.v. Then we study the estimation of a random variable by a linear combination of other random variables and then give the equation to calculate the best estimate of the filter's coefficients in the least squares sense. In the case of wide-sense stationary signals, a first example is the search for a Wiener filter providing the estimate of a random signal from the measurement of a second random signal which is correlated to it. The search for a filter in the form of a FIR filter requires the resolution of the Wiener–Hopf linear system of equations. An example of application to the case of an additive noise provides the coefficients of the FIR filter and allows quantifying the gain of the signal-to-noise ratio introduced by the filtering. A second important application is the prediction of the value of a signal from the previous measurements on a finite number of points. In the case of finding an IIR Wiener filter, two different situations arise. In the case of looking for a non-causal filter, the resolution is easy by processing in the Fourier domain. In the search for a more realistic causal filter, treatment is more difficult and requires the factorization of the z-spectral density which is difficult to carry out in practice. Kalman brought a breakthrough to this problem by searching recursively for the estimator. This formulation allows the treatment of non-stationary signals, and the recursive nature of the calculations allows for quick calculations using only the last immediate estimate and the last measured value. The applications are countless, in control systems and in the defense industry. We only focus here on the principle of the filter and its application to simple cases. However, the reader is equipped to extend his fundamental understanding of this technique to a wide range of advanced applications.

© Springer International Publishing Switzerland 2016   543
F. Cohen Tenoudji, *Analog and Digital Signal Analysis*,
Modern Acoustics and Signal Processing, DOI 10.1007/978-3-319-42382-1_27

27.1 Optimal Estimation

27.1.1 Stochastic Orthogonality

The scalar product of two complex random variables is defined as follows:

$$<x,y> \ = E\{xy^*\} \qquad (27.1)$$

For real variables, we have $<x,y> \ = E\{xy\}$.
The norm of a r.v. is defined by a square

$$\|x\|^2 = E\{xx^*\} = \ <x,x> . \qquad (27.2)$$

If this inner product is zero, we say that the two r.v. are orthogonal.

27.1.2 Optimal Least Squares Estimate

We seek to estimate a random variable y by a linear combination of other r.v. x_n.
We note

$$\hat{y} = \sum_{n=1}^{N} a_n x_n. \qquad (27.3)$$

The estimate is erroneous, we commit the error

$$e = y - \hat{y}. \qquad (27.4)$$

The square error is

$$\varepsilon = \|e\|^2 = E\left\{ \left| y - \hat{y} \right|^2 \right\} = E\left\{ \left| y - \sum_{n=1}^{N} a_n x_n \right|^2 \right\} \qquad (27.5)$$

We seek the coefficients a_n that minimize the error norm

$$\frac{\partial \varepsilon}{\partial a_n^*} = E\left\{ -2\left(y - \sum_{n=1}^{N} a_n x_n \right) x_n^* \right\} = 0,$$
$$E\{ex_n^*\} = 0. \qquad (27.6)$$

We note ε_{LS} the minimum square error. The coefficients a_n that minimize the error norm will be such that the error is orthogonal to the r.v. $\boldsymbol{x}_n$. ε_{LS} then becomes

$$\varepsilon_{LS} = E\left\{ \boldsymbol{e} \left(\boldsymbol{y} - \sum_{n=1}^{N} a_n \boldsymbol{x}_n \right)^* \right\}_{min} = E\{\boldsymbol{e}\boldsymbol{y}^*\}. \tag{27.7}$$

The filter is optimal in the sense of minimizing the quadratic error. The set of relationships

$$E\left\{ \left(\boldsymbol{y} - \sum_{n=1}^{N} a_n \boldsymbol{x}_n \right) \boldsymbol{x}_n^* \right\} = 0 \tag{27.8}$$

takes the following form structure:

$$\begin{aligned}
a_1 r_{11} + a_2 r_{12} \ldots + a_N r_{1N} &= r_{01} \\
a_1 r_{21} + a_2 r_{22} \ldots + a_w r_{2N} &= r_{02} \\
\ldots & \\
a_1 r_{N1} + a_2 r_{N2} \ldots + a_N r_{NN} &= r_{0N}
\end{aligned} \tag{27.9}$$

with $r_{0i} = E\{\boldsymbol{y}\boldsymbol{x}_i^*\}$ et $r_{ij} = E\{\boldsymbol{x}_i \boldsymbol{x}_j^*\}$.

The resolution of this linear system allows the determination of the coefficients of the optimal estimate of $\boldsymbol{y}$.

The Wiener FIR filter which is detailed now is an application to the temporal signals optimal estimation.

27.2   Wiener Optimal Filtering

27.2.1   FIR Wiener Filter

In this section, we seek an estimator $\hat{\boldsymbol{d}}[n]$ of a random signal $\boldsymbol{d}[n]$, assumed wide-sense stationary, from the values of a measured signal $\boldsymbol{x}[n]$.

The unknown impulse response of the FIR filter that minimizes the square error is noted $w[n]$. We note q the order of the moving average filter. Assuming a causal filter we can write

$$\boldsymbol{e}[n] = \boldsymbol{d}[n] - \hat{\boldsymbol{d}}[n] = \boldsymbol{d}[n] - \sum_{l=0}^{q-1} w[l] \boldsymbol{x}[n-l]. \tag{27.10}$$

The squared error will be minimum when the error vector is orthogonal to the approximation vectors:

$$E\{e[n]x^*[n-k]\} = 0; \quad (k = 0, 1, 2, \ldots, q-1),$$

so:

$$E\{e[n]x^*[n-k]\} = E\{d[n]x^*[n-k]\} - \sum_{l=0}^{q-1} w[l]E\{x[n-l]x^*[n-k]\} = 0.$$

This equation is

$$\sum_{l=0}^{q-1} w[l]r_{xx}[k-l] = r_{dx}[k]; \quad k = 0, 1, 2, \ldots, q-1, \tag{27.11}$$

This is the system of **Wiener–Hopf** equations that can be written in matrix form using the property of the correlation matrix having the Hermitian symmetry:

$$\begin{bmatrix} r_{xx}[0] & r_{xx}^*[1] & r_{xx}^*[2] & \cdots & r_{xx}^*[q-1] \\ r_{xx}[1] & r_{xx}[0] & r_{xx}^*[1] & \cdots & r_{xx}^*[q-2] \\ r_{xx}[2] & r_{xx}[1] & r_{xx}[0] & \cdots & r_{xx}^*[q-3] \\ \vdots & \vdots & \vdots & & \vdots \\ r_{xx}[q-1] & r_{xx}[q-2] & r_{xx}[q-3] & \cdots & r_{xx}[0] \end{bmatrix} \begin{bmatrix} w[0] \\ w[1] \\ w[2] \\ \vdots \\ w[q-1] \end{bmatrix}$$

$$= \begin{bmatrix} r_{dx}[0] \\ r_{dx}[1] \\ r_{dx}[2] \\ \vdots \\ r_{dx}[q-1] \end{bmatrix}, \tag{27.12}$$

or in condensed form:

$$R_{xx}w = r_{dx}. \tag{27.13}$$

As seen above, the minimum square error is written as

$$\varepsilon_{LS} = E\{e[n]d^*[n]\} = E\left\{ \left[d[n] - \sum_{l=0}^{q-1} w[l]x[n-l] \right] d^*[n] \right\}, \tag{27.14}$$

thus

$$\varepsilon_{LS} = r_{dd}[0] - \sum_{l=0}^{q-1} w[l]r_{dx}^*[l]. \tag{27.15}$$

$$\varepsilon_{LS} = r_{dd}[0] - r_{dx}^H w, \tag{27.16}$$

or

$$\varepsilon_{LS} = r_{dd}[0] - r_{dx}^H R_{xx}^{-1} r_{dx}. \tag{27.17}$$

Application. Filtering an additive noise

It is now assumed that the signal $d[n]$ is marred by a noise $v[n]$ with a zero expectation of noise and uncorrelated with the signal $d[n]$.

Writing $x[n]$ the measured signal, we have

$$x[n] = d[n] + v[n]. \tag{27.18}$$

Since $E\{d[n]v^*[n-k]\} = 0$, we can write

$$r_{dx}[k] = E\{d[n]x^*[n-k]\} = E\{d[n]d^*[n-k]\} = r_{dd}[k]. \tag{27.19}$$

Similarly

$$r_{xx}[k] = E\{x[n]x^*[n-k]\} = E\{d[n]d^*[n-k]\} + E\{v[n]v^*[n-k]\} = r_{dd}[k] + r_{vv}[k]. \tag{27.20}$$

In matrix form we note

$$[R_{dd} + R_{vv}]w = r_{dd}. \tag{27.21}$$

Example

$d[n]$ is a real random process with zero expectation with correlation function $r_{dd}[k] = \alpha^{|k|}$ (autoregressive first-order filter).

We denote $x[n]$ the measured signal as $x[n] = d[n] + v[n]$, where $v[n]$ is a white noise with power $r_{vv}[0] = \sigma_v^2$ that is added to the signal $d[n]$.

We look, for example, for an FIR filter limited to three elements. Its transfer function has the form:

$$W(z) = w[0] + w[1]z^{-1} + w[2]z^{-2}. \tag{27.22}$$

Wiener-Hopf equations are written as

$$\begin{bmatrix} r_{xx}[0] & r_{xx}[1] & r_{xx}[2] \\ r_{xx}[1] & r_{xx}[0] & r_{xx}[1] \\ r_{xx}[2] & r_{xx}[1] & r_{xx}[0] \end{bmatrix} \begin{bmatrix} w[0] \\ w[1] \\ w[2] \end{bmatrix} = \begin{bmatrix} r_{dx}[0] \\ r_{dx}[1] \\ r_{dx}[2] \end{bmatrix}, \tag{27.23}$$

or, taking into account Eqs. (27.19) and (27.20):

$$\begin{bmatrix} 1+\sigma_v^2 & \alpha & \alpha^2 \\ \alpha & 1+\sigma_v^2 & \alpha \\ \alpha^2 & \alpha & 1+\sigma_v^2 \end{bmatrix} \begin{bmatrix} w[0] \\ w[1] \\ w[2] \end{bmatrix} = \begin{bmatrix} 1 \\ \alpha \\ \alpha^2 \end{bmatrix}. \qquad (27.24)$$

The resolution of the previous system then gives the coefficients of the FIR Wiener filter.

Numerical application: Treat the case where $\alpha = 0.7$ and $\sigma_v^2 = 1$.

We solve the previous system using Matlab and we get

$$w[0] = 0.4189; \; w[1] = 0.1750; \; w[2] = 0.0811.$$

The minimum square error can be calculated using Eq. (27.15):

$$\varepsilon_{LS} = r_{dd}[0] - \sum_{l=0}^{p-1} w[l] r_{dx}^*[l] = 1 - (0.4189 + 0.1750 * 0.7 + 0.0811 * 0.49)$$
$$= 0.4189.$$

Signal-to-noise ratios:

- Before filtering: The signal power is $E\left\{ |v[n]^2| \right\} = \sigma_v^2 = 1$. The noise power is $E\left\{ |v[n]^2| \right\} = \sigma_v^2 = 1$. The signal-to-noise ratio is 1 or 0 dB.
- After filtering: The signal becomes $d'[n] = d[n] \otimes w[n]$. The output noise is $v'[n] = v[n] \otimes w[n]$.
- The signal power is

$$E\left\{ |d'[n]^2| \right\} = w^T R_{dd} w = [\, w[0] \quad w[1] \quad w[2]\,] \begin{bmatrix} 1 & \alpha & \alpha^2 \\ \alpha & 1 & \alpha \\ \alpha^2 & \alpha & 1 \end{bmatrix} \cdot \begin{bmatrix} w[0] \\ w[1] \\ w[2] \end{bmatrix}$$
$$= 0.3685$$

The noise power is

$$E\left\{ |v'[n]^2| \right\} = w^T w = [\, w[0] \quad w[1] \quad w[2]\,] \cdot \begin{bmatrix} w[0] \\ w[1] \\ w[2] \end{bmatrix} = 0.2127$$

The output signal-to-noise ratio:

$$10 \log_{10} \frac{0.3685}{0.2127} = 2.39\,\text{dB}.$$

Mean square deviation of the signal $x[n]$ relative to the signal $d[n]$:

$$E\{v[n]v^*[n]\} = \sigma_v^2 = 1.$$

Mean square deviation of the estimator $\hat{d}[n]$ relative to the signal $d[n]$:

$$\varepsilon_{LS} = 0.4189.$$

The decrease of the error power is $10 \log 10 \; (1/0.4189) = 3.78$ dB.

In Fig. 27.1 (Top), the power spectral density of the AR process is shown as a bold line and the frequency response of the Wiener filter (adjusted to have the same maximum) is in fine line. It is seen that the calculated filter favors the frequency interval containing the signal. In the bottom figure, we see the two zeros of the transfer function of this filter.

27.2.2   Linear Prediction of a Random Signal

Here we seek an estimator $\hat{x}[n+1]$ of the random signal $x[n+1]$. The signal values were measured until time n. The unknown impulse response of the FIR which minimizes the square error is noted $w[n]$. Assuming a causal filter we can write

$$\hat{x}[n+1] = \sum_{k=0}^{p-1} w[k]\, x[n-k]. \tag{27.25}$$

Fig. 27.1 Wiener modeling of an AR process; *top* moduli (Wiener model in *fine line*); *bottom* filter zeros positions

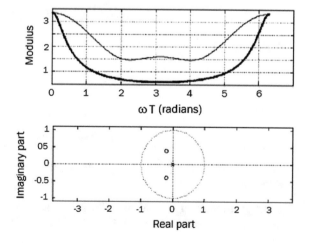

Previous results are used by searching

$$r_{dx}[k] = E\{d[n]x^*[n-k]\} = E\{x[n+1]x^*[n-k]\} = r_{xx}[k+1]. \qquad (27.26)$$

Wiener-Hopf equations then become

$$
\begin{bmatrix}
r_{xx}[0] & r_{xx}^*[1] & r_{xx}^*[2] & \cdots & r_{xx}^*[p-1] \\
r_{xx}[1] & r_{xx}[0] & r_{xx}^*[1] & \cdots & r_{xx}^*[p-2] \\
r_{xx}[2] & r_{xx}[1] & r_{xx}[0] & \cdots & r_{xx}^*[p-3] \\
\vdots & \vdots & \vdots & & \vdots \\
r_{xx}[p-1] & r_{xx}[p-2] & r_{xx}[p-3] & \cdots & r_{xx}[0]
\end{bmatrix}
\cdot
\begin{bmatrix}
w[0] \\
w[1] \\
w[2] \\
\vdots \\
w[p-1]
\end{bmatrix}
=
\begin{bmatrix}
r_{xx}[1] \\
r_{xx}[2] \\
r_{xx}[3] \\
\vdots \\
r_{xx}[p]
\end{bmatrix},
$$

$$(27.27)$$

and the minimum squared error is

$$\varepsilon_{LS} = r_{xx}[0] - \sum_{k=0}^{p-1} w[k] r_{xx}^*[k+1]. \qquad (27.28)$$

For example, we seek the predictive FIR filter with three coefficients of a first-order AR process with autocorrelation function $r_{xx}[k] = \alpha^{|k|}$ (with $\alpha = 0.7$). It is thus sought that

$$\hat{x}[n+1] = w[0]x[n] + w[1]x[n-1] + w[2]x[n-2].$$

Wiener–Hopf equations are written as

$$
\begin{bmatrix}
1 & \alpha & \alpha^2 \\
\alpha & 1 & \alpha \\
\alpha^2 & \alpha & 1
\end{bmatrix}
\cdot
\begin{bmatrix}
w[0] \\
w[1] \\
w[2]
\end{bmatrix}
=
\begin{bmatrix}
\alpha \\
\alpha^2 \\
\alpha^3
\end{bmatrix}.
$$

Using Matlab, we find

$$w[0] = 0.7; \; w[1] = 0; \; w[2] = 0.$$

Thus we get

$$\hat{x}[n+1] = 0.7x[n].$$

We thus find in this particular case the recurrence relation of the AR filter which defines $x[n]$.

Linear prediction in case of an additive noise
Here we seek an estimator $\hat{x}[n+1]$ of the random signal $x[n+1]$.

Measures $y[n]$ are noisy by an additive noise $v[n]$ that is assumed white with unit variance:

$$y[n] = x[n] + v[n]. \qquad (27.29)$$

The estimate takes the form:

$$\hat{x}[n+1] = \sum_{k=0}^{p-1} w[k]\, y[n-k]. \qquad (27.30)$$

$$e[n+1] = x[n+1] - \hat{x}[n+1] = x[n+1] - \sum_{l=0}^{p-1} w[l]\, y[n-l].$$

The quadratic error is minimal when the error vector is orthogonal to the approximation vectors:

$$E\{e[n+1]y^*[n-k]\} = 0; \quad (k = 0, 1, 2, \ldots, p-1),$$

so:

$$E\{e[n+1]y^*[n-k]\} = E\{x[n+1]y^*[n-k]\} - \sum_{l=0}^{p-1} w[l]E\{y[n-l]y^*[n-k]\} = 0.$$

This equation is

$$\sum_{l=0}^{p-1} w[l]r_{yy}[k-l] = r_{xx}[k+1]; \quad (k = 0, 1, 2, \ldots, p-1), \qquad (27.31)$$

We have noted

$$r_{xx}[k+1] = r_{dy}[k].$$

Or in condensed form

$$\left[R_{xx} + \sigma_v^2 I \right] w = r_{dy}. \qquad (27.32)$$

In the example discussed above, with $\alpha = 0.7$ we would have

$$\begin{bmatrix} 1+\sigma_v^2 & \alpha & \alpha^2 \\ \alpha & 1+\sigma_v^2 & \alpha \\ \alpha^2 & \alpha & 1+\sigma_v^2 \end{bmatrix} \cdot \begin{bmatrix} w[0] \\ w[1] \\ w[2] \end{bmatrix} = \begin{bmatrix} \alpha \\ \alpha^2 \\ \alpha^3 \end{bmatrix}. \qquad (27.33)$$

with $w[0] = 0.2932$, $w[1] = 0.1225$, $w[2] = 0.0568$ (Fig. 27.2).

Fig. 27.2 FIR linear
prediction of an AR process;
top moduli (Linear prediction
model in *fine line*); *bottom*:
filter zeros positions

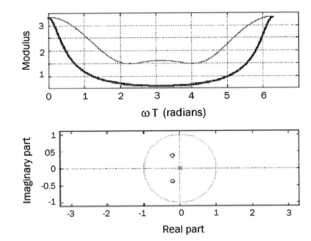

27.3   IIR Wiener Filter

27.3.1   *Non-Causal Filter*

We search again in this section an estimator $\hat{d}[n]$ of a random signal $d[n]$ from the
values of a measured signal $x[n]$. The unknown impulse response of the IIR filter
which minimizes the quadratic error is noted $h[n]$. We can write

$$e[n] = d[n] - \hat{d}[n] = d[n] - \sum_{l=-\infty}^{\infty} h[l]x[n-l]. \qquad (27.34)$$

The quadratic error is minimal when the error vector is orthogonal to the
approximation vectors

$$E\{e[n]x^*[n-k]\} = 0; \quad -\infty < k < \infty.$$

or

$$E\{e[n]x^*[n-k]\} = E\{d[n]x^*[n-k]\} - \sum_{l=-\infty}^{\infty} h[l]E\{x[n-l]x^*[n-k]\} = 0.$$

This equation is

$$\sum_{l=-\infty}^{\infty} h[l]r_{xx}[k-l] = r_{dx}[k]; \quad -\infty < k < \infty. \qquad (27.35)$$

These are the Wiener-Hopf equations for the non-causal filter with Infinite Impulse Response (IIR).

The solution is sought in the frequency domain. It was recognized in the first member of (27.35) a convolution product. Passing in the Fourier domain we have

$$S_{xx}(e^{j\omega T})H(e^{j\omega T}) = S_{dx}(e^{j\omega T}).$$

We deduce the filter's frequency response as

$$H(e^{j\omega T}) = \frac{S_{dx}(e^{j\omega T})}{S_{xx}(e^{j\omega T})}. \tag{27.36}$$

27.3.2 Causal Filter

This time we seek an estimator in the form of the output signal of a causal filter as follows:

$$\hat{d}[n] = \sum_{l=0}^{\infty} h[l]x[n-l]; \quad e[n] = d[n] - \hat{d}[n] = d[n] - \sum_{l=0}^{\infty} h[l]x[n-l].$$

The quadratic error is minimum when the error vector is orthogonal to the approximation vectors:

$$E\{e[n]x^*[n-k]\} = 0; \quad 0 \le k < \infty.$$

Wiener–Hopf equations are written in this case:

$$\sum_{l=0}^{\infty} h[l]r_{xx}[k-l] = r_{dx}[k]; \quad 0 \le k < \infty. \tag{27.37}$$

First, we study the particular case where the filter input signal $v[n]$ is white noise with unit variance. Note $g[n]$ the impulse response of this filter.

Equation (27.37) can be written in this case:

$$\sum_{l=0}^{\infty} g[l]\delta[k-l] = r_{dv}[k],$$

so, $g[k] = r_{dv}[k]$ for $0 \le k < \infty$, or, using the step function $U[k] : g[k] = r_{dv}[k]U[k]$.

Taking the z-transform:

$$G(z) = [S_{dv}(z)]_+, \tag{27.38}$$

The causal part of the transform of $r_{dv}[k]$ was noted $[S_{dv}(z)]_+$.

It is now assumed that the filter input signal $x[n]$ is regular with

$$S_{xx}(z) = \sigma_0^2 Q(z)Q^*(1/z^*), \tag{27.39}$$

where $Q(z)$ is minimum phase

$$Q(z) = 1 + q[1]z^{-1} + q[2]z^{-2} + \dots$$

If $x[n]$ is filtered by a filter with transfer function

$$F(z) = \frac{1}{\sigma_0 Q(z)}. \tag{27.40}$$

(This filter is system 1 in the diagram below).
The output $v[n]$ will have the spectral density:

$$S_{vv}(z) = S_{xx}(z)F(z)F^*(1/z^*) = 1.$$

It is seen that $F(z)$ is the transfer function of a whitening filter.
The optimum filter is the cascade of the two previous filters:

$$H(z) = G(z)F(z). \tag{27.41}$$

As $v[n]$ is formed by the filtering of $x[n]$ by the whitening filter $f[n]$, we can write

$$r_{dv}[k] = E\{d[n]v^*[n-k]\} = E\left\{d[n]\left[\sum_{l=-\infty}^{\infty} f[l]x[n-k-l]\right]^*\right\}$$

$$= \sum_{l=-\infty}^{\infty} f^*[l]r_{dx}[k+l],$$

whose z-Transform is written:

$$S_{dv}(z) = F^*(1/z^*)S_{dx}(z) = \frac{S_{dx}(z)}{\sigma_0 Q^*(1/z^*)}.$$

So $G(z)$ reads

$$G(z) = \frac{1}{\sigma_0}\left[\frac{S_{dx}(z)}{Q^*(1/z^*)}\right]_+. \tag{27.42}$$

Finally, we get

$$H(z) = G(z)F(z) = \frac{1}{\sigma_0^2 Q(z)} \left[\frac{S_{dx}(z)}{Q^*(1/z^*)} \right]_+ \tag{27.43}$$

The minimum square error is

$$\varepsilon_{LS} = r_{dd}[l] - \sum_{l=0}^{\infty} h[l] r_{dx}^*[l] = \frac{1}{\omega_e} \int_{-\frac{\omega_e}{2}}^{\frac{\omega_e}{2}} \left[S_{dd}(e^{j\omega T}) - H(e^{j\omega T}) S_{dx}^*(e^{j\omega T}) \right] d\omega. \tag{27.44}$$

27.4 Kalman Filter[1]

Up to this point in this chapter, the processed signals were wide-sense stationary, their first and second time moments were constant over time. This property is only little or not met in practice where the properties of a physical system under modeling vary slowly or quickly in time. Slowly, in the case, for example, for the drift of the components of an electronic circuit with temperature, quickly, in the case of the in-flight environment of an aircraft. The very notion of correlation function defined in an earlier chapter by of (24.7) does not apply. There needs to be an instantaneous signal processing in optimal seeking treatment. The Kalman filter operates an optimal recursive data processing. Since 1960, when R. Kalman published his article in the ASME transactions, this filtering is widely used in prediction problems of the state of a system in the most diverse fields as, for example, robotics or aeronautics. This is a subject of intense theoretical and applied research. Many books have been devoted to it. An exhaustive presentation of this subject is beyond the scope of this book. We simply expose its principle here.

The measurements on the system are performed successively in time and each new measure allows reevaluating the system state and predicting the state in a near future. The goal of the formulation is to integrate the new data in recursive formulas.

27.4.1 Recursive Estimate of a Constant State

For this presentation, we follow P. Maybeck supporting his explanation with an example. First, to simplify the interpretation of the recurrence relation, the system is assumed to be stationary, and each new measurement improves the knowledge of the system as described in the following.

---

[1]Kalman R.E., A new approach to linear filtering and prediction problems, J. Basic Eng., vol 82D, pp. 35–45, 1960

Suppose that we try to determine the position of an object standing still using successive measurements in a noisy environment. At time t the signal to be measured is noted $z(t) = x + w(t)$. The noise $w(t)$ is assumed white Gaussian, non-stationary.

At time t_1, the measurement result is $z(t_1) = z_1 = x + w(t_1)$. The variance of the measurement is assumed to be known; it is noted $\sigma_{z_1}^2$. At time t_2, the measurement result is $z(t_2) = z_2 = x + w(t_2)$. It is assumed that the measurement environment has changed. The variance of the measurement is now $\sigma_{z_2}^2$. Let us now detail how the second measure has improved the evaluation of the position of the object.

The statistical problem was treated as an exercise at the end of Chap. 22. The search of the best estimation was sought in the sense of maximum likelihood. Using that result, with the notations of the current paragraph, the best estimates of x at times t_1 and t_2 are

$$\hat{x}(t_1) = z_1, \quad \sigma_{\hat{x}(t_1)}^2 = \sigma_{z_1}^2; \tag{27.45}$$

$$\hat{x}(t_2) = z_1 \frac{\sigma_{z_2}^2}{\sigma_{z_1}^2 + \sigma_{z_2}^2} + z_2 \frac{\sigma_{z_1}^2}{\sigma_{z_1}^2 + \sigma_{z_2}^2}, \frac{1}{\sigma_{\hat{x}(t_2)}^2} = \frac{1}{\sigma_{z_1}^2} + \frac{1}{\sigma_{z_2}^2}. \tag{27.46}$$

The estimate of x and the variance of this estimate have varied between the first and the second evaluations. We can describe these changes in a recursive manner as follows:

Variation of the position x estimator as follows:

$$\hat{x}(t_2) = \hat{x}(t_1) + \frac{\sigma_{z_1}^2}{\sigma_{z_1}^2 + \sigma_{z_2}^2} (z_2 - z_1). \text{ Thus } \boxed{\hat{x}(t_2) = \hat{x}(t_1) + K(z_2 - z_1)}, \text{ with } K = \frac{\sigma_{z_1}^2}{\sigma_{z_1}^2 + \sigma_{z_2}^2}.$$

Variation of the estimator of the variance of the x estimator as follows:

$$\sigma_{\hat{x}(t_2)}^2 = \frac{\sigma_{z_1}^2 \sigma_{z_2}^2}{\sigma_{z_1}^2 + \sigma_{z_2}^2} = \sigma_{z_1}^2 - K\sigma_{z_1}^2, \text{ or, } \boxed{\sigma_{\hat{x}(t_2)}^2 = \sigma_{\hat{x}(t_1)}^2 - K\sigma_{\hat{x}(t_1)}^2}. \tag{27.47}$$

The two framed equations are written in the Kalman recursion form.

27.4.2   General Form of the Kalman Recursive Equation

Bold typeface emphasizes that the variables are vectors of parameters.

The operator has access to measurements whose values at time n constitute the vector $y[n]$. These measurements represent a linear modification by the factor $H[n]$ of the vector of interest $x[n]$ to which is added a white measurement noise with zero mean $v[n]$.

Thus, the measurement at time n is given by

$$y[n] = H[n]x[n] + v[n].\tag{27.48}$$

The noise vector is white:

$$E\{v[n]v^H[k]\} = R\delta[k-n] = \sigma_v^2\delta[k-n].\tag{27.49}$$

The system state evolves over time as

$$x[n] = \Phi[n-1]x[n-1] + w[n].\tag{27.50}$$

$\Phi[n-1]$ is the propagator which drives the system from state $x[n-1]$ to state $x[n]$.

$w[n]$ is a disruptive random white noise in the system evolution:

$$E\{w[n]w^H[k]\} = Q\delta[k-n] = \sigma_w^2\delta[k-n].\tag{27.51}$$

The estimator $\hat{x}[n]$ of $x[n]$ at time n uses the propagation by $\Phi[n-1]$ of the estimate $\hat{x}[n-1]$ of $x[n-1]$ at time $n-1$ and corrects this estimation with the term

$$K[n](y[n] - H[n-1]\Phi[n-1]\hat{x}[n-1]).\tag{27.52}$$

This correction term is proportional to the difference between the measure $y[n]$ at time n and the output multiplied by factor $H[n-1]$ of the propagation of $\hat{x}[n-1]$. $K[n]$ is the Kalman filter coefficient to be determined by optimization.

The recursive equation of the Kalman filter is

$$\hat{x}[n] = \Phi[n-1]\hat{x}[n-1] + K[n](y[n] - H[n-1]\Phi[n-1]\hat{x}[n-1]).\tag{27.53}$$

Here we follow the formulation of M.H. Hayes remarkable for its clarity.

The error at time n is the difference between $x[n]$ and the value of the estimator $\hat{x}[n]$, the measure $y[n]$ at time n having been integrated.

$$e[n|n] = x[n] - \hat{x}[n|n].\tag{27.54}$$

The error which does not take into account the measurement at time n is

$$e[n|n-1] = x[n] - \hat{x}[n|n-1].\tag{27.55}$$

The goal now is to find a recursive equation for the mean square error which is at time $x[n]$:

$$P[n|n] = E\{e[n|n]e^H[n|n]\}.\tag{27.56}$$

In the same manner, we note

$$P[n|n-1] = E\{e[n|n-1]e^{H}[n|n-1]\}. \tag{27.57}$$

Estimate of $x[n]$ without using the measurement $y[n]$:

$$\hat{x}[n|n-1] = \Phi[n-1]\hat{x}[n-1|n-1]. \tag{27.58}$$

We may write the error $e[n|n-1] = x[n] - \hat{x}[n|n-1]$ as

$$e[n|n-1] = \Phi[n-1]x[n-1] + w[n] - \Phi[n-1]\hat{x}[n-1|n-1]. \tag{27.59}$$

Using $e[n-1|n-1] = x[n-1] - \hat{x}[n-1|n-1]$, it comes

$$e[n|n-1] = \Phi[n-1]e[n-1|n-1] + w[n]. \tag{27.60}$$

We look for an unbiased estimator, so we impose a zero average of the error:

$$E\{e[n-1|n-1]\} = 0. \tag{27.61}$$

and also

$$E\{e[n|n-1]\} = 0. \tag{27.62}$$

We have $P[n|n-1] = E\{e[n|n-1]e^{H}[n|n-1]\}$

$$P[n|n-1] = E\{(\Phi[n-1]e[n-1|n-1] + w[n])(\Phi[n-1]e[n-1|n-1] + w[n])^{H}\}$$

$$P[n|n-1] = \Phi[n-1]P[n-1|n-1]\Phi^{H}[n-1] + Q[n] \tag{27.63}$$

We note the following, defining $K'[n]$:

$$\hat{x}[n|n] = K'[n]\hat{x}[n|n-1] + K[n]y[n]. \tag{27.64}$$

As $e[n|n] = x[n] - \hat{x}[n|n]$, we write:

$$e[n|n] = x[n] - K'[n]\hat{x}[n|n-1] - K[n]y[n]. \tag{27.65}$$

Thus

$$\begin{aligned}
e[n|n] &= x[n] - K'[n](x[n] - e[n|n-1]) - K[n](H[n]x[n] + v[n]) \\
&= (I - K'[n] - K[n]H[n])x[n] + K'[n]e[n|n-1] - K[n]v[n]
\end{aligned} \tag{27.66}$$

Since the two terms in the above expression are such that $E\{e[n|n-1]\} = 0$, and $E\{v[n]\} = 0$, the error $e[n|n]$ will be unbiased if

$$K'[n] = I - K[n]H[n] \tag{27.67}$$

Expression (27.66) becomes

$$e[n|n] = K'[n]e[n|n-1] - K[n]v[n],$$

or, using (27.67),

$$e[n|n] = (I - K[n]H[n])e[n|n-1] - K[n]v[n]. \tag{27.68}$$

Using expression (27.67) in (27.64) we write

$$\hat{x}[n|n] = (I - K[n]H[n])\hat{x}[n|n-1] + K[n]y[n].$$

We have obtained the recursion relation giving $\hat{x}[n|n]$:

$$\hat{x}[n|n] = \hat{x}[n|n-1] + K[n](y[n] - H[n]\hat{x}[n|n-1]). \tag{27.69}$$

The final stage of the demonstration is now to obtain the Kalman factor $K[n]$ by minimizing the least square error of the estimation.

We necessarily have

$$E\{e[n|n-1]v[n]\} = 0.$$

The square error $P[n|n] = E\{e[n|n]e^H[n|n]\}$ is

$$P[n|n] = (I - K[n]H[n])P[n|n-1](I - K[n]H[n])^H + K[n]R[n]K^H[n]. \tag{27.70}$$

We seek to minimize the average square error:

$$\xi[n] = \mathrm{tr}(P[n|n]).$$

We accept here the following mathematical results:

$$\frac{\mathrm{d}}{\mathrm{d}K}\mathrm{tr}(KA) = A^H \quad \text{and} \quad \frac{\mathrm{d}}{\mathrm{d}K}\mathrm{tr}(KAK^H) = 2KA. \tag{27.71}$$

It comes

$$\frac{\mathrm{d}}{\mathrm{d}K}\mathrm{tr}(P[n|n]) = -2(I - K[n]H[n])P[n|n-1]H^H + 2K[n]R[n] = 0. \tag{27.72}$$

We solve the above equation to obtain $K[n]$:

$$\boxed{K[n] = P[n|n-1]H^H[n]\left(H[n]P[n|n-1]H^H[n] + R[n]\right)^{-1}.}$$ (27.73)

To obtain the recursion relation on the mean square error we write Eq. (27.70) in the form:

$$P[n|n] = (I - K[n]H[n])P[n|n-1]$$
$$- \left((I - K[n]H[n])P[n|n-1]H^H[n] + K[n]R[n]\right)K^H[n].$$

The second part of the last equation is zero because of (27.72). Then

$$\boxed{P[n|n] = (I - K[n]H[n])P[n|n-1].}$$ (27.74)

The three framed equations, (27.69), (27.73), and (27.74) are the Kalman filter equations.

Recursivity requests to set the initial conditions. It is reasonable to take

$$\hat{x}[0|0] = E\{x[0]\} \text{ and } P[0|0] = E\{x[0]x^H[0]\}.$$

Application

Recursive estimation of the position of a stationary object with direct observation.
For this particular case, we have $\Phi[n] = 1$, $w[n] = 0$, $H[n] = 1$.
Equation (27.63) becomes $P[n|n-1] = P[n-1|n-1]$. We note it $P[n-1]$.
Equation (27.74) becomes

$$P[n] = (1 - K[n])P[n-1].$$

Equation (27.73) becomes

$$K[n] = \frac{P[n-1]}{P[n-1] + \sigma_v^2}.$$

Recursively we have, $P[1] = \frac{P[0]\sigma_v^2}{P[0] + \sigma_v^2}$, $P[2] = \ldots$, $P[n] = \frac{P[0]\sigma_v^2}{nP[0] + \sigma_v^2}$.
Thus

$$K[n] = \frac{P[0]}{nP[0] + \sigma_v^2}.$$

The recursive estimation Eq. (27.69) becomes

$$\hat{x}[n] = \hat{x}[n-1] + \frac{P[0]}{nP[0] + \sigma_v^2}(y[n] - \hat{x}[n-1]).$$

We find that as n increases indefinitely, the estimation becomes constant, toward the exact value, because the estimation is unbiased by construction.

Summary
We have studied in this chapter the modeling and filtering of noisy signals by
seeking an optimum estimator in the least squares sense. After defining stochastic
orthogonality of two r.v., we have studied the Wiener estimation of a random
variable by a linear combination of other random variables and then gave the
equation to calculate the best estimate of the filter's coefficients in the least squares
sense. In the case of wide-sense stationary signals, in a first example we have
exposed the search of a Wiener filter providing the estimate of a random signal from
the measurement of a second random signal which is correlated to it. In the choice
of a FIR filter, its determination passes by the resolution of the Wiener–Hopf linear
system of equations. We have shown on an example the improvement of the signal
to noise ratio brought by the filtering. A second important application is the prediction of the value of a signal from the previous measurements on a finite number
of points. In the case of finding an IIR Wiener filter, we studied the non-causal and
the more elaborate causal cases.

We have developed the recursive approach of Kalman to the least square estimation of the state of a physical system. It applies to non-stationary signals and
allows for quick calculations, using only the last immediate estimate and the last
measured value. Some applications to simple cases have been given.

Exercises
This exercise has been met in Chap. 22 and its solution was searched using the
Maximum likehood estimation. Here we search the solution with the least square
technique.

An unknown parameter x is measured with two equipments having different
precisions. The errors are random. The result of the first measurement is x_1; the
measurement standard error is σ_1. The second equipment delivers the value x_2 with
a standard error σ_2.

Give the least square error linear estimation of x. What is the standard error of
the estimation?

Solution:

We search the estimator under the form: $\hat{x} = a_1 x_1 + a_2 x_2$. The coefficients
should be such that $\varepsilon^2 = E\left\{ \|\hat{x} - x\|^2 \right\}$ is minimum. The partial derivatives
respective to the coefficients should be zero.

$$\frac{\partial \varepsilon^2}{\partial a_1} = \frac{\partial}{\partial a_1} E\left\{ (x - (a_1 x_1 + a_2 x_2))^2 \right\} = 0; \quad E\{(x - (a_1 x_1 + a_2 x_2))x_1\} = 0.$$

It comes

$$E\{xx_1\} + E\{-a_1 x_1 x_1\} + E\{-a_2 x_2 x_1\} = 0.$$
$$xE\{x_1\} - a_1 E\{x_1 x_1\} - a_2 E\{x_2 x_1\} = 0.$$

Likewise deriving ε^2 versus a_2 leads to

$$x^2 - a_1 E\{x_1 x_1\} - a_2 E\{x_2 x_1\} = 0.$$
$$x^2 - a_1 E\{x_1 x_2\} - a_2 E\{x_2 x_2\} = 0.$$

Taking the difference of the two last equations, it comes $-a_1 (E\{x_1^2\}$ $-E\{x_2 x_1\}) - a_2 (E\{x_2 x_1\} - E\{x_2^2\}) = 0$. The measurements are independent, then $E\{x_2 x_1\} = E\{x_2\}E\{x_1\} = x^2$. Since $E\{x_1^2\} - x^2 = \sigma_1^2$, we have $a_1 \sigma_1^2 - a_2 \sigma_2^2 = 0$, or $a_1 = a_2 \frac{\sigma_2^2}{\sigma_1^2}$.

As $E\{\hat{x}\} = a_1 E\{x_1\} + a_2 E\{x_2\} = x$, we have $a_1 + a_2 = 1$, Finally $a_1 = \frac{\sigma_2^2}{\sigma_1^2 + \sigma_2^2}$, $a_2 = \frac{\sigma_1^2}{\sigma_1^2 + \sigma_2^2}$.

The standard squared error of the estimation is $\varepsilon^2 = E\left\{ (x - (a_1 x_1 + a_2 x_2))^2 \right\}$.
The result should be valid for any value x and cannot depend on x.
To simplify we set $x = 0$.
$\varepsilon^2 = E\left\{ (a_1 x_1 + a_2 x_2)^2 \right\} = a_1^2 E\{x_1^2\} + 2a_1 a_2 E\{x_1 x_2\} + a_2^2 E\{x_2^2\}$.
As in this case $E\{x_1 x_2\} = E\{x_1\}E\{x_2\} = x^2 = 0$.

$$\varepsilon^2 = \sigma^2 = a_1^2 E\{x_1^2\} + a_2^2 E\{x_2^2\} = a_1^2 \sigma_1^2 + a_2^2 \sigma_2^2.$$

Finally, we have

$$\frac{1}{\sigma^2} = \frac{1}{\sigma_1^2} + \frac{1}{\sigma_2^2}.$$

The above system can be written in matrix form: $\boldsymbol{R}\boldsymbol{a} = x^2 \begin{pmatrix} 1 \\ 1 \end{pmatrix}$, with $\boldsymbol{R} = \begin{pmatrix} R_{11} & R_{12} \\ R_{12} & R_{22} \end{pmatrix}$, and $R_{11} = E\{x_1^2\}$, $R_{12} = E\{x_1 x_2\}$, $R_{22} = E\{x_2^2\}$, $\boldsymbol{a} = \begin{pmatrix} a_1 \\ a_2 \end{pmatrix}$.
$\boldsymbol{a} = x^2 \boldsymbol{R}^{-1} \begin{pmatrix} 1 \\ 1 \end{pmatrix} = \frac{x^2}{R_{11} R_{22} - R_{12}^2} \begin{pmatrix} R_{22} - R_{12} \\ R_{11} - R_{12} \end{pmatrix}$. We verify that $a_1 = a_2 \frac{\sigma_2^2}{\sigma_1^2}$

Appendix A
Functions of a Complex Variable

Only the essential concepts for the signal theory on the complex functions of the complex variable are presented here. Because of its brevity, this discussion is necessarily incomplete. Readers wishing to deepen these concepts are referred to the general mathematics courses and to the many books on this important subject.

A.1. Notions on Complex Variables

A.1.1. Notion of a Complex Number

A complex number consists of the sum of a real part and an imaginary part. We note $z = x + jy$, where $j = \sqrt{-1}$ (In mathematics, we write $i = \sqrt{-1}$. However, in electronics and signal analysis we note j the root of -1, reserving the letter i to name the current in a circuit).

A complex number z corresponds to a point M in the plane xOy called the complex plane (Fig. A.1). The real part of z corresponds to the abscissa of M, its imaginary part to the ordinate of M.

M is called image. We say that z is the affix of M.

As can be seen in the figure above, we have:

$$x = \rho \cos \theta \quad \text{and} \quad y = \rho \sin \theta. \tag{A.1}$$

We can write z as

$$z = \rho \cos \theta + j\rho \sin \theta. \tag{A.2}$$

The modulus ρ is given by $\rho = \sqrt{x^2 + y^2}$. The argument θ is given by

$$\theta = \operatorname{Arg} \frac{y}{x}. \tag{A.3}$$

© Springer International Publishing Switzerland 2016
F. Cohen Tenoudji, *Analog and Digital Signal Analysis*,
Modern Acoustics and Signal Processing, DOI 10.1007/978-3-319-42382-1

Fig. A.1 Complex number
z in the plane xOy.

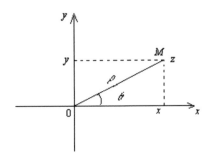

We admit the very important relationship (Euler formula):

$$e^{j\theta} = \cos\theta + j\sin\theta. \tag{A.4}$$

A complex number z can be written in exponential trigonometric form:

$$z = \rho e^{j\theta}. \tag{A.5}$$

A.1.2. Complex Function of a Complex Variable

A function of the complex variable z is noted $f(z)$. It is an application of a subset belonging to set $\mathbb{C}$ with value in $\mathbb{C}$.

Signal theory focuses on the simple case of uniform functions of the variable z, that is to say to functions having a single value for each value of z.

The concept of limit is particularly important for a function of the complex variable defined in an area surrounding a point z_0. By definition, we say that the function $f(z)$ tends towards the limit l as z approaches z_0 if

$$l = \lim_{z \to z_0} f(z). \tag{A.6}$$

z_0 being a point of the complex plane, the limit should be the same when z tends towards z_0 from any point z of the neighborhood of z_0 (Fig. A.2).

The function $f(z)$ is continuous in z_0 if

$$\lim_{z \to z_0} f(z) = f(z_0). \tag{A.7}$$

Fig. A.2 Limit of function
$f(z)$ at z_0.

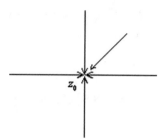

A.2. Complex Derivation

The derivative $f'(z)$ of the function $f(z)$ at any point z_0 is defined by

$$f'(z)|_{z=z_0} = \lim_{z \to z_0} \frac{f(z) - f(z_0)}{z - z_0}. \tag{A.8}$$

Here again, the limit must be the same starting from any point z in the neighborhood of z_0.

Example Let the function $f(z) = z^2$.

$$f(z) = (x + jy)^2 = x^2 + 2jxy - y^2. \tag{A.9}$$

The derivative calculated for y constant is taken on a horizontal path $z \to z_0$:

$$\frac{df(z)}{dz}\bigg|_{y=cte} = \frac{\partial f(z)}{\partial x} = 2x + 2jy = 2z. \tag{A.10}$$

The derivative calculated for x constant is taken along a vertical path $z \to z_0$:

$$\frac{df(z)}{dz}\bigg|_{x=cte} = \frac{\partial f(z)}{j \partial y} = 2x - \frac{1}{j}2y = 2(x + jy) = 2z. \tag{A.11}$$

We see that we could have derived formally $f(z)$ with respect to z:

$$\frac{df(z)}{dz} = \frac{dz^2}{dz} = 2z. \tag{A.12}$$

This geometric nature of complex functions defined in a planar domain gives them particular properties.

Only a few properties are given here:

- We write the function $f(z)$ as

$$f(z) = f_1(x, y) + jf_2(x, y); \qquad (A.13)$$

The following general property must be assessed:

$$\frac{\partial f_1(x, y)}{\partial y} = -\frac{\partial f_2(x, y)}{\partial x}. \qquad (A.14)$$

In the previous example,

$$\begin{aligned} f_1(x, y) &= x^2 - y^2 \\ f_2(x, y) &= 2xy. \end{aligned} \qquad (A.15)$$

The derivatives of functions $f_1(x, y)$ and $f_2(x, y)$ have satisfied the relationship (A.14).
- If the function $f(z)$ is defined, continuous at a point, it is infinitely differentiable at this point.
- A differentiable function at a point is analytic at this point, that is to say, it is developable in power series. Noting z_0 this point, we have

$$\begin{aligned} f(z) = f(z_0) + (z - z_0)\frac{df}{dz}\Big|_{z_0} + \frac{1}{2}(z - z_0)^2\frac{d^2f}{dz^2}\Big|_{z_0} \\ + \cdots + \frac{1}{n!}(z - z_0)^n\frac{d^nf}{dz^n}\Big|_{z_0} + \cdots \end{aligned} \qquad (A.16)$$

- Except for the constant function $f(z) = C$, a function $f(z)$ cannot be defined, analytic in the entire complex plane.
  For example, the function $f(z) = z$ is defined in the whole complex plane, except in $z = \infty$.

  The function $f(z) = \frac{1}{z}$ is not defined in $z = 0$. We say that it has a pole in $z = 0$.
  It is said that the function $f(z)$ has a pole of order n in $z = z_0$, if in a neighborhood of this point it behaves like $\frac{A}{(z-z_0)^n}$.

A.3. Complex Integration

The complex integral is a path integral defined on the path Γ by (see Fig. A.3)

Fig. A.3 Path integral in complex plane

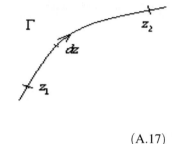

$$\int_\Gamma f(z)dz. \tag{A.17}$$

This integral on a path from z_1 to z_2 depends in the general case on the path followed from z_1 to z_2.

A.3.1. Cauchy Theorem

This theorem is one of the most important in analysis. It reads:
The integral of $f(z)$ on a closed contour enclosing an area within which the function is analytical is zero (see Fig. A.4):

$$\oint_\Gamma f(z)dz = 0. \tag{A.18}$$

We verify this theorem on a simple example: We assume that $f(z) = z$ and that the closed contour Γ consists of the sequence of segments as shown in Fig. A.5: $\Gamma = L_1 + L_2 + L_3 + L_4$. We choose to follow the contour counter-clockwise.
Then

$$\oint_\Gamma f(z)dz = \int_{L_1} zdz + \int_{L_2} zdz + \int_{L_3} zdz + \int_{L_4} zdz. \tag{A.19}$$

Calculating the first integral over L_1:

$$\int_{L_1} zdz = \int_0^2 (x+jy)dx = \left[\frac{x^2}{2} + jyx\Big|_{y=0}\right]_0^2 = \left[\frac{x^2}{2}\right]_0^2 = 2. \tag{A.20}$$

Fig. A.4 Integration on a closed path Γ

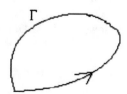

Fig. A.5 Integration contour
Γ

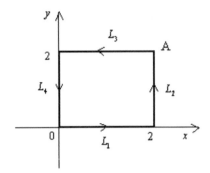

Calculating the second integral on L_2 where $x = cste = 2$:

$$\int_{L_2} z \, dz = \int_0^2 (2 + jy)j \, dy = j\left[2y + j\frac{y^2}{2}\right]_0^2 = 4j - 2. \qquad (A.21)$$

On L_3, $y = cste = 2$, the third integral becomes

$$\int_{L_3} z \, dz = \int_2^0 (x + jy) \, dx = \int_2^0 (x + 2j) \, dx = \left[\frac{x^2}{2} + 2jx\right]_2^0 = -2 - 4j. \qquad (A.22)$$

Calculating the last integral on L_4:

$$\int_{L_4} z \, dz = \int_2^0 (x + jy)\big|_{x=0} \ j \, dy = \int_2^0 jyj \, dy = -\frac{y^2}{2}\bigg|_2^0 = 2. \qquad (A.23)$$

The sum of these four integrals is zero as expected from Cauchy Theorem. The integral on the closed contour is zero. We say that the function $f(z)$ is holomorphic in the area inside the contour.

The condition that the integrand is analytic inside the integration contour is satisfied since the function $f(z) = z$ is regular in the whole complex plane except at infinity.

The integral on the path $L_1 + L_2$ leading from 0 to A is equal to $4j$.

If we traveled the path $-L_3 - L_4$ leading from 0 to A, the integral would be $-(-2 - 4j + 2) = 4j$ too.

The property that we find here is that the integral does not depend on the path followed if the inside area enclosed by the two paths contains no singularity.

A.3.2. Integration on a Closed Contour Surrounding a Pole

Integration of the function: $f(z) = z^n$:

We calculate the integral of the function $f(z) = z^n$ on the circle C centered at $z = 0$, and browsed in the forward direction (Fig. A1.6):

$$I_n = \oint_C z^n \, dz. \tag{A.24}$$

The situation is different depending on whether $n \geq 0$ or $n < 0$.

- If $n \geq 0$, the function $f(z) = z^n$ is analytic inside the circle, and the integral will be zero by Cauchy Theorem.
- If $n < 0$, $z = 0$ is a pole of order n. The function is not defined at all points of the disk inside C and the integral I_n may be different from zero.

We use the trigonometric form of z: $z = \rho e^{j\theta}$; $z^n = \rho^n e^{jn\theta}$.

On the integration circle centered in $z = 0$, ρ is constant and we have: $dz = \rho \, e^{j\theta} j \, d\theta$.

$$I_n = \oint_C z^n \, dz = \int_0^{2\pi} \rho^n e^{jn\theta} \rho e^{j\theta} j \, d\theta = j\rho^{n+1} \int_0^{2\pi} e^{j(n+1)\theta} \, d\theta.$$

If $n \neq -1$,

$$I_n = j\rho^{n+1} \frac{\left[e^{j(n+1)\theta} \right]_0^{2\pi}}{j(n+1)} = 0.$$

If $n = -1$,

$$I_{-1} = j\rho^0 \int_0^{2\pi} e^0 \, d\theta = j \int_0^{2\pi} d\theta = 2\pi j.$$

Fig. A.6 Integration circle for the function $f(z) = z^n$

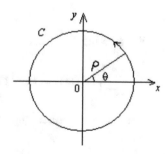

So we note that:

$$\oint_C z^n \, dz = \begin{vmatrix} 0 & \text{if } n \neq -1 \\ 2\pi j & \text{if } n = -1 \end{vmatrix}. \tag{A.25}$$

Integration on a circle around the origin of a function $f(z)$ having a pole of order n in $z = 0$:

$z = 0$ is a pole of order so n of $f(z)$ if

$$\lim_{z \to 0} z^n f(z) = A_n. \tag{A.26}$$

The general form of $f(z)$ valid in a neighborhood of $z = 0$ is:

$$f(z) = \frac{A_n}{z^n} + \frac{A_{n-1}}{z^{n-1}} + \cdots \frac{A_{-1}}{z} + A_0 + A_1 z + \cdots + A_n z^n \ldots. \tag{A.27}$$

We seek to evaluate the integral of $f(z)$ on the circle centered in $z = 0$ traveled in the forward direction $I = \oint_C f(z) \, dz$.

By integrating term by term the second member of (A.27), we have:

$$\begin{aligned} I &= \oint_C f(z) \, dz = \oint_C \frac{A_n}{z^n} \, dz + \oint_C \frac{A_{n-1}}{z^{n-1}} \, dz \\ &\quad + \cdots + \oint_C \frac{A_{-1}}{z} \, dz + \oint_C A_0 \, dz + \cdots + \oint_C A_n z^n \, dz + \cdots \end{aligned} \tag{A.28}$$

All these integrals are zero except $\oint_C \frac{A_{-1}}{z} \, dz$.

So we have:

$$I = \oint_C f(z) \, dz = \oint_C \frac{A_{-1}}{z} \, dz = 2\pi j A_{-1}. \tag{A.29}$$

A_{-1} is called the residue of the integral of the function $f(z)$ in $z = 0$.

The practical problem remaining to be solved is the evaluation of the residue A_{-1} of a function $f(z)$.

For this, equation (A.27) is multiplied by z^n.

$$z^n f(z) = A_n + A_{n-1} z + \cdots + A_{-1} z^{n-1} + A_0 z^n + \cdots \tag{A.30}$$

We take the $n - 1$ derivative of this equation

$$\frac{d^{n-1}}{dz^{n-1}} (z^n f(z)) = (n-1)(n-2) \ldots A_{-1} + n(n-1) \ldots A_0 z + \cdots \tag{A.31}$$

The second member tends toward $(n-1)! A_{-1}$ when $z \to 0$.

We have

$$A_{-1} = \frac{1}{(n-1)!} \lim_{z \to 0} \frac{d^{n-1}}{dz^{n-1}} (z^n f(z)). \tag{A.32}$$

More generally, let the function $f(z)$ having a pole of order n in $z = a$. The residue of this function in $z = a$ is given by:

$$A_{-1} = \frac{1}{(n-1)!} \lim_{z \to a} \frac{d^{n-1}}{dz^{n-1}} ((z-a)^n f(z)) \tag{A.33}$$

By applying Cauchy's theorem, it can be shown that the integral over a closed contour of any shape surrounding the poles is equal to the integral over a circle surrounding the pole if it is possible to deform this contour to reduce it to a circle without encountering poles.

Residue theorem: The integral over a closed contour Γ traveled counterclockwise of a function $f(z)$ having poles inside Γ is equal to the sum of residues within this contour multiplied by $2\pi j$:

$$I = \oint_{\Gamma} f(z) \, dz = 2\pi j \sum_i \text{Residues}_i \tag{A.34}$$

A.3.3. Jordan's Lemma

There are several Jordan's lemma. We will only consider here that lemma which relates to the problems most frequently encountered in signal analysis.

Consider a function $f(z)$ defined in an area of $1/2$ upper plane $y > 0$.

If $\lim_{|z| \to \infty} f(z) = 0$, the integral $\int f(z) e^{jz} \, dz$, extended to a circular arc centered in O with radius r contained in the upper half plane (Fig. A.7), tends toward 0 when $r \to \infty$.

Fig. A.7 Upper half circle of integration

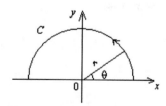

Indeed, we set $z = re^{j\theta}$ and let $M(r)$ the upper limit of $|f(z)|$ on the circular arc $|z| = r$. We have $|e^{jz}| = e^{-r\sin\theta}$, and on the circle $|dz| = |re^{j\theta}\,d\theta| = r|d\theta|$.

According to the general theorem stating that the integral modulus is less than or equal to the integral of the modulus, we can write

$$\left| \int f(z)e^{jz}\,dz \right| \leq \int |f(z)|e^{-r\sin\theta}|dz| \leq M(r) \int_0^\pi e^{-r\sin\theta}r\,d\theta \qquad (A.35)$$

The function $\sin\theta$ is symmetrical around $\theta = \frac{\pi}{2}$, we can write (Fig. A.8):

$$\int_0^\pi e^{-r\sin\theta}\,d\theta = 2\int_0^{\frac{\pi}{2}} e^{-r\sin\theta}\,d\theta.$$

It can be seen from the figure that for $0 \leq \theta \leq \frac{\pi}{2}$, $\sin\theta \geq \frac{2}{\pi}\theta$, as the sinus is above the line of slope $\frac{2}{\pi}$.

So

$$\left| \int_C f(z)e^{jz}\,dz \right| \leq rM(r)2\int_0^{\frac{\pi}{2}} e^{-\frac{2\theta r}{\pi}}\,d\theta = 2rM(r)\frac{[e^{-r}-1]}{\frac{-2r}{\pi}} = \pi M(r)[1 - e^{-r}].$$

$$(A.36)$$

This last expression tends towards 0 when $r \to \infty$.

Consequences of the previous Lemma:

1. If $\lim_{|z| \to \infty} f(z) = 0$, the integral $\int f(z)e^{-jz}\,dz$ over the arc in the 1/2 lower plane $y < 0$ tends to zero as $r \to \infty$.
2. Let α a real number. The integral $\int f(z)e^{-j\alpha z}\,dz$ tends toward zero as $r \to \infty$ on the circular arc of radius r

   a. in the lower 1/2 plane $y < 0$ if $\alpha > 0$.
   b. in the upper 1/2 plane $y > 0$ if $\alpha < 0$.

Fig. A.8 Plot of $\sin\theta$ and line with slope $\frac{2}{\pi}$

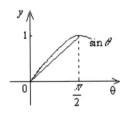

Appendix B
Linear Algebra

B.1. Vectors

A vector is an array of real or complex numbers or functions. The vectors are treated mainly as column vectors (in the example here with N elements):

$$x = \begin{bmatrix} x_1 \\ x_2 \\ \vdots \\ x_N \end{bmatrix}. \tag{B.1}$$

Transposed vector of a vector x (row vector):

$$x^{\mathrm{T}} = [x_1, \quad x_2, \quad \ldots, \quad x_N]. \tag{B.2}$$

The *Transposed Hermitian* vector is:

$$x^{\mathrm{H}} = \left(x^{\mathrm{T}}\right)^* = [x_1^*, \quad x_2^*, \quad \ldots, \quad x_N^*]. \tag{B.3}$$

Euclidean norm (length) of a vector:

$$\|x\|_2 = \left\{ \sum_{i=1}^{N} |x_i|^2 \right\}^{\frac{1}{2}}. \tag{B.4}$$

B.1.1. Linear Independence

Consider a set of n vectors $v_1, v_2, \ldots, v_n$. These vectors are called *linearly independent* if the relation

$$\alpha_1 v_1 + \alpha_2 v_2 + \cdots + \alpha_n v_n = 0 \text{ is verified only when } \alpha_i = 0 \text{ for any} i \tag{B.5}$$

© Springer International Publishing Switzerland 2016
F. Cohen Tenoudji, *Analog and Digital Signal Analysis*,
Modern Acoustics and Signal Processing, DOI 10.1007/978-3-319-42382-1

If we can find a set of non-zero coefficients α_i, such that the relationship (B.5) is verified, the vectors $v_1, v_2, \ldots, v_n$ are *linearly dependent*.

B.1.2. Basis of a Vector Space

If the vectors $v_1, v_2, \ldots, v_n$ are linearly independent, the set of linear combinations $w = \sum_{i=1}^{n} \alpha_i v_i$ is a vector space whose vectors $v_1, v_2, \ldots, v_n$ are a basis (not unique, though).

B.1.3. Scalar Product

Let two complex vectors with same dimensions m,

$$a = \begin{bmatrix} a_1 \\ a_2 \\ \vdots \\ a_m \end{bmatrix} \quad \text{and} \quad b = \begin{bmatrix} b_1 \\ b_2 \\ \vdots \\ b_m \end{bmatrix}.$$

Their inner product (or dot product) is defined by

$$\langle a, b \rangle = a^H b = \sum_{i=1}^{m} a_i^* b_i. \tag{B.6}$$

The *squared norm* of a vector a is defined by

$$\langle a, a \rangle = a^H a = \sum_{i=1}^{m} a_i^* a_i. \tag{B.7}$$

The norm of vector a is

$$\|a\|_2 = \{\langle a, a \rangle\}^{\frac{1}{2}}. \tag{B.8}$$

B.2. Matrices

A matrix is an array of numbers typically noted as follows:

$$A = \{a_{ij}\} = \begin{bmatrix} a_{11} & a_{12} & a_{13} & \cdots & a_{1n} \\ a_{21} & a_{22} & a_{23} & \cdots & a_{2n} \\ a_{31} & a_{32} & a_{33} & \cdots & a_{3n} \\ \vdots & \vdots & \vdots & & \vdots \\ a_{m1} & a_{m2} & a_{m3} & \cdots & a_{mn} \end{bmatrix}. \ A \text{ has } m \text{ rows and } n \text{ columns.} \quad (B.9)$$

Its *transpose* is obtained by interchanging the rows and columns of A:

$$A^{\mathrm{T}} = \begin{bmatrix} a_{11} & a_{21} & a_{31} & \cdots & a_{m1} \\ a_{12} & a_{22} & a_{32} & \cdots & a_{m2} \\ a_{13} & a_{23} & a_{33} & \cdots & a_{m3} \\ \vdots & \vdots & \vdots & & \vdots \\ a_{1n} & a_{2n} & a_{3n} & \cdots & a_{mn} \end{bmatrix}. \quad (B.10)$$

Its *Hermitian transpose* is the conjugate transpose:

$$A^{\mathrm{H}} = \begin{bmatrix} a_{11}^* & a_{21}^* & a_{31}^* & \cdots & a_{m1}^* \\ a_{12}^* & a_{22}^* & a_{32}^* & \cdots & a_{m2}^* \\ a_{13}^* & a_{23}^* & a_{33}^* & \cdots & a_{m3}^* \\ \vdots & \vdots & \vdots & & \vdots \\ a_{1n}^* & a_{2n}^* & a_{3n}^* & \cdots & a_{mn}^* \end{bmatrix}. \quad (B.11)$$

If the matrix A is square, its transpose is obtained by reflection on the diagonal.
If a square matrix is equal to its transpose, it is said to be *symmetric*.
If a square matrix is equal to its Hermitian transpose, it is called *Hermitian*.
Evidently we have $\left(A^{\mathrm{H}}\right)^{\mathrm{H}} = A$.

The *rank* $r(A)$ of a matrix is the number (common to rows and columns) of linearly independent row vectors and column vectors. We necessarily have:

$$r(A) \leq \min(m, n). \quad (B.12)$$

If

$$r(A) = \min(m, n), \text{ the matrix is said } full \ rank. \quad (B.13)$$

Product of two matrices
The product of a $(m \times n)$ matrix A by a $(n \times p)$ matrix B, is a $(m \times p)$ matrix C whose elements are given by:

$$c_{ij} = \sum_{k=1}^{n} a_{ik}b_{kj}. \tag{B.14}$$

Note that the number of columns of A must be equal to the number of rows of B.

Important properties

– Let $C = AB$ the product of matrices A and B.
  We have:

$$C^T = B^T A^T \quad \text{and} \quad C^H = B^H A^H. \tag{B.15}$$

– Let $R = A^H A$ the product of the Hermitian transpose A^H of a $(m \times n)$ matrix A and of matrix A. R is square $(n \times n)$.
  According to (B.15),

$$R = R^H. \tag{B.16}$$

The matrix R is Hermitian (symmetric if the elements are real).
The rank of R is that of A:

$$r(A) = r(A^H A) = r(AA^H). \tag{B.17}$$

– Similarly, let $S = AA^H$ the product of $(m \times n)$ matrix A and of its Hermitian transpose A^H.

$$S \text{ is square } (m \times m), \text{ Hermitian.} \tag{B.18}$$

Square matrices

If A is a square $n \times n$ full rank matrix, there is a single matrix A^{-1} called *inverse* of A such that:

$$A^{-1}A = AA^{-1} = I, \tag{B.19}$$

where $I = \begin{bmatrix} 1 & 0 & 0 & \dots & 0 \\ 0 & 1 & 0 & \dots & 0 \\ 0 & 0 & 1 & \dots & 0 \\ \vdots & \vdots & \vdots & & \vdots \\ 0 & 0 & 0 & \dots & 1 \end{bmatrix}$ is the square *identity matrix*. In this case A is said *invertible* or *nonsingular*.

If A is not full rank, $r(A) < n$, A has no inverse. It is said *noninvertible* or *singular*.

Determinant of a square matrix A:

$$\det(A_{nn}) = \sum_{i=1}^{n} (-1)^{i+j} a_{ij} \det(A_{ij}), \tag{B.20}$$

where A_{ij} is the $(n-1) \times (n-1)$ matrix obtained by removing the ith row and jth column of matrix A.

The determinant of a product of matrices is equal to the product of the determinants of these matrices:

$$\text{If } C = AB, \quad \det(C) = \det(A)\det(B). \tag{B.21}$$

A square $(n \times n)$ matrix A, is nonsingular if and only if

$$\det(A) \neq 0. \tag{B.22}$$

The *trace* of a square matrix is the sum of its diagonal elements

$$\text{tr}(A) = \sum_{i=1}^{n} a_{ii}. \tag{B.23}$$

B.3. Linear Systems

Partition of a matrix A in column vectors:

$$A = [c_1, \quad c_2, \quad \ldots, \quad c_n].$$

Partition of a matrix A in row vectors:

$$A = \begin{bmatrix} l_1 \\ l_2 \\ \vdots \\ l_m \end{bmatrix}.$$

B.3.1. Linear System Equation

We denote b the result of the multiplication of a $(m \times n)$ matrix A by a n vector x; $Ax = b$. The expanded form of this relationship is:

$$\begin{bmatrix} a_{11} & a_{12} & a_{13} & \cdots & a_{1n} \\ a_{21} & a_{22} & a_{23} & \cdots & a_{2n} \\ a_{31} & a_{32} & a_{33} & \cdots & a_{3n} \\ \vdots & \vdots & \vdots & & \vdots \\ a_{m1} & a_{m2} & a_{m3} & \cdots & a_{mn} \end{bmatrix} \begin{bmatrix} x_1 \\ x_2 \\ x_3 \\ \vdots \\ x_n \end{bmatrix} = \begin{bmatrix} b_1 \\ b_2 \\ b_3 \\ \vdots \\ b_m \end{bmatrix}. \tag{B.24}$$

Using the rules of the matrix product we have:

$$\begin{aligned} a_{11}x_1 + a_{12}x_2 + \cdots + a_{1n}x_n = b_1 \\ a_{21}x_1 + a_{22}x_2 + \cdots + a_{2n}x_n = b_2 \\ \vdots \\ a_{m1}x_1 + a_{m2}x_2 + \cdots + a_{mn}x_n = b_m \end{aligned} \tag{B.25}$$

We can check that

$$\begin{bmatrix} b_1 \\ b_2 \\ b_3 \\ \vdots \\ b_m \end{bmatrix} = x_1 \begin{bmatrix} a_{11} \\ a_{21} \\ a_{31} \\ \vdots \\ a_{m1} \end{bmatrix} + x_2 \begin{bmatrix} a_{12} \\ a_{22} \\ a_{32} \\ \vdots \\ a_{m2} \end{bmatrix} + x_3 \begin{bmatrix} a_{13} \\ a_{23} \\ a_{33} \\ \vdots \\ a_{m3} \end{bmatrix} + \cdots + x_n \begin{bmatrix} a_{1n} \\ a_{2n} \\ a_{3n} \\ \vdots \\ a_{mn} \end{bmatrix}. \tag{B.26}$$

The vector b appears as a linear combination of the columns of A. We can write:

$$b = \sum_{i=1}^{n} x_i c_i. \tag{B.27}$$

The vector b is a member of the m-dimensional space $\mathbb{C}^m$. If the matrix A and the vector x are real, $b \in \mathbb{R}^m$.

The column vectors of A generate a p-dimensional space, with $p \leq m$, called *column space* of A. This space $\mathbb{R}^p$ is a sub-space of $\mathbb{R}^m$.

B.3.2. Basis of Space $\mathbb{R}^m$

If, among the n column vectors of the matrix A, $p = m$ linearly independent vectors, can be found, these vectors form a basis of $\mathbb{R}^m$. To make it so, it is obviously necessary (but not sufficient) that $n \geq m$.

A m-dimensional vector space has an infinite number of bases.

B.3.3. Resolution of a Linear System

If the $(m \times n)$ matrix A and n vector x are given a-priori, a unique vector b may be found using the relationship (B.24). b is said to be the solution of a *direct, well-posed* problem.

If the $(m \times n)$ matrix A and m vector b are given a-priori, the relationship (B.24) appears as a system of equations to be resolved to find the n vector x solution. It may happen that there is no solution or there is an infinite number of solutions. The problem is said to be an *inverse, ill-posed* problem.

The resolution of this system is to search for the coefficients of the linear combination of the m dimensions column vectors of the matrix A which could be equal to the vector b. These coefficients of the linear combination are the components of the vector x (see B.26).

We look for the n vectors x, solutions of the linear equation:

$$Ax = b. \tag{B.28}$$

Equation (B.28) is a system of m linear equations with n unknown x_i with $i = 1, 2, \ldots, n$.

By construction, $b \in \mathbb{R}^m$, that is to say that the number of components of b are equal to the number of rows of the matrix A.

The solution of the equation $Ax = b$ depends on the elements of the vector b, on the respective values m and n, and on the rank of A.

The resolution of the system is the search of one or more vectors x which possibly verify this equation.

The first question that arises is: does b belong to the column space of A?

We can support the reasoning on writing the system in form (B.26) and discuss different cases occurring:

$\boxed{n < m}$: The number of column vectors is insufficient for generating all vectors of a m-dimensional space. In other words, the p-dimensional (with $p \leq n$) column space of the matrix cannot cover $\mathbb{R}^m$. An arbitrary vector b cannot always be represented as a linear combination of the column vectors of A. In other words, there are more equations than unknowns. The system is *overdetermined*.

- If the vector $b \in \mathbb{R}^p$, we can find the coefficients x_i and so find solutions for x. If $r = n$ the solution is unique.
  If $r < n < m$, the columns of A are not linearly independent; the equation $Ax = 0$ has an infinite number of solutions. It results that $Ax = b$ has an infinite number of solutions.
- If $b \notin \mathbb{R}^p$, the system has no solution. In some applications one is interested in an approximate solution. In these cases, we look for a vector x_0 which generates a vector $\hat{b} \in \mathbb{R}^p$ whose difference with b is minimal. Thus, we pose:

$$\hat{b} = \sum_{i=1}^{m} x_{0i} c_i = A x_0.$$

The minimum error vector is:

$$e = b - \hat{b} = b - A x_0. \tag{B.29}$$

One generally chooses to search for a minimum error in the least squares sense. The squared norm of the error $\|e\|^2 = \|b - Ax\|^2$ should be minimal. We recall that it is given by:

$$\|e\|^2 = \|b - Ax\|^2 = (b - Ax)^H (b - Ax) = (b^H - x^H A^H)(b - Ax). \tag{B.30}$$

The vector x should be such that:

$$\frac{\partial \|e\|^2}{\partial x^H} = -A^H (b - A x_0) = 0. \tag{B.31}$$

The notation x_0 appears in the last equation as being its solution.
So,

$$A^H e = 0. \tag{B.32}$$

This result means that the norm of the squared error is minimum when the vector e is orthogonal to each of the column vectors of A.
Equation (B.31) is written as:

$$A^H A x_0 = A^H b. \tag{B.33}$$

We saw above in (B.18) that the matrix $A^H A$ is square with $(n \times n)$ dimensions. If the columns of A are linearly independent (A has full rank, $r = n$), then the matrix $A^H A$ is invertible and by left multiplying equation (B.33) by $(A^H A)^{-1}$, we see that the solution in the least squares sense is:

$$x_0 = (A^H A)^{-1} A^H b. \tag{B.34}$$

So:

$$\hat{b} = A x_0 = A (A^H A)^{-1} A^H b = P_A b, \tag{B.35}$$

where

$$P_A = A\left(A^H A\right)^{-1} A^H.\tag{B.36}$$

P_A is called the *projection matrix* of a vector onto the space column of A.

$\boxed{n = m}$: If $r = m$, A is a nonsingular square matrix, A^{-1} exists and the solution of the system is: $x = A^{-1}b$. We may verify on (B.36) that

$$P_A = I.\tag{B.37}$$

If $r < m$, A is singular. The column vectors of A are not linearly independent. The discussion is similar to that conducted in the previous case. If the vector $b \in \mathbb{R}^r$, we can find a solution x (or an infinite number of solutions). Otherwise, a minimum error may be considered.

$\boxed{n > m}$: There are fewer equations than unknowns. The system is *underdetermined*.

The existence of a solution will depend on the rank of the matrix.

If $r = m$ there is a solution or an infinite number of solutions.

If $r < m$ there are 0 or an infinite number of solutions.

B.4. Special Forms of Matrices

A *diagonal* matrix is a matrix whose elements outside the main diagonal are zero. A square diagonal matrix has the form:

$$A = \begin{bmatrix} a_{11} & 0 & 0 & \cdots & 0 \\ 0 & a_{22} & 0 & \cdots & 0 \\ 0 & 0 & a_{33} & \cdots & 0 \\ \vdots & \vdots & \vdots & & \vdots \\ 0 & 0 & 0 & \cdots & a_{nn} \end{bmatrix}.$$

A rectangular diagonal matrix has the form:

$$A = \begin{bmatrix} a_{11} & 0 & 0 & 0 & 0 \\ 0 & a_{12} & 0 & 0 & 0 \\ \vdots & \vdots & \vdots & \vdots & \vdots \\ 0 & 0 & 0 & a_{mk} & 0 \end{bmatrix} \quad \text{or} \quad A = \begin{bmatrix} a_{11} & 0 & \vdots & 0 \\ 0 & a_{12} & \vdots & 0 \\ 0 & 0 & \vdots & 0 \\ 0 & 0 & \vdots & a_{kn} \\ 0 & 0 & \vdots & 0 \end{bmatrix}.$$

A *Bloc diagonal* matrix is a square matrix whose diagonal elements are square matrices:

$$A = \begin{bmatrix} A_{11} & 0 & 0 & \cdots & 0 \\ 0 & A_{22} & 0 & \cdots & 0 \\ 0 & 0 & A_{33} & \cdots & 0 \\ \vdots & \vdots & \vdots & & \vdots \\ 0 & 0 & 0 & \cdots & A_{kk} \end{bmatrix}.$$

Exchange matrix:

$$J = \begin{bmatrix} 0 & \cdots & 0 & 0 & 1 \\ 0 & \cdots & 0 & 1 & 0 \\ 0 & \cdots & 1 & 0 & 0 \\ \vdots & \vdots & \vdots & & \vdots \\ 1 & \cdots & 0 & 0 & 0 \end{bmatrix}. \tag{B.38}$$

This matrix is used to perform the inversion of terms of a vector:

$$J \begin{bmatrix} v_1 \\ v_2 \\ \vdots \\ v_n \end{bmatrix} = \begin{bmatrix} v_n \\ v_{n-1} \\ \vdots \\ v_1 \end{bmatrix}. \tag{B.39}$$

The left multiplication of a square matrix by J reverses the order of the terms in each column of the matrix:

$$\text{If } A = \begin{bmatrix} a_{11} & a_{12} & a_{13} \\ a_{21} & a_{22} & a_{23} \\ a_{31} & a_{32} & a_{33} \end{bmatrix}, \text{ then } J^\mathsf{T} A = \begin{bmatrix} a_{31} & a_{32} & a_{33} \\ a_{21} & a_{22} & a_{23} \\ a_{11} & a_{12} & a_{13} \end{bmatrix}. \tag{B.40}$$

The right multiplication of a square matrix by J inverses the terms of each row of the matrix:

$$AJ = \begin{bmatrix} a_{13} & a_{12} & a_{11} \\ a_{23} & a_{22} & a_{21} \\ a_{33} & a_{32} & a_{31} \end{bmatrix}. \tag{B.41}$$

Likewise:

$$J^\mathsf{T} A J = \begin{bmatrix} a_{33} & a_{32} & a_{31} \\ a_{23} & a_{22} & a_{21} \\ a_{13} & a_{12} & a_{11} \end{bmatrix}. \tag{B.42}$$

Property

$$J^2 = I.$$

A *square, matrix* A *is said* Toeplitz *if its elements along each diagonal are equal:*

$$a_{ij} = a_{i+1\,j+1} \quad \text{for all } i < n \text{ and } j < n.$$

A $(n \times n)$ *matrix* A *is said* Hankel *if its elements along each of the diagonals perpendicular to the main diagonal are equal:* $a_{ij} = \text{Cte}_{i+j}$ *for all* $i \leq n$ *and* $j \leq n$.

A *real* $(n \times n)$ *matrix* A *is called* orthogonal *if its columns (and rows) are orthonormal, namely, if* $A = [a_1, \quad a_2, \quad \ldots, \quad a_n]$ *and* $a_i^T a_j = \begin{cases} 1 & \text{for } i = j \\ 0 & \text{for } i \neq j \end{cases}$.

So:

$$A^T A = I \quad \text{and} \quad A^{-1} = A^T.$$

A *complex* $(n \times n)$ *matrix* A *is said* unitary *if its columns (and rows) are orthonormal:*

$$a_i^H a_j = \begin{cases} 1 & \text{for } i = j \\ 0 & \text{for } i \neq j \end{cases}. \quad \text{So: } A^H A = I \quad \text{and} \quad A^{-1} = A^H.$$

B.5. Quadratic and Hermitian Forms

The *quadratic form* of a real, $(n \times n)$ matrix A is the scalar defined by:

$$Q_A(x) = x^T A x = \sum_{i=1}^{n} \sum_{j=1}^{n} x_i a_{ij} x_j, \tag{B.43}$$

where $x^T = [x_1, \quad x_2, \quad \ldots, \quad x_n]$ is a vector made up of n real variables. The quadratic form is a quadratic function of the n variables $x_1, \quad x_2, \quad \ldots, \quad x_n$.

Example

$$A = \begin{bmatrix} 3 & 4 \\ 1 & 2 \end{bmatrix}; \quad Q_A(x) = 3x_1^2 + 5x_1 x_2 + 2x_2^2.$$

Similarly, the *Hermitian form* of a $(n \times n)$ matrix A is defined by:

$$Q_A(x) = x^H A x = \sum_{i=1}^{n} \sum_{j=1}^{n} x_i^* a_{ij} x_j. \tag{B.44}$$

If the quadratic form of a matrix A is positive for all non-zero vectors x, the matrix is said to be *positive definite*. If the quadratic form is non-negative, the matrix is called *positive semi-definite*.

B.6. Eigenvalues and Eigenvectors of a Square Matrix

Let A be a square $(n \times n)$, matrix (singular or not). We consider the linear equation $Av = \lambda v$ (where v is a vector solution with n elements and λ a complex constant).

This equation can also be written $(A - \lambda I)v = 0$. For this equation to have a non-zero vector solution v, it is necessary that the matrix $A - \lambda I$ is singular. Its determinant must be zero.

$p(\lambda) = \det(A - \lambda I)$ is the characteristic polynomial of order n. Its roots λ_i are the eigenvalues of the matrix A for $i = 1, 2, \ldots, n$. The corresponding vectors v_i are the eigenvectors: $A v_i = \lambda_i v_i$.

Properties

a. Nonzero eigenvectors $[v_1, \ v_2, \ \ldots, \ v_n]$ corresponding to different eigenvalues $\lambda_1, \lambda_2, \ldots, \lambda_n$ are linearly independent.

   To show this, we consider two eigenvectors v_1 and v_2: $A v_1 = \lambda_1 v_1$ and $A v_2 = \lambda_2 v_2$.

   The equation $(\alpha_1 v_1 + \alpha_2 v_2) = 0$ leads to $A(\alpha_1 v_1 + \alpha_2 v_2) = 0$ and, due to linearity:

   $$A(\alpha_1 v_1 + \alpha_2 v_2) = (\alpha_1 \lambda_1 v_1 + \alpha_2 \lambda_2 v_2) = 0.$$

   The first equation may be written $v_2 = -\frac{\alpha_1 v_1}{\alpha_2}$, the second, $v_2 = -\frac{\alpha_1 \lambda_1 v_1}{\alpha_2 \lambda_2}$.

   If $\lambda_1 \neq \lambda_2$ the only solution satisfying the two equations is $v_1 = v_2 = 0$.

   It follows that two non-null eigenvectors v_1 and v_2 that correspond to different eigenvalues are linearly independent.

b. The eigenvalues of a Hermitian matrix are real. Indeed, let v_i be an eigenvector and λ_i its associated eigenvalue given by $A v_i = \lambda_i v_i$.

   We multiply the equation on the left by v_i^{H}: $v_i^H A v_i = v_i^H \lambda_i v_i$.

   We take the Hermitian transpose of the previous equation $\left(v_i^H A v_i\right)^H = v_i^H A^H v_i = v_i^H \lambda_i^* v_i$.

The matrix A is assumed Hermitian, $A = A^H$, one must have: $v_i^H \lambda_i v_i = v_i^H \lambda_i^* v_i$, then

$$\lambda_i = \lambda_i^*. \tag{B.45}$$

c. The eigenvectors of a Hermitian matrix corresponding to different eigenvalues are orthogonal. In fact, consider two eigenvectors v_i and v_j corresponding to two distinct eigenvalues $\lambda_i \neq \lambda_j$. They verify the equations $Av_i = \lambda_i v_i$ and $Av_j = \lambda_j v_j$ with $\lambda_i \neq \lambda_j$.

Left multiplying the two equations respectively by v_j^H and v_i^H, we have:

$$v_j^H A v_i = \lambda_i v_j^H v_i \quad \text{and} \quad v_i^H A v_j = \lambda_j v_i^H v_j.$$

Taking the Hermitian conjugate of the last equation yields to:

$$v_j^H A^H v_i = \lambda_j^* v_j^H v_i.$$

The matrix is assumed Hermitian then $\lambda_i v_j^H v_i = \lambda_j^* v_j^H v_i$.

The eigenvalues being real, $\lambda_j^* = \lambda_j$, then $(\lambda_i - \lambda_j) v_j^H v_i = 0s$. As the eigenvalues are different by assumption, in order that this equation be verified, we must have necessarily, $v_j^H v_i = 0$.

Vectors v_i and v_j are orthogonal.

d. Any **invertible** $(n \times n)$ Hermitian matrix has n orthogonal eigenvectors.

e. An invertible Hermitian matrix A can be written as

$$A = U \Lambda U^H, \tag{B.46}$$

wherein the matrix U is formed of the eigenvectors of A with unit norms (in columns) and Λ is the diagonal matrix composed of the eigenvalues put in order relative to the eigenvectors of A. The matrix U is unitary.

Indeed:

$$AU = A \begin{pmatrix} \vdots & \vdots & & \vdots \\ v_1 & v_2 & \cdots & v_n \\ \vdots & \vdots & & \vdots \end{pmatrix} = \begin{pmatrix} \vdots & \vdots & & \vdots \\ \lambda_1 v_1 & \lambda_2 v_2 & \cdots & \lambda_n v_n \\ \vdots & \vdots & & \vdots \end{pmatrix} = U\Lambda. \tag{B.47}$$

By right multiplying Eq. (B.47) by U^H, we have $A\, U\, U^H = U\, \Lambda\, U^H$, and as the matrix U is unitary, $U\, U^H = I$ and so:

$$A = U\, \Lambda\, U^H. \tag{B.48}$$

As A is invertible, we have

$$A^{-1} = U \Lambda^{-1} U^{\mathrm{H}}. \tag{B.49}$$

Matrix A and Λ are called *similar*. They are related by the basis change given by U.

f. A unitary matrix U has the property $|\det(U)| = 1$.

g. In consequence

$$\det(A) = \det(\Lambda) = \prod_{i=1}^{n} \lambda_i. \tag{B.50}$$

Therefore a matrix is nonsingular (invertible) if all of its eigenvalues are different from 0.

h. We have also

$$\mathrm{tr}(A) = \mathrm{tr}(\Lambda) = \sum_{i=1}^{n} \lambda_i. \tag{B.51}$$

i. A Hermitian matrix is positive definite if and only if its eigenvalues are positive. A necessary and sufficient condition for A to be positive definite is that there is an invertible matrix C such that $A = C^H C$.

j. A square matrix A and its transpose A^{T} have the same set of eigenvalues.

k. Let A be an invertible Hermitian $(n \times n)$ matrix, $A = U \Lambda U^{\mathrm{H}}$ according to relationship (B.46). The **spectral theorem for a square matrix** states that

$$A = \sum_{i=1}^{n} \lambda_i u_i u_i^{\mathrm{H}}, \tag{B.52}$$

where λ_i and u_i are associated eigenvalues and eigenvectors. (The reader may verify the theorem as an exercise on a 2×2 matrix A).

We have then:

$$A^{-1} = \sum_{i=1}^{n} \frac{1}{\lambda_i} u_i^{\mathrm{H}} u_i. \tag{B.53}$$

B.7. Singular Value Decomposition (SVD)

We have seen above that a square, symmetric $(n \times n)$ matrix A can be decomposed as $A = U \Lambda U^{\mathrm{H}}$, where Λ is the diagonal matrix of its eigenvalues and U is a unitary matrix composed of the normalized relative eigenvectors.

We show in the following that any rectangular $(m \times n)$ matrix A can be decomposed as:

$$A = U\Sigma V^{\mathrm{H}}, \tag{B.54}$$

where U and V are unitary matrices composed of the eigenvectors respectively of AA^{H} and $A^{\mathrm{H}}A$, and Σ is a rectangular $(m \times n)$ matrix whose only non null elements are along the first diagonal. Relationship (B.54) is called the Singular Value Decomposition, and the non zero elements of Σ are called the singular values of A.

To demonstrate this property, let us search for a unitary matrix V such that:

$$AV = U\Sigma, \tag{B.55}$$

where U is unitary and Σ is diagonal (but non square if $(m \neq n)$). Is this possible?

We right multiply (B.55) by V^{H}: $AVV^{\mathrm{H}} = U\Sigma V^{\mathrm{H}}$. Thus

$$A = U\Sigma V^{\mathrm{H}}. \tag{B.56}$$

We left multiply (B.56) by A^{H}:

$$A^{\mathrm{H}}A = V\Sigma^{\mathrm{H}}U^{\mathrm{H}}U\Sigma V^{\mathrm{H}} = V\Sigma^{\mathrm{H}}\Sigma V^{\mathrm{H}}. \tag{B.57}$$

We know from (B.16) that $A^{\mathrm{H}}A$ is a square $(n \times n)$ Hermitian matrix. $\Sigma^{\mathrm{H}}\Sigma$ is also square $(n \times n)$, diagonal. This result is analogous to the decomposition in (B.48). V is the unitary matrix composed of the eigenvectors (n dimensions) of $A^{\mathrm{H}}A$; the elements (diagonal) of matrix $\Sigma^{\mathrm{H}}\Sigma$ are the corresponding eigenvalues (positives or zero).

Similarly, let us right multiply (B.56) by A^{H}:

$$AA^{\mathrm{H}} = U\Sigma V^{\mathrm{H}}V\Sigma^{\mathrm{H}}U^{\mathrm{H}} = U\Sigma\Sigma^{\mathrm{H}}U^{\mathrm{H}}. \tag{B.58}$$

AA^{H} is square $(m \times m)$, Hermitian. $\Sigma\Sigma^{\mathrm{H}}$ is also square $(m \times m)$, diagonal, and its elements are the eigenvalues of AA^{H} (which are the same as the eigenvalues of $A^{\mathrm{H}}A$). Its elements are positive or zero. U is the unitary matrix of the eigenvectors (m dimensions) of $A^{\mathrm{H}}A$.

$\Sigma\Sigma^{\mathrm{H}}$ being square, diagonal, we can assess from the property (B.21) that the product of diagonal matrices is diagonal, that we can find Σ, diagonal, whose non zero elements are the square roots of the elements of $\Sigma\Sigma^{\mathrm{H}}$.

Spectral theorem for a rectangular matrix A

$$A = \sum_{i=1}^{r} u_i \sigma_i v_i^{\mathrm{H}}. \tag{B.59}$$

where u_i and v_i are the eigenvectors respectively of AA^H and A^HA related to σ_i, the diagonal elements of Σ; r is the rank of A (the same as that of AA^H and A^HA).

In summary, any rectangular matrix A can be decomposed as $A = U\Sigma V^H$ where U and V are unitary matrices composed of the eigenvectors respectively of AA^H and A^HA, and Σ is diagonal such that the elements of $\Sigma\Sigma^H$ are the common eigenvalues of AA^H and A^HA.

The matrix Σ is rectangular, diagonal, its elements are called the *singular values* of the matrix A. The relation $AV = U\Sigma$ is verified.

The *condition number* of a matrix is the ratio of its largest to its smallest singular values.

Pseudo-inverse of a rectangular matrix

Let us revisit the resolution of system equation $Ax = b$.

Using SVD decomposition of A given in (B.56) we write: $Ax = U\Sigma V^H x = b$.

The pseudo-inverse of A is the $(n \times m)$ matrix defined as

$$A^\dagger = V\Sigma' U^H, \tag{B.60}$$

where Σ' is a matrix whose elements are inverse of those of Σ.

It may be shown that:

- if A is full column rank, that is to say, $r(A) = n \leq m$ (overdetermined system), $A^\dagger$ is a *left-inverse* of A (meaning $A^\dagger A = I_n$ with I_n, the square $(n \times n)$ identity matrix) with

$$A^\dagger = (A^H A)^{-1} A^H. \tag{B.61}$$

  We have seen above in (B.34) that $x_0 = A^\dagger b$ is the best estimate of x in the least square sense.

- if A is full row rank, that is to say, $r(A) = m \leq n$ (underdetermined system), $A^\dagger$ is a *right-inverse* of A (meaning $AA^\dagger = I_m$ with I_m, the square $(m \times m)$ identity matrix) with

$$A^\dagger = A^H (AA^H)^{-1}. \tag{B.62}$$

We have seen above that there are infinitely many solutions to the equation $Ax = b$.

The solution with minimal norm is $x_{\text{min norm}} = A^H z$ where z is a vector of the m-dimensional complex space.

$$Ax = AA^H z = b. \quad z = (AA^H)^{-1} b. \quad x_{\text{min norm}} = A^H z = A^H (AA^H)^{-1} b = A^\dagger b.$$

B.8. Signal Filtering and Linear Algebra

A time limited signal having values equal to zero outside an interval $\{0, N-1\}$ can be written as a vector:

$$x = \begin{bmatrix} x[0] \\ x[1] \\ \vdots \\ x[N-1] \end{bmatrix}. \tag{B.63}$$

The *energy* of vector x is its squared norm:

$$\|x\|^2 = \sum_{i=1}^{N} |x[n]|^2. \tag{B.64}$$

In some cases we will consider a set of vectors containing the values of the signal at the instants $n, n-1, n-N-1$:

$$x[n] = \begin{bmatrix} x[n] \\ x[n-1] \\ \vdots \\ x[n-N-1] \end{bmatrix}. \tag{B.65}$$

Let $h[n]$ the impulse response of a linear, causal, time invariant filter, whose impulse response is finite, of order $N-1$ and let $x[n]$ be the filter input signal. The filter output is given by the convolution:

$$y[n] = \sum_{m=0}^{N} h[m]x[n-m]. \tag{B.66}$$

$$\text{Posing } \boldsymbol{h} = \begin{bmatrix} h[0] \\ h[1] \\ \vdots \\ h[N-1] \end{bmatrix},$$

we have:

$$\boldsymbol{y}[n] = \boldsymbol{h}^T \boldsymbol{x}[n] = \boldsymbol{x}^T[n]\boldsymbol{h}. \tag{B.67}$$

The output of a digital filter is given by:

$$\boldsymbol{y}[n] = \boldsymbol{h}^T \boldsymbol{x}[n] = \boldsymbol{x}^T[n]\boldsymbol{h}. \tag{B.68}$$

If $x[n] = 0$ for $n < 0$, the output $y[n]$ of the filter can be written for $n \geq 0$:

$$\boldsymbol{y} = \boldsymbol{X}_0\boldsymbol{h}, \tag{B.69}$$

where $\boldsymbol{X}_0$ is a *convolution matrix* defined by:

$$\boldsymbol{X}_0 = \begin{bmatrix} x[0] & 0 & 0 & \dots & 0 \\ x[1] & x[0] & 0 & \dots & 0 \\ x[2] & x[1] & x[0] & \dots & 0 \\ \vdots & \vdots & \vdots & & \vdots \\ x[N-1] & x[N-2] & x[N-3] & & x[0] \\ \vdots & \vdots & \vdots & \vdots & \vdots \end{bmatrix}. \tag{B.70}$$

Appendix C
Computer Calculations

The professional programming software Matlab is widely used in signal analysis. It is simple to use and has great flexibility, its handling is fast, and it is well documented. It has Toolboxes on various subjects, particularly for signal analysis.

To learn about signal analysis, students can use a freeware, downloadable by Internet. One can find very good clones of Matlab as Octave or Scilab where the syntax is very similar to that of Matlab for the basic operations in signal analysis.

At the end of this appendix we give two small basic programs for a first contact with this type of programming. Having copied the examples in the software editor, one will launch the execution or make a copy and paste of a few lines in the software command window.

Practical works, signal samples to be analyzed, and corrections are available upon request at the email address: fcohentenoudji@yahoo.fr.

C.1. Notions in Matlab

A well-written program demands many comments describing the operations of ongoing calculations. This practice facilitates understanding at a later reading.

Use the % symbol to put the rest of a line in comments.

Matlab is a numerical computing software that operates on arrays (vectors or matrices).

The calculations are reduced to arithmetic operations on these arrays.

We note that there are 3 types of product of two vectors in Matlab:

The scalar and vector products use the operator *
The term by term product uses the operator . * (note the dot before the star).

Matlab is optimized for matrix calculation. A matrix product is calculated very quickly. In contrast the for loops are extremely slow. Use for loops as a last resort.

© Springer International Publishing Switzerland 2016
F. Cohen Tenoudji, *Analog and Digital Signal Analysis*,
Modern Acoustics and Signal Processing, DOI 10.1007/978-3-319-42382-1

Resizing an array in a loop by adding values is extremely detrimental in time. We should declare the array to its maximum size at the start of the program, by filling it with zeros for example.

A Matlab program is a series of calls to functions. For example the calculation of a square root is performed with the call to sqrt(x) function which is programmed into the function program sqrt.m.

In your work you will often see operations that you want to repeat. It is strongly recommended to use functions such as myfunction.m that you have programmed. These functions will have to be located in a directory accessible by the program.

Matlab manages the logic of the use of the operators and generates an error message in the case of an illegal operation. Understanding an error message generated by Matlab avoids wasting time when eliminating this error.

The most common errors are the basic confusion between a row vector and a column vector, poor integration of the dimension of a vector, the plot of a complex vector, this last operation providing strange results. We will have to trace the real part, imaginary, the modulus or the phase of a complex vector.

The following sections present some Matlab functions commonly used in signal analysis.

C.1.1. Miscellaneous

%: To put a program line in comment

clear: removes all variables from the calculation memory (It is recommended to start a program by this command to avoid confusion with variables already used in the same Matlab session)

;: puts a semicolon after the declaration of a variable or a calculation to avoid printing on screen the calculation result (this could result in a huge waste of time if the calculation involves several thousand items)

home: refreshes the command window by deleting old texts

who: provides information on the variables used

whos: provides information on the variables used with the dimensions of the arrays

disp('Text'): writes text in the command window when running a program

help function: questions Matlab on how to use a function

for s = 0.0 : 1 : 1000, end: increments the variable s in steps of 1

for s = 1.0 : −0.1 : 0, end: decrements s in steps of −0.1

Assigning values to elements of a matrix. Example:

```
for m = 1 : N ,
    for n = 1 : N ,
        a(m, n) = m +n;
    end
end
```

break: to exit a loop prematurely

Creating a loop with the while command. Example:

n = 0;

while n < 1025,

....

    n = n + 1;

end

C.1.2. Vectors, Matrices

$a = \begin{bmatrix} 1 & 2 & 3 \end{bmatrix}$: generates a real row vector

$e = \begin{bmatrix} 1+2i & 2+3i & 3+5i \end{bmatrix}$: generates a complex row vector

$b = \begin{bmatrix} 1+2i & 2+3i & 3+5i \end{bmatrix}'$: Transposition conjugate of a vector (we obtain a column vector b conjugate transposed of e)

$g = \begin{bmatrix} 1+2i & 2+3i & 3+5i \end{bmatrix}'$: Transposition of a vector: (the use of the point indicates that the complex conjugation is not done)

$zc = $ conj(z): complex conjugation of a vector

$x = $ real(z): extraction of the real part of a complex vector z

$y = $ imag(z): extraction of the imaginary part of a complex vector x

dim(x): returns the number of elements of vector x

sum(x): returns the value of the sum of vector x elements

prod(x): returns the value of the product of vector x elements

fliplr(a): reverses the order of elements of a row vector

flipud(b): reverses the order of elements of a column vector

$x = $ zeros$(1, n)$: creates a row vector with n zeros

$x = $ ones$(1, n)$: creates a row vector with n values 1

$v = $ diag(A): returns a vector composed of diagonal elements of matrix A

$t = $ trace(A): returns the trace (sum of diagonal elements) of matrix A.

Products of vectors

If $a = \begin{bmatrix} 1 & 2 & 3 \end{bmatrix}$,

$c = a * a' = 14$: inner product (a is a row vector)

$$d = a' * a = \begin{bmatrix} 1 & 2 & 3 \\ 2 & 4 & 6 \\ 3 & 6 & 9 \end{bmatrix}$$: outer product

$e = 2 * a = a * 2 = \begin{bmatrix} 2 & 4 & 6 \end{bmatrix}$: product of a vector by a scalar

$f = a .* a = \begin{bmatrix} 1 & 4 & 9 \end{bmatrix}$: multiplication of arrays term by term (the point before the operator * allows term by term multiplication)

$a = \begin{bmatrix} 1 : 1024 \end{bmatrix}$: creates an array: (row vector with 1024 elements consisting of integers from 1 to 1024)

$b = \begin{bmatrix} 1 : 0.1 : 1024 \end{bmatrix}$: (creates numbers in the range 1–1024 by increments of 0.1)

$c = a(12)$: value of an element of an array

Creation of a square matrix:

$$A = [1 \quad 2 \quad 0 \; ; \; 2 \quad 5 \quad 1 \; ; \; 4 \quad 10 \quad -1] \text{ gives } A = \begin{bmatrix} 1 & 2 & 0 \\ 2 & 5 & 1 \\ 4 & 10 & -1 \end{bmatrix}$$

$A(3,1)$: gives 4 (3rd row, 1st column)

rank(A): returns the rank of matrix A (the number of linearly independent rows or columns)

Square matrix d:

det(d): calculates the determinant of matrix d

inv(d): calculates the inverse matrix of matrix d

$t = $ eig(d): t is a column vector containing the eigenvalues of d

$[V,L] = $ eig (d): produces a diagonal matrix L containing the eigenvalues, and a matrix V whose columns are the corresponding eigenvectors of d (V is the modal matrix); the eigenvectors are normalized to have a norm 1. We have the relation: $d * V = V * L$.

Rectangular matrix X:

Singular value decomposition (SVD):

$[U,S,V] = $ svd(X): produces a diagonal matrix S, of the same dimensions as X and with nonnegative diagonal elements in decreasing order, and unitary matrices U and V so that

$X = U * S * V'$.

cond(X): returns the condition number of matrix X, the ratio of the largest to the smallest singular values; ratio of the largest to smallest eigenvalues if X is square.

$X = $ pinv(A): produces the pseudo-inverse matrix X of matrix A.

C.1.3. Graphics

figure(4): prepares a graph in Figure 4

plot(A): graphically shows the elements of the vector A

grid on: draws a grid

xlabel('Time (sec)'): writes a label on x axis

ylabel('Amplitude (V)'): writes a label on y axis

plot(b, '*'): traces the elements of vector b as stars

axis([xmin   xmax   ymin   ymax])): chooses the variation intervals of the graphic coordinates

title ('string'): displays the text string above the figure

legend('signal 1'): displays a small box combining the plot style to its identity

plot(x,y): represents the vector y as a function of the vector x. x and y must be 2 vectors with same dimension

stem(x): performs the bar graph of values of a vector x

zoom on: triggers a graphic zoom that one controls with a box drawn with the mouse by holding down the left button

zoom off: stops the ability to zoom with the mouse

plottools('on'): displays a dialog box that allows to select all parameters of a graph; fonts, line thickness, axis intervals, etc ..

plot(x,y1,x,y2): represents both y1 and y2 as a function of x

semilogy(x,y): graph with linear scale for x and logarithmic for y

subplot(m,n,p): allows an $m \times n$ matrix of graphs and prepares the graph plot p

num2str (x): converts number x to a string. To perform the reverse operation use sprintf()

surf(M) represents the values of a two-dimensional matrix as a 3D plot. For options refer to Help section (help surfing)

shading interp: represents a smoothed graph after calling the surf() function

close(n): closes figure number n

close all: closes all figures.

C.1.4. Polynomials

poly(v): v being a vector, gives a vector whose elements are the coefficients of the polynomial whose roots are the elements of v

roots(d): calculates the roots of the polynomial whose coefficients are elements of the vector d. If d has $n + 1$ components, the polynomial is:

$$d(1)x^n + \cdots + d(n)x + d(n+1)$$

polyval(p, x): calculates the value of the polynomial whose coefficients are the elements in p for the value of the variable x

zplane(z, p): traces in the complex plane the zeros specified in column vector z and the poles specified in column vector p in the current figure

C.1.5. Signal

[y, n] = max(x): searches the maximum y of a vector x and returns the index n of the position of this maximum

y = diff(x): calculates a vector y whose components are the differences of two consecutive elements of a vector x. For example: $y(1) = x(1) - x(1)$; $y(1) = x(2) - x(1)$; ... etc ...

index = find(x > 1.1): Returns the index vector whose elements are the indices of the elements x satisfying the condition $x > 1.1$

fft(x): computes the DFT of the vector x (fast if dim(x) is a power of 2, using the Cooley-Tukey algorithm)

fft(x, Nfft): calculates the DFT on Nfft points of a vector x. (if Nfft > dim(x) the vector is filled with zeros, if Nfft < dim(x) a truncated version of x is used)

ifft(x, Nfft): calculates the inverse DFT of the vector x on Nfft points

z = fftshift(y): shifts the vector y obtained by FFT by placing the center frequency in the middle of the vector

abs(x): calculates the x complex modulus

angle(x): calculates the complex x argument in radians

a = rand(1,1024): calculates 1024 random numbers according to an uniformly distributed law in the interval (0,1). a is a row vector; successive draws are independent

a = randn(1,1024): calculates 1024 random numbers according to a Gaussian distribution N (0,1). a is a row vector; successive draws are independent

Y = filter(B, A, X): filters the data in the vector X with the filter described by the vectors A and B to give the filtered data Y. With:

$$a(1)y(n) = b(1)x(n) + b(2)x(n-1) + \cdots + b(nb+1)x(n-nb) - a(2)y(n-1) - \cdots - a(na+1)y(n-na)$$

sound(x, fs): sends the signal corresponding to the values of the vector x to the computer sound card with the sampling frequency fs. The min and max values of the vector are expected between -1 and $+1$. If x is a matrix (N, 2) the output is in stereo

$[H,w]$ = freqz(B,A,N): returns the frequency vector and N-point complex frequency response. If N is not specified, it is 512. We have:

$$H(e^{j\omega}) = \frac{B(e^{j\omega})}{A(e^{j\omega})} = \frac{b(1) + b(2)e^{-j\omega} + \cdots + b(nb+1)e^{-jnb\omega}}{a(1) + a(2)e^{-j\omega} + \cdots + a(na+1)e^{-jna\omega}}$$

H = freqz(B, A, w): returns the frequency response for frequency values specified in w, in radians/ample; normally between 0 and π.

$[h,t]$ = impz(b,a): calculates the impulse response of the filter with the coefficients b in the numerator and a in denominator. The function chooses the number of samples and returns the response in the column vector h and time in the column vector t (with $t = [0, n-1]'$). n is the length of the impulse response.

c = xcorr(a, b): calculates the cross-correlation function of the vectors a and b. If M is the dimension of the vectors a and b, the dimension of c is $2*M-1$.

c = conv(a, b):   convolves   vectors a and b.   We   have

$c(n+1) = \sum_{k=0}^{N-1} a(k+1)b(n-k)$.

R = toeplitz(FuncR): builds a Toeplitz symmetric matrix having FuncR as its first row (used to pass from correlation function to correlation matrix).

$y = \text{chirp}(t, f_0, t_1, f_1)$: generates a chirp signal whose frequency varies linearly with the given instants in time t vector. The instantaneous frequency at time $t = 0$ is f_0. The final frequency at time t_1 is f_1

$x = \text{Hilbert}(x1)$: returns a complex vector x whose $x1$ is the real part and $x2$, the imaginary part, $x2$ is the Hilbert transform of $x1$

$b = \text{specgram}(a, \text{Nfft}, \text{fs}, w, \text{Noverlap})$: calculates the short-time Fourier transform on Nfft points of the vector a after multiplication by the window vector w, using an overlap of Noverlap points

$P = \text{periodogram}(a)$: calculates the power spectral density (PSD) of the signal a. The calculation is performed using an FFT on a number of points equal to the power of 2 immediately greater than the length of vector a

$Cxy = \text{mscohere}(x, y)$: calculates the coherence function of two signals x and y (of equal length)

wintool: opens a dialog box that allows you to set a window (Hann, Bartlett, etc.) that can be applied to the data

C.1.6. Wavelets

waveinfo('wname'): provides information on the family of wavelets whose name is 'wname'. For example 'haar' for the Haar wavelet, or 'dbx' for Daubechies wavelets

[Phi, Psi, xval] = wavefun('db2' iter): returns the scaling function Phi and wavelet Psi (here of the wavelet db2) on a grid of values of x on $2^{\wedge}\text{iter}$ points. The iter number is the number of iterations

[LO_D, HI_D, LO_R, HI_R] = wfilters('wname'): calculates the 4 analysis and reconstruction filters associated with the wavelet whose name is 'wname'

Coefs = cwt (S, SCALES, 'wname', 'plot'): calculates the continuous wavelet coefficients whose name is 'wname'. Traces the result

[C, L] = wavedec(X, N, 'wname'): returns the wavelet decomposition of the signal X

[CA, CD] = dwt(X, 'wname'): calculates the approximation coefficients CA and details CD of the vector X. the name of the wavelet is 'wname'

$X = \text{idwt}(AC, CD, \text{'wname'})$: inverse function of dwt()

C.1.7. File Management

save filename a: saving on disk the variable a to a file in Matlab format

save filename: saves to disk all current variables in a file in Matlab format

load filename: reads on the disk all the variables recorded in a file in Matlab format. Can also read files saved in txt format

M = dlmread(filename, delimiter) reads ASCII data in the file with the name filename. The delineation between data is created by the delimiter (by default, comma). Use '\t' for the TAB as the delimiter. The result is placed in the matrix M

fid = fopen(filename, permission) opens a file name filename and places the value of the file handle in the variable fid. Permission can be 'r' or 'w' or '...' for reading, writing or other

count = fwrite(fid, A, precision): writes the elements of the matrix A in the file specified by fid. The values are written in the accuracy that can be 'float', 'double' or the like

status = fclose(fid): closes the file identified by fid

status = fclose('all'): closes all open files

s = wavread(filename): reads a file in Wave format(.wav extension)

sound(s): plays the signal s (from a .wav file)

C.2. Examples of Matlab Programs

Programming a vector:

One can define a frequency vector as follows:

f = −1000: 0.1: 1000; (one defines here a vector composed of frequency values between −1000 and 1000 Hz with an increment of 0.1 Hz).

We could also operate as follows:

kmax = 10,000;
k = −kmax: 1: kmax;
frequencystep = 0.1; % 0.1 Hz
$f = k$ * frequencystep;

This second way seems longer but can make clearer further reading of the program.

Programming a function:

We assume that we want to write a function that calculates the module in decibels and the phase of a complex vector. We write the following program:

function[moddB, phase] = moduledBphase (x)
maxi = max (abs (x));
moddB = 20 * log10 (abs (x) / maxi);
phase = angle (x);

This program has to be recorded under the same name as the function: moduledBphase.m.

The call to the function in the program will be as follows:

n = [0: 1: 1023];
vector = exp (i * n / 100);
[moduledB, phase] = moduledBphase (vector);

```
figure (1)
subplot (1, 1, 2)
plot (moduledB)
subplot (1, 2, 2)
plot (phase)
```

Exercise: Type this MATLAB program:

```
clear
close all
fs = 20000;
disp ('The sampling frequency is fs = 20000 Hz');
% f0 frequency of the sine
f0 = input ('Enter the frequency of the sine to be displayed (in Hz) ');
n = [0:1:2047];
tn = n/fs;
sine = sin(2*pi*f0*tn);
figure (1)
plot (tn,sine);
title ('(fs = 20,000 Hz) Display of a sine')
grid on
xlabel ('time (seconds)')
ylabel ('sine ')
string = strcat ('signal frequency:' , num2str(f0), 'Hz');
legend(string)
% Spectrum calculation
k = [0:1:2047];
spectrum = fft(sine,2048);
% Here we chose the number of points of the FFT equal the number of signal
points.
fk = k*fs /2048;
omegak = 2*pi*fk;
figure (1)
plot (omegak(1:2048),abs(spectrum(1:2048)));
title ('FT of a sine')
grid on
xlabel ('omega (rad/s)')
ylabel ('Modulus in linear scale')
figure (2)
plot (fk(1:2048),abs(spectrum(1:2048)));
title ('FT of a sine function')
grid on
xlabel ('frequency (Hz)')
ylabel ('Modulus in linear scale')
% Now we represent the spectral amplitude in decibels:
Maxspectrum = max (abs(spectrum));
spectrumdB = 20 * log10(abs(spectrum)/maxspectrum);
```

figure (3)
plot (fk,spectrumdB(1:2048));
title ('FT of a sine')
grid on
xlabel ('frequency (Hz)')
ylabel ('Spectrum in decibels')

The following figures show the results for $f_0 = 400$ Hz:

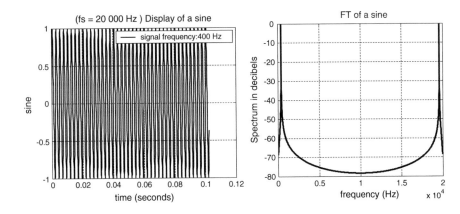

Bibliography

Signals and Systems

Anderson, B.D.O., Moore J.B.: Optimal Filtering. Prentice Hall, Englewood Cliffs; Dover Publication, New York

Bendat, J.S., Piersol, A.G.: Random Data Analysis and Measurement Procedures. Wiley, New York

Blackman, R.B., Tukey, J.W.: The Measurement of Power Spectra—From the Point of View of Communications Engineering. Dover Publications, New York

Bouvet, M.: Traitement des signaux pour les systèmes sonars, Masson Ed, (Collection CNET-Ens).

Byrne, C.L.: Signal Processing—A Mathematical Approach. A. K. Peters Ltd.

Combes, J.M., Grossmann A., Tchamitchian P.: Wavelets—Time-Frequency Methods and Phase Space, Springer, Berlin

Couch II, L.W.: Modern Communication Systems—Principles and Applications. Prentice Hall, Englewood Cliffs

Daubechies, I.: Ten lectures on Wavelets, CBMS-NSF Conferences in Applied Mathematics

Flandrin, P.: Temps-Fréquence, Hermes Ed, (Collection traitement du signal)

Haykin, S.: Communication Systems. Wiley, New York

Hsu, H.P.: Signals and Systems, Schaum's Outlines. McGraw-Hill, New York

Lacombe, A.: Analyse du signal, Editions de l'Ecole polytechnique de Lausanne.

Lacoume, J.L.: Théorie du signal, Que sais-je? PUF

Mallat S.: A Wavelet Tour of Signal Processing. Academic Press, London

Max, J.: Méthodes et techniques du traitement du signal et application aux mesures physiques, Masson (2 tomes)

Papoulis, A.: Signal Analysis, McGraw-Hill, New York

Papoulis, A.: Probability, Random Variables and Stochastic Processes. McGraw-Hill, New York

Phillips, C.L., Parr, J.M.: Signals, Systems and Transforms. Prentice Hall, Englewood Cliffs

Proakis, J.G., Salehi, M.: Communication Systems Engineering. Prentice Hall, Englewood Cliffs

Roubine, E.: Introduction à la théorie de la communication. Masson (3 tomes)

Singh, R.P., Sapre, S.D.: Communication Systems—Analog and Digital. Tata MacGraw-Hill, New Delhi

Soize, C.: Méthodes mathématiques en analyse du signal. Masson

Vetterli, M., Kovacevic, J.: Wavelets and Subband Coding. Prentice Hall, Englewood Cliffs

© Springer International Publishing Switzerland 2016
F. Cohen Tenoudji, *Analog and Digital Signal Analysis*,
Modern Acoustics and Signal Processing, DOI 10.1007/978-3-319-42382-1

Mathematics

Angot, A.: Compléments de Mathématiques, Editions de la revue d'Optique (Collection CNET)
Bass, J.: Cours de Mathématiques. Masson
Goertzel, G., Tralli, N.: Some Mathematical Methods of Physics. Mc Graw-Hill, New York
Lavoine, J.: Transformation de Fourier des pseudo-fonctions. CNRS
Lischutz, S.: Linear algebra, Schaum's Outlines. McGraw-Hill, New York
Rodier, J.: Transformation de Fourier et distributions. Ediscience
Schwartz, L.: Théorie des Distributions. Hermann Ed
Spiegel, M.R.: Complex Variables, Schaum's Outlines. McGraw-Hill, New York
Strang, G.: Introduction to Linear Algebra. Wellesley-Cambridge Press, Wellesley

Discrete Signals

Bellanger, M.: Traitement numérique du signal. Masson
Boite, R., Leich M.: Analyse et synthèse des filtres unidimensionnels. Masson
Claerbout, J. F.: Fundamentals of Geophysical Data Processing. Mc Graw-Hill, New York
Hayes, M.H.: Statistical digital signal processing and modeling. Wiley, New York
Kay, S.M.: Fundamentals of Statistical Signal Processing; Estimation Theory. Prentice Hall, Englewood Cliffs
Kundt, M.: Théorie du signal. Editions de l'Ecole Polytechnique de Lausanne.
Oppenheim, A.V., Schafer, R.W.: Digital Processing. Prentice Hall, Englewood Cliffs
Press, W.H., Teukolsky, S.A., Vetterling, W.T., Flannery, B.P.: Numerical recipes in C. Cambridge
Rabiner, L., Gold, B.: Theory and Application of Digital Signal Processing. Prentice Hall, Englewood Cliffs
Robinson, E.: Time Series Analysis and Applications. Prentice Hall, Englewood Cliffs
Robinson, E., Treitel, S.: Geophysical Signal Analysis. Prentice Hall, Englewood Cliffs

Probabilities and Statistics

Blanc-Lapierre, A., Fortet R.: Théorie des fonctions aléatoires. Masson
Doob, J.L.: Stochastic Processes. Wiley, New York
Fortet, R.: Eléments de la théorie des Probabilités. Editions du C N R S.
Hayes, M.H.: Statistical Digital Signal Processing and Modeling. Wiley, New York
Kay, S.M.: Fundamentals of Statistical Signal Processing—Estimation Theory. Prentice Hall Processing Series.
Kullback, S.: Information Theory and Statistics. Dover Publication, New York
Labarrère, M., Krief, J.P., Gimonet, B.: Le filtrage et ses applications. 2e Ed. Cepadues
Maybeck, P.S.: Stochastic Models, Estimation and Control. Academic Press, London
Porat, B.: Digital Processing of Random Signals. Theory and Methods. Prentice Hall Information and System Science Series
Taylor, H.M., Karlin, S.: An Introduction to Stochastic Modeling. Academic Press, London
Therrien, C.W.: Discrete Random Signals and Statistical Signal Processing. Prentice Hall Signal Processing Series
Wiener, N.: Extrapolation, Interpolation and Smoothing of Stationary Time Series. MIT Press, Cambridge

Internet

Course of G. Strang in linear algebra; Keywords: Strang Linear Algebra MIT Open
Google: Keywords: Signal analysis; Signal processing; Wikipedia; Fourier

Index

A

Adaptive filtering, 235, 375, 396
Aliasing, 227, 230, 265, 280, 288, 330, 339, 359
All-pole modeling, 375, 381, 392, 394, 396
Amplitude modulation, 193
Analog filters. *See* Analog systems
Analog systems, 1–9, 11, 13–16, 18–20, 22–24, 26–31, 33, 34, 159–163, 165–170, 172
Analog to digital conversion (ADC), 228
Analytic signal, 177, 184–193, 195–197
Anticausal system, 254, 255
Apodization, 122, 212, 263, 275–277, 279, 288
ARMA filters, 291, 305, 517
Artifacts of Fourier transform with computer, 264, 286, 287
Autocorrelation matrix, 394, 496
Autocorrelation method, 394
Autoregressive filters. *See* AutoRegressive discrete systems
Autoregressive discrete systems (AR), 292–294, 296–302, 304–310, 312–315, 483, 547

B

Bandpass filtering of amplitude modulated signal, 195
Bartlett method, 517, 522
Bartlett window, 107, 109, 113, 519
Bayes, 427, 441
Behavior at infinite of Fourier amplitude of a signal, 120, 177
Bessel, 91, 92, 159, 168–170, 172–174
Bode, 11, 16, 28, 32
Butterworth, 159–163, 170, 172

C

Capon, 517, 524
Cauchy, 48, 63, 79, 118, 140, 178, 567
Cauchy Principal Value, 60, 63, 78, 81
Causal signal, 149, 157, 177, 181, 237, 307, 375, 386, 401
Causal system, 153, 156, 157, 177, 292, 317, 532
Central limit theorem, 407, 433, 439, 445, 448, 507
Characteristic function, 419, 421, 425, 426, 430, 445
Chebyshev, 159, 163, 166, 411, 456
Chirp, 93, 101, 102, 106, 109, 110, 190, 207, 210, 212, 217, 223
Chi-square variable, 420, 445
Coherence function, 483, 493–495
Complex
  cepstrum, 330, 331
  derivation, 565
  integration, 566
  random variables, 435–437, 460, 543
Conditional probability density, 425, 429, 472
Continuous systems, 3
Convolution, 60, 70, 71, 93, 95–98, 100, 106, 108, 111, 113, 120, 121, 123–126, 128, 130, 133, 135, 253, 259, 359, 369, 398, 432, 433, 439, 440, 445, 503, 507, 509, 518, 520, 523, 535
Cooley-Tukey, 263, 281–283, 596
Correlation, 106, 358
  function (deterministic), 93, 99, 106, 359, 369, 376, 386, 389, 390, 393–395, 403
  function (random), 483, 484, 486, 511, 515, 517, 518, 521, 522, 528, 529, 532, 534, 535, 537, 539, 547, 549

© Springer International Publishing Switzerland 2016
F. Cohen Tenoudji, *Analog and Digital Signal Analysis*,
Modern Acoustics and Signal Processing, DOI 10.1007/978-3-319-42382-1

Correlation (*cont.*)
    of two complex r.v, 407, 436
    of two r.v, 421, 425, 431, 432, 434
Correlation and covariance matrices, 467, 468,
    476, 546
Couple of jointly Gaussian r.v., 428–430
Covariance, 424, 468, 476, 477, 513, 514, 520,
    521
Covariance method, 375, 395
Cramér-Rao, 445, 460, 462, 463
Cumulative distribution function, 407, 409,
    414, 423, 431, 434
    joint, 422

D

Daubechies, 344, 346, 353, 360–364, 367,
    371–373
Deconvolution, 325, 328, 330, 332, 333
Derivatives of Dirac distribution, 60, 73
Differential linear equations with constant
    coefficients, 4, 5, 7, 157
Dirac distribution, 59, 60, 67–69, 71–79, 87,
    92, 95, 99, 108
Direct Fourier transform, 77, 79, 81, 254
Dirichlet, 61, 73
Discrete Fourier Transform (DFT), 279, 281,
    288, 313, 314, 351
Double Side Band modulation (DSB), 190, 192

E

Eigenfunctions of a LTI system, 1, 3
Eigenvalue of an operator, 1, 6, 7, 50
Eigenvalues and eigenvectors of a Hermitian
    operator, 35, 49, 50, 56
Eigenvalues and eigenvectors of a matrix, 510
Energy spectral density, 86, 100
Ergodicity, 509, 511, 514, 517
Estimation of the
    correlation function, 474, 515, 517
    Correlation matrix from measurement, 474
    mean of a random signal, 486, 511, 512,
        515

F

Fast Fourier Transform (FFT), 330, 358
FFT. *See* Fast Fourier Transform
FIR filters. *See* Moving Average filters
Filter bank, 337, 338, 343
Filtering of a random signal, 483, 486
First order system
    analog, 11–20

Fourier series, 35
Fourier series decomposition, 35, 37, 261
Fourier transform and filtering by a LTI system
    (analog), 93, 95, 99, 101, 104, 105
Fourier transform and filtering by a LTI system
    (digital), 239
Fourier transform of
    a causal signal, 181
    a Dirac comb, 88
    a Gaussian function, 111, 115
    a Hanning window, 114
    a periodic function, 88
    a rectangular window, 111, 112
    a time-limited cosine (analog), 120
    a time-limited sine (digital), 275
    a triangular window (analog), 113
    a triangular window (digital), 272
    analog signals, 79
    digital signals, 253, 254
    Dirac distribution, 87
    the product of two functions (analog), 97
    the product of two functions (digital), 270
    trigonometric functions, 92
Frequency modulation, 203
Frequency response, 6, 9, 13–15, 17, 23, 28,
    32, 33, 77, 78, 92, 97, 137, 154, 155
Frequency spreading, 275, 279

G

Gabor, 102, 104, 106, 109, 134
Gauss distribution, 372
Gaussian probability density, 413, 418, 429
Geometric interpretation of the frequency
    response variation, 11, 17, 23
Gibbs, 35, 51–53, 56, 122–124, 133, 352
Goertzel, 158
Group delay, 160, 168, 170–172, 198, 199

H

Haar, 338
    function, 346, 350, 352
    transform, 338, 340, 343
Hanning window, 114
Harmonic generation, 53
Heaviside, 72, 74, 79, 128, 140, 151, 446
Heisenberg–Gabor, 93, 102, 104, 106, 109,
    134
Hermitian operator. *See* Self-adjoint operator
Hilbert spaces, 35, 48, 56, 82
Hilbert transform, 177, 182, 194, 201, 597

I

IIR filters. *See* Autoregressive filters
Improving Spectral Estimation, 522
Impulse response, 31, 34, 77, 78, 92, 95, 96,
    124, 137–142, 143, 145, 147, 148,
    151–154, 156, 157, 259, 260, 338
Independence in probability, 407, 439
Instantaneous frequency, 187, 189–191, 207
Instantaneous phase, 189
Instantaneous power, 42, 84, 504
Interpolation by zero padding, 284
Inverse Fourier transform (analog), 77, 78
Inverse Fourier transform (digital), 262

J

Joint probability density, 407, 422
Jordan Lemma, 65, 139, 140

K

Kalman filter, 543, 555
Karhunen-Loève, 467, 476, 478, 480, 481

L

Laplace transform, 13, 148–151, 153, 155, 157,
    158, 253
Least square methods, 376, 449, 544–546, 549,
    555, 557, 559, 561
LeGall-Tabatabai, 344, 345, 372
Likelihood function, 458, 459, 464
Linear
    algebra, 445, 451
    independence, 451
    prediction coding (LPC), 375, 376, 394,
        397
    system, 1
    time invariant system (LTI), 3, 6
    transformation of random vectors, 469
Linearity of physical systems, 3, 8

M

Mallat, Meyer, 361
Marginal stability, 154
Matched filtering, 483
Matlab, 591
Matrices. *See* Linear algebra
Maximum likelihood. *See* Likelihood function
Minimum phase system, 155, 321, 326
Modulation
    amplitude, 193, 201
    frequency, 203
Moments of a statistical distribution, 117
Morlet wavelet, 134
Mother wavelet, 349
Moving Average filters (MA), 229, 235–249

Moving window, 211
Multiresolution, 353, 372
Multivariate Gaussian Probability Densities,
    471

N

Nonlinearity of a system, 53
Non-parametric methods, 517
Normal law. *See* Gaussian law
Nyquist frequency, 244, 246, 247, 249, 273,
    311, 312

O

Optimal
    coefficients, 35, 45, 46, 56
    estimation, 544, 545, 549, 551, 553, 555

P

Padé, 375, 378, 381, 383, 385
Paley–Wiener, 529
Parametric
    estimation, 375, 378, 381, 383, 384, 445,
        455
    modeling, 375, 385, 532, 534, 535,
        537–539, 541, 543
Parseval, 42, 43, 56, 77, 84, 85, 92, 100, 102,
    109, 376
Periodogram, 518–522
Phase and group delays, 160, 168
Pisarenko, 511, 517, 526
Poisson, 77, 89, 92, 99
Power spectral density (PSD), 483, 488, 490,
    491, 496, 499, 503, 511, 529, 536–539
Primitive of the Dirac distribution, 72
Probability density of
    a Complex r.v., 435
    a function of a random variable, 407
    the sum of two r.v., 431
    two r.v, 434
Prony, 385, 387–390, 393, 395, 396

Q

Quadratic form and Hermitian form, 424, 473,
    583
Quality factor Q, 11, 30

R

Raw estimate of the power spectral density,
    518, 524, 539
Regularization, 445, 452–454
Regular random signal, 521, 532, 535, 536,
    538, 539
Residue theorem, 138, 145, 148, 253, 257

S

Sampling theorem, 224, 267
Scaling function, 346–348, 353, 356, 359, 362, 365, 367, 373
Second order system, 9, 11, 21, 23, 31, 137, 153
Self-adjoint operator, 49
Shanks, 375, 385, 389
Shannon, 359
Sharp resonance, 29, 31, 33, 34
Signal
    energy, 42, 84, 86, 102, 108, 109, 326
    power, 42, 84, 477, 485, 491, 499, 504, 548
    spread, 93, 102, 104, 106, 109
Signal-to-noise ratio, 483, 501, 503, 543, 548
Singular Value Decomposition (SVD), 454, 586
Spectral
    energy density, 86, 100
    estimation, 511, 517, 518, 522, 523, 539
Spectrogram, 104
Stationary systems, 2, 236
Stochastic orthogonality, 479
System stability, 138, 149, 153

T

Table of
    Fourier transforms, 111, 133, 135
    normal centered reduced, 442
    Laplace transforms, 158
Tikhonov. *See* Regularization
Time-frequency analysis, 207–224

Transfer Function, 6, 12, 21
Transition from analog to digital filters, 310–313

U

Uncertainty principle. *See* Heisenberg-Gabor

V

Variance of
    correlation function estimator, 515
    mean estimator, 515

W

Wavelets, 338, 344, 352, 353, 355, 373
Welch, 517
White centered noise, 486, 490, 491, 499
Whitening filter, 554
Wide sense stationary signal (WSS), 491, 496, 499, 503
Wiener, 529, 531, 535, 543, 545, 546, 548, 550, 553
Wiener-Hopf, 543, 546, 547, 550, 553
Wigner-Ville distribution, 207, 212–217

Y

Yule-Walker, 529, 532, 534, 535, 537

Z

Z-transform, 253
Z-transform of the Product of Two Functions, 258

<barcode>||| || || ||||||||| || |||| |||| | ||||| || ||| || || |||</barcode>

Printed in the United States
By Bookmasters